Lecture Notes in Mathematics 1968

Editors:
J.-M. Morel, Cachan
F. Takens, Groningen
B. Teissier, Paris

T0216631

Rainer Weissauer

Endoscopy for GSp(4) and the Cohomology of Siegel Modular Threefolds

 Springer

Rainer Weissauer
Universität Heidelberg
Mathematisches Institut
Im Neuenheimer Feld 288
69120 Heidelberg
Germany
weissauer@mathi.uni-heidelberg.de

ISBN: 978-3-540-89305-9   e-ISBN: 978-3-540-89306-6
DOI: 10.1007/978-3-540-89306-6

Lecture Notes in Mathematics ISSN print edition: 0075-8434
                              ISSN electronic edition: 1617-9692

Library of Congress Control Number: 2008940147

Mathematics Subject Classification (2000): 11Fxx, 11Gxx, 11Exx, 32Nxx, 57M99

Cover design: SPi Publishing Services

Printed on acid-free paper

springer.com

Preface

This volume grew out of a series of preprints which were written and circulated between 1993 and 1994. Around the same time, related work was done independently by Harder [40] and Laumon [62]. In writing this text based on a revised version of these preprints that were widely distributed in summer 1995, I finally did not pursue the original plan to completely reorganize the original preprints. After the long delay, one of the reasons was that an overview of the results is now available in [115]. Instead I tried to improve the presentation modestly, in particular by adding cross-references wherever I felt this was necessary. In addition, Chaps. 11 and 12 and Sects. 5.1, 5.4, and 5.5 were added; these were written in 1998.

I will give a more detailed overview of the content of the different chapters below. Before that I should mention that the two main results are the proof of Ramanujan's conjecture for Siegel modular forms of genus 2 for forms which are not cuspidal representations associated with parabolic subgroups (CAP representations), and the study of the endoscopic lift for the group $GSp(4)$. Both topics are formulated and proved in the first five chapters assuming the stabilization of the trace formula. All the remaining technical results, which are necessary to obtain the stabilized trace formula, are presented in the remaining chapters.

Chapter 1 gathers results on the cohomology of Siegel modular threefolds that are used in later chapters, notably in Chap. 3. At the beginning of Chap. 1, important facts from [19] on the Hodge structure and l-adic cohomology of the Siegel modular varieties $S_K(\mathbb{C})$ are reviewed. In the case of genus 2, the Siegel modular varieties $S_K(\mathbb{C})$ define algebraic varieties of dimension 3. They are the Shimura varieties attached to the group of symplectic similitudes $G = GSp(4, \mathbb{Q})$. One can define coefficient systems E_λ on these threefolds associated with irreducible finite-dimensional algebraic representations λ of the group $GSp(4)$, which are defined over $\mathbb{Q}$. The most interesting cohomology groups of these coefficient systems are the cohomology groups $H^i(S_K(\mathbb{C}), E_\lambda)$ in the middle degree $i = 3$. The group $G(\mathbb{A}_{fin})$ acts on the direct limit of the cohomology groups $H^i(S_K(\mathbb{C}), E_\lambda)$, where the limit is over the adelic compact open level groups $K \subseteq G(\mathbb{A}_{fin})$, and this defines an admissible automorphic representation of the group $GSp(4, \mathbb{A})$ for the adele ring $\mathbb{A} = \mathbb{R} \times \mathbb{A}_{fin}$. Since the Siegel moduli spaces are not proper, the cohomology of these varieties is not pure. Besides the interior cohomology, which is the image of the cohomology with compact supports in the ordinary cohomology, there occur representations of Eisenstein type. The automorphic representations of $G(\mathbb{A}_{fin})$ defined by the Eisenstein constituents are by definition isomorphic to constituents of induced representations $Ind_{P(\mathbb{A}_{fin})}^{G(\mathbb{A}_{fin})}(\sigma)$, where σ is an automorphic representation of the Levi group of a $\mathbb{Q}$-rational parabolic subgroup of G. It is well known that the Eisenstein cohomology is rarely pure. But also some part of the cuspidal cohomology, which is a subspace of the interior cohomology, fails to be pure. In fact, some of the irreducible cuspidal automorphic representations behave similarly to Eisenstein representations. These are called CAP representations π. By definition, an irreducible automorphic representation $\pi = \otimes_v \pi_v$ is a CAP representation if there exists some global automorphic representation σ of a Levi group of some proper $\mathbb{Q}$-rational parabolic subgroup of G such that π_v and $Ind_{P_v}^{G_v}(\sigma_v)$ have the same spherical constituent for almost all non-Archimedean places v. For the group $GSp(4)$ the CAP representations were completely classified by Piatetski-Shapiro [69] and Soudry [95], and the Ramanujan conjecture (purity) does not hold for them. The main result obtained in Chap. 1 (Sect. 1.3), Theorem 1.1, states that except for the cohomology degree $i = 3$ all the irreducible automorphic representations, which occur as irreducible constituents of the representations $\lim_K H^i(S_K(\mathbb{C}), E_\lambda)$, are either CAP representations or belong to the Eisenstein cohomology. This, in principle, allows a complete description of the cuspidal part of the cohomology representations in degree $i \neq 3$ by the classification of CAP representations. Even for the degree $i = 3$ the CAP representations occur, indeed those defined by the Saito–Kurokawa lift. Sections 1.3–1.7 contain the proof of Theorem 1.1, which is based on the study of the Lefschetz map and a weak form of the Ramanujan conjecture. This eventually shows that for irreducible automorphic cuspidal representations π, which occur as constituents in degree $i \neq 3$, a certain L-function must have poles at specific points, which forces π to be a CAP representation.

In Chap. 2, we consider the topological trace formula of Goresky and MacPherson for general reductive groups G. We do not consider this for the spaces $S_K(\mathbb{C})$ themselves, but for spaces $S_K(G)$ with a slightly better behavior with respect to "parabolic induction." This suffices for our purposes, since $S_K(\mathbb{C})$ is a finite unramified covering of $S_K(G)$:

$$S_K(\mathbb{C}) \twoheadrightarrow S_K(G).$$

For a Hecke operator f the topological trace formula computes the alternating sum $\sum_i (-1)^i tr(f; H^i(S_K(G), E_\lambda)$ of its traces. Its computation is considerably simplified if one discards all contributions from CAP representations and Eisenstein representations, which we abbreviate by the notion "CAP localization." The corresponding simplified formula obtained by CAP localization still allows us to compute the alternating sum of the dimensions of generalized eigenspaces $H^i(S_K(G), E_\lambda)(\pi)$ for an irreducible cuspidal automorphic representation π, which is not CAP (see page 47). To do this for a single fixed π, we construct suitable Hecke operators f

in Sects. 2.7 and 2.8, called good projectors, whose elliptic trace $T_{ell}^G(f)$ computes the alternating sum of these virtual dimensions (Theorem 2.1). For the construction of good projectors, it is essential that all the irreducible representations of $GSp(4, \mathbb{A}_{fin})$, which arise as constituents of the cohomology, are automorphic. This was shown by Schwermer for the group $GSp(4)$ and by Franke in the general case. In Sect. 2.9, we compare the formula thus obtained with Arthur's L^2-trace formula, which has the property that these elliptic traces coincide with the elliptic part of the geometric side of the Selberg trace formula. Now assume π is a cuspidal irreducible representation of $G(\mathbb{A}_{fin})$, which are not CAP. For a prime p, let us denote $\pi^p = \otimes_{v \neq \infty, p} \pi_v$. Then the results obtained in Chap. 2 in the special case $G = GSp(4)$ combined with the results obtained in Chap. 1 give a simple trace formula for the action of Hecke operators on generalized eigenspaces $\lim_K H^3(X, E_\lambda)(\pi^p)$ of the middle cohomology for either $X = S_K(G)$ or $X = S_K(\mathbb{C})$ for large enough primes p. Furthermore, this simple formula also allows us to compute the action of the nth powers of the geometric Frobenius substitution $Frob_p$ at the prime p in terms of certain Hecke operators $h_p = h_p^{(n)}$. See Theorem 2.2 and its applications in Chap. 4.

In Chap. 3, the simple localized topological trace formula for $GSp(4, \mathbb{Q})$ is compared with the Grothendieck–Verdier–Lefschetz trace formula for $S_K(\overline{\mathbb{F}}_p)$, which computes the traces of the Frobenius homomorphism attached for certain l-adic sheaves attached to the coefficient systems E_λ. But unlike the CAP localized form of the topological trace formula, this other trace formula, studied by Langlands, Kottwitz, and Milne, is not a stable trace formula. To compare both trace formulas, one has to stabilize it [53, 59]. This requires certain local identities at the non-Archimedean places – the so-called fundamental lemma and certain variants of it. More precisely, since the Grothendieck–Verdier–Lefschetz trace formula can only be explicitly computed for sufficiently high powers $Frob_p^n$ of the Frobenius $Frob_p$, one needs for each such n a twisted version of the fundamental lemma [51]. With use of these local assumptions, which are considered in the later chapters, one obtains without effort the main formula (Corollary 3.4) which expresses the Frobenius traces as a sum of two terms. One of these two terms is the trace of a suitably defined Hecke operator on the cohomology. The other term is the so-called endoscopic term, which is related to an automorphic lift. This lift is implicitly defined by the trace formula; however, it is not yet properly understood at this point of the discussion. This lift will eventually be constructed in several steps by a bootstrap argument using repeated comparison of traces. At this stage of the discussion we are therefore content with the following weak characterization: a cuspidal automorphic representation is a weak endoscopic lift if its L-functions are the L-function of an irreducible automorphic form $\sigma = (\sigma_1, \sigma_2)$ for $Gl(2, \mathbb{A}) \times Gl(2, \mathbb{A})$ at almost all places, provided σ_1 and σ_2 are cuspidal automorphic representations of $Gl(2, \mathbb{A})$ such that they have the same central character. In fact σ can be viewed as an irreducible cuspidal automorphic representation of the nontrivial elliptic endoscopic group $M \cong Gl(2)^2/\mathbb{G}_m$ of G. In the situation of Corollary 3.4, a preliminary condition at the Archimedean place has also been added. This temporarily relevant definition of a weak lift, involving some technical conditions, can be found at the beginning of Chap. 3. At the end of Chap. 3, we prove the Ramanujan conjecture (purity) for the cuspidal representations π (at the spherical places), which are neither CAP nor a weak lift, and which occur as constituents in the cohomology of degree $i = 3$. We then indicate why this should yield four-dimensional $\overline{\mathbb{Q}}_l$-adic representations of the absolute Galois group of $\mathbb{Q}$ attached to such irreducible automorphic representations. In fact, the trace formula stated in Sect. 3.6 finally leads to this [115]. We neglect the discussion of the cohomology in degree different from 3. In fact, only CAP representations contribute in these degrees, and the CAP representations have all been classified for the group $GSp(4)$; hence, it is not difficult to determine their contributions to the Hasse–Weil zeta function by weight considerations. So there is no need to exploit the trace formula in these cases. Nevertheless this is still worthwhile since it gives refined formulas for the local automorphic representations of CAP representations (see the notion of "Arthur packets" in Sect. 4.11), and it can be done using the formulas of Chap. 2. However, we have not included this discussion.

In Chap. 4, we take up the study of weak lifts. For this discussion we fix a cuspidal irreducible representation $\sigma = (\sigma_1, \sigma_2)$ of $M(\mathbb{A})$ for the unique nontrivial elliptic endoscopic group M of $G = GSp(4, \mathbb{Q})$. In Chap. 4, we then consider irreducible cuspidal automorphic representations π, which are weak lifts attached to σ but which are not CAP. This is the general assumption of Chap. 4. Since only the group $G(\mathbb{A}_{fin})$ acts on the cohomology, it is natural to ask for the Archimedean component π_∞ of the automorphic representations $\pi = \pi_\infty \otimes \pi_{fin}$. For this we fix some coefficient system E_λ. Then π_∞ necessarily must belong to the discrete series of $GSp(4, \mathbb{R})$, and π_∞ is almost determined by the condition that π_{fin} defines a nontrivial generalized eigenspace on the direct limit $\lim_K H^3(S_K(\mathbb{C}), E_\lambda)$. More precisely, this means that π_∞ belongs to a local L-packet of discrete series representations in the sense of Shelstad [91]. This L-packet is uniquely determined by the irreducible representation λ, which defines the coefficient system. This L-packet contains two equivalence classes of irreducible representations. One of the representations, $\pi_{-,\infty} = \pi_\infty^H(\lambda)$, of this L-packet belongs to the holomorphic/antiholomorphic discrete series; the other representation, $\pi_{+,\infty} = \pi_\infty^W(\lambda)$, has a Whittaker model. Let $m_1(\pi_{fin})$ and $m_2(\pi_{fin})$ be the automorphic multiplicities of the cuspidal representations $\pi = \pi_{-,\infty} \otimes \pi_{fin}$ and $\pi = \pi_{+,\infty} \otimes \pi_{fin}$, respectively. The multiplicity of π_{fin} in the generalized eigenspace $\lim_K H^3(S_K(\mathbb{C}), E_\lambda)(\pi_{fin})$ is $2m_1(\pi_{fin}) + 2m_2(\pi_{fin})$. In fact, the semisimplification of the $\overline{\mathbb{Q}}_l$-adic representations of the absolute Galois group of $\mathbb{Q}$ on the corresponding eigenspace defined for the l-adic cohomology is $m_1\rho_1 \oplus m_2\rho_2$. Here, ρ_1 and ρ_2 are the two-dimensional irreducible $\overline{\mathbb{Q}}_l$-representations attached to $\sigma_i, i = 1, 2$, by Deligne. Indeed, if some weak lift π of σ contributes nontrivially to the cohomology in degree 3, then the two cuspidal representations σ_1 and σ_2 are irreducible automorphic representations of $Gl(2, \mathbb{A})$, whose Archimedean component again belongs to the discrete series. Such automorphic representations σ_i are related to elliptic holomorphic new forms of weights r_i. The weights are not arbitrary. They must be different, so we can assume $r_1 > r_2$. This being said, there is the finer result (Lemma 4.2)

$$\lim_K H^3(S_K(\mathbb{C}), E_\lambda)(\pi_{fin})^{ss} \cong m_1 \cdot \rho_1 \oplus m_2 \cdot (\nu_l^{k_2-2} \otimes \rho_2),$$

where ν_l denotes the cyclotomic character and k_2 is an integer determined by the underlying coefficient system E_λ. The trace formula comparison of Chap. 3 provides a formula which expresses

$$m_1(\pi_{fin}) - m_2(\pi_{fin})$$

in terms of local data, but in which only the local non-Archimedean components π_v of the representations $\pi_{fin} = \prod_{v \neq \infty} \pi_v$ enter. More precisely, this formula is

$$m_1(\pi_{fin}) - m_2(\pi_{fin}) = - \prod_{v \neq \infty} n(\sigma_v, \pi_v),$$

where the coefficients $n(\sigma_v, \pi_v)$ are complex numbers obtained by a distribution formula

$$\chi_{\sigma_v}^G = \sum_{\pi_v} n(\sigma_v, \pi_v) \chi_{\pi_v}.$$

Here, χ_{σ_v} and χ_{π_v} denote characters of admissible irreducible representations, and $\chi_{\sigma_v}^G$ is the endoscopic distribution lift of the character χ_{σ_v} (see page 83). The definition of this local lift for distributions requires the existence of matching functions, where certain transfer factors have been fixed. Transfer factors will be discussed in Chaps. 6 and 7. It should be mentioned that for the group $GSp(4, F_v)$ the existence of matching functions was established by Hales [36]. The existence of a character expansion $\chi_{\sigma_v}^G = \sum_{\pi_v} n(\sigma_v, \pi_v) \chi_{\pi_v}$ is then derived from the trace comparisons studied in Chap. 3 and the first sections of Chap. 4. Most of the content of Sects. 4.5–4.12 are devoted to proving that this sum is finite and that the transfer coefficients $n(\sigma_v, \pi_v)$ are integers. In fact this finally defines the endoscopic lift $r : R_{\mathbb{Z}}[M_v] \to R_{\mathbb{Z}}[G_v]$ between the integral Grothendieck groups of irreducible admissible representations of $G = GSp(4)$ and $M = Gl(2)^2/\mathbb{G}_m$ for non-Archimedean p-adic fields. In the real case such formulas are known in general from the work of Shelstad [90, 91]. The final result is stated in Sect. 4.11. With use of the classification of representations, the results obtained by Moeglin, Rodier, Sally, Shahidi, Soudry, Tadic, Vigneras, and Waldspurger, this is reduced to establish the existence of r for local non-Archimedean admissible irreducible representations σ_v of M_v, which belong to the discrete series. For these representations, it turns out that the local character lift has the form

$$r(\sigma_v) = \pi_+(\sigma_v) - \pi_-(\sigma_v)$$

for two irreducible admissible representations $\pi_\pm(\sigma_v)$ of the group G_v. We furthermore show that $\pi_+(\sigma_v)$ does have a Whittaker model, whereas $\pi_-(\sigma_v)$ does not. Finally, we use global theta series to describe $\pi_\pm(\sigma_v)$ in terms of local theta lifts similarly to the case of the group $Sp(4)$ studied by Howe and Piatetski-Shapiro [41]. In fact this study is continued in Chap. 5. Indeed, some results obtained in Chap. 5 are already used in Sect. 4.12. Besides these local results, studied in Sects. 4.5–4.12, we consider in Sect. 4.4 rationality questions, i.e., questions concerning the field of definition. Using some properties of the endoscopic transfer factor, defined in the later chapters, we can describe the numbers $m_1(\pi_{fin})$ and $m_2(\pi_{fin})$ in terms of Hodge theory. In fact $2m_1(\pi_{fin})$ turns out to be the multiplicity of π_{fin} in the holomorphic/antiholomorphic part and $2m_2(\pi_{fin})$ turns out to be the nonholomorphic contribution.

In Chap. 5, we continue the discussion of Chap. 4, but return to global questions. The main result obtained in this chapter is Theorem 5.2, which is the final version of the preliminary multiplicity formula for weak lifts given in Sects. 4.1–4.3. We obtain the formula $m_1(\pi_{fin}) + m_2(\pi_{fin}) = 1$; hence, one of the global multiplicities $m_i(\pi_{fin})$ is 1 and the other is 0. Which of them does not vanish depends only on the non-Archimedean components of π_{fin}. The essential argument is the principle of exchange, which controls exchange in $\otimes_v \pi_v$ of a representation π_v within its local packet at one specific place v. The final formula is, of course, a special case of Arthur's conjecture [3], which originated from considering the special case of the group $GSp(4, \mathbb{Q})$. However we only consider this formula in the case of weak lifts, which are not CAP. Nevertheless, the multiplicity formula for the cohomology groups in the case of the Saito–Kurokawa lift can be derived along the same line of arguments (although this is not carried through explicitly).

Chapter 5 also contains sections in which the global results on the endoscopic lift are extended to the case $G = Res_{F/\mathbb{Q}}(GSp(4))$ for an arbitrary totally real number field F, and also applies to representations which do not necessarily appear in the cohomology of Shimura varieties. This is contained in Sects. 5.4 and 5.5. The arguments here use Arthur's trace formula instead of the topological trace formula, and they are subtler and more technical than the arguments involving the topological trace formula. The analogous local results, which extend those obtained in Chap. 4 to arbitrary local fields of characteristic 0, are considered in Sect. 5.1.

In Chaps. 6 and 7, the fundamental lemma for the group $GSp(4, F_v)$ over a local non-Archimedean field F_v of residue characteristic different from 2 is proved. This fundamental lemma (Theorems 6.1 and 7.1) is an identity between local orbital integrals $O_\eta^{G_v}(f_v)$ and $O_t^{M_v}(f^{M_v})$ for the groups M_v and G_v. This identity involves a transfer factor $\Delta(\eta, t)$. Here, the elements $\eta \in GSp(4, F_v)$ and $t \in M(F_v)$ are sufficiently regular semisimple elements and η and t are related by a norm mapping. η is an element whose conjugacy class over the algebraic closure is determined by the conjugacy class of t in $M(F_v)$. This does not determine the $G(F_v)$-conjugacy class of η uniquely. In the case under consideration, there are one or two such conjugacy classes in the stable conjugacy class. The κ-orbital integral is the difference $O_\eta^{G_v, \kappa}(f_v) = O_\eta^{G_v}(f_v) - O_{\eta'}^{G_v}(f_v)$ of orbital integrals $O_\eta^{G_v}(f_v) = \int_{G_{\eta,v} \backslash G_v} f_v(g^{-1}\eta g) dg/dg_\eta$ in the case where there are two such classes. Since there is no canonical choice, which might privilege η or η', one has to make a choice for the definition of $O_\eta^{G_v, \kappa}(f_v)$. The dependence on this choice is compensated for by a transfer factor $\Delta(\eta, t)$, which depends on the class of η chosen. Then the fundamental lemma is the statement that there exists a homomorphism $b : f_v \mapsto f^{M_v}$ between the spherical Hecke algebras (prescribed by the principles of Langlands functoriality) with the matching condition

$$\Delta(\eta, t) O_\eta^{G_v, \kappa}(f_v) \;=\; SO_t^{M_v}(f_v^{M_v})$$

for all sufficiently regular t and the corresponding η. Here, SO denotes the stable orbital integral on M_v. A couple of remarks are in order. First, although the fundamental lemma is later used for (G, M)-regular elements t, it is enough to prove the fundamental lemma for sufficiently regular elements by a degeneration argument. Second, it is enough to prove the fundamental lemma for the unit elements of the Hecke algebras, and for almost all primes (see Hales [35] for the case of ordinary endoscopy). See also Chap. 10, where the reduction to the case of the unit element $f_v = 1$ of the spherical Hecke algebra is discussed in the slightly greater generality of twisted base change. Third, for our purposes it is important that the transfer factors have certain nice properties. One of these properties is the product formula (global property)

$$\prod_v \Delta(\eta_v, t_v) = 1$$

for global elements η and t, where the product is over all Archimedean and Non-archimedean places. For the product formula above, it is essential for us that the formula holds precisely with the Archimedean transfer factor $\Delta(\eta_\infty, t_\infty)$ used by Shelstad. Concerning this, we show in Chap. 8 that for our choice of transfer factor the product formula holds, and that our chosen Archimedean transfer factor is the same as the one defined by Langlands and Shelstad over the field $\mathbb{R}$. Unfortunately, already in the case of the group $GSp(4)$, this amounts to a lengthy and tedious unraveling of the definitions, which are based on the cohomological reciprocity pairings of local class field theory [60]. The proof of the fundamental lemma is done by an explicit calculation. We distinguish two cases dealt with in Chaps. 6 and 7, respectively. In fact, the computation gives the local orbital integrals explicitly, not only the κ-orbital integral. This turned out to be useful for later computations in the twisted case done by Flicker [26]. The explicit calculation of the orbital integrals hinges on an approach which in the case of the group $Gl(2)$ is used in the book by Jaquet and Langlands [42] on $Gl(2)$ and which in an implicit form is based on some double coset computations in the group $G(F_v)$ for the group $G = Gl(2)$. For me, an analogous double coset decomposition for the group $G = GSp(4)$, due to Schröder, suggested this approach. A special case was been carried out by Schröder [81]. It seems that nice representatives for double cosets $H(F_v) \backslash G(F_v) / K$ of this type exist for reductive, hyperspecial maximal compact subgroups K and maximal proper reductive F_v-subgroups H of G quite generally, in the sense that they should define a generalization of the classical theory of genera of quadratic forms. In classical genus theory, H is the orthogonal group contained in the linear group G. The maximal subgroups of reductive groups are well known, and new types of genus theory mainly arise from considering the inclusions $H \hookrightarrow G$ of centralizers $H = G_s$ of semisimple elements s in reductive groups G. In Chap. 12 we consider this situation for the group $G = GSp(2n)$, where we generalize the result obtained by Schröder to the case $n = 2$. Similar computations can be made in the case of classical groups [116]. The case of the exceptional case group G of type G_2 was considered by Weselmann [117].

Chapter 9 considers the fundamental lemma for twisted base change endoscopy. This type of fundamental lemma is needed for the trace comparison theorems in Chap. 3. We show that these twisted endoscopic fundamental lemmas can be reduced to the ordinary fundamental lemmas of standard endoscopy. Such a reduction can be carried out quite generally, except that we consider the global trace formula arguments only in the case of the group $GSp(4)$. However, the argument can be extended to the general case, and this will be considered elsewhere. As a side result, one gets a variant of the fundamental lemma (see Lemma 9.7) where the transfer factors are defined in a slightly different form, which is needed for Chap. 3. This is based on some explicit formulas for the Langlands reciprocity map as given in Kottwitz [48] or Schröder and Weissauer [82]. An entirely local proof was given later by Kaiser [43].

In Chap. 10, we verify that the twisted endoscopic fundamental lemma is a consequence of the special case of the fundamental lemma for unit elements, as one expects from the untwisted case [35], and that it is enough to know it for almost all primes. This reduces the fundamental lemmas needed for Chap. 3 to the statements given in Chaps. 6 and 7. The argument uses the method of Labesse [57]; hence, it is based on elementary functions and can be further generalized [113] to the twisted adjoint cases. For standard endoscopy this reduction was proved in Hales [35] by a different argument. Finally, Chaps. 11 and 12 contain some prerequisite material needed in Chaps. 6–10.

Contents

Chapter 1
An Application of the Hard Lefschetz Theorem

In the later parts of this chapter, we will use the hard Lefschetz theorem to show that an irreducible cuspidal automorphic representation $\pi = \pi_\infty \pi_{fin}$ of the group $GSp(4, \mathbb{A})$, whose Archimedean component π_∞ belongs to the discrete series, gives a contribution to a cohomology group of some canonical associated locally constant sheaves on Siegel modular threefolds only if the cohomology degree is 3, provided π is not a cuspidal representation associated with parabolic subgroup (CAP representation). In other words, under the action of the adele group $GSp(4, \mathbb{A}_{fin})$, all irreducible constituents occur in the middle degree, if one discards so-called CAP representations. Since CAP representations are well understood for the group $GSp(4)$, this result is important for the analysis of the supertrace of Hecke operators acting on the cohomology of Siegel modular threefolds, and hence for the proof of the generalized Ramanujan conjecture for holomorphic Siegel modular forms of genus 2 and weight 3 or more.

1.1 Review of Eichler Integrals

In this section, we collect facts on the Hodge structure on the cohomology of Siegel modular threefolds. For this we first define the underlying locally constant coefficient systems. Although our main interest will be the l-adic cohomology groups defined by these coefficient systems, the Hodge structures defined by them will play a vital role later. The Hodge structure helps to improve upon results on the l-adic representations obtained by the topological trace formula in Chap. 4, since it allows us to obtain an interpretation of certain multiplicities m_1, m_2 which are defined via l-adic representations. See, for instance, Corollaries 4.1 and 4.4.

Notation. Let Γ be a neat subgroup of finite index in the symplectic group $Sp(4, \mathbb{Z})$ (e.g., a principal congruence subgroup of level 3 or more). Let H denote the Siegel upper half-space $\{Z = Z' \in M_{2,2}(\mathbb{C}) \mid Im(Z) > 0\}$ of genus 2. Matrices $M = \left(\begin{smallmatrix} A & B \\ C & D \end{smallmatrix}\right)$ in $GSp(4, \mathbb{R})$ act on H in the usual way by $M\langle Z \rangle = (AZ+B)(CZ+D)^{-1}$.

R. Weissauer, *Endoscopy for GSp(4) and the Cohomology of Siegel Modular Threefolds*, Lecture Notes in Mathematics 1968, DOI: 10.1007/978-3-540-89306-6_1,

Let $\overline{X}$ denote a suitable toroidal compactification of the Siegel modular variety defined by the quotient $X = \Gamma \backslash H$. Let $j : X \hookrightarrow \overline{X}$ denote the inclusion map.

Representations of $GSp(4,\mathbb{Q})$.   To any irreducible algebraic representation

$$\lambda : GSp(4, \mathbb{Q}) \to Gl(V)$$

of the symplectic group of similitudes $GSp(4, \mathbb{Q})$ on a complex vector space V one can associate a locally constant sheaf V_λ on $X = \Gamma \backslash H$ whose fibers are isomorphic to V. The mixed Hodge structure of the cohomology of these fiber bundles is described in [19], Sect. VI.5.

Except for the similitude character $\nu : GSp(4, \mathbb{Q}) \to \mathbb{C}^*$, there are two fundamental representations of the group $GSp(4, \mathbb{Q})$. One of them is the four-dimensional standard representation on $\mathbb{C}^4$. The other is contained in $\Lambda^2(\mathbb{C}^4)$, which decomposes into a one-dimensional representation and the five-dimensional fundamental representation. These representations are defined over $\mathbb{Q}$. Up to a twist by $\nu^n, n \in \mathbb{Z}$, all irreducible representations occur in the decomposition of tensor products of the two fundamental representations.

An irreducible representation λ of $GSp(4, \mathbb{Q})$ as above remains irreducible after restriction to the subgroup $Sp(4, \mathbb{Q})$. The isomorphism class of the restriction to $Sp(4, \mathbb{Q})$ is determined by its highest weight. In our case this may be made more explicit: The $Sp(4)$ dominant weights may be identified with pairs of integers

$$\lambda = (\lambda_1, \lambda_2) \text{ such that } \lambda_1 \geq \lambda_2 \geq 0.$$

For the four-dimensional standard representation $(\lambda_1, \lambda_2) = (1, 0)$, and for the five-dimensional fundamental representation $(\lambda_1, \lambda_2) = (1, 1)$. The irreducible representation λ of $GSp(4, \mathbb{Q})$ is indeed uniquely determined by its central character and the integers λ_1, λ_2. A character of the center of $GSp(4, \mathbb{Q})$, identified with the scalar matrices $t \cdot id$, has the form $t \mapsto t^n$ for an integer $n \in \mathbb{Z}$. For fixed λ_1, λ_2 the possible values of n are the integers $n \equiv \lambda_1 + \lambda_2$ modulo 2. In fact, two irreducible algebraic representations of $GSp(4, \mathbb{Q})$ with isomorphic restriction to the symplectic group, i.e., coinciding parameters (λ_1, λ_2), are obtained from each other by a twist with a power of the similitude character ν.

Algebraic Normalization.   In the following we attach to integers $\lambda_1 \geq \lambda_2 \geq 0$ the unique class of irreducible algebraic representation λ of the group $GSp(4, \mathbb{Q})$, whose central character is $t \mapsto t^{\lambda_1 + \lambda_2}$ and whose restriction to $Sp(4, \mathbb{Q})$ has highest weight (λ_1, λ_2). By abuse of notation, we write $\lambda = (\lambda_1, \lambda_2)$ both for this representation of $GSp(4, \mathbb{Q})$ and for its highest weight vector. With use of this notation, the dual representation $\lambda^\vee$ attached to λ is $\lambda \otimes \nu^{-\lambda_1 - \lambda_2}$.

Let λ be an irreducible algebraic representation of the group $GSp(4, \mathbb{Q})$. Consider the coefficient system $V_\lambda = \mathbf{V}(\lambda)$ on X attached to the representation λ as in [19], p. 232. Put $E_\lambda = \mathbf{V}(\lambda^\vee)$ for the dual representation $\lambda^\vee \cong \lambda \otimes \nu^{-\lambda_1 - \lambda_2}$. This notation is motivated by the fact that for the four-dimensional standard representation λ of $GSp(4, \mathbb{Q})$ this gives

$$E_\lambda = R^1 \pi_* \mathbb{C},$$

where $\pi : A \to X$ is "the universal family" of principally polarized Abelian varieties of genus 2 over X [19], p. 234.

Algebraic and Unitary Normalizations. If $n = 2r$ is even, our normalization above normalized the representation λ such that the central character becomes trivial – by twisting with a suitable multiple of the similitude character. Even if n is odd, one might proceed in this way by twisting formally with a half-integral power of the similitude character. If $n = 2r$ is odd and $r \in \frac{1}{2}\mathbb{Z}$, this "unitary" normalization can be introduced only in a purely formal manner (see [19], p. 231) except that one then has to deal with the slight inconvenience of having half-integral F-weights in Hodge filtrations or half-integral Tate twists. However, as this is done purely formally, it presents no difficulties. Following Chai and Faltings [19], we will use this notation.

This normalization of the representation is suitable if we consider L-series and spectral theory. We refer to it as the unitary normalization. However, one has to be careful. The operation of the Hecke algebra and also the Galois action, defined later, essentially depend on the central character of the representation λ of $GSp(4, \mathbb{Q})$. For this our previous *algebraic* normalization is better suited.

In contrast, the unitary normalization formally leads to Hodge structures of half-integral weights. See [19], p. 235, Remark (1). Although it implies that the corresponding l-adic sheaves will formally become pure of weight 0 and self-dual, half-integral Tate twists are not well defined over the base $Spec(\mathbb{Q})$. Nevertheless, half-integral Tate twists are well defined in characteristic p, namely, in the sense of Weil sheaves. Notice the unitary normalization has the advantage that all automorphic representations arising in the cohomology groups then have trivial central character on some open subgroup of $Z_G(\mathbb{A})/Z_G(\mathbb{Q})$; hence, the central characters become unitary of finite order. Here Z_G denotes the center of G. For the study of the Hodge structure or the study of Galois representations on etale cohomology groups, the artificial unitary normalization can easily be removed by an "r-fold" formal Tate twist for $r = \frac{\lambda_1 + \lambda_2}{2} \in \frac{1}{2}\mathbb{Z}$.

The Dual Bernstein–Gelfand–Gelfand (BGG) Complex. It is shown in [19] that the direct image complex $Rj_*(E_\lambda)$ on $\overline{X}$ defined for the inclusion $j : X \to \overline{X}$ can be resolved using the dual BGG complex. This means that the complex is quasi-isomorphic to complex $\overline{K}_\lambda^\bullet \cong Rj_*(E_\lambda)$ on $\overline{X}$, where

$$\overline{K}_\lambda^p = \bigoplus_\mu \overline{W}_\mu^*$$

and where the complex maps are given by certain differential operators [19], p. 232, Proposition 5.4(1). The vector bundles $\overline{W}_\mu$, and their duals $\overline{W}_\mu^*$, which occur in this sum, are prolongations to $\overline{X}$ of certain vector bundles W_μ on X, being defined by automorphic cocycles attached to a rational irreducible representation μ of the group $Gl(2, \mathbb{Q})$.

A spectral Sequence. By abuse of notation, let μ denote the representation and also the corresponding $\mathbf{M}^0 = Gl(2, \mathbb{Q})$-dominant weight. Notice that owing to our normalization of λ (either algebraic or unitary) we can replace the group $\mathbf{M} = Gl(2, \mathbb{Q}) \times \mathbb{G}_m$ considered in [19], p. 222f, by the subgroup $\mathbf{M}^0 = Gl(2, \mathbb{Q})$. In the following, let $P = \mathbf{M}^0 U$ be the Siegel parabolic subgroup of upper block triangular matrices in $Sp(4, \mathbb{Q})$ with the notation as in [19]. This gives a canonical isomorphism $\mathbf{M}^0 \cong P/U \cong Gl(2, \mathbb{Q})$. Fixing a Borel subgroup in P containing the diagonal matrices as in [19], pp. 222 and 228f, induces a Borel subgroup of $\mathbf{M}^0$. Then the weight ρ (half sum of the positive roots of $Sp(4)$) is defined and becomes $\rho = (2, 1)$. In this sense the sum $\bigoplus_\mu \overline{W}_\mu^*$ above extends over all those $Gl(2)$-dominant weights μ, where

$$\mu = w(\lambda + \rho) - \rho$$

holds for some element w of length $l(w) = p$ in the Weyl group of $Sp(4, \mathbb{Q})$. According to Chai and Faltings [19], Theorem 5.5(i), p. 233, the spectral sequence

$$E_1^{p,q} = H^q(\overline{X}, \overline{K}_\lambda^p) \Rightarrow H^{p+q}(X, E_\lambda)$$

degenerates at the 1-level.

In our case let us make this more explicit.

Given the dominant weight $\lambda = (\lambda_1, \lambda_2)$, such that $\lambda_1 \geq \lambda_2 \geq 0$, of the group $Sp(4)$, there are four substitutions w in the Weyl group of $Sp(4)$, such that $\mu = w(\lambda + \rho) - \rho$ is $Gl(2)$-dominant. These are the Weyl group elements denoted id, σ, τ, ω of length 0, 1, 2, and 3, respectively, where $\sigma(u, v) = (u, -v)$, $\tau(u, v) = (v, -u)$, and $\omega(u, v) = (-v, -u)$. For these substitutions $\mu = w(\lambda + \rho) - \rho$ is given by $(\lambda_1, \lambda_2), (\lambda_1, -\lambda_2 - 2), (\lambda_2 - 1, -\lambda_1 - 3)$, and $(-\lambda_2 - 3, -\lambda_1 - 3)$, respectively.

Since $\overline{W}_{(\mu_1, \mu_2)}^* = \overline{W}_{(-\mu_2, -\mu_1)}$ [19], p. 229, we obtain

$$\overline{K}_\lambda^0 = \overline{W}_{(-\lambda_2, -\lambda_1)}, \qquad \overline{K}_\lambda^1 = \overline{W}_{(\lambda_2 + 2, -\lambda_1)},$$

$$\overline{K}_\lambda^2 = \overline{W}_{(\lambda_1 + 3, 1 - \lambda_2)}, \qquad \overline{K}_\lambda^3 = \overline{W}_{(\lambda_1 + 3, \lambda_2 + 3)}.$$

Example 1.1. For $(\lambda_1, \lambda_2) = (0, 0)$ we get the sheaves of differentials $\overline{K}_\lambda^p = \overline{\Omega}_X^p = \Omega_X^p(log)$ with log poles at the boundary. In fact $\overline{\Omega}_X^1 = \overline{W}_{(2,0)}$.

Example 1.2. For $\lambda = (2, 0)$ we have $H^1(X, E_\lambda) = H^1(\overline{X}, \Omega_X(log)^*)$.

Hodge Structure (weights). Let us first review certain facts about the mixed Hodge structure defined on $H^3(X, E_{\lambda_1, \lambda_2})$. The pure part of this Hodge structure, which contains the image of the space of holomorphic cusp forms or the image of the cohomology with compact support, has weight $3 + \lambda_1 + \lambda_2$ (algebraic normalization) or formal weight 3 (unitary normalization of E_λ). For details see [19], Theorems 5.5 and 6.2 and p. 237.

F-filtration. On $H_!^3(X, E_{\lambda_1, \lambda_2}) = image(H_c^3(X, E_{\lambda_1, \lambda_2}) \to H^3(X, E_{\lambda_1, \lambda_2}))$ we have a pure Hodge structure, since this space is contained in the restriction of the cohomology of the smooth toroidal compactification. According to [19], Theorem 5.5, the decomposition of the weight 3 Hodge structure on $H^3_!(X, E_{\lambda_1, \lambda_2}) = image(H^3_c(X, E_{\lambda_1, \lambda_2}) \to H^3(X, E_{\lambda_1, \lambda_2}))$ into (p, q)-types yields "formal" Hodge types with p expressed as

$$p = \rho(H) - w(\lambda + \rho)(H),$$

where $H = (1/2, 1/2, -1/2, -1/2)$ holds (everything is now in the unitary normalization [19], pp. 231 and 232). Notice that the weight (μ_1, μ_2) maps $(x_1, x_2, -x_1, -x_2)$ (an element from the Lie algebra of the diagonal matrices in $Sp(4)$) to $\mu_1 x_1 + \mu_2 x_2$; hence, $(\mu_1, \mu_2)(H) = \frac{\mu_1 + \mu_2}{2}$. For the Weyl group elements $1, \sigma, \tau, \omega$ this gives $p = -\frac{1}{2}(\lambda_1 + \lambda_2)$, $p = 1 - \frac{1}{2}(\lambda_1 - \lambda_2)$, $p = 2 + \frac{1}{2}(\lambda_1 - \lambda_2)$, and $p = 3 + \frac{1}{2}(\lambda_1 + \lambda_2)$, respectively.

Let us return for a while to the more familiar algebraic normalization. Removing the formally defined Tate twist $\mathbb{Q}(r)$ on

$$H^3_!(X, E_{\lambda_1, \lambda_2}) = image\big(H^3_c(X, E_{(\lambda_1, \lambda_2)}) \to H^3(X, E_{(\lambda_1, \lambda_2)})\big),$$

for the half-integer $r = \frac{\lambda_1 + \lambda_2}{2}$, we obtain the algebraic normalization with the Hodge types $(p, q)_{alg} = (p, q)_{unit} + (\frac{\lambda_1 + \lambda_2}{2}, \frac{\lambda_1 + \lambda_2}{2})$, where p_{alg} ranges now over the integers $0, 1 + \lambda_2, 2 + \lambda_1, 3 + \lambda_1 + \lambda_2$. This defines an "honest" Hodge structure on

$$V = H^3_!(X, E^{unit}_{(\lambda_1, \lambda_2)}(-r)) = H^3_!(X, E^{alg}_{(\lambda_1, \lambda_2)}),$$

which is pure of weight $\lambda_1 + \lambda_2 + 3 = k_1 + k_2 - 3$. Its decomposition into (p, q)-types is given by

$$\boxed{V = V^{(k_1 + k_2 - 3, 0)} \oplus V^{(k_1 - 1, k_2 - 2)} \oplus V^{(k_2 - 2, k_1 - 1)} \oplus V^{(0, k_1 + k_2 - 3)}}$$

The new notation $k_1 = \lambda_1 + 3$ and $k_2 = \lambda_2 + 3$ is introduced for the following reason.

Links with Holomorphic Modular Forms. Consider the complex analytic situation. Then the space of global sections $H^0(\overline{X}, \overline{W}_{(k_1, k_2)})$ can be identified with the space $[\Gamma, (k_1, k_2)]$ of classical Siegel holomorphic vector-valued modular forms of weight (k_1, k_2). The degeneration of the spectral sequence [19], Theorem 5.5, gives inclusions from the edge homomorphisms

$$[\Gamma, (\lambda_1 + 3, \lambda_2 + 3)] = H^0(\overline{X}, \overline{W}_{\lambda_1 + 3, \lambda_2 + 3}) = H^0(\overline{X}, \overline{K}^3_\lambda) \hookrightarrow H^3(X, E_{(\lambda_1, \lambda_2)}).$$

Hence, vector-valued holomorphic modular forms of weight $k_1 \geq k_2 \geq 3$ can be embedded into the cohomology of suitable canonical coefficient systems E_λ. These embeddings are compatible with the action of Hecke operators. The case $k_1 = k_2 = k$ corresponds to classical Siegel modular forms of weight k.

For $k_1 \geq k_2 \geq 3$ the space of holomorphic cusp forms $[\Gamma, (k_1, k_2)]$ maps to the subspace $V^{(k_1 + k_2 - 3, 0)}$ of the Hodge decomposition.

Remark 1.1. More generally for $k_1 \geq k_2 > 0$ one could ask which holomorphic cusp forms can be detected by higher cohomology groups. From the list above we see that the space of vector-valued holomorphic modular forms of type (k_1, k_2) can be embedded into the cohomology of a canonical coefficient system if and only if $\rho + (-k_2, -k_1)$ is not $Sp(4)$-singular. In other words in all cases where $k_2 \neq 2$ and $(k_1, k_2) \neq (2, 1)$ or $(k_1, k_2) \neq (1, 1)$.

1.2 Automorphic Representations

For the group $G = GSp(4, \mathbb{Q})$ of symplectic similitudes we consider the corresponding Shimura varieties

$$S_K(\mathbb{C}) = G(\mathbb{Q}) \backslash G(\mathbb{A})/Zentr(h)_\infty K,$$

where K is a compact open subgroup of $G(\mathbb{A}_{fin})$, and where $Zentr(h)_\infty$ is contained in the product of a maximal compact subgroup K_∞ and the topologically connected component $A_G(\mathbb{R})^0 = Z^0_{G,\infty}$ of the center $Z_{G,\infty}$ of $G_\infty = G(\mathbb{R})$ (we also refer to [110]). As an analytic space each $S_K(\mathbb{C})$ is a disjoint union of Siegel modular varieties $X = \bigsqcup_i \Gamma_i \backslash H$ for certain Γ_i depending on K, as considered in Sect. 1.1. The topologically connected components $\pi_0(S_K(\mathbb{C})) = \pi_0(G(\mathbb{Q}) \backslash G(\mathbb{A})/Zentr(h)_\infty K)$ can be identified with

$$(\prod_p \mathbb{Z}_p^*)/\nu(K),$$

where $\nu : G \to G^{ab} \cong \mathbb{G}_m$ is the similitude character of G. Hence, the results of Sect. 1.1 also apply for the analytic spaces $S_K(\mathbb{C})$. In particular, we can consider the embedding of the spaces of holomorphic modular forms for $X = S_K(\mathbb{C})$ and $H^\bullet(X, E_\lambda)$ as follows.

The adele group $GSp(4, \mathbb{A}_{fin})$ acts on the direct limit

$$H^\bullet(S(\mathbb{C}), E_\lambda) = \varinjlim_K H^\bullet(S_K(\mathbb{C}), E_\lambda)$$

by an admissible representation. The automorphic representation of the group $GSp(4, \mathbb{A}_{fin})$ on the direct limit over K of the spaces $\bigoplus_i [\Gamma_i, (k_1, k_2)]$ of holomorphic vector-valued modular forms of type (k_1, k_2) with fixed $k_1 \geq k_2 \geq 3$ (holomorphic discrete series) embeds into the representation $H^3(S(\mathbb{C}), E_{k_1-3, k_2-3})$, as explained in Sect. 1.1.

Furthermore, it is well known (see [13] or again [19], p. 233) that the subspace of cusp forms within this space of holomorphic modular forms maps under this embedding into the cohomology subspace

$$H^3_!(X, E_\lambda) = Image(H^3_c(X, E_\lambda) \to H^3(X, E_\lambda)).$$

Let us fix $k_1 \geq k_2 \geq 3$; hence, $\lambda = (k_1 - 3, k_2 - 3)$ and $E = E_\lambda$. Attached to (k_1, k_2) there exists the holomorphic discrete series representation $\pi_\infty^{hol}(k_1, k_2)$ of $GSp(4, \mathbb{R})$ of type (k_1, k_2). Then from what has been said above we get

Lemma 1.1. *The subspace defined by the direct limit of holomorphic cusp forms is injected into $H_!^3(S(\mathbb{C}), E_{k_1-3, k_2-3})$, and is isomorphic to*

$$\bigoplus_{\pi = \pi_{fin} \otimes \pi_{\infty, hol}(k_1, k_2)} \pi_{fin}$$

as a module under the group $GSp(4, \mathbb{A}_{fin})$. The sum extends over all irreducible cuspidal automorphic representations $\pi = \pi_\infty \otimes \pi_{fin}$, such that π_∞ belongs to the holomorphic discrete series of type (k_1, k_2).

Fixing notation. For an irreducible cuspidal automorphic representation π of $GSp(4, \mathbb{A})$, let S be a finite set of non-Archimedean places, which includes the ramified places of π_{fin}. Choose a compact open small level group $K = K^S K_S \subseteq G(\mathbb{A}_{fin})$ such that π^K is nonzero, where $K^S = GSp(4, \mathbb{Z}^S)$ is the maximal compact subgroup of unimodular symplectic similitudes at the non-Archimedean places outside S.

Canonical Models. The Shimura varieties $X = S_K(\mathbb{C})$ have canonical models over the field $\mathbb{Q}$. For the group K_S there are two particular choices of interest.

The First Choice. $K_S = K_N \subseteq GSp(4, \mathbb{Z}_S)$ is a principal congruence subgroup of level N, i.e., the subgroup of all matrices $g \in GSp(4, \mathbb{Z}_S)$ with the congruence condition $g \equiv id \bmod N$ for the integer $N = \prod_{l \in S} l^{v_l(N)}$. In fact, for this choice of K, the Shimura variety can be identified with the complex analytic space attached to the moduli space of principally polarized Abelian varieties with level N structure. Let $\mathcal{M}$ temporarily denote its canonical model over $Spec(\mathbb{Q}(\zeta_N))$.

The Second Choice. Within the set S of bad non-Archimedean places we assume $K_S = K_N'$ to be a "modified" principal congruence subgroup (denoted $K_N^{(2)}$ in [110], p. 102). K_N' is the group generated by the matrices $M \equiv id \bmod N$ in $GSp(4, \mathbb{Z}_f)$ in K_S together with the matrices

$$\begin{pmatrix} E & 0 \\ 0 & \lambda \cdot E \end{pmatrix}, \qquad \lambda \in \mathbb{Z}_f^*.$$

Obviously, $\Delta = K_N'/K_N \cong (\mathbb{Z}/N\mathbb{Z})^*$. Let the model temporarily denote $\mathcal{M}'$.

For both choices the corresponding Shimura variety $S_K(\mathbb{C})$ has a canonical model defined over $Spec(\mathbb{Q})$. For the second choice this model $\mathcal{M}'$ becomes geometrically connected, which is convenient from the point of view of classical algebraic geometry, whereas the canonical model $\mathcal{M}$ considered as a scheme over

$Spec(\mathbb{Q})$ by restriction of scalars has many geometric components. However, using Galois descent (see [24], p. 286, Lemma 2.7.5), these two models $\mathcal{M}$ and $\mathcal{M}'$ are related by Remark 1.2, which easily allows us to pass from one point of view to the other.

Remark 1.2. For fixed N with the notation above the canonical $\mathbb{Q}$-model $\mathcal{M}$ is obtained from the $\mathbb{Q}$-model $\mathcal{M}'$ by the Cartesian product $\mathcal{M} = \mathcal{M}' \times_{Spec(\mathbb{Q})} Spec(\mathbb{Q}(\zeta_N))$.

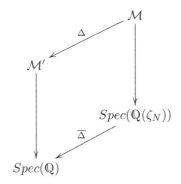

The action of K'_N on $\mathcal{M}$ factorizes over the quotient $\Delta = (\mathbb{Z}/N\mathbb{Z})^*$, and this action is compatible with the Galois action on $Spec(\mathbb{Q}(\zeta_N))$ provided the reciprocity law has been suitably normalized.

The space $\mathcal{M}'$ is a quotient of $\mathcal{M}$. We will later consider certain ample divisors on the quotient space $\mathcal{M}'$, which pull back to an ample divisor of the geometrically nonconnected scheme $\mathcal{M}$.

For places p outside S, $\mathcal{M}$ has a smooth model $\mathcal{X}$ over $Spec(\mathbb{Z}_p)$ for level $N \geq 3$ and $(p, N) = 1$ defined by the moduli space of principal polarized Abelian varieties with level N structure.

Generalized Eigenspaces. Let H_S denote the Hecke algebra of all K^S bi-invariant functions on $G(\mathbb{A}^S)$ with compact support. The Hecke algebra H_S acts on the cohomology groups $H^\bullet(X, E)$ and $H_c^\bullet(X, E)$ via correspondences. Let us fix some further notation which will be used frequently. Let S be a set of places as above containing all bad places. Consider some non-Archimedean place p not in S. This will be a prime of good reduction for the model $\mathcal{X}$ of $S_{K_N K^S}(\mathbb{C})$. Then we write

$$\mathbb{A} = \mathbb{R} \times \mathbb{A}_{fin} = \mathbb{R} \times \mathbb{A}_S \times \mathbb{Q}_p \times \mathbb{A}_f,$$

where f means finite places away from S and some distinguished prime p, also called the Frobenius prime later. In particular, we will consider the tensor decomposition

$$\pi = \pi_\infty \otimes \pi_{fin}$$

for

$$\pi_{fin} = \pi_S \otimes \pi_p \otimes \pi_f.$$

Hence, in this setting we always let S be some fixed finite set including all ramified places of π, and p will be some distinguished unramified place for π (i.e., not in S). At the remaining non-Archimedean places the spherical representation π_f is then uniquely determined by the eigenvalues of the action of the Hecke algebra H_f on its spherical vector. Any finite-dimensional module V of the commutative H_f algebra uniquely decomposes into a direct sum of generalized eigenspaces (Jordan blocks) $V = \oplus V(\chi)$, on each of which the semisimplification of the image of H_f in $End(V)$ acts by a character χ of H_f. If χ arises from the spherical representation π_f, we also write $V(\chi) = V(\pi_f)$ for this generalized eigenspace. In the following we consider the following *generalized* eigenspaces

$$H^\bullet(X, E(\pi_f)) := H^\bullet(X, E)(\pi_f)$$

on the finite-dimensional vector space $H^\bullet(X, E)$

$$H^\bullet(X, E) = \bigoplus_{\pi_f} H^\bullet(X, E(\pi_f))$$

for $K = K_N K^S$ and $K^S = K_p K_f = GSp(4, \mathbb{Z}^S)$. Furthermore, we consider the related spaces $H_c^\bullet(X, E(\pi_f))$ and $H_!^\bullet(X, E(\pi_f))$.

1.3 CAP and Toric Representations

For a reductive group G over $\mathbb{Q}$ let π be any irreducible cuspidal automorphic representation of $G(\mathbb{A})$. For a parabolic subgroup $P \subseteq G$ for almost all places v the following holds:

- There exists an irreducible representation τ_v, with central character denoted ω_v, of the Levi component of P_v such that π_v is a subquotient of $Ind_{P_v}^{G_v}(\tau_v)$.

Definition 1.1. The cuspidal representation π will be called CAP [69] if for a proper parabolic subgroup $P \subseteq G$ over $\mathbb{Q}$ there exists a finite set of places T such that the representation $\otimes_{v \notin T} \tau_v$ of the Levi group of P is automorphic or it will be called toric if its central character $\otimes_{v \in T} \omega_v$ is automorphic (viewed as a character of the center of the Levi component).

Being CAP or toric is, of course, a property of the weak equivalence class of irreducible cuspidal automorphic representations π of G; hence, we will not distinguish between π and π_f in using this notation.

Recall. Two irreducible admissible representations are called weakly equivalent if their local admissible representations coincide at almost all places.

Obviously CAP implies toric. It seems plausible that a converse form also holds, i.e., an irreducible cuspidal automorphic representation should be toric if and only if it is a CAP representation. See [112] for some results in this direction in the case of the group $G = GSp(4)$. The CAP representations for the group $G = GSp(4)$ have been classified by Piatetski-Shapiro [69] and Soudry [97].

Theorem 1.1. *Let us maintain the notation from above. For an irreducible cuspidal automorphic representation π whose infinite component π_∞ belongs to a holomorphic discrete series representation of type (k_1, k_2), consider the spherical representation π_f as above and the sheaf $E = E_\lambda$ for $\lambda = (k_1 - 3, k_2 - 3)$. Then:*

1a. The generalized π_f-eigenspace $H^3_!(X, E(\pi_f)) \subseteq H^3(X, E(\pi_f))$ is nontrivial for some $X = S_K(\mathbb{C})$ for some $K = \prod_{v \neq \infty} K_v$ with $K_v = GSp(\mathbb{Z}_v)$ for all v, where π is spherical.

1b. If π not is CAP (or toric) $\left(H^3(S_K(\mathbb{C}), E)/H^3_!(S_K(\mathbb{C}), E) \right)(\pi_f) = 0$ for all K.

2. If π is not CAP, for all K the generalized eigenspaces

$$H^i(S_K(\mathbb{C}), E(\pi_f)) = 0, \quad i \neq 3$$

vanish in degree different from $i = 3$.

Corollary 1.1. *Suppose π is not toric. Then for all Hecke operators $f_p \in C^\infty_c(G(\mathbb{Q}_p)//K_p)$ and $X = S_K(\mathbb{C})$ the following holds:*

$$-Trace(\pi_p(f_p); H^3(X, E(\pi_f))) = \sum_{i=0}^{6} (-1)^i Trace(\pi_p(f_p); H^i(X, E(\pi_f))).$$

Proof of Theorem 1.1. By our assumptions regarding π and E, the assertion (1a) follows from Lemma 1.1. In particular,

$$H^3(S(\mathbb{C}), E_{k_1-3, k_2-3}) \neq 0.$$

By assumption π is cuspidal, but not CAP. Hence, all generalized eigenspaces

$$H^\bullet(S_K(\mathbb{C}), E)(\pi_f)$$

are contained in the subspaces $H^\bullet_!(S_K(\mathbb{C}), E)$, since from [83] the cohomology is spanned by the subspaces $H^\bullet_!(S_K(\mathbb{C}), E)$ and the Eisenstein cohomology. We remark that this result in [83] uses the Borel–Serre compactifications. Since π is not CAP, the generalized eigenspace of π_f on the Eisenstein cohomology is obviously trivial; hence, this implies (1b), and for (2) it is enough to consider the cohomology groups $H^i_!(X, E)$. In other words, one has to show the vanishing of the $H^i_!(S_K(\mathbb{C}), E(\pi_f))$ for $i \neq 3$.   □

This will be done in the remaining sections of this chapter with arguments that are a slight modification of those in [112, 114]. They use the weak form of the Ramanujan conjecture proven in Chai and Faltings [19], p. 264. The point will be that if π_f contributes nontrivially to the third cohomology of E and to some other cohomology group of E in a different degree, then the L-series of π_f will have a pole, which forces π to be CAP, contradicting the assumptions.

1.4 The *l*-Adic Sheaves

Remark 1.3 concerns the sheaves E. For the following fix some auxiliary prime l, say, contained in S. Furthermore, fix an isomorphism ι between the algebraic closure of $\mathbb{Q}_l$ and $\mathbb{C}$. The comparison isomorphism allows us to identify Betti cohomology and $\overline{\mathbb{Q}}_l$-adic cohomology. We now consider the l-adic version of the coefficient system E.

In fact, the sheaves $E = E_\lambda$ can already be defined as an l-adic smooth Weil sheaf of a model over $\mathbb{Q}$ of the algebraic variety $X = S_K(\mathbb{C})$. They are of geometric origin and occur in the direct image of products of the universal Abelian variety

$$A^s \to S_K(\mathbb{C}), \qquad A^s = A \times_{S_K(\mathbb{C})} A \times_{S_K(\mathbb{C})} \ldots \times A.$$

For $\lambda = (1,0)$ the basic example is the sheaf $E_\lambda = R^1\pi_*(\overline{\mathbb{Q}}_l)$. All the other l-adic coefficient systems E_λ can be produced from such direct images in a similar way. See [19], p. 235.

The construction also works over $Spec(\mathbb{Z}) \setminus S$. Consider a prime not in S and let T be the spectrum of the strict henselization of $\mathbb{Z}$ with respect to this prime. There exists a smooth model $\pi : \mathcal{X}_T \to T$ of the moduli space defined by $S_K(\mathbb{C})$ (for $N \geq 3$) together with a universal Abelian variety $\mathcal{A}$ (or products of it over $\mathcal{X}_T$), being smooth over $\mathcal{X}_T$. This defines coefficient systems $\mathcal{E} = \mathcal{E}_\lambda$.

Lemma 1.2. *The construction above defines smooth sheaves $\mathcal{E}$ on $\mathcal{X}_T$ such that the sheaves $R^\bullet\pi_!(\mathcal{E})$ and $R^\bullet\pi_*(\mathcal{E})$ are smooth on T and commute with arbitrary base change $T' \to T$.*

Proof of Lemma 1.2. Since the Leray spectral sequence for the morphism π degenerates at the E_2-level by the Liebermann trick (cohomology decomposes by the different eigenvalues of the multiplication map), this leads us to consider the case of the constant sheaf $\overline{\mathbb{Q}}_l$ and its higher direct images with respect to the composed morphism from the Cartesian product $f : \mathcal{A}^s \to \mathcal{X}_T \to T$. This map is smooth and admits a proper smooth extension $\overline{f} : \overline{\mathcal{A}^s} \to T$, where the boundary divisor is a relative divisor with normal crossing [19], p. 195. Hence, according to [72], theorem of finitude (App. 1.3.3(ii)), the pair $(f, \overline{\mathbb{Q}}_l)$ is cohomologically proper. Therefore, the higher direct image sheaves $Rf_*(\overline{\mathbb{Q}}_l)$ commute with arbitrary base change $T' \to T$. So, according to [72], theorem of finitude (App. 2.4), the specialization and the cospecialization maps are isomorphisms, because f is smooth and therefore locally acyclic. Therefore, the direct image sheaves $R^i f_*(\overline{\mathbb{Q}}_l)$ are smooth sheaves.

Without restriction of generality, the morphism f can be compactified by a projective morphism $\overline{f}$ [19], p. 207; therefore, the relative Poincaré duality theorem holds for the morphism f

$$Rf_* R\mathcal{H}om(K, f^!(\overline{\mathbb{Q}}_l)) \cong R\mathcal{H}om(Rf_!(K), \overline{\mathbb{Q}}_l).$$

$f^!(\overline{\mathbb{Q}}_l) = \overline{\mathbb{Q}}_l[2d](d)$, since f is smooth, where d is the relative dimension. Hence, the cohomology sheaves of the complex $R\mathcal{H}om(Rf_!(\overline{\mathbb{Q}}_l), \overline{\mathbb{Q}}_l)$ are smooth sheaves

on T. By the theorem of finitude (Theorem 4.3) this complex is quasi-isomorphic
to $D(Rf_!(\overline{\mathbb{Q}}_l))$; hence, $Rf_!(\overline{\mathbb{Q}}_l) \cong D(D(R_!f(\overline{\mathbb{Q}}_l))) = D(R_*f(\overline{\mathbb{Q}}_l))$. A spectral
sequence is now obtained from $Rf_!(\overline{\mathbb{Q}}) = D(R_*f(\overline{\mathbb{Q}}_l))$, which computes $R^if_!(\overline{\mathbb{Q}})$
in terms of the $R^if_*(\overline{\mathbb{Q}}_l)$. It has smooth E_2-terms by the result above; therefore, no
higher $\mathcal{E}xt$ sheaves occur. Hence, the spectral sequence degenerates. Thus, also the
sheaves $R^if_!(\overline{\mathbb{Q}}_l)$ are smooth sheaves on T. $\square$

Positive Characteristics. Fix some prime $p \notin S$ and different from l. Lemma 1.2
now allows us to consider the situation in characteristic p using base change. Let
now X and $\overline{X}$, respectively, denote the moduli scheme and a smooth proper toroidal
compactification over the algebraic closure $\overline{\mathbb{F}}_p$ of the prime field of characteristic p.
(Of course, the sheaf E need not extend to a locally constant coefficient system
on $\overline{X}$.) First of all, $H_c^\bullet(X, E)$ is independently defined using comparison isomorphisms, whether we consider $X = S_K(\mathbb{C})$ to be the Shimura variety over $\mathbb{C}$ or use
its smooth reduction mod p of a model over T, considered as a scheme over $\overline{\mathbb{F}}_p$.
Namely, these spaces will be identified using the specialization map. This follows
from the base change assertions of Lemma 1.2. By Poincaré duality, the same holds
for cohomology with compact support and therefore also for $H_!^\bullet(X, E)$. These identifications are equivariant with respect to the action of the Hecke correspondences
on these cohomology groups.

Reduction Mod p. To simplify comparison with automorphic L-series later, we
choose a square root $p^{1/2}$ of p in $\overline{\mathbb{Q}}_l$ and an isomorphism $\iota : \overline{\mathbb{Q}}_l \cong \mathbb{C}$ and normalize
the Weil sheaves E_λ by a half-integral Tate twist to become ι-pure Weil sheaves of
weight 0 (well defined over the residue field $\mathbb{F}_p$ after reduction mod p). For example,
$E_{(1,0)} = R^1\pi_*(\overline{\mathbb{Q}})_l(\frac{1}{2})$. This sheaf is self-dual as a Weil sheaf. The same is true for
the other E_λ by tensor calculus.

Over the algebraic closure $\overline{\mathbb{F}}_p$ of the residue field the Frobenius automorphism acts
on the cohomology groups commuting with the action of the Hecke correspon-
dences. The eigenvalues of the Frobenius on the spaces $H_!^i(X, E)$ are pure of weight
i. To take into account that our spaces are not compact, let $\overline{E}$ be the sheaf complex
on $\overline{X}$ defined by the perverse intermediate extension of the sheaf E to the compactification $\overline{X}$ considered in characteristic p. This is a pure self-dual perverse Weil
sheaf complex of weight 0. Its cohomology groups are pure, and hence so are the
subquotients $H_!^\bullet(X, E)$.

1.5 Lefschetz Maps

By the hard Lefschetz theorem for the pure perverse sheaf $\overline{E}$ on $\overline{X} = \overline{S_K(\mathbb{C})}$ we
get for all $r \geq 0$ isomorphisms

$$L^r : H^{3-r}(\overline{X}, \overline{E}) \to H^{3+r}(\overline{X}, \overline{E})(r),$$

where L^r is the map induced from the iterated cup product with the fundamental class of some ample line bundle $\mathcal{L}$ on the toroidal compactification $\overline{S_K(\mathbb{C})}$.

There are corresponding maps on the cohomology groups and cohomology groups with compact support. We have a commutative diagram of morphisms

$$
\begin{array}{ccc}
H_c^\bullet(X, E) & \to & H^\bullet(\overline{X}, \overline{E}) \\
\downarrow & & \downarrow res \\
H_!^\bullet(X, E) & \hookrightarrow & H^\bullet(X, E)
\end{array}
$$

compatible with the Lefschetz maps L^r. We consider these both in characteristic p and characteristic 0. In positive characteristic L^r commutes with the Frobenius action up to some factor (coming from the r-fold Tate twist). It is enough to consider the case $\overline{E}$.

Hecke operators. Next we show that L^r commutes with the action of Hecke operators. Using the base change and smoothness (Lemma 1.2), the specialization map is an isomorphism outside S. It therefore suffices to consider the question in characteristic 0. We can also replace the moduli variety X of level N by some geometrically connected component X' using the remark of Sect. 1.2. For the variety X' choose $\mathcal{L}$ as in [110]. Then, as explained in [110], pp. 104 and 108, the Lefschetz map $L^r : H_!^{3-r}(X', E) \to H_!^{3+r}(X', E)(r)$ is defined by the iterated cup product with respect to the translation-invariant harmonic Kähler 2-form on $G(\mathbb{R})$. The same then holds by pullback for X. Hence, L commutes with the action of Hecke operators, because it already does so on the level of differential forms. We therefore obtain induced maps preserving the generalized eigenspaces:

$$
L^r : H_!^{3-r}(X, E(\pi_f)) \to H_!^{3+r}(X, E(\pi_f))(r).
$$

Lemma 1.3. *Suppose π_f is a representation which is not weakly isomorphic to an Eisenstein representation attached to a proper parabolic subgroup. Then the restriction of L^r to the generalized π_f eigenspace defines an isomorphism*

$$
L^r : H_!^{3-r}(X, E(\pi_f)) \to H_!^{3+r}(X, E(\pi_f))(r).
$$

Proof. Poincaré duality applied to the self-dual sheaf E on X proves $\dim H_!^{3-r}$ $(X, E) = \dim H_!^{3+r}(X, E)(r)$. Use $Im(\phi)^* \cong Im(\phi^*)$ for the dual ϕ^* of a linear map $\phi : V \to W$ between finite-dimensional $\overline{\mathbb{Q}}_l$ vector spaces. Apply this to the natural map ϕ from cohomology with compact support to ordinary cohomology. Observe that $\phi = \phi^*$ is self-dual with respect to Poincaré duality. Hence, it is enough to show injectivity of L^r.

At this point of the proof we use a transcendental argument. According to [83] every cohomology class has an automorphic representative in our case (for the general case, this holds by a theorem of J. Franke). By our assumption regarding π_f the representatives have to be cuspidal. This means that the generalized π_f eigenspaces are contained in the cuspidal cohomology. Hence, they can be represented by harmonic differential forms (with respect to the translation-invariant metric) that have rapid decay on the Siegel space $\Gamma \backslash H$. From [13], p. 47, the natural map from harmonic rapidly decreasing forms $\mathcal{H}_{rd}^{\bullet}(E)$ to the cohomology $H^{\bullet}(\Gamma \setminus H, E)$ is injective (coefficients of E are assumed to be complex). As explained in the section on Hecke operators above, the Lefschetz map L^r is realized by iterated cup products with the translation-invariant harmonic Kähler 2-form ω. Hence, L^r preserves the space $\mathcal{H}_{rd}^{\bullet}(E)$ of harmonic rapidly decreasing forms:

$$
\begin{array}{ccc}
\mathcal{H}_{rd}^{3-r} & \lhook\joinrel\longrightarrow & H^{3-r}(X, E) \\
{\scriptstyle \wedge\omega^r}\big\downarrow & & \big\downarrow{\scriptstyle L^r} \\
\mathcal{H}_{rd}^{3+r} & \lhook\joinrel\longrightarrow & H^{3+r}(X, E)
\end{array}
$$

This shows that our hypothesis regarding π_f implies that the natural map

$$
H^{3-r}(X, E(\pi_f)) \to H^{3+r}(X, E(\pi_f))(r)
$$

is injective. By the dimension argument at the beginning of the proof it is therefore an isomorphism. $\square$

Remark 1.3. For the constant sheaf $E = \overline{\mathbb{Q}}_l$ the map

$$
L^r : H_!^{3-r}(X, E) \to H_!^{3+r}(X, E)(r)
$$

defines an isomorphism. See [114]. It would be interesting to know whether this holds in general.

1.6 Weak Ramanujan

A weak form of the Ramanujan conjecture, proved in [19], p. 264, implies that the graded generalized eigenspace $H_!^{\bullet}(X, E)(\pi_f)$, when considered in positive characteristic p, is preserved by the Frobenius action. Furthermore, the Frobenius $Frob_p$ can have at most four different eigenvalues on this subspace, and – although not stated explicitly in [19] – the possible eigenvalues are given by

$$
p^{3/2}\alpha_0, p^{3/2}\alpha_0\alpha_1, p^{3/2}\alpha_0\alpha_2, p^{3/2}\alpha_0\alpha_1\alpha_2,
$$

where these numbers define the local spinor L-function $L_p(\pi_f, s)$

$$
L_p(\pi_f, s)^{-1} = (1 - \alpha_0 p^{-s})(1 - \alpha_0\alpha_1 p^{-s})(1 - \alpha_0\alpha_2 p^{-s})(1 - \alpha_0\alpha_1\alpha_2 p^{-s})
$$

attached to π. Notice the shift by $\frac{3}{2}$ (unitary normalization). The complex numbers $\alpha_0, \alpha_1, \alpha_2$ are the Satake parameters of the unramified local representation π_p attached to π_f at the prime p. Owing to the unitary normalization, the value $(\alpha_0)^2\alpha_1\alpha_2$, which defines the central character, has absolute value 1.

Low Degrees. Concerning the cohomology groups in degrees $0, 1, 5, 6$ it is well known that cuspidal representations do not contribute to these degrees. To see this, it is enough to consider degrees 0 and 1 using Poincaré duality. By the vanishing theorems for the continuous L^2-cohomology groups $H^q(\Gamma, V)_{(2)}$, the cohomology groups $H^i_{cusp}(X, E)$ (sheaves with coefficients over $\mathbb{C}$) vanish for $i \leq rk_{\mathbb{R}}(G) - 1$. Here $rk_{\mathbb{R}}(G) = 2$ denotes the real rank of the group $GSp(4)$ ([13], pp. 40–42, in particular Theorem 4.4(i).2). If, therefore, π_f is not CAP, the generalized eigenspaces $H^i(X, E(\pi_f))$, etc. vanish in degree $i = 0, 1, 5, 6$. Hence, for the remaining part of this chapter – in particular for the proof of assertion (2) of Theorem 1.1 – we can restrict ourselves to discussing non-CAP representations π_f, whose generalized eigenspaces are nontrivial on the cohomology groups $H^i_!(X, E)$ both in degree 3 and in degree 2 (or degre 4). By Lemma 1.3 π_f contributes both to degree 2 and to degree 4.

Multiplicity 1. Let us now suppose π_f is cuspidal automorphic, but not CAP, and contributes to cohomology in the degrees $i = 2, 3, 4$. If β_1, β_2 are two different eigenvalues of the Frobenius on $H^2_!(X, E)(\pi_f)$ and $p\beta_1, p\beta_2$ are two different eigenvalues on $H^4_!(X, E)(\pi_f)$, corresponding vice versa under the Lefschetz isomorphism L, then by purity of the eigenvalues of the Frobenius on $H^i_!(X, E)$ all these four eigenvalues would be distinct from each other, having weights 2 and 4, respectively. Then, by the weak form of the Ramanujan conjecture stated above, these four eigenvalues must all be eigenvalues. This implies a contradiction, since π_f also contributes to degree 3. Hence, at least one of the eigenvalues of the Frobenius must have absolute value of weight 3. Therefore, at most one eigenvalue of the Frobenius $Frob_p$ occurs in $H^2_!(X, E)$. Then the same holds for $H^4_!(X, E)$ by the Lefschetz isomorphism.

1.7 Residues of L-Series

For a cuspidal representation π_f which is not CAP and violates assertion (2) of Theorem 1.1, we conclude that the Frobenius at the prime p has at most one eigenvalue on the generalized π_f-eigenspace of the cohomology in degree 2. If the action of the absolute Galois group $Gal(\Gamma_{\mathbb{Q}})$ of $\mathbb{Q}$ on $H^2_!(X, E(\pi_f))$ is semisimple, then the corresponding Galois representation has to be a multiple of a fixed Abelian character. The same is also true in degree 4.

To show semisimplicity we proceed as follows.

First we reduce to the case of the scheme X' such that $X = X' \times_{Spec(\mathbb{Q})} Spec(\mathbb{Q}(\zeta_N))$. Then according to [110], p. 104, there exists an ample divisor $\overline{Y}$ in $\overline{X}$ defined over $\mathbb{Q}$ described below such that $\overline{Y} = \overline{Y}_1 \cup (\overline{X} \setminus X)$. Put $Y = X \cap \overline{Y}_1$. Then the restriction map on the right of

$$H^2_!(X, E) \leftarrow H^2_c(X, E) \hookrightarrow H^2_c(Y, E|Y)$$

is injective. This holds since the kernel is a quotient of $H_c^2(\overline{X} \setminus \overline{Y}, E)$, which is zero by the affine vanishing theorem of Artin–Grothendieck. Notice that $X \setminus Y = \overline{X} \setminus \overline{Y}$ is affine of dimension 3. The divisor $\overline{Y}$ is singular, a union of products of two modular curves together with the components of the boundary divisor $\overline{X} \setminus X$, which itself is a union of Shioda surfaces. Passing to the normalization of Y, one can control the cohomology groups $H_c^2(Y, E|Y)$ in terms of the cohomology groups of modular curves and canonical coefficient systems on them as in [110]. Again as in [110] this implies for $H_c^2(X, E)$, and hence for the quotient $H_!^2(X, E)$, semisimplicity and furthermore that all one-dimensional Galois representations, which occur in the (l-adic) cohomology, are grössen characters viewed as characters of the absolute Galois group of $\mathbb{Q}$.

That the Hecke algebra acts by a single character on $H_!^2(X, E(\pi_f))$ and $H_!^4(X, E(\pi_f))$ (multiplicity 1) now easily implies that the absolute Galois group $\Gamma_{\mathbb{Q}}$ of $\mathbb{Q}$ acts by a multiple of some grössen character $\chi(-1)$ on $H_!^2(X, E)(\pi_f)$ for a Dirichlet character χ. Similarly, in degree 4, it acts by the corresponding multiple $\chi(-2)$.

We still assume that π_f is cuspidal, but not a CAP representation, and contributes to degree 3, however not exclusively. Then both $p\chi(p)$ and $p^2\chi(p)$ must occur as eigenvalues of the Frobenius $Frob_p$ on the generalized eigenspaces $H_!^\bullet(X, E(\pi_f))$. These two numbers occur among the four "roots" of the spinor L-series at the prime p. Since the Weyl group acts transitively on the four "roots" of the local spinor L-function $L_p(\pi_p, s)$, we can assume $p^{3/2}\alpha_0 = p\chi(p)$ without restriction of generality. Furthermore, the Weyl group allows us to interchange $\alpha_0\alpha_1$ and $\alpha_0\alpha_2$. So we can assume $p^2\chi(p)$ to be either $p^{3/2}\alpha_0\alpha_1$ or $p^{3/2}\alpha_0\alpha_1\alpha_2$. Hence, for all primes $p \notin S$ (S is finite, fixed, and large enough) one of the two cases holds:

- Case 1. $\alpha_1 = p$
- Case 2. $\alpha_1\alpha_2 = p$

Lemma 1.4. *$\alpha_1\alpha_2 = p$ holds for all $p \notin S$, and $|\alpha_1| = |\alpha_2| = p^{1/2}$. Furthermore, $p^{3/2}\alpha_0 = p\chi(p)$, and the central character of π is $\omega_\pi = \chi^2$.*

Proof. Assume to the contrary $\alpha_1 = p$. The unitary central character ω_π of π is related to the value $\alpha_0^2\alpha_1\alpha_2$, which has absolute value 1. Since $p^{3/2}\alpha_0 = p\chi(p)$ and $\alpha_1 = p$, this would imply $|\alpha_2| = 1$. Notice $p^{3/2}\alpha_0\alpha_1\alpha_2$ then has absolute value p^2; hence, at least one of the values $p^{3/2}\alpha_0\alpha_1$ or $p^{3/2}\alpha_0\alpha_2$ occurs as an eigenvalue of the Frobenius in the third cohomology with absolute value $p^{3/2}$. Since $\alpha_1 = p$, $|\alpha_2| = 1$, and $\alpha_0 = \chi(p)p^{-1/2}$, this is impossible; hence, only the second case $\alpha_1\alpha_2 = p$ is possible. Then $\alpha_0 = \chi(p)p^{-1/2}$ and $\alpha_0\alpha_1\alpha_2 = \chi(p)p^{1/2}$. Again there must be an eigenvalue $p^{3/2}\alpha_0\alpha_1$ or $p^{3/2}\alpha_0\alpha_2$ of absolute value $p^{3/2}$ coming from the third cohomology. Therefore, $|\alpha_1| = |\alpha_2| = p^{1/2}$. Finally $\omega_{\pi_f}(q) = \alpha_0 \cdot \alpha_0\alpha_1\alpha_2 = p^{-1/2}\chi(p) \cdot p^{1/2}\chi(p)$ proves the last claim. This proves the lemma. $\square$

The Final Contradiction. Consider the partial L-series for the primes $p \notin S$

$$L^S(\pi, s) = \prod_{p \notin S} L_p(\pi, s).$$

Since $(1 - \alpha_0\alpha_1\alpha_2 p^{-s})^{-1} = (1 - \chi(p)p^{\frac{1}{2}-s})^{-1}$ by Lemma 1.4 and $\alpha_0 = \chi(p)p^{-1/2}$, the estimates for the Satake parameters $|\alpha_1| = |\alpha_2| = p^{1/2}$ obtained above imply for any Dirichlet character $\tilde{\chi}$

$$Res_{s=\frac{3}{2}}L^S(\pi \otimes \tilde{\chi}, s) = c \cdot Res_{s=1}L^S(\chi\tilde{\chi}, s), \qquad c \neq 0,$$

where c is some nonvanishing constant arising from an absolutely convergent product. Since the partial Dirichlet L-series $L^S(\chi\tilde{\chi}, s)$ has a pole at $s = 1$ for $\tilde{\chi}\chi = 1$, the partial L-series $L^S(\pi \otimes \tilde{\chi}, s)$ of the representation $\pi \otimes \chi^{-1}$ has a pole at $s = 3/2$. Furthermore, π has central character χ^2; hence, $\pi \otimes \chi^{-1}$ has trivial central character by Lemma 1.4. The characterization of CAP representations of $PGSp(4, \mathbb{A})$ in [69] thus forces $\pi \otimes \chi^{-1}$ to be a CAP representation of the group $PGSp(4, \mathbb{A})$ of the Saito–Kurokawa type. That π is a CAP representation of $GSp(4, \mathbb{A})$, however, was excluded by assumption. This proves Theorem 1.1.

Corollary 1.2. *Let* $\pi = \pi_\infty \pi_{fin}$ *be an irreducible cuspidal automorphic representation with infinite component* π_∞ *belonging to the L-packet of the holomorphic discrete series of type* (k_1, k_2), *and let* $E = E_{(k_1-3, k_2-3)}$ *and* π_f *be as above. If* π *is not a CAP representation, then it contributes only to cohomology in degree 3, in the sense that* $H^3(X, E(\pi_f)) \neq 0$ *but* $H^i(X, E(\pi_f)) = 0$ *for* $i \neq 3$.

Remark 1.4. The CAP representations which appear for the classical holomorphic Siegel cusp forms of weight $k \geq 3$ ($k_1 = k_2 = k$) are just those given by the Saito–Kurokawa lift. See [69]. The situation is reversed for weights $k = 1$, where the CAP forms studied by Soudry appear. See [111].

Chapter 2
CAP Localization

In this chapter, we will express the π-isotypic Lefschetz numbers of Hecke operators acting on the cohomology of symmetric spaces $S_K(G)$ attached to reductive groups G in terms of so-called elliptic traces T_{ell}, provided the underlying representation π is not a cuspidal representation associated with a parabolic subgroup (CAP representation) of $G(\mathbb{A})$. In the following two chapters we derive from these formulas all the essential information required.

For a connected reductive group G over $\mathbb{Q}$, let K_∞ be a maximal compact subgroup of $G_\infty = G(\mathbb{R})$ and let $A_G(\mathbb{R})^0$ be the topologically connected component of the maximal $\mathbb{Q}$-split component A_G of the center Z_G of G. Then $X_G = G_\infty/\tilde{K}_\infty$ for $\tilde{K}_\infty = K_\infty \cdot A_G(\mathbb{R})^0$ will be called the connected symmetric space attached to G. For a compact open subgroup $K \subseteq G(\mathbb{A}_{fin})$

$$S_K(G) = G(\mathbb{Q}) \setminus \big(X_G \times G(\mathbb{A}_{fin})/K\big) = G(\mathbb{Q}) \setminus \big(X_G \times G(\mathbb{A}_{fin})\big)/K$$

is a disjoint union of arithmetic quotients of X_G.

Example 2.1. For $G = GSp(4, \mathbb{Q})$ we have $X_G = H \cup -H$ for the Siegel upper half-space H of genus 2. Hence, $S_K(G)$ does not coincide with the Shimura variety $S_K(\mathbb{C})$, which is an unramified covering of $S_K(G)$.

Assumption Regarding G_{der}. In this chapter assume that the derived group G_{der} of G is simply connected. This property is inherited by the Levi subgroups L of G.

Proof: $G = G_{der}Z(G)$ and $Z(G) \subseteq L$ implies $L_{der} = (G_{der} \cap L)_{der}$. $L \cap G_{der}$ is a Levi group of G_{der}, since this holds for the Lie algebras by characterizing Levi subgroups as centralizers. So it is enough to consider the semisimple case to see that L_{der} is simply connected. For this case see [99], Lemma 5.3 or Theorem 5.8, p. 208, which proves the claim. Since all groups L_{der} are simply connected implies that the centralizers L_γ of semisimple elements in the Levi groups L are connected reductive groups.

Lefschetz Numbers. An irreducible complex representation of the group $G_\infty = G(\mathbb{R})$ with highest weights λ restricts us to a representation of $G(\mathbb{Q})$, which defines

R. Weissauer, *Endoscopy for GSp(4) and the Cohomology of Siegel Modular Threefolds*,   19
Lecture Notes in Mathematics 1968, DOI: 10.1007/978-3-540-89306-6_2,
© Springer-Verlag Berlin Heidelberg 2009

a coefficient system[1] V_λ on $S_K(G)$. The cohomology groups $H^\nu(S_K(G), V_\lambda)$ are modules under the Hecke algebra of K-bi-invariant functions on $G(\mathbb{A}_{fin})$ with compact support. Assume $K = \prod_v K_v$. Fix a finite set S of non-Archimedean places such that for all non-Archimedean places $v \notin S$ the group K_v is a special, good maximal compact subgroup of G_v. Let

$$\pi^S = \otimes_{v \notin S} \pi_v$$

be an irreducible spherical automorphic representation of $G(\mathbb{A}_f^S)$. The π^S-isotypic generalized eigenspace of the νth cohomology group

$$H^\nu(S_K(G), V_\lambda)(\pi^S)$$

is a module under the Hecke algebra $\mathcal{H}_{S,K} \subseteq \mathcal{H}_S$, defined by the locally constant K_S-bi-invariant functions on $G(\mathbb{A}_S)$ with compact support. A simple formula for the trace of Hecke operators $f_S \in \mathcal{H}_S = \otimes_{v \in S} \mathcal{H}_v$ in the subspace $\mathcal{H}_{S,K}$ of the Hecke algebra (see Appendices 1 and 2) defined by

$$tr_s(f_S) = \sum_\nu (-1)^\nu tr\left(f_S, H^\nu(S_K(G), V_\lambda)(\pi^S) \right)$$

is provided by the topological trace formula of Goresky and MacPherson. Assume that the unramified automorphic spherical representation π^S of $G(\mathbb{A}_f^S)$ is not isomorphic to a subquotient of an induced representation $Ind_{P^S}^{G^S}(\sigma^S)$ for all proper parabolic subgroups $P \neq G$ with Levi component L, and all irreducible automorphic representations σ^S of $L(\mathbb{A}_f^S)$. In this case π^S is cuspidal, and π^S is not a CAP representation in the sense of [69, 97]. With these assumptions, the formula for the trace of f_S is further simplified (Sects. 2.6, 2.8).

Of special interest is the case where G_∞ has discrete series representations (Sect. 2.9). In this case the formula for the trace becomes the following (see Corollary 2.6): If the group $K = \prod_v K_v$ is small and π^S is not CAP, the trace $Tr_s(f_S)$ of f_S is equal to

$$d(G) \cdot \sum_{\gamma \in G(\mathbb{Q})/\sim}^{\prime} \tau(G_\gamma) O_\gamma^{G(\mathbb{A})}(f_S f_{\pi^S} f_\infty).$$

The sum is over all strongly elliptic semisimple conjugacy classes in $G(\mathbb{Q})$ (see page 46); G_γ denotes the centralizer of γ in G, which by our assumptions is a connected reductive group. The coefficients $O_\gamma^{G(\mathbb{A})}$ are adelic orbital integrals. Measures are such that $vol_{dg_f}(K) = 1$ and $vol_{dg_\infty dg_f}\left(G(\mathbb{Q}) \setminus G(\mathbb{A})\right) = \tau(G)$ is the Tamagawa number. The function f_{π^S} is a suitable chosen good π^S-projector depending on the fixed f_S (see Sect. 2.8), and f_∞ is a suitable linear combination of pseudocoefficients of discrete series representations with respect to the measure dg_∞ (see Sect. 2.9). The corresponding L-packet is determined by the representation

---

[1] In this chapter we consider the dual V_λ of the coefficient system E_λ of Chap. 1.

λ defining the coefficient system V_λ. $d(G)$ denotes the number of discrete series representations in this L-packet.

Remark 2.1. If $\tilde{K}_\infty$ is replaced by a subgroup U of finite index such that $G_\infty/U \to G_\infty/\tilde{K} = X_G$ is a finite unramified covering of degree d, then the trace formula also holds for $G(\mathbb{Q}) \setminus (G_\infty/U \times G(\mathbb{A}_{fin})/K$ except that the formula above has to be multiplied by the degree d of the covering. This applies for Shimura varieties (G, h) attached to a reductive $\mathbb{Q}$-group, for which $Z(G)/A_G$ is $\mathbb{R}$-anisotropic, since in this case the centralizer $Z(h)$ of the structure homomorphism h of the Shimura variety is a subgroup of finite index in $\tilde{K} = K_\infty \cdot A_G(\mathbb{R})^0$. See page 53.

2.1 Standard Parabolic Subgroups

Fix a minimal $\mathbb{Q}$-parabolic subgroup P_0. For a $\mathbb{Q}$-rational parabolic subgroup $P = LN$ containing P_0, and $\gamma \in P(\mathbb{Q})$ let γ_L denote the image of γ under the projection $P(\mathbb{Q}) \to L(\mathbb{Q})$ to the Levi component.

Contractive Elements. A semisimple element $\gamma \in L(\mathbb{Q})$, which is contained in a real torus T of L, which is $\mathbb{R}$-anisotropic modulo A_L is called *P-contractive*, if $|\gamma_L^\sigma|_\infty > 1$ holds for all simple roots σ (over the algebraic closure), which occur in the Lie algebra of the nilpotent radical of P, restricted to the maximal $\mathbb{Q}$-split torus A_L (in the center of L). In fact, it does not matter if we consider the absolute root system or the $\mathbb{Q}$-root system. Since $\gamma_L = a_\infty \cdot x_\infty k_\infty x_\infty^{-1}$ for $a_\infty \in A_L(\mathbb{R})^0, k_\infty \in K_{L,\infty}$, this implies $|\gamma_L^\sigma|_\infty = |a_\infty^\sigma|_\infty$ for all roots σ. Hence, γ_L is P-contractive if and only if the central component a_∞ is P-contractive and this notion depends only on the $L(\mathbb{Q})$-conjugacy class of the element γ. Suppose $P = P_\theta = L_\theta N_\theta$ is a $\mathbb{Q}$-rational standard parabolic subgroup defined by a subset θ of the simple positive $\mathbb{Q}$-roots. Then by definition $|\alpha(\gamma_L)|_\infty = |a_\infty^\alpha|_\infty = 1$ holds for all simple roots $\alpha \in \theta$. Since the roots in $Lie(N_P)$ are the positive roots which are not linear combinations of the roots in θ, the condition defining the notion P-contractive may be replaced by the *stronger condition*: $|\gamma_L^\alpha|_\infty \geq 1$ holds for all positive roots in Φ^+, and $|\gamma_L^\alpha|_\infty = 1$ holds if and only if α is a root which occurs in $Lie(L_P)$, or alternatively this could also be replaced by the condition $|a_\infty^\alpha|_\infty > 1$ for all simple $\mathbb{Q}$-roots $\alpha \notin \theta$.

The Set W'. Let $\Phi_G = \Phi = \Phi^+ \cup \Phi^-$ be the decomposition into the positive and negative roots of the absolute root system. Define W' as a subset of the absolute Weyl group W (considered over the algebraic closure) to consist of the elements $w \in W$ for which $\Phi^+ \cap w\Phi^- \subseteq \Phi(Lie(N_P))$ [33], p. 474, or equivalently $w\Phi^- \cap \Phi_L^+ = \emptyset \Leftrightarrow w^{-1}(\Phi_L^+) \subseteq \Phi_G^+$. Then $W' = W^P$ is the set of all $w \in W$ such that $w^{-1}(\alpha) > 0$ holds for all $\alpha \in \theta$. By a result obtained by Kostant, W is the disjoint union of the cosets $W_L \cdot w$ for representatives $w \in W^P$; hence $|W^{P_\theta}| = |W|/|W_{L_\theta}|$. Here W_L denotes the absolute Weyl group of L, considered as a subgroup of $W = W_G$. The representatives $w \in W^P$ are uniquely characterized as the representatives of minimal length in the W_L left cosets of W_G.

Inductivity. Notice the following inductive property of the sets $W^P \subseteq W$. Let $P = P_{\theta_1} \subseteq Q = P_{\theta_2} \subseteq G$ be standard parabolic subgroups, corresponding to $\theta_1 \subseteq \theta_2$. Let $L = L_{\theta_2}$ be the standard Levi component of P_{θ_2}. Then $P_{\theta_1} \cap L = P'$ is a standard parabolic subgroup of L with Levi component $L' = L_{\theta_1}$. In particular, $W^{P'} \subseteq W_L$ is defined. Then

$$W^{P'} \cdot W^Q = W^P.$$

In fact $w_1 \cdot w_2 = w_1' \cdot w_2'$ with $w_1, w_1' \in W^{P'}$ and $w_2, w_2' \in W^Q$ implies $W_{L_{\theta_2}} w_2 = W_{L_{\theta_2}} w_2'$; hence, $w_2 = w_2'$ and therefore also $w_1 = w_1'$. By the above-mentioned formula for the cardinalities it is enough to show that the product set on the left side is contained in the right side. But this is clear. Every w_1^{-1} for $w_1 \in W^{P'}$ maps $\Phi(L_{\theta_1})^+$ to $\Phi(L_{\theta_2})^+$ and every w_2^{-1} for $w_2 \in W^{P_{\theta_2}}$ maps $\Phi(L_{\theta_2})^+$ to $\Phi^+ = \Phi(G)^+$; thus, $w_1 w_2 \in W^{P_{\theta_1}}$.

Characters. For a dominant weight λ of L let ψ_λ denote the character of the finite-dimensional irreducible complex representation of L with highest weight λ. Let ρ_G denote half of the sum of the roots in Φ^+. Similarly define ρ_L for the reductive group L. Put $\rho_P = \rho_G - \rho_L$ as characters of L. If λ is a dominant weight for G, then $w(\rho_G + \lambda) - \rho_G$ is dominant for L (see [15], Sect. III.1.4 and Sect. III.3.1, and [45]). Using the Coxeter lengths $l(w)$, define

$$\Psi(\gamma, \lambda) = \sum_{w \in W^P} (-1)^{l(w)} \psi_{w(\lambda + \rho_G) - \rho_G}(\gamma_L^{-1}).$$

Since $-\rho_G + \rho_P = -\rho_L$, we have for $\gamma \in L(\mathbb{Q})$

$$|\gamma|_\infty^{-\rho_P} \cdot \Psi(\gamma, \lambda) = \sum_{w \in W^P} (-1)^{l(w)} \psi_{w(\lambda + \rho_G) - \rho_L}(\gamma_L^{-1}).$$

The Function $r(\gamma)$. Let $\mathbb{A}$ denote the ring of adeles of $\mathbb{Q}$ and $\mathbb{A}_{fin}$ the ring of finite adeles. Let $K = \prod_{v \; finite} K_v$ be a compact open subgroup of $G(\mathbb{A}_{fin})$. For a $\mathbb{Q}$-rational parabolic $P = LN$ and for semisimple $\gamma \in P(\mathbb{Q})$ define $\Gamma = G(\mathbb{Q}) \cap K$, $\Gamma_N = \Gamma \cap N$, $\Gamma' = \Gamma \cap \gamma^{-1} \Gamma \gamma$, $\Gamma_N' = \Gamma' \cap N$. Then

$$r = r(\gamma) = [\Gamma_N : \Gamma_N'] = [\Gamma_N : \Gamma_N \cap \gamma^{-1} \Gamma_N \gamma],$$

$$s = s(\gamma) = [\gamma^{-1} \Gamma_N \gamma : \Gamma_N'] = [\Gamma_N : \gamma \Gamma_N' \gamma^{-1}]$$

satisfy $s(\gamma) = [\Gamma_N : \gamma(\Gamma_N \cap \gamma^{-1}\Gamma_N\gamma)\gamma^{-1}] = [\Gamma_N : \gamma\Gamma_N\gamma^{-1} \cap \Gamma_N] = r(\gamma^{-1})$; hence,

Lemma 2.1. $s(\gamma) = r(\gamma^{-1})$, *which only depends on* γ_L.

Lemma 2.2. $s(\gamma)/r(\gamma) = \gamma^{2\rho_P}$ *or* $|\gamma^{\rho_P}|_\infty r(\gamma) = |\gamma^{-\rho_P}|_\infty r(\gamma^{-1})$.

Proof. The quotient $[\Gamma_N : \Gamma_N \cap \gamma \Gamma_N \gamma^{-1}] / [\Gamma_N : \Gamma_N \cap \gamma^{-1} \Gamma_N \gamma]$ is the virtual index

$$[\Gamma_N \cap \gamma^{-1}\Gamma_N\gamma : \Gamma_N \cap \gamma\Gamma_N\gamma^{-1}] = [\Gamma_N \cap \gamma^{-1}\Gamma_N\gamma : \gamma(\Gamma_N \cap \gamma^{-1}\Gamma_N\gamma)\gamma^{-1}] = |\gamma^{2\rho_P}|_\infty. \quad \Box$$

2.2 The Adelic Reductive Borel–Serre Compactification

As a set, the adelic reductive Borel–Serre compactification is

$$(S_K^G)^+ = G(\mathbb{Q}) \setminus [\bigcup_P X_L \times (G(\mathbb{A}_{fin})/K)] = G(\mathbb{Q}) \setminus [\bigcup_P X_L \times G(\mathbb{A}_{fin})]/K,$$

a disjoint union over all $\mathbb{Q}$-rational parabolic subgroups P of G. $X_L = L_\infty/\tilde{K}_{L,\infty}$ is the connected symmetric domain attached to L, i.e., $\tilde{K}_\infty = K_{L,\infty} A_L(\mathbb{R})^0$, where $K_{L,\infty}$ denotes a maximal compact subgroup of L_∞ and $A_L(\mathbb{R})^0$ the topologically connected component of the maximal $\mathbb{Q}$-split subtorus A_L in the center $Z(L)$ of L. Elements $g \in G(\mathbb{A}_{fin})$ act on the projective limit $(S^G)^+ = \lim_K (S_K^G)^+ = G(\mathbb{Q}) \setminus [\bigcup_P X_L \times G(\mathbb{A}_{fin})]$ by $x \mapsto xg^{-1}$. This defines a left action of $G(\mathbb{A}_{fin})$ on $(S^G)^+$, which induces a right action on cohomology groups. Now consider

$$T(g^{-1}) : \ G(\mathbb{Q})x_\infty x_{fin} \mapsto G(\mathbb{Q})x_\infty (x_{fin}g^{-1}).$$

Here $x_\infty \in \bigcup_P X_L$ and $x_{fin} \in G(\mathbb{A}_{fin})$. On the quotients $(S_K^G)^+$ this defines Hecke correspondences. Put

$$K' = K \cap g^{-1}Kg.$$

Then the induced Hecke correspondence is given by two maps $c_1 = T(1)$ and $c_2 = T(g^{-1})$ (see Appendix 1)

$$(S_{K'}^G)^+ \ \overset{c_1}{\underset{c_2}{\rightrightarrows}} \ (S_K^G)^+ \ .$$

The action of $G(\mathbb{Q})$ on the $\mathbb{Q}$-parabolic subgroups by conjugation is transitive on the minimal $\mathbb{Q}$-parabolic subgroups. Fixing a minimal parabolic P_0, every $\mathbb{Q}$-parabolic is conjugate over $\mathbb{Q}$ to one and only one standard parabolic $\mathbb{Q}$-subgroup P with respect to P_0. Since the stabilizer of P under conjugation with $G(\mathbb{Q})$ is $P(\mathbb{Q})$, $(S_K^G)^+$ is a union over the finitely many standard $\mathbb{Q}$-parabolic subgroups $P = P_\theta$ containing P_0:

$$(S_K^G)^+ = \bigcup_{P_0 \subseteq P} S_K^P, \quad \text{where} \quad S_K^P = P(\mathbb{Q}) \setminus [X_L \times G(\mathbb{A}_{fin})]/K.$$

Goresky and MacPherson [33] deduced a formula for the alternating trace

$$tr_s(T(g^{-1}); H^\bullet(S_K(G), V_\lambda))$$

from the Grothendieck–Verdier–Lefschetz fixed-point formula which they applied for the reductive Borel–Serre compactification $(S_K^G)^+$ of $S_K(G)$. They used the property that the cohomology groups $H^\bullet(S_K(G), V_\lambda)$ coincide with the cohomology groups $H^\bullet((S_K^G)^+, i_*V_\lambda)$ of the reductive Borel–Serre compactification $(S_K^G)^+$, where $i : S_K(G) \hookrightarrow (S_K^G)^+$ is the inclusion. As in [33], Theorem (version 0), the Lefschetz fixed-point theorem of Grothendieck, Verdier, and Illusie therefore expresses the Lefschetz number as a sum of "local" contributions $LC(F)$

$$\sum_P \sum_F LC(F)$$

for the connected components F of the intersection of the fixed-point set of the correspondence with the boundary strata S_K^P attached to the rational parabolic group P.

Rational Hecke Correspondences. We say a double coset KgK or the corresponding Hecke correspondence is rational if a representative g can be chosen to be $g = \gamma_{fin}$ for some $\gamma = \gamma_\infty \gamma_{fin} \in G(\mathbb{Q})$. In this case the correspondence $T(g^{-1})$ defined on $(S^G)^+ = G(\mathbb{Q}) \setminus [\bigcup_P X_L \times G(\mathbb{A}_{fin})]$ satisfies $G(\mathbb{Q})x_\infty \mapsto G(\mathbb{Q})x_\infty \gamma_{fin}^{-1} = G(\mathbb{Q})\gamma_\infty x_\infty$; hence, it induces the Hecke correspondence considered in [33], p. 467, defined by $c_1(\Gamma'y) = \Gamma y$ and $c_2(\Gamma'y) = \Gamma\gamma_\infty y$.

2.2.1 Components

First consider the connected components of S_K^P. Since X_L is topologically connected, the topologically connected components h of the stratum S_K^P are the fibers of the map

$$S_K^P = P(\mathbb{Q}) \setminus [X_L \times (G(\mathbb{A}_{fin})/K)] \longrightarrow \pi_0(S_K^P) = P(\mathbb{Q}) \setminus G(\mathbb{A}_{fin})/K.$$

For each component $h = P(\mathbb{Q})x_{fin}K$ in $\pi_0(S_K^P)$ put

$$\Gamma_{P_h} = P(\mathbb{Q}) \cap x_{fin}Kx_{fin}^{-1} \text{ and } \Gamma_{N_h} = N(\mathbb{Q}) \cap x_{fin}Kx_{fin}^{-1}.$$

For $K_N(h) := N(\mathbb{A}_{fin}) \cap x_{fin}Kx_{fin}^{-1}$ and $K_N'(h) := N(\mathbb{A}_{fin}) \cap x_{fin}K'x_{fin}^{-1}$ then obviously $[K_N(h) : K_N'(h)] = [\Gamma_{N_h} : \Gamma_{N_h}']$, where $\Gamma_{N_h}' := K_N'(h) \cap N(\mathbb{Q})$.

Fixed Components. Now consider the connected components F of the fixed-point locus of a Hecke correspondence within S_K^P, for fixed P. Then

$$F \subseteq h$$

for some unique component h of S_K^P. If F is fixed, then h is also fixed. So we first determine the fixed components h of the Hecke correspondence, and then the fixed components F in h.

2.2.2 Fixed Components h

The component $h = P(\mathbb{Q})(X_L \times \{x_{fin}\})K$ is fixed

$$T(g^{-1})h = h,$$

if and only if $x_{fin}g^{-1}K = \gamma x_{fin}K$ holds for some $\gamma \in P(\mathbb{Q})$ and some $k \in K$.

Recall $gK'g^{-1} \subseteq K$. Hence, $\gamma^{-1}x_{fin}g^{-1} = x_{fin}k$ implies $\gamma^{-1}x_{fin}K'x_{fin}^{-1}\gamma \subseteq x_{fin}Kx_{fin}^{-1}$. Thus,

$$\gamma^{-1}\Gamma'_{P_h}\gamma := P(\mathbb{Q}) \cap \gamma^{-1}x_{fin}K'(\gamma^{-1}x_{fin})^{-1} \subseteq P(\mathbb{Q}) \cap x_{fin}Kx_{fin}^{-1} =: \Gamma_{P_h}.$$

$x_{fin}K'x_{fin}^{-1}=x_{fin}Kx_{fin}^{-1}\cap x_{fin}g^{-1}K(x_{fin}g^{-1})^{-1}=x_{fin}Kx_{fin}^{-1}\cap\gamma(x_{fin}Kx_{fin}^{-1})\gamma^{-1}$
again using $x_{fin}g^{-1}=\gamma x_{fin}k$. Hence, the intersection with $P(\mathbb{Q})$ is $\Gamma'_{P_h}=\Gamma_{P_h}\cap\gamma\Gamma_{P_h}\gamma^{-1}$. In particular, $\Gamma'_{N_h}=(\Gamma_{N_h}\cap\gamma\Gamma_{N_h}\gamma^{-1})$. Hence, the fixed equation $T(g^{-1})h=h$ given by $\gamma^{-1}x_{fin}g^{-1}=x_{fin}k$ implies.

Lemma 2.3. $[K_N(h) : K'_N(h)] = [\Gamma_{N_h} : \Gamma'_{N_h}] = [\Gamma_{N_h} : (\Gamma_{N_h} \cap \gamma\Gamma_{N_h}\gamma^{-1})] = r(\gamma^{-1})$. *For fixed g, K this number only depends on P and the coset $P(\mathbb{A}_{fin})x_{fin}$.*

Rationality. To simplify the notation we now replace K by $x_{fin}Kx_{fin}^{-1}$, and g^{-1} by $x_{fin}g^{-1}x_{fin}^{-1}$, which allows us to assume $x_{fin} = 1$ without restriction of generality. Then the fixed-component equation becomes $\gamma \in g^{-1}K$. Hence, the coset $g^{-1}K \subseteq Kg^{-1}K$ has a rational point, and the Hecke correspondence defined by $Kg^{-1}K = K\gamma K$ is rational. For a fixed component h one can thus reduce the local computations of the local term $LC(F)$ for F to the classical setting considered in [33].

2.2.3 Another Formulation

The action of $G(\mathbb{Q})$ on the $\mathbb{Q}$-parabolic subgroups by conjugation is transitive on the minimal $\mathbb{Q}$-parabolic subgroups. Hence, choosing a minimal $\mathbb{Q}$-parabolic P_0, every $\mathbb{Q}$-parabolic is conjugate over $\mathbb{Q}$ to one and only one standard parabolic $\mathbb{Q}$-subgroup P with respect to P_0. Since the stabilizer of P under conjugation with $G(\mathbb{Q})$ is $P(\mathbb{Q})$, $(S_K^G)^+$ is a union over the finitely many standard $\mathbb{Q}$-parabolic subgroups $P = P_\theta$ containing P_0,

$$(S_K^G)^+ = \bigcup_{P_0 \subseteq P} P(\mathbb{Q}) \setminus [X_L \times G(\mathbb{A}_{fin})]/K.$$

Since $gKg^{-1} \cap N_P(\mathbb{A}_{fin})$ is open in $N_P(\mathbb{A}_{fin})$ for $P = L_P N_P$, for the strata $S_K^P = P(\mathbb{Q}) \setminus [X_L \times G(\mathbb{A}_{fin})]/K$ of

$$(S_K^G)^+ = \bigcup_{P_0 \subseteq P} S_K^P$$

an easy density argument gives the formula $S_K^P = L_P(\mathbb{Q}) \setminus [X_L \times (N_P(\mathbb{A}_{fin}) \setminus G(\mathbb{A}_{fin})/K)]$ or

$$S_K^P = L_P(\mathbb{Q})N_P(\mathbb{A}_{fin}) \setminus [L_\infty \times G(\mathbb{A}_{fin})]/\tilde{K}_\infty K.$$

Hence, $\overline{x} \in (S_K^G)^+$ is a double coset represented by some $x = x_\infty x_{fin} \in L_\infty \times G(\mathbb{A}_{fin})$. Iwasawa decomposition $G(\mathbb{A}_{fin}) = P(\mathbb{A}_{fin}) \cdot \Omega$ for some maximal compact group Ω containing the group K gives a finite decomposition $G(\mathbb{A}_{fin}) = \bigcup_g P(\mathbb{A}_{fin})gK$. Therefore, the set $\pi_0(S_K^P)$ of the topologically connected components is finite, since by a result obtained by Borel and Harish-Chandra [14], $M(\mathbb{Q}) \setminus M(\mathbb{A}_{fin})/K_M$ is finite for any reductive $\mathbb{Q}$-group M and any compact open group $K_M \subseteq M(\mathbb{A}_{fin})$. Of course we may choose the representatives elements

$$x_{fin} = k \in \Omega.$$

2.2.4 Small Groups

Consider compact open subgroups $K \subseteq G(\mathbb{A}_f)$ and $\tilde{K}_\infty = K_\infty Z_{G,\infty}^0$, where K_∞ is maximal compact in G_∞. $K \subseteq G(\mathbb{A}_{fin})$ will be called *small* if

$$x^{-1}n\gamma x \in K\tilde{K}_\infty Z_{L,\infty}$$

for $x \in G(\mathbb{A}), n \in N(\mathbb{A}), \gamma \in P(\mathbb{Q})$ and any $\mathbb{Q}$-parabolic $P = L \cdot N$ with unipotent radical N implies $\gamma_L \in Z_L(\mathbb{Q})$ (image in the Levi component is contained in the center) and in addition implies $\gamma_L = 1$ if γ_L is a torsion element.

Remark 2.2. Of course it is enough to demand this for all standard parabolic groups containing a fixed P_0.

Remark 2.3. "Small" implies "neat" in the sense that $L(\mathbb{Q})_{tor} \cap (xKx^{-1} \cap P)_L = 1$.

Small-level groups K exist: $G(\mathbb{A})$ is a finite union of cosets $P(\mathbb{A})kKK_\infty$ for $k \in G(\mathbb{A}_{fin})$. This allows us to replace K by some conjugate K_k, and x by some $p \in P(\mathbb{A})$, and gives equations $p^{-1}n\gamma p \in K_k$ for $p \in P(\mathbb{A})$ instead of $x \in G(\mathbb{A})$. Equivalently, $m^{-1}\gamma_L m \in (K_k \cap P)_L$ for $m \in L(\mathbb{A})$, where the index L indicates projection from P to the Levi component L. γ_L is semisimple since modulo the center it is contained in a maximal compact subgroup of L_∞. The groups L and $L_{ad} = L/Z_L$ are connected reductive groups; hence, by embedding L_{ad} into some linear group and using for L_{ad} the argument at the beginning of the proof [44], Proposition 8.2, one can show that only finitely many $L(\overline{\mathbb{Q}})$ conjugacy classes of semisimple elements γ_L in $L_{ad}(\mathbb{Q})$ meet $(K_k \cap P)_L$. Shrinking K leaves us, considering eigenvalues, with the unique $\overline{\mathbb{Q}}$ conjugacy class $\{1\}$. Thus, $\gamma_L \in Z_L(\mathbb{Q})$. Finally, $Z_L(\mathbb{Q})_{tor}$ is finite (consider a splitting field of Z_L). Since it is enough to consider the finitely many standard parabolic groups P and for each finitely many cosets k, shrinking K therefore allows us to assume $Z_L(\mathbb{Q})_{tor} \cap (K_k \cap P)_L = \{1\}$ for the finitely many relevant cases.

2.3 Fixed Points

Now we want to determine the fixed points $\overline{x}$ of the Hecke correspondence $T(g^{-1})$ in the reductive Borel–Serre compactification $(S_{K'}^G)^+$. They are described by the equations $c_1(\overline{x}) = T(1)\overline{x}$ and $c_2(\overline{x}) = T(g^{-1})\overline{x}$ in $(S_K^G)^+$. The unique component h containing $\overline{x}$ is necessarily a fixed component. h is contained in a stratum S_K^P. Now fix the standard parabolic $P = L_P N_P$, or $P = LN$ for short.

Suppose $\overline{x} \in S_{K'}^P$ is represented by $x = x_\infty x_{fin} \in L_\infty \times G(\mathbb{A}_{fin})$. Then $\overline{x}$ is a fixed point of $T(g^{-1})$ if and only if $xg^{-1} = \gamma \cdot x \cdot k$ holds for some $\gamma \in P(\mathbb{Q})$ and $k \in \tilde{K}_{L,\infty}K$, or equivalently if and only if

$$x^{-1}\gamma x \in g^{-1}K\tilde{K}_{L,\infty}.$$

We may replace x by another representative $\delta^{-1}x$, $\delta \in P(\mathbb{Q})$. Then instead of γ its conjugate $\delta\gamma\delta^{-1}$ appears in the fixed-point equation. Moreover

Lemma 2.4. *The element γ is semisimple and $\mathbb{R}$-elliptic. For small K the $L(\mathbb{Q})$-conjugacy class of the image γ_L of γ in $L(\mathbb{Q})$ is uniquely determined by the fixed point $\overline{x} \in S_{K'}^P$.*

Proof. The equation $x_\infty^{-1}\gamma_L x_\infty \in \tilde{K}_{L,\infty}$ implies that $\gamma_L \in L(\mathbb{Q})$ is $\mathbb{R}$-elliptic, hence semisimple. Now choose an equivalent representative $\delta nxk'$ for $\overline{x}$ for some $n \in N(\mathbb{A}_{fin}), \delta \in P(\mathbb{Q}), k' \in \tilde{K}_{L,\infty}K'$. Suppose $xg^{-1} = \gamma_1 xk_1$ and $(\delta nxk')g^{-1} = \gamma_2(\delta nxk')k_2$ holds for $\gamma_i \in P(\mathbb{Q})$, $k_i \in \tilde{K}_{L,\infty}K$. Replacing k_2 by $k'k_2(gk'g^{-1})^{-1}$ allows us to assume $k' = 1$. Replacing γ_2 by $\delta^{-1}\gamma_2\delta$ allows us to assume $\delta = 1$. Hence, $\gamma_1 xk_1 = xg^{-1} = n^{-1}\gamma_2 nxk_2$. Since K is small, this implies $\gamma_L \in Z_L(\mathbb{Q})$ for $\gamma = \gamma_1^{-1}\gamma_2$ and hence γ commutes with x_∞, which then implies $\gamma_L \in K_\infty K_L$, where K_L is the image of $K \cap P(\mathbb{A}_{fin})$ in $L(\mathbb{A}_{fin})$. Thus, $\gamma_L \in Z_L(\mathbb{Q}) \cap \tilde{K}_{L,\infty}K_L$. Looking at the Archimedean place and the non-Archimedean places separately, this forces γ_L to be a torsion element. Therefore, $\gamma_L = 1$, since K is small.

This lemma gives a decomposition of the fixed-point set in the stratum $S_{K'}^P$ according to the conjugacy classes $\gamma_L \in L(\mathbb{Q})/\sim$. $\square$

Fixing $\gamma_L/\sim$. We want to determine the set $Fix(\gamma_L)$ of all fixed points $\overline{x} \in S_{K'}^P$ of $T(g^{-1})$, where in the fixed-point equation for some representative an element γ appears whose projection to $L(\mathbb{Q})$ belongs to the fixed conjugacy class $\gamma_L/\sim$. To unburden the notation we also write $Fix(\gamma)$ instead of $Fix(\gamma_L)$,

$$Fix(\gamma) \subseteq S_{K'}^P = L_P(\mathbb{Q}) \setminus [(L_\infty/\tilde{K}_{L,\infty}) \times (N(\mathbb{A}_{fin}) \setminus G(\mathbb{A}_{fin})/K')].$$

For $x = x_\infty x_{fin} \in L_\infty \times G(\mathbb{A}_{fin})$ the double coset $\overline{x} = L(\mathbb{Q})N(\mathbb{A}_{fin})x\tilde{K}_{L,\infty}K'$ is in $Fix(\gamma)$ if and only if there exist $n \in N(\mathbb{A}_f), \delta \in P(\mathbb{Q}), k \in \tilde{K}_{L,\infty}K, \gamma' \in N(\mathbb{Q})$ such that $n(\delta\gamma'\gamma\delta^{-1})xk = xg^{-1}$ holds, or equivalently if and only if there exist $n \in N(\mathbb{A}_f), \delta \in P(\mathbb{Q}), k \in \tilde{K}_{L,\infty}K$, such that

$$(*) \qquad x^{-1}n\delta\gamma\delta^{-1}x \in g^{-1}\tilde{K}_{L,\infty}K,$$

since we are free to replace n by $n\delta\gamma'\delta^{-1}$.

By abuse of notation we do not distinguish between global elements γ, δ in $G(\mathbb{Q})$ and their images in G_v or $G(\mathbb{A}_{fin})$. Since x is considered in $S_{K'}^P = P(\mathbb{Q}) \setminus [X_L \times G(\mathbb{A}_{fin})]/K'$, we may replace x by $\delta^{-1}x$ and n by $\delta^{-1}n\delta$, which simplifies the equations for $Fix(\gamma)$. Hence, we get

Lemma 2.5. *We have* $Fix(\gamma) \cong L(\mathbb{Q}) \setminus \left(L(\mathbb{Q}) \cdot \prod_v Sol_v(\gamma) \right)$, *where*

$$Sol_v(\gamma) = \{x_v \in N(\mathbb{Q}_v) \setminus G(\mathbb{Q}_v)/K'_v \mid x_v^{-1}n\gamma x_v \in g_v^{-1}K_v \text{ holds for some } n \in N_v\}$$

at the non-Archimedean places v *and* $K'_v = K_v \cap g_v^{-1}K_v g_v$, *and where*

$$Sol_\infty(\gamma) = \{x_\infty \in L_\infty(\mathbb{R})/\tilde{K}_{L,\infty} \mid x_\infty^{-1}\gamma_L x_\infty \in \tilde{K}_{L,\infty} \}$$

at the Archimedean place ∞.

By abuse of notation we write L_γ for the centralizer L_{γ_L} of the element γ_L in L, which is a connected reductive group by the assumption that G_{der} is simply connected.

Corollary 2.1. *For small* K *we obtain* $Fix(\gamma) = L_\gamma(\mathbb{Q}) \setminus Sol(\gamma)$ *for* $Sol(\gamma) = \prod_v Sol_v(\gamma)$.

Proof. $L(\mathbb{Q})$-equivalent solutions in $Sol(\gamma)$, say, $x^{-1}n_1\gamma x$ and $x^{-1}n_2\delta^{-1}\gamma\delta x$ in $g^{-1}K\tilde{K}_\infty$ for suitable $n_1, n_2 \in N(\mathbb{A}_{fin})$ and $\delta \in P(\mathbb{Q})$, satisfy $x^{-1}n_2\delta^{-1}\gamma\delta\gamma^{-1}$ $n_1x \in K\tilde{K}_\infty$ for some $n \in N(\mathbb{A})$. We may then assume $n_1 = 1$, and since K is small, this implies $\delta_L^{-1}\gamma_L\delta_L\gamma_L^{-1} \in Z_L(\mathbb{Q})$. Since the commutator $\delta_L^{-1}\gamma_L\delta_L\gamma_L^{-1}$ is in $L_{der}(\mathbb{Q})$, and since $Z_L(\mathbb{Q}) \cap L_{der}(\mathbb{Q})$ is finite, the commutator is a torsion element, and hence is 1 since K is small. This implies $\delta_L \in L_\gamma(\mathbb{Q})$ and completes the proof. □

2.3.1 Archimedean Place

$Sol_\infty(\gamma) \cong L_{\gamma,\infty}/(L_{\gamma,\infty} \cap \tilde{K}_{L,\infty})$ by the corollary in Appendix 2, unless it is empty. If it is nonempty, the Archimedean fixed-point condition shows that γ_L is L_∞-conjugate to a point in $\tilde{K}_{L,\infty}$. To determine $L_{\gamma,\infty}$ we may therefore assume $\gamma_L \in \tilde{K}_{L,\infty}$ without restriction of generality. Hence, the centralizer $L_{\gamma,\infty}$ becomes θ-stable for the Cartan involution θ (see Appendix 2). Therefore, $K_\infty \cap L_{\gamma,\infty}$ is a maximal compact subgroup $K_{L_\gamma,\infty}$ of $L_{\gamma,\infty}$. Since $A_L(\mathbb{R})^0 \subseteq L_{\gamma,\infty}$, $Sol_\infty(\gamma) = L_{\gamma,\infty}/(K_{L_\gamma,\infty}A_L(\mathbb{R})^0)$ admits a smooth surjective map to the symmetric space $X_{L_\gamma} = L_{\gamma,\infty}/(K_{L_\gamma,\infty}A_{L_\gamma}(\mathbb{R})^0)$ of the centralizer L_γ, which defines a trivial fibration by the Euclidean space $A_{L_\gamma}(\mathbb{R})^0/A_L(\mathbb{R})^0$, and hence a homotopy equivalence. See the Remark 2.15 in Appendix 2.

2.3.2 Non-Archimedean Places

Recall that $T(g^{-1})$ and $\gamma/\sim$ are now fixed. Let Ω_v be the stabilizer of a special point in the Bruhat–Tits building. Special points always exist, and Ω_v is a maximal compact subgroup of G_v. We now assume $K_v \subseteq \Omega_v$; hence, $K'_v \subseteq \Omega_v$. Then by the Iwasawa decomposition $G_v = P_v \cdot \Omega_v$ (see [103], Sect. 3.3.2). For $k \in \Omega_v$ put $g_k^{-1} = kg^{-1}k^{-1}$, $K_k = kK_vk^{-1}$, and $K'_k = kK'_vk^{-1}$. Elements $x_v \in Sol_v(\gamma)$ may be written $x_v = p \cdot k$ for $p \in P_v$ and $k \in \Omega_v$. The coset $(P_v \cap \Omega_v)kK'_v$ is uniquely determined by x_v, and $G/K' = \bigcup_{k \in P_v \cap \Omega_v \backslash \Omega_v / K'_v} P_v/(P_v \cap kK'_vk^{-1})$. Therefore,

$$Sol_v(\gamma) = \bigcup_{k \in P_v \cap \Omega_v \backslash \Omega_v / K'_v} Sol_v(\gamma, k).$$

Here $Sol_v(\gamma, k) = \{p \in N_v \backslash P_v/(P_v \cap kK'_vk^{-1}) \mid p_L^{-1}\gamma_L p_L \in (g_k^{-1}K_k \cap P)_L\}$ or

$$Sol_v(\gamma, k) \cong \mathcal{S}_v(\gamma, k)/K'(k)_v$$

for $K'(k)_v := (P_v \cap kK'_vk^{-1})_L$ and

$$\mathcal{S}_v(\gamma, k) = \left\{ m \in L_v \mid m^{-1}\gamma_L m \in (g_k^{-1}K_k \cap P_v)_L \right\} = \biguplus_{\xi_v}(L_\gamma)_v \cdot \xi_v \cdot K'(k)_v.$$

This is a finite (possibly empty) union over representatives $\xi_v \in L_v$. From [53], Propositions 7.1 and 8.2, there is only one representative $\xi_v = 1$ for almost all v.

2.3.3 Globally

With this notation

$$Sol(\gamma, k) = \mathcal{S}(\gamma, k)/K'(k)_{\mathbb{A}} \quad \text{for} \quad K'_{\mathbb{A}}(k) = \tilde{K}_{L,\infty} \prod_{v \, fin} K'(k)_v,$$

where $\mathcal{S}(\gamma, k) = \left\{ m \in L(\mathbb{A}) \mid m^{-1}\gamma_L m \in \tilde{K}_{L,\infty}\left(g_k^{-1}K_k \cap P(\mathbb{A}_{fin})\right)_L \right\}. L_\gamma(\mathbb{A})$ acts on $\mathcal{S}(\gamma, k)$ from the left. Choose a decomposition

$$\mathcal{S}(\gamma, k) = \biguplus_{\xi} L_\gamma(\mathbb{A}) \cdot \xi \cdot K'(k)_{\mathbb{A}}$$

with representatives $\xi \in L(\mathbb{A})$, where representatives $\xi = \prod_v \xi_v$ are chosen to be products of corresponding local non-Archimedean representatives ξ_v for $L_\gamma(\mathbb{Q}_v) \backslash \mathcal{S}_v(\gamma, k)/K'(k)_v$, and $\xi_\infty = 1$. Then

Lemma 2.6. *For small K the contribution of a fixed conjugacy class $\gamma/\sim$ in $L(\mathbb{Q})$ to the fixed-point locus of $T(g^{-1})$ is*

$$Fix(\gamma) \cong \biguplus_{k \in P(\mathbb{A}_{fin}) \cap \Omega \backslash \Omega / K'} Fix(\gamma, k),$$

$$Fix(\gamma, k) \cong \biguplus_{\xi} L_\gamma(\mathbb{Q}) \backslash L_\gamma(\mathbb{A}) / (\xi K'(k)_\mathbb{A} \xi^{-1} \cap L_\gamma(\mathbb{A})).$$

Of course $L_\gamma(\mathbb{Q}) \backslash L_\gamma(\mathbb{A}) / (\xi K'(k)_\mathbb{A} \xi^{-1} \cap L_\gamma(\mathbb{A})) = \biguplus F_\nu$ is a finite union of arithmetic quotients F_ν.

2.4 Lefschetz Numbers

The Lefschetz number becomes

$$\sum_P \sum_{\gamma/\sim} \sum_k \sum_\xi \sum_\nu LC(F_\nu),$$

where $k \in P \cap \Omega \backslash \Omega / K'$. Put $F = F_\nu$. For the local terms $LC(F)$ Goresky and MacPherson gave an explicit description as a product $\chi(F) r(\gamma_F) \Psi(\gamma_F, \lambda)$ if γ_F is P-contractive, and it vanishes otherwise. See [33], pp. 470–471 and Theorem (version 3a), p. 474. Here $\gamma_F = \gamma^{-1}$ is the characteristic element defined in [33], p. 469, which is the inverse of the element γ defined in Lemma 2.4. Hence, if it is nonvanishing, the local number $LC(F)$ is the product of:

- The Euler characteristic $\chi(F)$
- $|\gamma_F|_\infty^{\rho_P} \cdot r(\gamma_F, k)$
- $|\gamma_F|_\infty^{-\rho_P} \cdot \Psi(\gamma_F, \lambda) = \sum_{w \in W^P} (-1)^{l(w)} \psi_{w(\lambda + \rho_G) - \rho_L}(\gamma_F^{-1})$

2.4.1 Euler Characteristics

We may sum the terms $\sum_\nu \chi(F_\nu)$ for fixed $P, \gamma/\sim, k, \xi$, which gives the Euler characteristic

$$\chi(L_\gamma(\mathbb{Q}) \backslash L_\gamma(\mathbb{A}) / (\xi K'(k)_\mathbb{A} \xi^{-1} \cap L_\gamma(\mathbb{A}))).$$

To compute it we may replace $L_{\gamma,\infty} / \tilde{K}_{L,\infty}$ by $X_{L_\gamma} = L_{\gamma,\infty} / \tilde{K}_{L_\gamma,\infty}$. See page 44. Notice $L_\gamma(\mathbb{Q}) \cap \xi \tilde{K}_{L_\gamma,\infty} K'(k)_\mathbb{A} \xi^{-1}$ is contained in the center of L_γ, since K is small. Hence, the intersection is discrete and compact, and hence we have a finite group. By our assumption K is small; hence, the intersection is trivial. Thus, we obtain

$$\sum_\nu \chi(F_\nu) = \frac{\chi(L_\gamma, dg_{fin})}{vol_{dg_{fin}}(\xi K'(k)_{\mathbb{A}_{fin}}\xi^{-1} \cap L_\gamma(\mathbb{A}_{fin}))}$$

for a constant $\chi(L_\gamma) = \chi(L_\gamma, dg_{fin})$ depending only on L_γ and on the choice of the Haar measure dg_{fin} on $L_\gamma(\mathbb{A}_{fin})$.

Remark 2.4. Observe that γ is $\mathbb{R}$-elliptic, and hence is $\mathbb{Q}$-elliptic. For the ambiguity of this notion, see [44], p. 392. We show that in our situation this ambiguity does not cause problems, since the Euler characteristic of the corresponding summands in the trace formula vanishes unless both notions agree. Consider the group L or better L/A_L. The center $Z(L_\gamma)$ is $\mathbb{Q}$-anisotropic modulo A_L. If the quotient were not anisotropic over $\mathbb{R}$, the corresponding global quotient space X would be a nontrivial torus fibration, whose Euler characteristic would therefore vanish. Similarly the Euler characteristic vanishes for locally symmetric arithmetic quotients of semisimple groups unless the $\mathbb{R}$-rank of the maximal compact subgroup equals the $\mathbb{R}$-rank of the group. Considering the map $L_{der} \to L$, we can assume that $(L_\gamma/A_L)(\mathbb{R})$ contains an $\mathbb{R}$-anisotropic torus of maximal rank, or otherwise the Euler characteristic vanishes and the corresponding summand does not contribute to the trace formula.

Definition 2.1. Call $\gamma \in L(\mathbb{Q})$ *strongly elliptic* if γ is $L(\mathbb{R})$-conjugate to an element in $K_{L,\infty} \cdot A_L(\mathbb{R})^0$ such that the Euler characteristic $\chi(L_\gamma)$ does not vanish.

Remark 2.5. For connected reductive groups L over $\mathbb{Q}$, for which the connected component of the center modulo A_L is anisotropic over $\mathbb{R}$, one also wants to compare $\chi(L, dg_{fin})$ with the Tamagawa number. At the moment we do not need to carry through this comparsion. When we need it later, it can be obtained directly from a comparison between the topological L^2-trace formula and Arthur's L^2-trace formula. On the other hand, it should not be difficult to obtain it by reduction to the case of semisimple groups (Harder's theorem [37]) adapting the argument of [68], pp. 129–131, with a z-extension $T' \to L^* \to L$ replacing the sequence (V), and $\tilde{L} = (L^*)_{der} \to L^* \to T$ replacing the sequence (H) in [68].

Only (semisimple) strongly elliptic elements γ contribute to the Lefschetz number. Let χ_P^G be the characteristic function of the P-contractive elements. We obtain for the Lefschetz number the expression

$$\sum_P \sum_{\gamma/\sim} \chi(L_\gamma, dg_{fin})\chi_P^G(\gamma_\infty^{-1}) \sum_{w \in W^P} (-1)^{l(w)} \psi_{w(\lambda+\rho_G)-\rho_L}(\gamma) \cdot O_\gamma$$

$$= \sum_P \sum_{\gamma/\sim} \chi(L_\gamma, dg_{fin})\chi_P^G(\gamma_\infty) \sum_{w \in W^P} (-1)^{l(w)} \psi_{w(\lambda+\rho_G)-\rho_L}(\gamma^{-1}) \cdot O_{\gamma^{-1}},$$

where

$$O_\gamma = \sum_k \sum_\xi \frac{|\gamma^{-1}|_\infty^{\rho_P} \cdot r(\gamma^{-1}, k)}{vol_{dg_{fin}}(\xi K'(k)_{\mathbb{A}_{fin}}\xi^{-1} \cap L_\gamma(\mathbb{A}_{fin}))}.$$

Here we used $r(\gamma, k) = r(\delta\gamma\delta^{-1}, k)$ for $\delta \in L(\mathbb{Q})$. To show this, recall $k = x_{fin}$ describes the component h of the P-stratum, which contains F. Recall $r(\gamma, x_{fin}) = r(\delta\gamma\delta^{-1}, \delta x_{fin})$ for $\delta \in P(\mathbb{Q})$. Since $r(\gamma, x_{fin})$ depends only on P and the coset $P(\mathbb{A}_{fin})x_{fin}$ (Lemmas 2.1 and 2.3), replacing γ by a conjugate does not change $r(\gamma, k)$.

2.4.2 Computation of O_γ

Notice $|\gamma^{-1}|_\infty^{\rho_P} r(\gamma^{-1}) = \prod_{v \neq \infty} |\gamma|_v^{\rho_P}[N_v \cap K_k : N_v \cap K_v']$ by Lemma 2.3. Since

$$\sum_{k \in (P_v \cap \Omega_v)\backslash \Omega_v/K_v'} f(k) = \sum_{k \in \Omega_v/K_v'} \frac{f(k)}{[(P_v \cap \Omega_v) : (P_v \cap K_k')]},$$

this allows us to write O_γ as a product $\prod_{v \neq \infty} O_{\gamma,v}$ of non-Archimedean local terms

$$O_{\gamma,v} = \sum_{k \in \Omega_v/K_v'} \sum_{\xi_v} \frac{|\gamma|_v^{\rho_P} \cdot [N_v \cap K_k : N_v \cap K_v']}{[(P_v \cap \Omega_v) : (P_v \cap K_k')] \cdot vol_{dg_v}(\xi_v K'(k)_v \xi_v^{-1} \cap L_{\gamma,v})}.$$

Since

$$0 \to N_v \cap K_k' \to P_v \cap K_k' \to K'(k)_v \to 0$$

is exact, this gives

$$O_{\gamma,v} = \sum_{k \in \Omega_v/K_v'} \sum_{\xi_v} \frac{|\gamma|_v^{\rho_P} \cdot vol_{N_v}(N_v \cap K_k) \cdot vol_{L_v}(K'(k)_v)}{vol_{dg_v}(\xi_v K'(k)_v \xi_v^{-1} \cap L_{\gamma,v})},$$

where measures are normalized such that $vol(\Omega_v \cap P_v) = 1$ and $vol(\Omega_v \cap N_v) = 1$. In Sect. 2.5 we show that this expresses O_γ as an orbital integral

$$O_\gamma = O_\gamma^L(\overline{f}^{(P)})$$

of the characteristic function f of the set $Kg^{-1}K$ up to a normalization factor.

2.4.3 Conclusion

The computations in Sects. 2.4.1 and 2.4.2 describe the right action of $1_{KgK}/vol_\Omega(K)$ on the cohomology. Any K-bi-invariant function f is a linear combination of functions f as above. However, we should keep in mind that so far we have used a left action of $G(\mathbb{A}_{fin})$ on S^G, where $g \in G(\mathbb{A}_{fin})$ acts by the formula on page 23; hence, the cohomology becomes a right module under the Hecke algebra.

Theorem 2.1. *Assume the derived group of G is simply connected and K is small. Then the Lefschetz number of the right action of a K-bi-invariant Hecke operator $f \in C_c^\infty(G(\mathbb{A}_{fin}))$ on the cohomology $H^\bullet(S_K(G), V_\lambda)$ is given by*

$$L(f, V_\lambda) = \sum_P \sum_{\gamma \in L(\mathbb{Q})/\sim} \chi(L_\gamma, dg_{fin}) O_\gamma^L(\overline{f}^{(P)}) \cdot |\gamma|_\infty^{-\rho_P} \Psi(\gamma, \lambda).$$

The sum extends over all standard $\mathbb{Q}$-parabolic subgroups $P = LN$ containing the fixed minimal $\mathbb{Q}$-parabolic P_0 and all $L(\mathbb{Q})$-conjugacy classes $\gamma \in L(\mathbb{Q})/\sim$ of semisimple, strongly elliptic elements in $L(\mathbb{Q})$ with P-contractive representatives.

Example 2.2. $G = \mathbb{G}_m$ for the representation $x \mapsto x^r$ of weight $\lambda = r$ on $V = \mathbb{C}$, and K maximal compact. Then S_K is a single point. The element $g = (g_v) \in \mathbb{A}_{fin}^*$, where $g_p = p$ and $g_v = 1$ for $v \neq p$, acts on V_λ by $1 \times 1 \mapsto 1 \times g^{-1} \simeq p^r \times 1$ in $V \times \mathbb{A}_{fin}^*$. It acts on the cohomology via multiplication by p^{-r} (right action on cohomology) or p^r (left action on cohomology).

Remark 2.6. Notice we used

$$O_{\gamma^{-1}}^L(\overline{f}^{(P)}) = O_\gamma^L(\overline{f^-}^{(P)}).$$

For the comparison of trace formulas with those in [64] in Chap. 3, we may turn the right action of the Hecke algebra on the cohomology groups into a left action by the substitution $f(x) \mapsto f^-(x) = f(x^{-1})$. This makes the formula compatible with that in [64], p. 197.

Remark 2.7. The factor $\chi(L_\gamma, dg_{fin}) O_\gamma^L(.)$ does not depend on the choice of the fixed Haar measure dg_{fin} on $L_\gamma(\mathbb{A}_{fin})$; therefore, we do not mention the choice of dg_{fin} in the following.

Remark 2.8. The condition imposed in Theorem 2.1 that $\gamma \in L(\mathbb{Q})/\sim$ contains a P-contractive representative $\gamma \in P(\mathbb{Q})$ can be replaced by the stronger condition that $|\alpha(\gamma)|_\infty \geq 1$ holds for all positive roots α of G and $|\alpha(\gamma)|_\infty = 1$ holds if and only if α is a root from L as explained after the definition of contractiveness. Of course it is enough to consider $\mathbb{Q}$-roots, since G and P are defined over $\mathbb{Q}$ and γ is a $\mathbb{Q}$-rational element. Therefore, the condition in Theorem 2.1 can be replaced by the condition $|\alpha(\gamma)|_{fin} \leq 1$ holds for all positive $\mathbb{Q}$-roots α and $|\alpha(\gamma)|_{fin} = 1$ holds if and only if α is a root from L.

Remark 2.9. For a standard $\mathbb{Q}$-parabolic group $P \supseteq P_0$ with Levi decomposition $P = LN$ let $X^*(P)_\mathbb{Q} = X^*(L)_\mathbb{Q} = Hom_{\mathbb{Q}-alg}(L, \mathbb{G}_m)$ be the group of characters defined over $\mathbb{Q}$. Then $\mathcal{X}_L = Hom(X^*(L)_\mathbb{Q}, \mathbb{R})$ can be canonically identified with the Lie algebra of A_L, and hence with $A_L(\mathbb{R})^0$ by the exponential map. One defines the Harish-Chandra homomorphism

$$H_P : L(\mathbb{A}) \to \mathcal{X}_L$$

by $exp(\langle H_P(l), \chi \rangle) = \|\chi(l)\|$, where $\|.\| : \mathbb{A}^* \to \mathbb{R}^*$ is the idele norm, and $\chi \in X^*(L)_{\mathbb{Q}}$. For the minimal $\mathbb{Q}$-parabolic $P = P_0$ we write $\mathcal{X}_P = \mathcal{X}$. Let Δ be a basis of the simple $\mathbb{Q}$-roots. Then the standard $\mathbb{Q}$-parabolic subgroups $P = P_\theta$ correspond uniquely to the subsets $\theta \subseteq \Delta$. The roots in F are the roots of the Levi component L_θ, and the simple roots in the Lie algebra of the unipotent radical N_θ are the roots in $\Delta \setminus F$. In fact, since γ is strongly elliptic, the condition for $\gamma \in L(\mathbb{Q}) = L_\theta(\mathbb{Q})$, given in Remark 2.8, could also be replaced by the condition

$$|\alpha(H_P(\gamma))|_{fin} < 1$$

for all simple $\mathbb{Q}$-roots α not in θ.

The Decomposition $\mathcal{X} = \mathcal{X}_L \oplus \mathcal{X}_L^{\perp}$ [1]. Let β_j denote the dual roots such that $\langle \beta_j, \alpha_i \rangle = \delta_{ij}$, both considered as elements of $\mathcal{X}$. Then there exists a natural orthogonal decomposition $\mathcal{X} = \mathcal{X}_L \oplus \mathcal{X}_L^{\perp}$ such that $\mathcal{X}_L$ is the span $\sum_{j \notin F} \mathbb{R}\beta_j$ and $\mathcal{X}_L^{\perp} = \sum_{i \in F} \mathbb{R}\alpha_i$. The projection $pr_L : \mathcal{X} \to \mathcal{X}_L$ is $pr(\sum_{j \notin F} x_j\beta_j + \sum_{i \in F} y_i\alpha_i) = \sum_{j \notin F} x_j\beta_j$. The image under pr_L of the open positive Weyl chamber $\mathcal{X}^+ = \sum_{j \in \Delta} \mathbb{R}_{>0}\beta_j \subseteq \mathcal{X}$ defines the open Weyl chamber $\mathcal{X}_M^+$ in $\mathcal{X}_M$; the image of the obtuse Weyl chamber ${}^+\mathcal{X} = \sum_i \mathbb{R}_{>0}\alpha_i \subseteq \mathcal{X}$ defines the obtuse open Weyl chamber in $\mathcal{X}_L$. Obviously ${}^+\mathcal{X}_L = \sum_{j \notin F} \mathbb{R}_{>0}\alpha_j$. Then $\mathcal{X}_L^+ = pr(\mathcal{X}^+) = \sum_{j \notin F} \mathbb{R}_{>0}\beta_j$, since $\langle \alpha_i, \alpha_j \rangle \leq 0$ for $i \neq j$ and $\langle \beta_i, \beta_j \rangle \geq 0$. In fact $\sum_{j \notin F} x_j\beta_j + \sum_{i \in F} y_i\alpha_i \in \sum_{i \in F} \mathbb{R}_{>0}\beta_i$ therefore implies $y_i \geq 0, i \in F$; hence, $x_j > 0, j \notin F$. Also $\mathcal{X}^+ \subseteq {}^+\mathcal{X}$; therefore, $\mathcal{X}_L^+ \subseteq {}^+\mathcal{X}_L$. Finally notice $\mathcal{X}^+ \cap -{}^+\mathcal{X} = \{0\}$.

2.5 Computation of an Orbital Integral

We write the terms O_γ in the formula for the Lefschetz numbers as an orbital integral $O_\gamma^L(\overline{f}^{(P)})$. This is done in steps 1–3. The final result is formulated in step 4.

Step 1. Assume measures are normalized by $vol_G(\Omega) = 1$. Recall $K' = K \cap g^{-1}Kg$ and $g \in G(\mathbb{A}_{fin})$ is fixed. The characteristic function $1_{g^{-1}K}(y)$ of the set $g^{-1}K$ is then K'-bi-invariant. Furthermore, $k^{-1}xk \in g^{-1}K \Longleftrightarrow x \in kg^{-1}k^{-1}kKk^{-1} =: g_k^{-1}K_k$. Hence,

$$\int_\Omega 1_{g^{-1}K}(k^{-1}xk)dk = [\Omega : K']^{-1} \cdot \sum_{k \in \Omega/K'} 1_{g^{-1}K}(k^{-1}xk)$$

$$= vol_\Omega(K') \cdot \sum_{k \in \Omega/K'} 1_{g_k^{-1}K_k}(x).$$

$\int_K 1_{g^{-1}K}(k^{-1}xk)dk = \int_K 1_{g^{-1}K}(k^{-1}x)dk = vol_\Omega(g^{-1}Kg \cap K)1_{Kg^{-1}K}(x) = vol_\Omega(K')1_{Kg^{-1}K}(x)$ holds for $x \in G(\mathbb{A}_{fin})$. Hence, $\int_\Omega = vol_\Omega(K)^{-1} \int_\Omega \int_K$ implies

$$\int_{\Omega} 1_{g^{-1}K}(k^{-1}xk)dk = vol_{\Omega}(K)^{-1}\int_{\Omega} vol_{\Omega}(K')1_{Kg^{-1}K}(k^{-1}xk)dk.$$

Comparison of the right sides thus gives for the Ω-average

Definition 2.2. $\overline{f}(x) = \int_{\Omega} f(k^{-1}xk)dk$, $\quad x \in G(\mathbb{A}_{fin})$, of the normalized characteristic function.

Definition 2.3. $\boxed{f(x) = vol_{\Omega}(K)^{-1}1_{Kg^{-1}K}(x)}$.

Lemma 2.7. $\overline{f}(x) = \sum_{k\in\Omega/K'} 1_{g_k^{-1}K_k}(x).$

Step 2. For $x \in P(\mathbb{A}_{fin})$ in a standard parabolic subgroup $P = LN$ by Lemma 2.7

$$\int_{N(\mathbb{A}_{fin})} \overline{f}(xn)dn = \sum_{k\in\Omega/K'} \int_{N(\mathbb{A}_{fin})} 1_{g_k^{-1}K_k}(xn)dn.$$

If x_L is not in $(g_k^{-1}K_k \cap P(\mathbb{A}_{fin}))_L$, the corresponding integral on the right side is zero. Otherwise $xn_0 = g_k^{-1}k_0$ holds for some $n_0 \in N(\mathbb{A}_{fin})$ and $k_0 \in K_k$, and in this case the integral becomes $\int_{N(\mathbb{A}_{fin})} 1_{g_k^{-1}K_k}(g_k^{-1}k_0 n)dn = vol(N(\mathbb{A}_{fin})\cap K_k)$. Hence, for $x \in L(\mathbb{A}_{fin})$ we get

Lemma 2.8. $\phi(x) := \int_{N(\mathbb{A}_{fin})} \overline{f}(xn)dn = \sum_{k\in\Omega/K'} vol\big(N(\mathbb{A}_{fin}) \cap K_k\big) \cdot 1_{(g_k^{-1}K_k\cap P(\mathbb{A}_{fin}))_L}(x).$

Step 3. Next consider the orbital integral of the function ϕ defined on $L(\mathbb{A}_{fin})$

$$O_{\gamma}^{L}(\phi) = \int_{L_{\gamma}(\mathbb{A}_{fin})\backslash L(\mathbb{A}_{fin})} \phi(m^{-1}\gamma m)dm.$$

By the definition of ϕ the value of $O_{\gamma}^{L}(\phi)$ is

$$\sum_{k\in\Omega/K'} vol\big(N(\mathbb{A}_{fin}) \cap K_k\big) \cdot \int_{L_{\gamma}(\mathbb{A}_{fin})\backslash L(\mathbb{A}_{fin})}$$

$$char\Big\{m \,\Big|\, m^{-1}\gamma_L m \in \big(g_k^{-1}K_k \cap P(\mathbb{A}_{fin})\big)_L\Big\}dm,$$

or by Sect. 6.16 and the decomposition $\mathcal{S}(\gamma, k) = \bigcup L_{\gamma}(\mathbb{A}_{fin}) \cdot \xi_{fin} \cdot K'(k)_{\mathbb{A}_{fin}}$

$$\sum_{k\in\Omega/K'} vol\big(N(\mathbb{A}_{fin})\cap K_k\big) \cdot \sum_{\xi_{fin}} \frac{vol_{L(\mathbb{A}_{fin})}\big(K'(k)_{\mathbb{A}_{fin}}\big)}{vol_{L_{\gamma}(\mathbb{A}_{fin})}\big(\xi_{fin}K'(k)_{\mathbb{A}_{fin}}\xi_{fin}^{-1} \cap L_{\gamma}(\mathbb{A}_{fin})\big)}.$$

Step 4. To put things together. The function

$$\overline{f}(x) = \int_{\Omega} f(k^{-1}xk)dk, \qquad f(x) = vol_{\Omega}(K)^{-1}1_{Kg^{-1}K}(x)$$

is K-bi-invariant on $G(\mathbb{A}_{fin})$. Define

$$\overline{f}^{(P)}(m) = |m|_{fin}^{\rho_P} \int_{N(\mathbb{A}_{fin})} \overline{f}(mn)dn, \qquad m \in L(\mathbb{A}_{fin})$$

$$O_\gamma^L(\overline{f}^{(P)}) = \int_{L_\gamma(\mathbb{A}_{fin})\backslash L(\mathbb{A}_{fin})} \overline{f}^{(P)}(m^{-1}\gamma m)dm,$$

assuming that the integrals are normalized by the conventions $vol_{P(\mathbb{A}_{fin})}(\Omega \cap P(\mathbb{A}_{fin})) = 1$, $vol(\Omega) = 1$, and $vol_{N(\mathbb{A}_{fin})}(\Omega \cap N(\mathbb{A}_{fin})) = 1$. See also [16], p. 144. The measure on $L_\gamma(\mathbb{A}_{fin})$ is dg_{fin}. Then the computation above proves that

$$O_\gamma^L(\overline{f}^{(P)}) = O_\gamma.$$

2.6 Elliptic Traces

Recall G is a connected reductive group over $\mathbb{Q}$ whose derived group is simply connected. Define elliptic "traces"

$$\boxed{T_{ell}^G(f,\tau) = \sum_{\gamma \in G(\mathbb{Q})/\sim} \chi(G_\gamma)O_\gamma^G(f) \cdot tr\big(\tau(\gamma^{-1})\big)}$$

for a finite-dimensional complex representations τ of $G(\mathbb{Q})$ and $f \in C_c^\infty(G(\mathbb{A}_{fin}))$. If τ is an irreducible complex representation defined by a highest weight λ, we also write $T_{ell}^G(f,\lambda)$ instead of $T_{ell}^G(f,\tau)$. Hence, we do not distinguish between representations and their highest weights. The sum defining $T_{ell}^G(f,\tau)$ extends over the $G(\mathbb{Q})$-conjugacy classes of semisimple, strongly elliptic elements in $G(\mathbb{Q})$. The integrals $O_\gamma^G(f)$ in this sum are orbital integrals with respect to the group of finite adeles for functions $f \in C_c^\infty(G(\mathbb{A}_{fin}))$. The same definition defines elliptic traces T_{ell}^L for the Levi subgroups L of all standard $\mathbb{Q}$-parabolic subgroups $P = LN_P$ of G.

Let $\chi_P^G = \tau_P^G \circ H_P$ be defined by the characteristic function τ_P^G of the open positive Weyl chamber of $\mathcal{X}_L = X_*(A_L)_{\mathbb{Q}} \otimes \mathbb{R}$, lifted to a function on $L(\mathbb{A}_{fin})$ via the Harish-Chandra homomorphism $H_L : L(\mathbb{A}_{fin}) \to \mathcal{X}_L$. Then Theorem 2.1 and the remarks following it imply

Lemma 2.9.
$$L(f, V_\lambda) = \sum_{P_0 \subseteq P \subseteq G} T_{ell}^P(\overline{f}^{(P)}\chi_P^G, \lambda),$$

where

$$T_{ell}^P(h,\lambda) = \sum_{w \in W^P} (-1)^{l(w)} \cdot T_{ell}^L(h, w(\lambda + \rho_G) - \rho_L)$$

for $h \in C_c^\infty(L(\mathbb{A}_{fin}))$, and where $P = LN_P$ runs over the $\mathbb{Q}$-rational standard parabolic subgroups of G.

Let $\hat{\chi}_P^G = \hat{\tau}_P^G \circ H_P$ be the characteristic function $\hat{\tau}_P^G$ of the open obtuse Weyl chamber in $\mathcal{X}_L$, considered as a function on $L(\mathbb{A}_{fin})$. Notice $\chi_P^G \leq \hat{\chi}_P^G$.

Lemma 2.10. *Let the situation be as in Lemma 2.9. Then*

$$T_{ell}^G(f, \lambda) = \sum_{P_0 \subseteq Q \subseteq G} (-1)^{rang(Q) - rang(G)} L^Q(\overline{f}^{(Q)} \hat{\chi}_Q^G, \lambda),$$

where the sum is over the standard $\mathbb{Q}$-parabolic subgroups of G, where $rang(Q)$ denotes the $\mathbb{Q}$-split rank of the Levi subgroup of Q, and the Lefschetz number L^Q for $h \in C_c^\infty(L(\mathbb{A}_{fin}))$ and the parabolic group Q is defined by

$$L^Q(h, V_\lambda) = \sum_{w \in W^Q} (-1)^{l(w)} \cdot L^L(h, V_{w(\lambda + \rho_G) - \rho_L}),$$

and $L^L(h, .)$ is the Lefschetz number attached to coefficient systems for the symmetric space attached to the Levi subgroup L of the (standard) parabolic subgroup Q.

Lemmas 2.9 and 2.10 were expected by Harder [39], pp. 144–145.

Proof of Lemma 2.10. Lemma 2.9 applied to the Levi subgroup L of $Q = LN$ gives

$$L^Q(\overline{f}^{(Q)} \hat{\chi}_Q^G, V_\lambda) = \sum_{w \in W^Q} (-1)^{l(w)} \cdot L^L(\overline{f}^{(Q)} \hat{\chi}_Q^G, V_{w(\lambda + \rho_G) - \rho_L})$$

$$= \sum_{w \in W^Q} (-1)^{l(w)} \sum_{P_0 \cap L \subseteq P' = L'N' \subseteq L} T_{ell}^{P'}\left(\overline{(\overline{f}^{(Q)} \hat{\chi}_Q^G)}^{(P')} \chi_{P'}^L, w(\lambda + \rho_G) - \rho_L\right)$$

$$= \sum_{P_0 \cap L \subseteq P' = L'N' \subseteq L} \sum_{w' \in W^{P'}} \sum_{w \in W^Q} (-1)^{l(w) + l(w')} \cdot$$
$$T_{ell}^{L'}\left(\overline{(\overline{f}^{(Q)} \hat{\chi}_Q^G)}^{(P')} \chi_{P'}^L, w'w(\lambda + \rho_G) - \rho_{L'}\right).$$

$P \subseteq Q$ induces the parabolic group $P' = P \cap L$ in the Levi component L of Q, and all standard Q parabolic groups P' are obtained in this way from the standard $\mathbb{Q}$-parabolic subgroups $P \subseteq Q$ such that the Levi components L' of P' and P coincide. Since $sn(w) = (-1)^{l(w)}$ satisfies $sn(w')sn(w) = sn(w'w)$, the inductivity $W^{P'}W^Q = W^P$ and the formula $\overline{f}^{(P)} \hat{\chi}_Q^G = \overline{(\overline{f}^{(Q)} \hat{\chi}_Q^G)}^{(P')}$ implies that the sum simplifies to

$$L^Q(\overline{f}^{(Q)} \hat{\chi}_Q^G, V_\lambda) = \sum_{P_0 \subseteq P = L'N_P \subseteq Q} T_{ell}^P(\overline{f}^{(P)} \hat{\chi}_Q^G \chi_{P'}^L, \lambda).$$

The sum is over all $\mathbb{Q}$-rational standard parabolic subgroups P of G contained in Q. Notice in the formula above $\chi_{P'}^L$ is a function on $\mathcal{X}_{L'}$, whereas $\hat{\chi}_Q^G$, which is defined as a function on $\mathcal{X}_L$, is tacitly considered as a function on $\mathcal{X}_{L'}$ via the canonical projection map $pr : \mathcal{X}_{L'} \to \mathcal{X}_L$. Summing these formulas over the standard parabolic groups Q, with the additional factors $(-1)^{rang(G)-rang(Q)}$,

$$\sum_{P_0 \subseteq Q} (-1)^{rang(Q)-rang(G)} \cdot L^Q(\overline{f}^{(Q)} \hat{\chi}_Q^G, V_\lambda),$$

gives the desired result, by interchanging the order of summation. Fixing P, the sum over all Q with $P \subseteq Q \subseteq G$ gives zero except for $P = G$. Indeed for fixed $P \subseteq G$ the sum $\sum_{P \subseteq Q \subseteq G} (-1)^{rang(G)-rang(Q)} \hat{\chi}_Q^G \chi_{L'}^L$ is zero except for $P = L'N_P = G$, where it is 1 instead. This is a well known result obtained by Arthur [1]. For the convenience of the reader we include the argument. $\square$

Proof. Let $F' \subseteq F \subseteq \Delta$ define $P' \subseteq P \subseteq G$. Since the support $Supp_F$ of the characteristic function $\hat{\tau}_Q^G \tau_{L'}^L$ of the subset $\sum_{i \notin F} \mathbb{R}_{>0} \alpha_i + \sum_{j \in F \setminus F'} \mathbb{R}_{>0} \beta_j$ of $\mathcal{X}_{L'}$ is contained in $^+\mathcal{X}_{L'} = \sum_{i \notin F'} \mathbb{R}_{>0} \alpha_i$ (if $F' \neq \Delta$), $Supp_F = {}^+\mathcal{X}_{L'} \cap \bigcap_{j \in F \setminus F'} \{H \mid \alpha_j(H) > 0\}$ follows as an immediate consequence of the inequalities $(\alpha_i, \alpha_j) \leq 0$ for $i \neq j$ and $(\beta_i, \beta_j) \geq 0$. For $H \in \mathcal{X}$ let Δ_H denote the set of $\alpha_i \notin F'$, for which $\langle \alpha_i, H \rangle > 0$. Δ_H is nonempty for $H \in {}^+\mathcal{X}$, since $-\overline{\mathcal{X}^+} \cap {}^+\mathcal{X} = \{0\}$. Hence, $\sum_{P \subseteq Q \subseteq G} (-1)^{rang(G)-rang(Q)} \hat{\tau}_Q^G \tau_{L'}^L(H) = 0$ follows from $\sum_{T \subseteq \Delta_H} (-1)^{|T|} = 0$. $\square$

Corollary 2.2. *The elliptic trace* $T_{ell}^G(f, \lambda)$ *is*

$$\boxed{\sum_{P_0 \subseteq Q \subseteq G} \sum_{w \in W^Q} (-1)^{rang(Q)-rang(G)+l(w)} \cdot tr_s(\overline{f}^{(Q)} \hat{\chi}_Q^G; H^\bullet(S_{L_Q}, V_{w(\lambda+\rho_G)-\rho_L}))}$$

Corollary 2.3. *The Lefschetz number* $L(f, V_\lambda)$ *is*

$$\boxed{\sum_{P_0 \subseteq P \subseteq G} \sum_{w \in W^P} (-1)^{l(w)} \cdot T_{ell}^L(\overline{f}^{(P)} \chi_P^G, w(\lambda + \rho_G) - \rho_L)}$$

2.7 The Satake Transform

For a connected reductive group G over a non-Archimedean local field F_v let A be a maximal F_v-split torus in the center of G. Let G^{ab} be the maximal Abelian quotient of G. Write $G_v = G(F_v)$, etc.

ord_G. There is a canonical homomorphism $ord_G : G_v \to X_*(G) = Hom_{F_v - alg}(G, \mathbb{G}_m)$ (see [16], p. 134). We also write ord_G for the induced homomorphism $G_v \to \mathcal{X}_{G_v} = X_*(G) \otimes \mathbb{R}$, and ^0G for the kernel. The homomorphism ord_G is functorial in G and induces the field valuation in the case $G = \mathbb{G}_m$. It factorizes over the quotient G^{ab}, and is trivial on compact subgroups. The kernel of the canonical map $A_v \to G_v^{ab}$ is contained in the maximal compact subgroup ^0A_v. Hence, the quotient group $A_v/^0A_v$, which can be identified with the F_v-rational cocharacter lattice $X_*(A)$ of the torus A, is injected into $G_v^{ab}/^0G_v^{ab}$ as a subgroup of finite index. Hence, the canonical maps $\mathcal{X}_{A_v} \to \mathcal{X}_{G_v} \to \mathcal{X}_{G_v^{ab}}$ induce isomorphisms, which allows us to identify these vector spaces.

The Map S. Now assume $G_v = G(F_v)$ to be quasisplit, and split over a finite unramified extension field of F_v such that the derived group is simply connected. Let Ω_v be a good maximal compact subgroup and $P = MN$ be a minimal F_v-rational parabolic subgroup of G such that $G_v = P_v \cdot \Omega_v$. To be precise, we demand Ω_v to be admissible relative to M_v in the sense of [7], p. 9. The Ω_v-bi-invariant functions on G_v with compact support define the spherical Hecke algebra $\mathcal{H}(G_v, \Omega_v)$ of G_v. Put $\Lambda = {}^0M_v \setminus M_v$. For $f_v \in C_c^\infty(G_v)$ define $\overline{f}_v^{(P_v)}(m) = |m|_v^{\rho_{P_v}} \int_{N_v} \overline{f}(mn)dn$ as on page 36 now locally for F_v. For elements f_v in the spherical Hecke algebra of G_v the Satake transform S is defined by (see [16], p. 146, formula (19))

$$f_v \mapsto S(f_v) = \overline{f}_v^{(P_v)}$$

and defines a function $S(f_v)$ on $M_v(\mathbb{Q}_v)/M_v(\mathbb{Q}_v) \cap \Omega_v = \Lambda$. The group Λ is a lattice, which contains and is commensurable with the cocharacter lattice $X_*(A)$ of the torus A (see [16], p. 135) in $\mathcal{X}_{A_v} = X_*(A) \otimes \mathbb{R}$. The Satake transform defines an isomorphism between the spherical Hecke algebra of the group G_v and the algebra $\mathbb{C}[\Lambda]^W$ (W-invariants in the group ring $\mathbb{C}[\Lambda]$ [16], Theorem 4.1). Furthermore, for γ regular in M_v the Satake transform S is given by the orbital integral up to a normalization factor

$$S(f_v)(\gamma) = D_G(\gamma)^{1/2}O_\gamma^{G_v}(f_v).$$

For an arbitrary function $\chi : \mathcal{X}_{G_v} \to \mathbb{R}$ multiplication by χ determines a $\mathbb{C}$-linear endomorphism $f_v(x) \mapsto \chi(ord_L(x))f_v(x)$ of the Hecke algebra of G_v, which preserves the spherical Hecke algebra such that for the orbital integral

$$O_\gamma^{G_v}(f_v\chi) = \chi(ord_G(\gamma)) \cdot O_\gamma^{G_v}(f_v)$$

holds, and also for the Satake transform $S(\chi f_v)(m) = \chi(ord_G(\gamma)(m))S(f_v)(m)$.

Standard F_v-parabolic Groups. Let Q be a F_v-rational standard parabolic subgroup of G with Levi component L. Let A_Q be the maximal F_v-split torus in Q. The natural map $A_v \to L_v \to \mathcal{X}_L$ factorizes over the quotient $A_v/^0A_v$, and hence induces a canonical $\mathbb{R}$-linear map

$$pr : \mathcal{X}_{M_v} \to \mathcal{X}_{L_v}.$$

The following two properties characterize the projection pr. Firstly, the embedding $A_{Q_v} \hookrightarrow A_v$ induces a canonical embedding $i : \mathcal{X}_{L_v} = X_*(A_Q) \otimes \mathbb{R} \hookrightarrow \mathcal{X}_{M_v} = X_*(A) \otimes \mathbb{R}$ such that $pr : \mathcal{X}_{M_v} \to \mathcal{X}_{L_v}$ restricts us to the identity map on the subspace $\mathcal{X}_{L_v} \subseteq \mathcal{X}_{M_v}$. Secondly pr is zero on the subspace $X_*(A'_Q) \otimes \mathbb{R} \subseteq \mathcal{X}_{M_v}$, where A'_Q denotes the split torus $L_{der} \cap A$.

This gives the following formulation in terms of the Killing form. Let $\alpha_i \in \Delta(G_v, A_v)$ denote the simple F_v-roots attached to $P_v \subseteq G_v$, let $\langle\ ,\ \rangle$ denote the Killing form, and let β_j denote the dual basis $\langle \alpha_i, \beta_j \rangle = \delta_{ij}$. Use the Killing form to identify $X_*(A) \otimes \mathbb{R}$ with its dual $X^*(A) \otimes \mathbb{R}$. The F_v-rational standard parabolic subgroups are in one-to-one correspondence with the subsets $F \subseteq \Delta(P_v, A_v)$. For $Q = Q_F$ the space $\mathcal{X}_{L_v} = X_*(A_Q) \otimes \mathbb{R}$ is given in $\mathcal{X}_{M_v} = X_*(A) \otimes \mathbb{R}$ by the equations $\langle ., \alpha_i \rangle = 0, \alpha_i \in F$ (or $i \in F$ by abuse of notation) for a subset F of the simple roots. $\mathcal{X}_{M_v}$ splits into the orthogonal direct sum of the two subspaces $\mathcal{X}_{L_v} = \sum_{j \notin F} \mathbb{R}\beta_j$ and the orthocomplement $\sum_{i \in F} \mathbb{R}\alpha_i$. pr is the orthogonal projection defined by $pr(\sum_{j \notin F} x_j \beta_j + \sum_{i \in F} y_i \alpha_i) = \sum_{j \notin F} x_j \beta_j$.

Transitivity. Let $Q = LN$ be an F_v-rational parabolic subgroup of G. Let σ_v be an irreducible admissible representation of L_v. The Hecke algebra $C_c^\infty(G_v)$ of locally constant functions with compact support on G_v acts by convolution on the unitary normalized induced representation $\pi_v = Ind_{Q_v}^{G_v}(\sigma_v)$ such that (for measures suitably normalized) the adjunction formula (see, e.g., [44], Sect. 2, Lemma 1, the slightly different definition involving f_v^* in the pairing in loc. cit. has no effect) holds

$$tr\ Ind_{Q_v}^{G_v}(\sigma_v)(f_v) = tr\ \sigma_v(\overline{f}_v^{(Q)}),$$

where $f_v \in C_c^\infty(G_v)$ and by definition $\overline{f}_v^{(Q)}(m) = |m|_v^{\rho_{Q_v}} \int_{N_v} \overline{f}(mn)dn$.

The group $\Omega_v \cap L_v = (\Omega_v \cap Q_v)_{L_v}$ is a good maximal compact subgroup of L_v, i.e., admissible with respect to M_v (see [7], p. 9). L_v is again quasisplit and splits over a unramified extension field. Hence, the spherical Hecke algebra $\mathcal{H}(L_v, \Omega_v \cap L_v)$ is defined. For $f_v \in \mathcal{H}(G_v, \Omega_v)$ the function $S_L^G(f_v) = \overline{f}_v^{(Q)}$ is bi-invariant under $\Omega_v \cap L_v$, and hence the partial Satake transform $S = S_M^G : \mathcal{H}(G_v, \Omega_v) \to \mathcal{H}(M_v, {}^0 M_v)$ factorizes over the spherical Hecke algebra $\mathcal{H}(L_v, \Omega_v \cap L_v)$

$$S = S_M^G = S_M^L \circ S_L^G.$$

Absolute Support. In the following, a cone C in Euclidean space is understood to be an open submonoid stable under multiplication by $\mathbb{R}_{>0}$ which does not contain a real line.

Lemma 2.11. *Fix an arbitrary nonempty open cone $C \subseteq \mathcal{X}_M$, which is contained in the positive Weyl chamber attached to P_v. Let $\pi_v = Ind_{P_v}^{G_v}(\sigma_v)$ be an unramified induced representation attached to an unramified character σ_v of M_v with spherical constituent π_v^0. Choose $x_0 \in C$. Then there exist spherical Hecke operators f_v with the properties:*

1. *$tr\ \pi_v(f_v) = tr\ \pi_v^0(f_v) = 1$.*
2. *The support of the Satake transform $S(f_v)$ of f_v is contained in the Weyl group orbit $\bigcup_{w \in W} w(x_0 + C)$ of the translated cone $x_0 + C$.*

Proof. It suffices to find $f_v \in \mathcal{H}(G_v, \Omega_v)$ with $tr\ \pi_v(f_v) \neq 0$ such that (2) holds. $tr\ \pi_v$, considered as a functional on the spherical Hecke algebra $\mathbb{C}[\Lambda]^W$, is a finite sum of characters on the group Λ. Up to a twist by $\delta^{1/2}(x)$ these characters are in the W-orbit of the character σ_v. This character sum is conjugation-invariant, and hence W-invariant. If the assertions of the lemma were false, there would exist finitely many different characters $\chi_i, i = 1, \ldots, r$, of Λ and $n_i \in \mathbb{C}$ such that

$$\sum_{i=1}^{r} n_i \chi_i(x) = 0, \qquad (n_1, \cdots, n_r) \neq 0$$

holds for all $x \in \Lambda \cap (x_0 + C)$. To see that this is impossible we can assume $x_0 = 0$, changing the coefficients n_i to $n_i \chi_i(x_0)$, and then use induction on r. Since $x, y \in C$ implies $x + y \in C$ we can lower the length r of such a nontrivial character relation on C by considering $\sum_i n_i(\chi_i(y) - \chi_1(y))\chi_i(x) = 0$, provided there exists $y \in C$ with $\chi_r(y) \neq \chi_1(y)$ if, say, $n_r \neq 0$. Because $\chi = \chi_r/\chi_1$ is a nontrivial character on Λ, such a y exists, since otherwise χ vanishes on $C \cap \Lambda$, and hence on the generated group $(C \cap \Lambda) - (C \cap \Lambda)$. However, $(C \cap \Lambda) - (C \cap \Lambda) = \Lambda$ holds for any nonempty open cone of $\mathcal{X}$. This proves the lemma. $\square$

Relative Support. For $f_v \in C_c^\infty(G_v)$ consider the support Σ of the orbital integral $O_\gamma^L(\overline{f}_v^{(Q)})$ as a function of $\gamma \in L_v$. Notice the support of $\overline{f}_v^{(Q)}$ itself is contained in Σ. The image of Σ in $\mathcal{X}_L$ of the regular, semisimple subset of this support under $ord_L : L_v \to \mathcal{X}_{L_v}$ will be called the *relative support* of f_v with respect to Q_v. The relative support contains the image of the support of $\overline{f}_v^{(Q)}$ in $\mathcal{X}_{L_v}$ under the map ord_L. Since the regular semisimple elements are dense in Σ, and since the maximal compact subgroup of L_v is in the kernel of ord_L, one could replace the support Σ by the regular, semisimple support of $O_\gamma^L(\overline{f}_v^{(Q)})$ for the definition of relative support above.

The relative support of f_v with respect to Q_v is a finite subset of the vector space $\mathcal{X}_L$. Notice that $\overline{f}_v^{(Q)}$ has compact support on L_v, and ord_L is invariant under conjugation. Hence, the image $ord_L(\Sigma)$ is relatively compact in $\mathcal{X}_{L_v}$. On the other hand $ord_L(L_v)$ is contained in a sublattice of $\mathcal{X}_{L_v}$.

Lemma 2.12. *Let $f_v \in \mathcal{H}(G_v, \Omega_v)$ be a spherical function. Let $Q = LN_Q$ be an F_v-rational standard parabolic subgroup of G containing the minimal F_v-parabolic subgroup $P = MN$. Then $x \in \mathcal{X}_{L_v}$ is in the relative support of $O_\gamma^L(\overline{f}_v^{(Q)})$ if and only if x is in the image of the support of the Satake transform $S(f_v) \in \mathcal{H}(M_v, {}^0 M_v)$ under the map $pr \circ ord_M$, where $pr : \mathcal{X}_{M_v} \to \mathcal{X}_{L_v}$ is the canonical projection.*

Proof. Let $\chi_x(\lambda)$ be the function on $\mathcal{X}_L$, which is not zero for $\lambda = x$ and is zero otherwise. Then by definition the following statements are equivalent. By abuse of notation we consider χ_x as a function on L_v using the map ord_L. Then $x \in \mathcal{X}_L$ is in the relative support of f_v if and only if

$$\chi_x(\gamma) O_\gamma^{L_v}(\overline{f}_v^{(Q)}) = O_\gamma^{L_v}(\chi_x \cdot \overline{f}_v^{(Q)})$$

does not vanish identically for all semisimple, regular elements $\gamma \in L_v$. Since f_v is a spherical function on G_v, $\overline{f}_v^{(Q)} = S_L^G(f_v)$ is spherical on L_v; hence, $\chi_x \cdot \overline{f}_v^{(Q)} = \chi_x \cdot S_L^G(f_v)$ is again spherical on L_v. If $O_\gamma^{L_v}(\chi_x \overline{f}_v^{(Q)})$ does not vanish identically for all semisimple, regular elements $\gamma \in L_v$, then $\chi_x \overline{f}_v^{(Q)}$ does not vanish identically on L_v. Since $\chi_x \overline{f}_v^{(Q)}$ is spherical, this implies $S_M^L(\chi_x \overline{f}_v^{(Q)}) \neq 0$; hence, $O_\gamma^{L_v}(\chi_x \overline{f}_v^{(Q)})$ does not vanish identically for all semisimple, regular elements $\gamma \in M_v \subseteq L_v$. In other words, $x \in \mathcal{X}_L$ is in the relative support of f_v if and only if $S_{M_v}^{L_v}(\chi_x \overline{f}_v^{(Q)}) \neq 0$. Obviously $S_{M_v}^{L_v}(\chi_x \overline{f}_v^{(Q)}) = \chi_x S_{M_v}^{L_v}(\overline{f}_v^{(Q)}) = (\chi_x \circ pr \circ ord_M) \cdot S_M^L(S_L^G(f_v)) = (\chi_x \circ pr \circ ord_M) \cdot S(f_v)$. This does not vanish identically if and only if x is in the image of the support of $S(f_v)$ in $\mathcal{X}_{M_v}$ under pr. This proves the lemma. $\square$

2.7.1 Subdivision of the Weyl Chambers

Suppose $Q = LN_Q$ is an F_v-rational standard parabolic subgroup $Q = Q_F$ defined by $F \subseteq \Delta(G_v, A_v)$, containing the minimal F_v-parabolic group $P = MN$. Then an element $x = \sum_{i \in \Delta(G_v, A_v)} x_i \alpha_i$ in $\mathcal{X}_{M_v}$ is contained in the support of the function

$$\hat{\chi}_{Q_F}^G = \hat{\tau}_{Q_F}^G \circ pr \circ ord_M$$

if and only if its projection $pr(x) = \sum_{i \notin F} x_i \alpha_i \in \mathcal{X}_{L_v}$ is in the obtuse Weyl chamber $^+\mathcal{X}_{L_v} = \sum_{i \notin F} \mathbb{R}_{>0} \alpha_i$, which means $x_i = \langle x, \beta_i \rangle > 0$ for all $i \notin F$.

The equations $\alpha_i(x) = 0$ and $\beta_i(x) = 0$ for $\alpha_i \in \Delta(G_v, A_v)$ define hyperplanes in $\mathcal{X}_{M_v}$. The images of these hyperplanes under the action of the Weyl group on $\mathcal{X}_{M_v}$ define finitely many hyperplanes. The complement of these hyperplanes in $\mathcal{X}_{M_v}$ is a union of open connected cones. Each of these cones is the image under the Weyl group of a subcone of the open Weyl chamber $\mathcal{X}_{M_v}^+$. Pick one of these cones C.

Example 2.3. For $G_v = Sl(3, F_v)$ the positive Weyl chamber contains two such cones.

Support Conditions. Suppose f_v is a spherical function on G_v such that its Satake transform is contained in the W-orbit of $x_0 + C \subseteq \mathcal{X}_{M_v}$ for some $x_0 \in C$, as in Lemma 2.11. Then a regular semisimple element γ is in the support of $O_\gamma^L(\overline{f}_v^{(Q)} \hat{\chi}_Q^G)$ if and only if $x = ord_L(\gamma)$ is in $pr(\bigcup_{w \in W}(x_0 + C))$. If this is the case then $x_i = \beta_i(x) > 0$ for all $i \notin F$. But then moreover, by our specific choice of the cone, we even get $x_i > const(x_0) > 0$ for all $i \notin F$. Similarly, if γ is not in the support of $O_\gamma^L(\overline{f}_v^{(Q)} \hat{\chi}_Q^G)$, then $x_i < -const(x_0)$ holds for at least one $i \notin F$. The constant $const(x_0)$ which appears in these formulas of course depends on the choice of $x_0 \in C$. By a suitable choice of x_0 it can be made arbitrarily large. A similar statement holds for the condition that $x = ord(\gamma) \in \mathcal{X}_{L'_v}$ is in the support of $O^{L'}(\overline{f}^{(P)} \hat{\chi}_Q^G \chi_{L'}^L)$ for $L' \subseteq L$, $P = L'N_P$, and $Q = LN_Q$. In fact all values

$$\alpha_i(w(x))\,,\ \beta_j(w(x)) \qquad w \in W,\ i,j \in \Delta(G_v, A_v)$$

are different from zero, and either $> const(x_0)$ or $< -const(x_0)$.

Preferred Places S'. These facts can now be used in the global context to concentrate the effect of the adelic cutoff functions $\hat{\chi}_Q^G$, as they appear in the formula of Corollary 2.2, to a finite set S' of "preferred" local non-Archimedean places in the sense that

$$tr_s(\overline{f}^{(Q)}\hat{\chi}_Q^G; H^\bullet(S_{L_Q}, V)) = tr_s(\overline{f}^{(Q)}(\hat{\chi}_Q^G)_{S'}; H^\bullet(S_{L_Q}, V))$$

holds (in a suitable context). For this it would suffice to know that

$$T_{ell}^{L'}(\overline{f}^{(P)}\hat{\chi}_Q^G\chi_{P'}^L, .) = T_{ell}^{L'}(\overline{f}^{(P)}(\hat{\chi}_Q^G\chi_{P'}^L)_{S'}, .)$$

holds for all $L' \subseteq L$, where L' is a Levi component of $P = L'N_P \subseteq Q = LN_Q$ (Corollary 2.2). Alternatively (Corollary 2.3) it would be enough to know that

$$O_\gamma^{L'}(\overline{f}^{(P)}\hat{\chi}_Q^G\chi_{P'}^L) = O_\gamma^{L'}(\overline{f}^{(P)}(\hat{\chi}_Q^G\chi_{P'}^L)_{S'}).$$

Before we explain under which conditions this holds, we first recall certain definitions.

2.7.2 Global Situation

For $\mathbb{Q}$-rational parabolic subgroups P and Q of T the global cutoff function $\hat{\chi}_Q^G\chi_{P'}^L$ on $L'(\mathbb{Q})$, which occurs in Corollary 2.3, was defined for $P = L'N_P$ using the Harish-Chandra map H_P via

$$L'(\mathbb{Q}) \lhook\joinrel\longrightarrow L'(\mathbb{A}) \xrightarrow{\ H_P\ } \mathcal{X}_{L'} .$$

In fact, by the product formula $H_P(\gamma) = log|\gamma_\infty|_\infty - \sum_{v \neq \infty} q_v \cdot ord_{L'}(\gamma_v)$, the global cutoff condition can be written as the condition on the point

$$\sum_{v \neq \infty} q_v \cdot ord_{L'}(\gamma_v) \in \mathcal{X}_{L'}$$

to lie in the support of $\hat{\tau}_Q^G\tau_{P'}^L$.

Notation: $\gamma = (\gamma_v)_v \in L(\mathbb{A}_{fin})$. q_v denotes the cardinality of the residue field, and $ord_{L'}(\gamma_v)$ the image of the local element $ord_{L'}(\gamma_v) \in \mathcal{X}_{L'_v}$ in $\mathcal{X}_{L'}$ under the natural projection map $\mathcal{X}_{L'_v} \to \mathcal{X}_{L'}$ (notice that locally the maximal F_v-split torus may be larger than the maximal $\mathbb{Q}$-split torus $A_{L'}$).

Assumptions. To be more specific about the concentration at specific places, let us assume $f = \prod_{v \neq \infty} f_v$. Furthermore, suppose there are two finite disjoint sets S and S' of non-Archimedean places such that f_v is the unit element of the spherical Hecke algebra for all $v \notin S \cup S'$. Suppose $f_S = \prod_{v \in S} f_v$ has support in a fixed compact subset of $G(\mathbb{A}_S)$. Finally, suppose that all f_v for $v \in S'$ are spherical such that the Satake transform $S(f_v)$ has the following property.

Property $(*)$. For all roots α and all dual roots β in the set of $\mathbb{Q}$-rational simple roots of (G, P_0) and all elements $w \in W$ the absolute value of the linear forms $\alpha \circ w$ and $\beta \circ w$ on

$$\sum_{v \in S'} q_v \cdot ord_{L'}(\gamma_v) \ \in \mathcal{X}_{L'}$$

is larger than a fixed constant $c > 0$.

If c is sufficiently large compared with the support of f_S, we obviously get

Lemma 2.13. *Under the assumptions above, if the constant c is large enough depending only on the support of f_S, the truncation condition concentrates on the places in S'*

$$O_\gamma^{L'}(\overline{f}^{(P)} \hat{\chi}_Q^G \chi_{P'}^L) = O_\gamma^{L'}(\overline{f}^{(P)} (\hat{\chi}_Q^G \chi_{P'}^L)_{S'}).$$

Notation. Let $\mathcal{E}_\nu$ denote the set of irreducible constituents $\rho = \rho_{S'} \otimes \rho^{S'} \in \mathcal{E}_\nu$ of the admissible representation of $G(\mathbb{A}_{fin})$ on the cohomology group $H^\nu(S_L, V)$.

Corollary 2.4. *Let the situation be as in Lemma 2.13. Then the truncated Lefschetz number $tr_s(\overline{f}^{(Q)} \hat{\chi}_Q^G; H^\bullet(S_L, V))$ is given by $tr_s(\overline{f}_S^{(Q)} (\hat{\chi}_Q^G)_{S'}; H^\bullet(S_L, V))$, or alternatively by a sum*

$$\sum_\nu (-1)^\nu \cdot \sum_{\rho \in \mathcal{E}_\nu} tr\left(f^{S'}; Ind_{L(\mathbb{A}_S)}^{G(\mathbb{A}_S)}(\rho_S)\right) \cdot tr\left(\overline{f}_{S'}^{(Q)} (\hat{\chi}_Q^G)_{S'}; \rho_{S'}\right),$$

where now $f^{S'} = f_S \prod_{w \notin S', w \neq \infty} 1_w$.

Proof. The first statement follows from Corollary 2.3 together with Lemma 2.13, which implies $T_{ell}^L(f\hat{\chi}_Q^G, \tau) = T_{ell}^L(f_S \cdot (f^S(\hat{\chi}_Q^G)_{S'}), \tau)$. The second formula then follows from the first assertion via the adjunction formula. $\square$

2.8 Automorphic Representations

Fix λ and a compact open subgroup $K = \prod_{v \neq \infty} K_v \subseteq \Omega$ of $G(\mathbb{A}_{fin})$, which defines the "level," the level group. The $G(\mathbb{A}_{fin})$-module given by the limit $H^\bullet(S(G), V_\lambda)$ is an admissible representation of $G(\mathbb{A}_{fin})$. Only finitely many irreducible constituents π with the property $\pi^K \neq 0$ occur. The same holds for the finitely many Levi subgroups L, the induced level groups $K_L = (K \cap P(\mathbb{A}_{fin})_L$, and the induced coefficient systems attached to the highest weights $\lambda' = w(\lambda + \rho_G) - \rho_L$. Thus, the admissible representation

$$\Pi(\lambda) = \bigoplus_{P_0 \subseteq Q \subseteq G} \bigoplus_{w \in W^Q} \bigoplus_i Ind_Q(\mathbb{A}_{fin})^{G(\mathbb{A}_{fin})}(H^i(S(L), V_{w(\lambda+\rho_G)-\rho_L})),$$

the "halo" of the $G(\mathbb{A}_{fin})$-module $H^\bullet(S(G), V_\lambda)$, again contains only finitely many irreducible $G(\mathbb{A}_{fin})$-constituents π with the property $\pi^K \neq 0$. Let $\mathcal{P}$ be the set of equivalence classes of these representations of level K.

Remark 2.10. $\Pi(\lambda)$ should be considered as a superspace whose grading is induced by the sign defined by the parity of the sum of the number $rank(G) - rank(Q)$, the length $l(w)$ for $w \in W^P$, and the degree i.

Let S_0 be the set of places for which $K_v \neq \Omega_v$ (level primes). Outside S_0 representations in $\mathcal{P}$ are unramified. Fix a prime $p \notin S_0$, the "Frobenius" prime. For π in $\mathcal{P}$ consider the representation π^p of $G(\mathbb{A}_{fin}^p)$ defined by $\pi = \pi^p \otimes \pi_p$. The set of places S_0 can be enlarged to a finite set S of places not containing p such that $\pi_1^p \cong \pi_2^p \Longleftrightarrow (\pi_1)_S \cong (\pi_2)_S$. There exists $f_S \in C_c^\infty(G(\mathbb{A}_S))$, so $tr\ \pi_S(f_S) = 0$ holds for all representations π' in $\mathcal{P}$ for which $(\pi')^p$ is not isomorphic to π^p, where π is some fixed representation in $\mathcal{P}$. Furthermore, we can assume $tr\ \pi_S(f_S) = 1$. For a suitable choice of K (in a cofinal system, where K^S is a product of special good maximal compact open subgroups), one can assume in addition that f_S is K_S-bi-invariant (see the Remark 4.3 on page 79). Now fix the π^p-projector f_S. For a non-Archimedean place $v \notin S$ consider functions

$$f = f_S \cdot h_p \cdot f_v \cdot \prod_{w \neq \infty\ else} 1_w$$

in $C_c^\infty(G(\mathbb{A}_{fin}))$, where h_p and f_v are suitable functions in the spherical Hecke algebra $\mathcal{H}(G_p, \Omega_p)$, respectively, $\mathcal{H}(G_v, \Omega_v)$. f_v is chosen subject to the conditions:

- Property $(*)$ (see the assumptions preceding Lemma 2.13) holds for $S' = \{v\}$ with respect to the fixed function f_S or more precisely its fixed support in $G(\mathbb{A}_S)$.
- $tr\pi_v(f_v) = 1$ holds for the unramified component π_v of our fixed representation $\pi^p = \otimes_{w \neq p, \infty} \pi_w$.

Such functions f_v exist, as explained on page 42, as a consequence of Lemma 2.11 choosing x_0 in the cone C to be sufficiently large. The function h_p is chosen to be either:

- $h_p = 1_p$ (unit element of $\mathcal{H}(G_p, \Omega_p)$) or
- $h_p^{(n)} = b(\phi_n)$ (the local cyclic base change of the Kottwitz function ϕ_n on $G(E_p)$ of [51] under the unramified base change map homomorphism b of spherical

Hecke algebras for some unramified local field extension $E_p/\mathbb{Q}_p$ of degree $[E_p : \mathbb{Q}_p] = n)$ in the context where G is attached to a Shimura variety as in [51] with reflex field $\mathbb{Q}$ (for simplicity)

We claim that either for $h_p = 1$, $S' = \{v\}$, or for $h_p = h_p^{(n)}$, $S' = \{p, v\}$, and for sufficiently large $n \gg 0$, the assumptions preceding Lemma 2.13 are satisfied. For $h_p = 1$ this has already been explained. The case $h_p = h_p^{(n)}$ and $n \gg 0$ can be reduced to the ensuing Lemma 2.14. We leave this as an exercise. So taking this for granted, now assume $n \gg 0$ or $h_p = 1$.

Then we get from Corollary 2.4 an expression for the truncated Lefschetz numbers

$$tr_s(\overline{f}^{(Q)}\hat{\chi}_Q^G, H^\bullet(S_L, V))$$

in terms of

$$\sum_\nu \sum_{\rho \in \mathcal{E}_\nu} (-1)^\nu \cdot tr\left(f^{S'}, Ind_{L(\mathbb{A}_S)}^{G(\mathbb{A}_S)}(\rho^{S'})\right) \cdot tr\left(\overline{f}_{S'}^{(Q)}(\hat{\chi}_Q^G)_{S'}, \rho_{S'}\right).$$

This allows us to apply a theorem of Franke [27] which states that all irreducible representations ρ of $L(\mathbb{A}_{fin})$ which occur in $\mathcal{E}_\nu$ as constituents of the cohomology group $H^\nu(S_L, V)$ are automorphic representations of $L(\mathbb{A}_{fin})$. Hence, all induced representations in

$$Ind_{L(\mathbb{A}_S)}^{G(\mathbb{A}_S)}(\rho^{S'})$$

are automorphic representations of $G(\mathbb{A}^{S'})$, and are Eisenstein representations for $L \neq G$.

Therefore, if the fixed representation $\pi \in \mathcal{P}$ is cuspidal and not CAP, π^p does not occur as a constituent in $\mathcal{P}$ from these induced representations in the case $L \neq G$. Since f and $\overline{f}$ are K_S-bi-invariant, the trace of f on $\Pi(\lambda)$ involves only constituents in $\mathcal{P}$, i.e., for the fixed level K. Since f_S is a projector for π^p among the representations in $\mathcal{P}$, this implies $tr\left(f_S, Ind_{L(\mathbb{A}_S)}^{G(\mathbb{A}_S)}(H^\nu(S_L, V))\right) = 0$. Hence, the truncated Lefschetz numbers

$$tr_s\left(\overline{f}^{(Q)}\hat{\chi}_Q^G, H^\bullet(S_L, V)\right)$$

all vanish except for the case $G = Q$, where the truncated Lefschetz number is the trace $tr_s(\overline{f}, H^\bullet(S_L, V))$ of $\overline{f}$ on the cohomology $H^\bullet(S_L, V)$. Notice $\overline{f} = \overline{f}^{(G)} \neq f$ in general. However, f and $\overline{f}$ have the same trace on every irreducible admissible representation. This follows from $O^G(\overline{f}) = O^G(f)$, since $vol(\Omega) = 1$. But then we can replace $\overline{f}$ by f. Then, since f is K_S-bi-invariant, the remaining Lefschetz number is the trace of f on the finite-dimensional space $H^\bullet(S_K(G), V)$ for fixed level K, and it only involves the representations in $\mathcal{P}$. Since f_S is a π^p-projector, the trace of $h_p f_v$ on this space is the trace of $h_p f_v$ on the generalized π^p-eigenspace of the cuspidal cohomology. Since $tr\,\pi_v(f_v) = 1$, this simplifies the formula for $T_{ell}^G(f_S f_v h_p, \lambda)$ of Corollary 2.2, and leaves only the term for $Q = G$ and $w = 1$. This proves

Theorem 2.2. *Suppose π is an irreducible cuspidal representation of $G(\mathbb{A}_{fin})$ and not CAP (see [69, 97]). Then for $f = f_S f_v h_p \prod_{w \neq \infty \text{ else}} 1_w$, where f_S, f_v, and $h_p = 1$ or $h_p = h_p^{(n)}$ and $n \gg 0$ is chosen as above, we get for the trace on the π^p-constituents*

$$tr_s\left(h_p, H^\bullet(S_K(G), V_\lambda)(\pi^p)\right) = T_{ell}^G\left(f_S f_v h_p, \lambda\right).$$

This theorem will be used in Chap. 3 for the "Frobenius" prime p.

Notice that $f_S f_v$ again is a projector on π^p among the representations in $\mathcal{P}$. We write $f_S f_v = f_{\pi^p}$, and call it a *"good π^p-projector."*

Lemma 2.14. *Let $C \subseteq V$ be a cone in Euclidean space, W a finite group acting on V, and $L_1, \ldots, L_r$ a W-stable set of linear forms in V^* nonvanishing on C. For $x_0 \in C$ and $x \in V$ and a bounded set $M \subseteq V$, there exists a integer m depending on M and an integer N depending on m and M such that the following holds. Suppose $v = v_1 + v_2 + v_3$ for $v_1 \in \bigcup_{w \in W}\{n \cdot w(v) \mid n \geq N\}$, $v_2 \in \bigcup_{w \in W} w(m \cdot x_0 + C)$, and $v_3 \in M$. Then $L_i(v) > 0$ for some $i = 1, \ldots, r$ holds if and only if $L_i(v_1 + v_2) > 0$ holds.*

Proof. Obvious. $\square$

Remark 2.11. In the case $h_p = 1$, we may also omit the auxiliary prime p or choose p to be large so that the formula in Theorem 2.2 becomes

$$\sum_\nu (-1)^\nu dim_\mathbb{C}(H^\nu(S_K(G), V_\lambda)(\pi)) = T_{ell}^G(f_\pi, \lambda)$$

for a good π-projector $f_\pi \in C_c^\infty(G(\mathbb{A}_{fin}))$.

Remark 2.12. In the Hermitian symmetric case there exists a formula analogous to Theorem 2.2 for the L^2-cohomology instead of the Betti cohomology. In this case the L^2-cohomology is finite-dimensional, so one can define the traces of Hecke operators on the L^2-cohomology. Using the results in [33], one obtains a formula for the L^2-Lefschetz numbers analogous to the one of Corollary 2.3. The relevant change in this case amounts to a subtler substitute of W^P, which in the case of L^2-cohomology also depends on the elements γ. In fact one obtains the following formula for the L^2-Lefschetz number:

$$\sum_{P_0 \subseteq P = LN \subseteq G} \sum_{w \in W^P} (-1)^{\#I(w)} \cdot T_{ell}^L(\overline{f}^{(P)}\chi_P^G(w), w(\lambda + \rho_G) - \rho_L),$$

where the cutoff functions $\chi_P^G(w)$ now depend on $w \in W^P$. They are defined as follows: $\chi_P^G(w)$ is the characteristic function of the set of all $\gamma \in L(\mathbb{Q})$, which satisfy $I(\gamma) = I(w)$, for certain finite sets $I(w)$ depending only on w, P, G, and λ

(see [33], p. 474), and where $I(\gamma)$ is the set of simple roots α of A_P in N_P such that $|\alpha(\gamma)|_f^{-1} = |\alpha(\gamma)|_\infty \leq 1$ (see [33], p. 471). Let $\chi_{P,i}$ denote the characteristic function of the set of all $\gamma \in L(\mathbb{R})$ such that $|\alpha_i(\gamma)|_\infty > 1$ for the simple root α_i. Then the characteristic function $\chi_P^G(w)$ can be expressed in the form $\prod_{i \notin I} \chi_{P,i} \prod_{j \in I}(1 - \chi_{P,j})$ for $I = I(w)$. It can therefore be expanded into a finite linear combination of the functions $\chi_K = \prod_{i \in K} \chi_{P,i}$ for $K \subseteq \Delta(G, A)$. For these finitely many global cutoff functions χ_K, which now appear in the L^2-Lefschetz formula, the effect of the cutoff can now be concentrated at some preferred non-Archimedean places $v \in S'$ by the choice of a suitable good π^S-projector, modified at a single place $S' = \{v\}$ as in the discussion above using a variant of Lemma 2.13. This implies

Corollary 2.5. *Suppose G_∞ is of Hermitian symmetric type. Suppose π is an irreducible cuspidal representation of $G(\mathbb{A}_{fin})$ and not CAP. Then there exists a good π-projector f_{π^S} such that the L^2-Lefschetz number of the π-constituents is*

$$tr_s(H_{(2)}^\bullet(S(G), V_\lambda)(\pi)) = T_{ell}^G(f_\pi, \lambda).$$

In particular, the alternating sums of the π-multiplicities on the cohomology and the L^2-cohomology coincide.

2.9 The Discrete Series Case

This is the case considered in [4]. Suppose G is a connected reductive group over $\mathbb{Q}$, G_{der} is simply connected, and G contains a maximal $\mathbb{R}$-torus B, for which $B(\mathbb{R})/A_G(\mathbb{R})^0$ is compact (see [4], p. 262).

Notation. Let $2q(G)$ denote the real dimension of the symmetric domain attached to G_∞ and $d(G)$ the cardinality of the packets of discrete series representations of G_∞. Let τ be an irreducible complex representation of $G(\mathbb{Q})$ defined by the highest weight $\lambda \in X^*(B)_\mathbb{C}$. λ defines a representation of $G(\mathbb{C})$, and hence of the compact inner form $\overline{G}$ of G over the field $\mathbb{R}$. Let τ^* denote the contragredient representation. Attached to the representation τ of $\overline{G}$ is a packet $\Pi_{disc}(\tau)$ of discrete series representations π_∞. Let π_∞^* denote the contragredient. Attached to τ and λ is the function

$$f_\infty = \frac{f_\lambda}{d(G)},$$

where $f_\lambda \in \mathcal{H}_{ac}(G_\infty, \xi_\lambda^{-1})$ (in the notation in [4], Lemma 3.1) is the stable cuspidal function (i.e., supported in discrete series, see [4], Sect. 4) defined by Clozel and Delorme. f_∞ is compactly supported modulo $A_G(\mathbb{R})^0$ and is $\tilde{K}_\infty$-invariant. Then using the notation in [4], p. 271, formula (4.3), $tr \, \rho^*(f_\infty) = tr \, \rho^*(f_\infty) = (-1)^{q(G_\infty)} tr \, \pi_\infty^*(\frac{f_\lambda}{d(G)})$ for $\pi_\infty \in \Pi_{disc}(\rho)$ becomes $d(G)^{-1}$ if $\pi_\infty^* \in \Pi_{disc}(\tau)$ and is zero otherwise (see [4], Lemma 3.1). Notice $\pi_\infty^* \in \Pi_{disc}(\tau)$ if and only if $\rho^* \cong \tau$. Hence,

$$tr\ \rho^*(f_\infty) = d(G)^{-1}$$

if $\rho \cong \tau^*$, and $tr\ \rho^*(f_\infty) = 0$ otherwise.

The orbital integral $O_\gamma^G(f) = \int_{G_\gamma(\mathbb{R})\backslash G(\mathbb{R})} f(x^{-1}gx)dx$, considered for fixed $\gamma \in G(\mathbb{R})$ as a distribution on $\mathcal{H}_{ac}(G(\mathbb{R}), \xi_\lambda^{-1})$, is denoted $\Phi_G(\gamma, f)$ in [4], p. 269, and in [5], p. 325. Theorem 5.1 in [4] gives a formula valid for all $\gamma \in G(\mathbb{R})$ which expresses the orbital integral of *stable cuspidal* functions $f_\infty \in \mathcal{H}_{ac}(G_\infty, \xi_\lambda^{-1})$ in terms of the distributions $\rho^*(f)$ discussed above,

$$O_\gamma^G(f_\infty) = (-1)^{q(G)}d(G_\gamma)vol(\overline{G}_{\gamma,\infty}/A_G(\mathbb{R})^0)^{-1}\sum_\rho \Phi_G(\gamma, \rho) \cdot tr\ \rho^*(f_\infty),$$

for certain coefficients $\Phi_G(\gamma, \rho)$. The sum runs over irreducible representations ρ of $\overline{G}(\mathbb{R})$ in $\Pi(\overline{G}(\mathbb{R}), \xi_\lambda)$. In particular, $O_\gamma^G(f)$ is zero unless γ is semisimple and $\gamma \in T(\mathbb{R})$ for some maximal $\mathbb{R}$-torus of G such that $(\mathbb{R})/A_G(\mathbb{R})^0$ is compact. Notice $T(\mathbb{R}) \cong B(\mathbb{R})$. On the regular part $T_{reg}(\mathbb{R})$ the function is $\Phi_G(\gamma, \rho) = tr\ \rho(f)$ (see [4], p. 271). Since $\Phi_G(\gamma, \rho)$ extends to a continuous function on $T(\mathbb{R})$ (see [4], Lemma 4.2), this holds for all $\gamma \in T(\mathbb{R})$. Hence, if $O_\gamma^G(f)$ does not vanish a priori, one has $\gamma \in T(\mathbb{R})$, where T is a maximal $\mathbb{R}$-torus in G such that $T(R)/A_G(\mathbb{R})$ is compact. And for all $\gamma \in T(\mathbb{R})$ one has the formula

$$O_\gamma^G(f_\infty) = (-1)^{q(G)}d(G_\gamma)vol(\overline{G}_{\gamma,\infty}/A_G(\mathbb{R})^0)^{-1}tr\ \tau^*(\gamma)d(G)^{-1},$$

since only $\rho \cong \tau^*$ contributes to the sum over all ρ. Notice $tr\ \tau(\gamma^{-1}) = tr\ \tau^*(\gamma)$ for the contragredient representation. Next, from the formula for the Euler numbers (see, e.g., [4], p. 281, formula (6.3), and also p. 282)

$$\chi(G, dg_f) = (-1)^{q(G)} \cdot d(G) \cdot vol(G(\mathbb{Q})A_G(\mathbb{R})^0\backslash G(\mathbb{A})) \cdot vol(\overline{G}(\mathbb{R})/A_G(\mathbb{R})^0)^{-1},$$

one obtains for $f_{fin} \in C_c^\infty(G(\mathbb{A}_{fin}))$

$$\chi(G_\gamma) \cdot tr\ \tau(\gamma^{-1}) \cdot O_\gamma^G(f_{fin})$$

$$= (-1)^{q(G_\gamma)}d(G_\gamma)vol(\overline{G}_\gamma(\mathbb{R})/A_G(\mathbb{R})^0)^{-1}\tau(G_\gamma) \cdot tr\ \tau^*(\gamma) \cdot O_\gamma^G(f_{fin})$$

$$= d(G)\tau(G_\gamma)O_\gamma^{G_\infty}(f_\infty)O_\gamma^G(f_{fin})$$

$$= d(G)\tau(G_\gamma)O_\gamma^{G(\mathbb{A})}(f_{fin}f_\infty),$$

provided the measure dg_∞ is chosen such that $dg_\infty dg_{fin}$ is the Tamagawa measure on $G(\mathbb{A})$. Hence, from the definition of $T_{ell}^G(f_{fin})$ we obtain

Lemma 2.15.

$$T_{ell}^G(f_{fin}, \tau) = d(G) \sum_{\gamma \in G(\mathbb{Q})/\sim}' \tau(G_\gamma)O_\gamma^{G(\mathbb{A})}(f_{fin}f_\infty).$$

The summation is over all semisimple, strongly elliptic conjugacy classes of $G(\mathbb{Q})$. Here $\tau(G_\gamma)$ is the Tamagawa number $vol(G_\gamma A_G(\mathbb{R})^0 \backslash G_\gamma(\mathbb{A}))$, where the measure dg_∞ is chosen such that $dg_\infty dg_{fin}$ is the Tamagawa measure on $G(\mathbb{A})$.

Corollary 2.6. *With the assumptions and the notation used in Theorem 2.2 we get for $f_{fin} = h_p f_{\pi^p}$*

$$tr(h_p, H^\bullet(S_K(G), V_\lambda)(\pi^p)) = d(G) \sum_{\gamma \in G(\mathbb{Q})/\sim}^{'} \tau(G_\gamma) O_\gamma^{G(\mathbb{A})}(h_p f_{\pi^p} f_\infty).$$

The summation is over all semisimple, strongly elliptic conjugacy classes of $G(\mathbb{Q})$. The measures defining the orbital integrals are assumed to be Tamagawa measures on $G(\mathbb{A})$ and $G_\gamma(\mathbb{A})$.

Remark 2.13. The term $O_\gamma^G(f_S f_{\pi^S} f_\infty)$ is independent of the chosen measures dg_f and dg_∞ provided $dg_\infty dg_f$ is the Tamagawa measure on $G(\mathbb{A})$. This follows from the definition of f_{fin} and f_∞. Hence, in applications we are now free to normalize the measures dg_f and dg_∞, e.g., such that $vol_{dg_f}(K) = 1$ following the convention of [51].

Remark 2.14. Assume that Z_G/A_G is anisotropic over $\mathbb{R}$. If one considers a Shimura variety attached to G (as in [51]) one replaces $S_K(G) = G(\mathbb{Q}) \backslash G(\mathbb{A})/\tilde{K}_\infty K$ by $G(\mathbb{Q}) \backslash G(\mathbb{A})/Zentr(h)_\infty K$, where h is the underlying structure homomorphism of the Shimura variety. For small K this multiplies the trace by the index $[\tilde{K} : Zentr(h)_\infty]$. See also the remark on page 21 In fact $\gamma\epsilon_\infty \in KZentr(h)_\infty$ for $\epsilon_\infty \in \tilde{K}_\infty$, and $\gamma \in G(\mathbb{Q})$ implies $\gamma \in Z_G(\mathbb{Q})$ (K is small) and $\gamma \in K\tilde{K}_\infty$. Hence, γ is finite, and hence is 1 (K is small). Therefore, $\epsilon_\infty \in Zentr(h)_\infty$.

Appendix 1

Let G be a reductive connected group over $\mathbb{Q}$. Let $K \subseteq G(\mathbb{A}_{fin})$ be a compact open subgroup. For $g \in G(\mathbb{A}_{fin})$ put $K' = K_g = g^{-1}Kg \cap K \subseteq K$. Consider $M = G(\mathbb{Q}) \backslash G(\mathbb{A})$, or some compactification, with continuous $G(\mathbb{A}_{fin})$ left action $m \mapsto mg^{-1}, g \in G(\mathbb{A}_{fin})$ together with the maps $p(m) = m$ and $p'(m) = mg^{-1}$

$$p : M/K' \to M/K$$

$$p' : M/K' \to M/K.$$

The map p (or the map p') is equivariant with respect to the map q (or the map q') from $K' = K_g$ to K, defined by $k \mapsto k$ or $k \mapsto gkg^{-1}$. Two points mK and $m'K$ in M/K are related by the correspondence underlying p, p' if there exists a point $m''K' \in M/K'$ such that

$$p(m''K') = mK \text{ and } p'(m''K') = m'K \text{ in } M/K.$$

This means that there exist $k, k' \in K, k'' \in K'$ such that $mkk'' = m''$ and $m''g^{-1} = m'(k')^{-1}$ holds. Hence, $mkk''g^{-1}k' = m'$. Stated in other terms, $m' = mx^{-1}$ for some $x \in KgK$. There exists a finite decomposition $KgK = \biguplus_j Kg_j$. Hence,

$$m'K = mg_j^{-1}K$$

for some j. Conversely, suppose $m'K = mkg^{-1}K$ for some $k \in K$. Then for $m'' := mkK_g$, we tget $p(m''K_g) = mkK = mK$ and $p'(m''K_g) = mkg^{-1}K = m'K$.

Put $\Gamma = K \cap G(\mathbb{Q})$. In general for $\gamma \in G(\mathbb{Q})$ the double coset $\Gamma\gamma\Gamma = \biguplus_i \Gamma\gamma_i$ decomposition gives $K\gamma\Gamma = \biguplus_i K\gamma_i$, again a disjoint union. Since $k_1\gamma_1 = k_2\gamma_2$ implies $k_2^{-1}k_1 = \gamma_2\gamma_1^{-1} \in G(\mathbb{Q}) \cap K = \Gamma$, we get $\Gamma\gamma_1 = \Gamma\gamma_2$. Passing to the closure defines the subset $K\gamma\overline{\Gamma} = \biguplus_i K\gamma_i$ of $K\gamma K = \biguplus_j Kg_j$, which might be smaller than KgK if $\overline{\Gamma} \neq K$. Therefore, to relate fixed points of the adelic correspondence to its classical analogue, one has to ensure that fixed points belong to cosets gK of the form γK for some $\gamma \in G(\mathbb{Q})$ and in particular $KgK = K\gamma K$. However, this is the case (see page 24). Only rational cosets γK contribute to the fixed points of the Goresky–MacPherson trace formula for the Lefschetz numbers.

Appendix 2

Let G_∞ be the group of real points of a reductive group over $\mathbb{R}$. Let K_∞ be a maximal compact group, and let $V_1 \subseteq Z_\infty$ be a vector group in the center Z_∞.

Claim 2.1. Then for every $y \in K_\infty \cdot V_1$, the set $\mathcal{S}$ of all $x \in G_\infty$, such that $x^{-1}yx \in K_\infty \cdot V_1$, is either empty or

$$\mathcal{S} = G_{y,\infty} \cdot K_\infty.$$

Here $G_{y,\infty}$ denotes the centralizer of y in G_∞.

Proof. The proof of this assertion is easily reduced to the case $V_1 = 1$. In fact, $G_\infty = {}^0G \cdot V$, where V is the maximal vector group in the center of G_∞ and ${}^0G$ is the normal subgroup of G_∞ with ${}^0G \cap V = \{e\}$ chosen as in [98], p. 19. Notice $K_\infty \subseteq {}^0G_\infty$.

This allows us to reduce the proof to the case where $y \in K_\infty$ and x satisfies the equation $x^{-1}yx \in K_\infty$. In fact, if $x_0^{-1}yx_0 = k \cdot v_1$ holds for some $x = x_0$ and $k \in K_\infty, v \in V_1$, we simply replace y by $y_1 = x_0yv_1^{-1}x_0^{-1} \in K_\infty$ and x by $x_1 = x_0^{-1}x$. Then $x_1^{-1}y_1x_1 \in K_\infty V_1$ is equivalent to $x^{-1}yx \in K_\infty \cdot V_1$. However $x_1^{-1}y_1x_1 \in K_\infty V_1$ if and only if $x_1^{-1}y_1x_1 \in {}^0G_\infty \cap (K_\infty V_1) = K_\infty$. So we assume $y \in K_\infty$ and $x^{-1}yx \in K_\infty$.

Choose a Cartan involution θ of G_∞ such that $g \in K_\infty$ if and only if $\theta(g) = g$ (see [98], Proposition 5). For x as above, the element $z = \theta(x)x^{-1}$ is in $G_{y,\infty}$, and satisfies $\theta(z) = z^{-1}$. One can write $x = s \cdot \kappa$ for $\kappa \in K_\infty$ and $s = exp(\sigma)$ and $\theta(\sigma) = -\sigma \in Lie(G_\infty)$ (follows from [98], Proposition 5). Then $z = exp(-2\sigma) \in G_{y,\infty}$. Since $y \in K_\infty$, y and hence also $G_{y,\infty}$ is θ-stable. Therefore, there exists a symmetric one-parameter subgroup in $G_{y,\infty}$ passing through z. See, e.g., [98], p. 20. In other words we find a symmetric root $r = exp(-\sigma) \in G_{y,\infty}, \theta(r) = r^{-1}$ of $z = r^2$ for $\theta(\sigma) = -\sigma \in Lie(G_{y,\infty})$. We conclude $1 = r^{-1}\theta(x)x^{-1}r^{-1} = \theta(rx)(rx)^{-1}$. Thus, $rx = k \in K_\infty$ and $x = r^{-1}k \in G_{y,\infty} \cdot K_\infty$, which proves the claim. $\square$

Corollary 2.7. $S/(K_\infty \cdot V_1)$ *is either empty or* $G_{y,\infty}/(G_{y,\infty} \cap K_\infty)V_1$, *where* $G_{y,\infty} \cap K_\infty$.

Proof. Notice $V_1 \subseteq G_{y,\infty}$. $\square$

Remark 2.15. Finally, there exists a diffeomorphism $G_\infty/(K_\infty \cdot V_1) \cong (^0G_\infty/K_\infty) \times V/V_1$. In particular for $V_1 \subseteq A_G(\mathbb{R})^0$, we see that $G_\infty/(K_\infty \cdot V_1)$ is homotopic to $X_G = G_\infty/(K_\infty \cdot A_G(\mathbb{R})^0)$.

Chapter 3
The Ramanujan Conjecture for Genus 2 Siegel Modular Forms

In this chapter we apply the results obtained in Chaps. 1 and 2 and work of Kottwitz in the special case of the symplectic group of similitudes $GSp(4)$. The main result obtained in this chapter is the proof of the Ramanujan conjecture for $GSp(4)$ for cohomological automorphic forms which are not cuspidal representations associated with parabolic subgroups (CAP). We will assume certain statements regarding the trace formula, the proof of which occupies Chaps. 6–10.

For the moment let G be a connected reductive group over $\mathbb{Q}$, whose derived group is simply connected. Assume that the maximal $\mathbb{Q}$-split torus A_G in the center coincides with the maximal $\mathbb{R}$-split torus in the center. Let $K \subseteq G(\mathbb{A}_{fin})$ be a sufficiently small compact open subgroup (see Sect. 2.2).

Shimura Varieties. Let

$$h : R_{\mathbb{C}/\mathbb{R}}(\mathbb{G}_m) \to G_{\mathbb{R}}$$

be the structure homomorphism of a Shimura variety attached to G (i.e., satisfying the conditions of [65] and [51], p. 162). Let $X_\infty = G(\mathbb{R})/Zentr(h)_\infty$ denote the $G(\mathbb{R})$-conjugacy class of the homomorphism h. The Shimura variety

$$S_K(\mathbb{C}) = G(\mathbb{Q}) \setminus (X_\infty \times G(\mathbb{A}_{fin})/K$$

is a smooth algebraic variety over $\mathbb{C}$. Let K_∞ denote a maximal compact subgroup in $G_\infty = G(\mathbb{R})$ and put $\tilde{K}_\infty = A_G(\mathbb{R})^0 K_\infty$. Then $Zentr(h)_\infty \subseteq \tilde{K}_\infty$ is a subgroup of finite index by our assumptions.

The Manifolds $S_K(G)$. Let $X_G = G_\infty/\tilde{K}_\infty$ denote the connected symmetric space attached to G. There is a finite covering map $X_\infty \to X_G$. The cardinality of the fibers is equal to the index $[\tilde{K}_\infty : Zentr(h)_\infty]$. At the beginning of Chap. 2 we defined the topological space

$$S_K(G) = G(\mathbb{Q}) \setminus (X_G \times G(\mathbb{A}_{fin})/K).$$

R. Weissauer, *Endoscopy for GSp(4) and the Cohomology of Siegel Modular Threefolds*, Lecture Notes in Mathematics 1968, DOI: 10.1007/978-3-540-89306-6_3, © Springer-Verlag Berlin Heidelberg 2009

Obviously
$$S_K(\mathbb{C}) \to S_K(G)$$
is a covering map of degree $[\tilde{K}_\infty : Zentr(h)_\infty]$. The trace formulas in Chap. 2 carry over to the spaces $S_K(\mathbb{C})$ instead of $S_K(G)$, except that they have to be multiplied by this index (see the remark on page 21 and the second remark on page 50).

Example 3.1. For $G = GSp(4, \mathbb{Q})$ $X_\infty = H$ is the Siegel upper half-space.

Example 3.2. For $G = GSp(4)$ and the standard choice of structure morphism h this index is $[\tilde{K}_\infty : Zentr(h)_\infty] = 2$. In fact, $\tilde{K}_\infty$ is generated by $Zentr(h)_\infty$ and the block diagonal matrix
$$\epsilon_\infty = diag(id, -id).$$
Hence,
$$X_\infty = G(\mathbb{R})/Zentr(h)_\infty = X_G \cup -X_G \quad , \quad X_G = H.$$

The spaces $S_K(\mathbb{C})$ are the (nonconnected) Siegel modular varieties of genus 2. The canonical model of the Shimura variety $S_K(\mathbb{C})$ over $\mathbb{Q}$ can be identified with the moduli space over $\mathbb{Q}$ of principally polarized Abelian varieties of level N for a suitable choice of K (Sect. 1.2). In the following we restrict ourselves to studying these particular cases, since the corresponding groups define a cofinal system of compact open subgroups $K \subseteq G(\mathbb{A}_{fin})$.

Relevant Constants. For $G = Sp(2g, \mathbb{Q})$ the number of discrete series representations $d(G)$ in a fixed L-packet for the group G_∞ is 2^g; for $G = GSp(2g, \mathbb{Q})$ this number is 2^{g-1}. In particular, $d(Gl(2, \mathbb{Q})) = 1$; hence, $d(M) = 1$ for the group $M = Gl(2, \mathbb{Q})^2/Z_1$. The group $Z_1 \cong \mathbb{G}_m$ is embedded via $t \mapsto (t \cdot id, t^{-1} \cdot id)$.

Tamagawa Numbers. There is the general formula $\tau(G) = |\pi_0(Z(\hat{G}))^\Gamma|/|ker^1(\mathbb{Q}, Z(\hat{G}))|$, which holds for all connected reductive groups G, whose derived group is simply connected. For the groups $G = GSp(2g)$ we have $Z(\hat{G}) = \mathbb{C}^*$, considered with trivial Galois action; hence,
$$\tau(G) = 1.$$

In the case $g = 2$, we also consider the group $M = Gl(2)^2/Z_1$, where $Z_1 \cong \mathbb{G}_m$ is embedded in $Gl(2)^2$ via the map $x \mapsto (x \cdot id, x^{-1} \cdot id)$. The derived group $M_{der} = Sl(2)^2/(\pm 1)$ is not simply connected. Oesterle [67], p. 40, showed that $\tau(M)\tau(\mathbb{Q}^*)/\tau(Gl(2, \mathbb{Q})^2)$ is the product of
$$\#Koker(Hom_{\mathbb{Q}-alg}(Gl(2)^2, \mathbb{G}_m) \to Hom_{\mathbb{Q}-alg}(Z_1, \mathbb{G}_m))$$
and $[M(\mathbb{A})/M(\mathbb{Q})Gl(2, \mathbb{A})^2]$ divided by
$$\#Ker(ker^1(\mathbb{Q}, Z_1) \to Ker^1(\mathbb{Q}, Gl(2)^2)).$$

The first factor is 2 and the second and third factors are 1 by Hilbert theorem 90. Thus,
$$\tau(M) = 2.$$

The Endoscopic Factor $i(G,M)$. An endoscopic datum $(M, {}^L M, s, \xi)$ for G is a quasisplit connected reductive $\mathbb{Q}$-group M (called an endoscopic group for G), an element s in the center of the dual group $\hat{M}$, and an embedding $\xi : \hat{M} \to \hat{G}$ of groups, satisfying the following conditions: (a) $\xi(\hat{M}) = (\hat{G}_{\xi(s)})^0$; (b) the $\hat{G}$-conjugacy class of ξ is fixed by the Galois group, which defines an extension ${}^L M \to {}^L G$ also denoted ξ; and (c) the image of s in $Z(\hat{M})/Z(\hat{G})$ is fixed by the Galois action and is contained in the group $\mathcal{K}(M/\mathbb{Q})$, i.e., has a locally trivial image in $\pi_0((Z(\hat{M})/Z(\hat{G}))^{\Gamma_\mathbb{Q}})$. Such an endoscopic datum is called elliptic if $\xi(Z(\hat{M})^{\Gamma_\mathbb{Q}})^0$ is contained in $Z(\hat{G})$. For further details and the notion of equivalence of endoscopic data we refer to [54], p. 17ff.

It is not hard to see that up to equivalence there exists a unique nontrivial elliptic endoscopic datum $(M, {}^L M, s, \xi)$ for the group $G = GSp(4, \mathbb{Q})$, namely,

$$
\xi : \quad \hat{M} = \left\{ \begin{pmatrix} * & 0 & * & 0 \\ 0 & * & 0 & * \\ * & 0 & * & 0 \\ 0 & * & 0 & * \end{pmatrix} \right\} \hookrightarrow \hat{G} = GSp(4, \mathbb{C})
$$

for the choice

$$
s = diag(1, -1, 1, -1).
$$

See also Sect. 6.3. The group M is the group $Gl(2)^2/Z_1$ defined above. For an elliptic endoscopic datum there is the constant

$$
i(G, M) = \tau(G)\tau(M)^{-1}\lambda^{-1},
$$

where λ is the order of a group $\Lambda(M, s, \eta) = Aut(M, s, \eta)/M_{ad}(\mathbb{Q})$ attached to the elliptic endoscopic group M. $Aut(M, s, \eta)$ is the group of those automorphisms φ of M which preserve the embedding η up to $\hat{G}$ equivalence and the element $s \in \mathcal{K}(M/\mathbb{Q})$ under the map $\mathcal{K}(M/\mathbb{Q}) \to \mathcal{K}(M/\mathbb{Q})$ induced by φ. See [60] and [53], p. 630, for these definitions and further details.

The case $G = GSp(4)$. Any automorphism of $PGl(2)$ is an inner automorphism; therefore, an automorphism ρ of M which does not switch the two factors of M_{ad} can be modified by an inner automorphism such that it induces the identity on M_{ad}. Since ρ preserves the center $Z_M \cong \mathbb{G}_m$, the restriction ρ_0 of ρ to Z_M is either the identity or the inversion. Automorphisms $\rho : M \to M$, which induce the identity on M_{ad} and induce some fixed homomorphism ρ_0 on the center Z_M, are of the form $g \mapsto g\psi(g)$ for an algebraic character $\psi \in Hom_{alg}(M, Z_M) \cong \mathbb{Z}$ (generated by $(g_1, g_2)/(t \cdot id, t^{-1} \cdot id) \mapsto det(g_1 g_2)^{-1}$) such that $\psi(z) = \rho_0(z)z^{-1}$ holds for $z \in Z_M$; hence, ρ_0 uniquely determines ρ. For $\rho_0 = id$, therefore, $\psi(z) = 1$ and $\rho = id$. For $\rho_0(z) = z^{-1}$ the automorphism ρ is $(g_1, g_2)/ \sim \mapsto (g_1 det(g_1)^{-1}, g_2 det(g_2)^{-1})/ \sim$. Its dual $\hat{\rho}$ is not induced by an inner automorphism of $\hat{G}$. There it does not contribute to $\Lambda(M, s, \xi)$. Hence, up to an inner automorphism, there is a unique further automorphism of M given by $(g_1, g_2)/ \sim \mapsto (g_2^{-t}, g_1^{-t})/ \sim$, which contributes nontrivially to $\Lambda(M, s, \xi)$ and switches the two factors of M_{ad}. The induced map on $\hat{M}$ is the restriction of an inner automorphism of G, which stabilizes s modulo the center of $Z(\hat{G})$. Therefore, $\lambda = 2$. Since $\tau(G) = 1$ and $\tau(M) = 2$, this implies

$$i(G, M) = \frac{1}{4}$$

for the nontrivial elliptic endoscopic datum of the group $GSp(4)$.

3.1 Results Obtained by Kottwitz, Milne, Pink, and Shpiz

In the following we consider principal congruence subgroups $K = K_N$ in $GSp(4, \mathbb{A}_{fin})$ for N sufficiently large. Then the moduli space of principally polarized Abelian varieties of level N structure defines a scheme S_K over $Spec(\mathbb{Z}[1/N])$, which gives the canonical model denoted $\mathcal{M}$ on page 7. Fix a prime p such that $(p, N) = 1$. Then p is a prime of good reduction for the Siegel modular variety S_K. We write as usual $K = K_p K^p$, where $K_p = GSp(\mathbb{Z}_p)$. By abuse of notation we write $S_K(\overline{\mathbb{F}}_p)$ for the variety over $\overline{\mathbb{F}}_p$ obtained by the extension of scalars from $\mathbb{F}_p$ to the algebraic closure $\overline{\mathbb{F}}_p$ of the reduction of S_K modulo p. The Hecke correspondences and the Frobenius act on the etale cohomology groups of the variety $S_K(\overline{\mathbb{F}}_p)$.

Counting Points. In [51], (3.1), Kottwitz states a conjectural formula for the "$P = G$-part" of the Lefschetz number (or supertrace) of a correspondence

$$f \times Frob_p^n$$

on the etale intersection cohomology groups $IH^\bullet(S_K(\overline{\mathbb{F}}_p), V_\lambda)$. The sheaves V_λ extend to smooth l-adic sheaves on the reduction of the Shimura variety, temporarily viewed as a variety over the algebraic closure $\overline{\mathbb{F}}_p$ of the prime field of characteristic p. See Sect. 1.4. In this chapter we use the *unitary* normalization (see page 3). In other words, we normalize these sheaves by a half-integral Tate twist so that they become Weil sheaves on $S_K(\overline{\mathbb{F}}_p)$, which are pure of weight zero. In this formula the Hecke operator $f = \prod_v f_v$ is supposed to be a product of local Hecke operators coming from the primes outside p. In fact, there is no other restriction except that $f = f_p f^p$ is K-bi-invariant and induces at the place p the local Hecke operator $f_p = 1_p$, which is the unit element of the local spherical Hecke algebra at p. See loc. cit. p. 163. We simply write f^p instead of $f^p 1_p$. To provide evidence for his formula, Kottwitz computed in loc. cit. Sect. 4 the cardinality of the number of fixed points of F_{p^n}-rational points of $S_K(\overline{\mathbb{F}}_p)$, and showed that this cardinality is given by his predicted formula (3.1) in the special case $f \times Frob_p^n = Frob_p^n$, where the Hecke correspondence is trivial, and where the coefficient system is $V_\lambda = \overline{\mathbb{Q}}_l$. Hence, by the Weil conjectures this computation of Kottwitz gives a formula for the Lefschetz number of $Frob_p^n$ on the etale cohomology $H_c^\bullet(S_K(\overline{\mathbb{F}}_p), V_\lambda)$.

Conjecture of Deligne. The immediate question is whether this calculation extends to cover the cases where V_λ is nontrivial and where $f = 1_p \otimes f^p$ is a Hecke operator, which has trivial component at p. In this case the supertrace $tr_s(f \times Frob_p^n)$ on $H_c^\bullet(S_K(\overline{\mathbb{F}}_p), V_\lambda)$ can be calculated by the Lefschetz trace formula of Grothendieck and Verdier. It computes this Lefschetz number as a sum over the fixed points in terms of contributions, which are defined purely locally. However, in general, these local terms of the Grothendieck–Verdier fixed-point formula may be complicated, and in addition there appear possible additional "boundary contributions" owing to the fact that the variety is not proper over the base field [65], p. 258f. On the other hand, Pink [73] has shown that for fixed f^p and all large enough powers $n \gg 0$ of the Frobenius $Frob_p^n$, the Lefschetz number of the correspondence $f^p \times Frob_p^n$ on the cohomology with compact support can be calculated by a "naive" Lefschetz number, in which the local terms at the fixed points of the correspondence are the obvious local traces. A similar result was obtained, independently, by Shpiz [65], Theorem C.4. Hence, again one can compute the Lefschetz number by a counting argument involving the fixed points of the correspondence [65], Theorem C.6. In fact the computation of Kottwitz in [51], Sect. 12, gives a stronger result. This was further refined by Milne [64] or [65], (0.1). As explained in [65], p. 157, his refined result [65], (0.1), gives a formula for the "naive" Lefschetz number of $f^p \times Frob_p^n$ on the etale cohomology $H_c^\bullet(S_K(\overline{\mathbb{F}}_p), V_\lambda)$, stated in [65], (0.5). In fact, this formula coincides with the formula (3.1) in [51]. Hence, combined with [65], Theorem C.4, this implies

Theorem 3.1 (Kottwitz–Milne). *For $f = 1_p f^p$ and $K_p = GSp(\mathbb{Z}_p)$ the following holds: there exists n_0 such that for all $n \geq n_0$*

$$tr_s(f^p \times Frob_p^n; H_c^\bullet(S_K(\overline{\mathbb{F}}_p), V_\lambda)) = \sum_{\gamma_0} \sum_{(\gamma, \delta)} c(\gamma_0; \gamma, \delta) O_\gamma^G(f^p) TO_\delta(\phi_n) tr\, \tau(\gamma),$$

subject to the summation conditions and the notation in [51], p. 171.

For the formula of the last theorem we assume that the Hecke operators f act on the cohomology groups by a left action, following Milne's convention. This differs from the conventions in the last chapter. To make a comparison with the last chapter, where the right action was used, simply replace $f = f(x)$ by $f^- = f(x^{-1})$.

Remark 3.1. In the following we do not need to know the rather complicated description of the terms in this formula. It is enough for us to know that the term $TO_\delta(\phi_n)$ is a local twisted orbital for the group $G(E)$, where E is the unramified extension of $\mathbb{Q}_p$ of degree n. A warning concerning different normalizations in the literature is in order. In [51] the function ϕ_n denotes the local spherical function defined by the characteristic function of the double coset $GSp(o_E) \cdot a \cdot GSp(o_E)$ on $GSp(E)$ for $a = \mu_h(\pi_F^{-1})$, where $\mu_h : (\mathbb{G}_m)_{\mathbb{C}} \to G_{\mathbb{C}}$ is defined by h as usual. In [64], p. 202, however it is the characteristic function $GSp(o_E) \cdot a \cdot GSp(o_E)$ for the inverse $a = \mu_h(\pi_F)$, where the action of Hecke operators is induced by a right action of $G(\mathbb{A}_{fin}^p)$ on the variety and the induced left action on cohomology, which was also our choice above.

3.1.1 Stabilization

In [51] Kottwitz rearranged the summation terms in the formula [51], (3.1), which appears in Theorem 3.1 under the assumption of certain conjectures on (twisted) endoscopy. These conjectures are described in loc. cit. p. 178 (existence of matching functions; an extension of matching to (G, M)-regular elements; the fundamental lemma for the unit element at almost all places) and on p. 180 (twisted fundamental lemma for the function ϕ_n). These assumptions are needed to construct the required functions $h^p \in C_c^\infty(H(\mathbb{A}_{fin}^p))$ and h_p in the spherical local Hecke algebra of $H(\mathbb{Q}_p)$, which match, respectively, with the function f^p (endoscopic matching condition) and ϕ_n (twisted endoscopic matching[1]), where H denotes the elliptic endoscopic groups of G. For $G = GSp(4)$, therefore, either $H = G$ or $H = M$. The required twisted versions of the fundamental lemma follow from the untwisted fundamental lemma, which is proven in Chaps. 6–10.

Archimedean Case. Similarly, using the theory of characters of the discrete series, Kottwitz constructed functions h_∞ on $H(\mathbb{R})$ in loc. cit. pp. 182–187. In the case $H = G$ the construction is as follows. Consider the L-packet $\Pi(\varphi)$ of discrete series representations of $G(\mathbb{R})$, having the same[2] central and infinitesimal character as $\tau = \tau_\lambda$:

$$h_\infty^G = (-1)^{q(G)} h(\varphi_G) = (-1)^{q(G)} d(G)^{-1} \sum_{\pi_\infty \in \Pi(\varphi)} f_{\pi_\infty},$$

where f_{π_∞} is a pseudocoefficient for π_∞. This slightly differs from [51], top of p. 185. The choice of Kottwitz coincides with the function f_∞ defined on page 48. We use the dual since we follow the conventions of Milne [65], p. 214(c).

Concerning the Assumptions. One further remark is in order. The matching conditions, among others, depend on the choice of transfer factors Δ_v. In [51], p. 188, Kottwitz assumes the global hypothesis, i.e., the product formula $\Delta = \prod_v \Delta_v = 1$ for global elements (Sect. 7.15). Furthermore, the other assumptions are not unrelated. The fundamental lemma at almost all places implies the fundamental lemma (Hales [35]) at all places. See also Sect. 10, where we give a proof of this which includes the twisted endoscopic case, using the method of Labesse [57] and Clozel [20]. In the case of the group $GSp(4)$ the existence of matching functions was shown by Hales [36], Sect. 6, Corollary 3. In general, it was shown by Waldspurger [108] that the existence of matching functions is already a consequence of the fundamental lemma. (Finally, it is possible to extend statements from regular to (G, M)-regular elements along the lines of [76], Proposition 8.1.3 and [50].)

---

[1] This matching condition is a priori not the same as the twisted endoscopic matching condition of [54]. The condition is described in Lemma 9.7. However, it is equivalent to the twisted endoscopic condition in general, and can be deduced from the ordinary endoscopic fundamental lemma. See Corollary 9.4.

[2] Well defined also for the unitary normalization.

Therefore, it enough to prove the fundamental lemma in the regular cases. Hence, since these assumptions are therefore fulfilled in our case of the group $GSp(4)$, Theorem 7.2 of Kottwitz [51] implies

Theorem 3.2. *Under the assumptions above and of Theorem 3.1, the following formula holds for $f = 1_p f^p$ and $K_p = GSp(\mathbb{Z}_p)$ and all $n \gg 0$:*

$$tr_s\left(f^p \times Frob_p^n; H_c^\bullet(S_K(\overline{\mathbb{F}}_p), V_\lambda)\right) = \sum_{\mathcal{E}} i(G, M) ST_e^*(h).$$

This requires some explanations:

1. First, $\mathcal{E}$ refers to a set of representatives of elliptic endoscopic triples. For example, in the case of the group $G = GSp(4, \mathbb{Q})$, the cardinality of $\mathcal{E}$ is 2. There is the trivial triple $(G, 1, id)$ and the triple (M, s, η) considered above.
2. Second, the constants $i(G, M)$ are the rational numbers defined already. For $G = GSp(4)$ they are 1 and $\frac{1}{4}$, respectively.
3. Third, the symbol $ST_e^*(\ )$ stands for the (G, M)-regular, $\mathbb{Q}$-elliptic part of the stable trace formula

$$ST_e^*(h) = \tau(M) \sum_{\gamma_M} |(M_{\gamma_M}/M_{\gamma_M}^0)(\mathbb{Q})|^{-1} SO_{\gamma_M}^M(h).$$

Summation is over a set of representatives for the (G, M)-regular semisimple $\mathbb{Q}$-elliptic stable conjugacy classes in $M(\mathbb{Q})$ [51], p. 189. The terms $SO_{\gamma_M}^M(h)$ are the corresponding stable orbital integrals. A semisimple element $\gamma_M \in M(\overline{\mathbb{Q}})$ is (G, M)-regular if the corresponding element $\gamma \in G(\overline{\mathbb{Q}})$ satisfies the following: $\gamma^\alpha = 1$ for a root α of G implies that α is already a root of M. $\mathbb{Q}$-elliptic means that Z_{G_γ}/Z_G is $\mathbb{Q}$-anisotropic.

Recall that the functions $h = \prod_v h_v$ have been carefully attached to the given correspondence $f^p \times Frob_p^n$ for each M. In the following we often write $h_v = h_v^M$ for the corresponding function f_v. Notice $h_v = f_v$ for $M = G$ except for $v \in \{p, \infty\}$. We write $f_p := h_p$ in the case $H = G$. This defines (!)

$$h^G = h_\infty f^p f_p \quad \text{and} \quad h^M = h_\infty^M h_p^M h^{M,p}.$$

Remark 3.2. Suppose $G = \mathbb{G}_m$, $K = \prod_v \mathbb{Z}_v$, and $\lambda = r \in \mathbb{Z}$. Then the Frobenius $Frob_p$ acts on $H_c^0(S(\overline{\mathbb{F}}_p), V_\lambda) \cong H^0(S(\overline{\mathbb{F}}_p), V_\lambda) \cong \mathbb{Q}_l$ in the same way as the Hecke operator (with left action on cohomology) defined by $g = (1, \ldots, p, 1 \ldots) \in \mathbb{A}_{fin}^*$, namely, by p^r. This seems to be in discrepancy with [40], p. 55. But notice that this depends on certain conventions, such as the normalization of the reciprocity law, the reciprocity law for the Shimura variety, and in particular on whether the absolute Galois group of the reflex field is considered to act by a left or a right action on the base field extension to $\overline{\mathbb{Q}}$ of the canonical model.

Let us Specialize for The Case $G=GSp(4,\mathbb{Q})$. Assuming the local conjectures on the endoscopic lifts, which will be proved in Chaps. 6–9, we get the formula

$$tr_s\left(f^p \times Frob_p^n; H_c^\bullet(S_K(\overline{\mathbb{F}}_p), V_\lambda)\right) = ST_e^*(h^G) + i(G,M)ST_e^*(h^M).$$

Let us consider the first term on the right-hand side. Since G_{der} is simply connected, it is

$$ST_e^*(h^G) = \tau(G) \cdot \sum_{\gamma_G} SO_{\gamma_G}^G(h^G).$$

Summation is over representatives of the semisimple $\mathbb{Q}$-elliptic stable conjugacy classes of $G(\mathbb{Q})$. The condition of (G,G)-regularity is void.

3.2 Destabilization

Theorem 9.6 of Kottwitz [53] gives the stabilization of the elliptic part of the trace formula – again using certain endoscopic conjectures [53], pp. 380 and 384, which for $G = GSp(4)$ are covered by the results proved in the chapters on the fundamental lemma. Notice that in these chapters only regular elements are considered. This will cause no problems since the extension to the (G,M)-regular case can be obtained along the lines of [76], Proposition 8.1.3 and [50]. So this allows us to destabilize the last formula for the Lefschetz number. Destabilization [53] gives

$$ST_e^{**}(h^G) = T_e^*(h^G) - i(G,M)ST_e^{**}\left((h^G)^M\right).$$

Here, following the notation of Kottwitz [53, p. 393], the double index $**$ instead of the single index $*$ should remind us that in the case $G = M$ one should not consider the (G,M)-regular stable conjugacy classes, but should exclude those coming from central elements. Notice (G,G)-regularity is the empty condition. We need not go through the full process of destabilization to observe that the second term $ST_e^{**}\left((h^G)^M\right)$ vanishes.

The Vanishing Term. In fact, $ST_e^{**}\left((h^G)^M\right)$ is the product over the local stable orbital integrals on $M(\mathbb{A})$. These are, up to some transfer factors, equal to the so-called κ-orbital integrals $O_\gamma^{G,\kappa}(h^H)$ of the functions h^G on $G(\mathbb{A})$. For the vanishing term it is therefore enough to see that at the Archimedean place the global κ-orbital integral vanishes, unless $\kappa = 1$. By a continuity argument similar to that in [51], p. 186, one should reduce this to the regular case. Since the sum defining $ST_e^{**}((h^G)^M)$ is indexed by stable conjugacy classes, all of which can be assumed to have $\mathbb{R}$-elliptic representatives γ_M ([51], p. 188), G_γ is an $\mathbb{R}$-anisotropic torus modulo $A_G(\mathbb{R})$. Hence, the global element $\kappa \in \mathcal{K}(G_\gamma/\mathbb{Q})$ is trivial if and only if its image in $\mathcal{K}(G_\gamma/\mathbb{R})$ is trivial. In fact both groups are cyclic of order 2 and are generated by $\kappa = \kappa_s$, where κ_s is attached to the endoscopic element s. But this is enough since for nontrivial $\kappa \in \mathcal{K}(G_\gamma/\mathbb{R})$ the Archimedean κ-orbital of h_∞^G vanishes, simply because

$$h_\infty^G = (-1)^{q(G)} d(G)^{-1} \sum_{\pi \in \Pi(\varphi)} f_{\pi_\infty}$$

is the sum over the pseudocoefficients f_{π_∞} of discrete series representations π_∞ in the Archimedean L-packet attached to an irreducible (finite-dimensional) representation of the compact form $\overline{G}$. Hence, h_∞^G is stable ([51], Sect. 7), and therefore has vanishing κ-orbital integrals at the Archimedean place for $\kappa \neq 1$. We conclude

$$ST_e^{**}(h^G) = T_e^*(h^G),$$

$$T_e^*(h^G) = \sum_{\gamma_G} |(G_{\gamma_G}/G_{\gamma_G}^0)(\mathbb{Q})|^{-1} \tau(G_{\gamma_G}^0) O_{\gamma_G}^G(h^G).$$

The sum is over representatives of the $G(\mathbb{Q})$-conjugacy classes of semisimple, $\mathbb{Q}$-elliptic elements in $G(\mathbb{Q})$, however with central elements removed. See [53], p. 391ff, in particular Theorem 9.6. In these formulas the global orbital integrals are defined using the canonical measures on $G(\mathbb{A})$, respectively, $G_{\gamma_G}^0(\mathbb{A})$ (i.e., the volumes on the adele quotients give the Tamagawa numbers).

Adding the Central Terms. In the stable trace $ST_e^{**}(h^G)$ the omitted summands are of the form $\tau(G) h^G(\gamma_G)$ for central elements γ_G. Also in $T_e^*(h^G)$ the omitted terms are the terms $\tau(G) h^G(\gamma_G)$ for central elements γ_G. This implies $ST_e^*(h^G) = T_{ell}^G(h^G)$, i.e.,

$$ST_e^*(h^G) = \sum_{\gamma_G} |(G_{\gamma_G}/G_{\gamma_G}^0)(\mathbb{Q})|^{-1} \tau(G_{\gamma_G}^0) O_{\gamma_G}^G(h^G) = \sum_{\gamma_G} \tau(G_{\gamma_G}) O_{\gamma_G}^G(h^G),$$

where now the summation is over representatives of all $G(\mathbb{Q})$-conjugacy classes of semisimple $\mathbb{Q}$-elliptic elements in $G(\mathbb{Q})$. Notice $G_{\gamma_G} = G_{\gamma_G}^0$, since G_{der} is simply connected. In the situation of Theorem 3.2 we therefore get from the result at the end of Sect. 3.1

$$\boxed{tr_s\left(f^p \times Frob_p^n; H_c^\bullet(S_K(\overline{\mathbb{F}}_p), V_\lambda) \right) = T_{ell}^G(h^G) + i(G, M) ST_e^*(h^M)}.$$

Notice the orbital integral over h_∞^G at the Archimedean place is zero unless the element γ_G is $\mathbb{R}$-elliptic [51], p. 182. This allows us to compare the summands on the right side of this formula with the terms in the topological trace formula of Goresky and MacPherson. In Sects. 3.3 and 3.4, we discuss the two terms on the right side of this formula.

3.3 The Topological Trace Formula

Assume that the level group K is small in the sense of Chap. 2. In the following, all measures are again supposed to be canonical measures. So the measure on $G(\mathbb{A})$ is the canonical measure. We may in addition assume the Haar measure to be suitably normalized on $G(\mathbb{A}_{fin})$. Now we apply Theorem 2.2. To be compatible with Milne's conventions, we have to convert the action of the Hecke algebra $C_c^\infty(\mathbb{A}^{p,\infty})$ on the cohomology from a right action to a left action. This is done by replacing f_{π^p} by $f_{\pi^p}^-$. However, to come closer to the formulas in [65], it is better to make the additional substitution $\gamma \mapsto \gamma^{-1}$ for the summation index. Notice $O_{\gamma^{-1}}^G(f^-) = O_\gamma^G(f)$ for $f = f^p$. At the infinite place the effect is $O_{\gamma^{-1}}^G(f_\infty) = O_\gamma^G(f_\infty^-)$, where now the function f_∞ becomes the function $f_\infty^- = h_\infty^G$. Recall that f^p is a good π-projector for an irreducible automorphic representation π^p of $GSp(4, \mathbb{A}_{fin}^p)$ with respect to the fixed level K in the sense of Chap. 2, Theorem 2.2.

If π is not CAP in the sense of Piatetski-Shapiro, we get from Theorem 2.2 the following expression for $T_{ell}^G(h^G)$:

$$T_{ell}^G(h^G) = (d(G)[\tilde{K}_{G,\infty} : Zentr(h)])^{-1} \sum_\nu (-1)^\nu tr\big(h_p^G; H^\nu(S_K(\mathbb{C}), V_\lambda)(\pi^p)\big)$$

$$= \frac{1}{4} \cdot \sum_\nu (-1)^\nu tr\big(h_p^G; H^\nu(S_K(\mathbb{C}), V_\lambda)(\pi^p)\big).$$

This needs some explanation. First, observe $h_v^G = f_v$ for all v not in $\{p, \infty\}$. At the archimedean place the definition of h_∞^G has already been given. The function h_p^G depends on the Frobenius power n and can be understood in terms of a cyclic base change of degree n [51, 52]. For the moment it is enough to remember the dependence of h_p^G on n. Hence, we will sometimes write $h_p^G = h_p^{(n)}$.

On the other hand, for trivial reasons the choice of a good π^p-projector (π not CAP) also localizes the etale trace formula involving the Frobenius trace. Hence,

Lemma 3.1. *Let $G = GSp(4, \mathbb{Q})$ and let π be irreducible cuspidal, but not CAP. Let K be a fixed small level subgroup with corresponding set S of primes of bad reduction. Then for $n \gg 0$ (depending on K and p) the Kottwitz functions $h_p^G = h_p^{(n)}$ define spherical Hecke operators such that under the assumptions of Theorem 3.2*

$$4 \cdot tr_s\big(Frob_p^n; H_{c,et}^\bullet(S_K(\overline{\mathbb{F}}_p), V_\lambda)(\pi^p)\big) - tr_s\big(h_p^{(n)}; H_{Betti}^\bullet(S_K(\mathbb{C}), V_\lambda)(\pi^p)\big)$$

$$= ST_e^*(h_\infty^M h_p^M f_{\pi^p}^M).$$

Proof. Multiply the identity above by 4 and observe that the two coefficients were both equal to $\frac{1}{4}$. For that see the introductory remarks on constants in this chapter. $\square$

Remark 3.3. By Theorem 1.1(1b) and its dual, our assumption that π is not CAP implies that all generalized π^p-eigenspaces of the cohomology groups (with or without compact support) with respect to π^p can be replaced by the corresponding spaces $H_!^\bullet(S_K, V_\lambda)(\pi^p)$, where $H_!^\nu(S_K, V_\lambda)$ is the image of $H_c^\nu(S_K, V_\lambda)$ in $H^\nu(S_K, V_\lambda)$. This shows that we might drop the index c and replace both cohomology groups by the cohomology groups $H_!^\nu(S_K, V_\lambda)(\pi^p)$

$$H_c^\nu(S_K, V_\lambda)(\pi^p) \xrightarrow{\sim} H_!^\nu(S_K, V_\lambda)(\pi^p) .$$

Furthermore, by Lemma 1.2 the two cohomology groups in characteristic p (coefficients in the etale $\overline{\mathbb{Q}}_l$-sheaf V_λ) and characteristic zero (coefficients in the complex coefficient system V_λ), which appear in Corollary 3.1, can be identified by the choice of an isomorphism $\iota : \overline{\mathbb{Q}}_l \cong \mathbb{C}$.

Corollary 3.1. *Under the assumptions of Lemma 3.1 we obtain*

$$4 \cdot tr_s\big(Frob_p^n; H_{!,et}^\bullet(S_K(\overline{\mathbb{F}}_p), V_\lambda)(\pi^p)\big) - tr_s\big(h_p^{(n)}; H_!^\bullet(S_K(\mathbb{C}), V_\lambda)(\pi^p)\big)$$

$$= ST_e^*(h_\infty^M h_p^M f_{\pi^p}^M).$$

For further applications of Corollary 3.1, it remains to simplify the righthand side of its formula. For that we should first understand the meaning of the terms which were removed by the condition (*), namely, the condition of (G, M)-regularity.

3.4 (G, M)-Regularity Removed

What does it mean for an $\mathbb{R}$-elliptic, semisimple element t of $M(\mathbb{Q})$ to be (G, M)-regular? Of course it is enough to deal with this over the field $\mathbb{R}$ of real numbers. Suppose t is represented by (t_1, t_2) modulo $Z_1(\mathbb{R})$ in the notation in Chap. 6, page 217. Not to be (G, M)-regular then means (see page 277) $x = x^\sigma$ or $x' = (x')^\sigma$, where $x = t_1 t_2$ and $x' = t_1(t_2)^\sigma$, and where σ denotes the nontrivial $\mathbb{R}$-automorphism of $\mathbb{C}$. In other words, t is not (G, M)-regular if and only if $t_1/t_1^\sigma = (t_2/t_2^\sigma)^{\pm 1}$ holds.

Lemma 3.2. $SO_\gamma^M(h_\infty^M) = 0$ *for all $\mathbb{R}$-elliptic, semisimple γ which are not (G, M)-regular.*

Proof. h_∞^M is defined in [51], Sect. 7. Remember that M is of adjoint type; hence, $d(M) = 1$. Each packet of discrete series representations of $M(\mathbb{R})$ has cardinality 1. Furthermore, $q(G) = 3$. Therefore, according to [51], p. 186, and Lemma 8.4

$$h_\infty^M = \langle \mu_h, s \rangle \cdot \sum_{\omega_* \in \Omega_M \backslash \Omega} det(\omega_*(\phi_M))(-1)^{q(G)} h(\phi_M)$$

or in our case $h_\infty^M = -\langle \mu_h, s \rangle \cdot \sum_{\omega_* \in \{id, w\}} det(\omega_*(\phi_M), X^*(T)) h(\phi_M)$. The term in front of the sum is a sign factor, and id and w are representatives of the cosets of the complex Weil groups Ω_M and Ω of M and G, respectively. w interchanges the two factors of the group M. If $\phi_M : W_{\mathbb{R}} \to {}^L M$ is a Langlands parameter for M for the discrete series representation π_M on $M(\mathbb{R})$ determined by λ, $h(\pi_M)$ denotes one of its pseudocoefficients. Hence, up to a sign determined later in Lemma 8.4,

$$\pm h_\infty^M = h(\pi_{r_1} \otimes \pi_{r_2}) - h(\pi_{r_2} \otimes \pi_{r_1}), \qquad (r_1 > r_2).$$

Here, by abuse of notation we regard the discrete series representation π_M as given by a pair of discrete series representations π_{r_1}, π_{r_2} for the group $Gl(2, \mathbb{R})$. The weights r_1 and r_2 (highest K-types) are numbers greater than or equal to 2 determined by the assumption that the composite map $W_{\mathbb{R}} \to {}^L M \to {}^L G$ gives the discrete series packet on $G(\mathbb{R})$ corresponding to τ, i.e., to the coefficient system E_λ under consideration. Similarly for the central character. For example, if V_λ corresponds to classical Siegel modular forms of weight k – which means $\lambda_1 = \lambda_2 = k - 3$ – then we get $r_1 = 2k - 2$ and $r_2 = 2$. More generally, for $\lambda_1 \geq \lambda_2 \geq 0$ and $k_i = \lambda_3 + 3$ with $k_1 \geq k_2 \geq 3$ [58], p. 212,

$$r_1 = k_1 + k_2 - 2, \quad r_2 = k_1 - k_2 + 2.$$

The stable orbital integral on $M(\mathbb{R})$ is essentially the product of two orbital integrals on the group $Gl(2, \mathbb{R})$.

Consider an $\mathbb{R}$-elliptic element in $M(\mathbb{R})$. Notice γ is conjugate to an element which is up to some central element contained in a fixed maximal compact subgroup of $M(\mathbb{R})$. Suppose $\gamma = (\gamma_1, \gamma_2) \mod Z_1(\mathbb{R})$. If $det(\gamma_i) < 0$, then $O_{\gamma_i}^{Gl(2,\mathbb{R})}(h_{\pi_r}) = 0$ for all r. In particular, $SO_\gamma^M(h_\infty^M)$ vanishes. For this recall that the restriction of an irreducible discrete representation π of $Gl(2, \mathbb{R})$ decomposes into the direct sum of two irreducible subrepresentations $\pi^+ \oplus \pi^-$, after restriction to $Gl(2, \mathbb{R})^+ = \{g \in Gl(2, \mathbb{R}) \mid det(g) > 0\}$. Furthermore, an irreducible representation π' of $Gl(2, \mathbb{R})$ is isomorphic to π if and only if the restriction of π' to $Gl(2, \mathbb{R})^+$ contains π^+. Therefore, if f^+ is a pseudocoefficient of the discrete series representation π^+ of G^+, extend f^+ to a function on $Gl(2, \mathbb{R})$ which is zero on the complement of $Gl(2, \mathbb{R})^+$. Then, evidently, f becomes a pseudocoefficient for the discrete series representation π of $Gl(2, \mathbb{R})$. Since the support of f is in G^+, our claim follows.

Hence, to prove the lemma, we can assume that γ_1, γ_2 are both contained in $Gl(2, \mathbb{R})^+$, and by conjugation

$$\gamma_i = \lambda_i \cdot \begin{pmatrix} cos(\alpha_i) & sin(\alpha_i) \\ -sin(\alpha_i) & cos(\alpha_i) \end{pmatrix}, \qquad \lambda_i \in \mathbb{R}^*$$

is contained in a fundamental torus, and $t_i/t_i^\sigma = exp(2i\alpha_i)$. Since now γ_i are contained in an elliptic torus, up to some constant $(-1)^{q(M_\gamma^0)} d(M_\gamma^0)/d(M) vol(Z_{M_{\gamma,\infty}}^0 \setminus \overline{M}_{\gamma,\infty}^0)^{-1}$ (Sect. 2.9), the orbital integral $O_{\gamma_i}^{Gl(2)}(h_{\pi_r})$ can be expressed by

$$\big(tr \, (\tau_{r_1 - 2} \otimes \tau_{r_2 - 2}) - tr \, (\tau_{r_2 - 2} \otimes \tau_{r_1 - 2}) \big)(\gamma),$$

where τ_r is a finite-dimensional irreducible complex representation of $Gl(2, \mathbb{C})$.
In fact $\tau_r = Symm^r(st)$, and $tr\ \tau_r(\gamma_i) = \omega(\lambda_i) \cdot P(t_i)$ for $P \in \mathbb{C}[X, X^{-1}]$ with
$P(X) = P(X^{-1})$ and $P(-X) = (-1)^r P(X)$. That γ is not (G, M)-regular means
$t_1^2 = t_2^{\pm 2}$, or equivalently $t_1 = \pm t_2$ or $\pm t_2^{-1}$. Therefore, $tr\ \tau_{r_1-2}(t_1) \cdot tr\ \tau_{r_2-2}(t_2) -$
$tr\ \tau_{r_2-2}(t_2) \cdot tr\ \tau_{r_1-2}(t_1) = 0$, since $r_1 - r_2 = 2k_2 - 4$ is an even integer. This
completes the proof of the lemma. $\square$

Corollary 3.2. *The term* $ST_e^*(h_\infty^M h_p^M f_{\pi^p}^M)$ *on the right-hand side in the formula of
Corollary 3.1 is*

$$T_{ell}^M(h_\infty^M h_p^M f_{\pi^p}^M) = \sum_{\gamma_M}[M_\gamma : M_\gamma^0]^{-1}\tau(M_\gamma^0)O_\gamma^M(h_\infty^M h_p^M f_{\pi^p}^M).$$

*Summation is over all conjugacy classes of $\mathbb{R}$-elliptic, semisimple elements of
$M(\mathbb{Q})$.*

Proof. M does not have nontrivial elliptic endoscopic groups, and the notions of conjugacy and stable conjugacy coincide; hence, $ST_e^*(h_\infty^M h_p^M f_{\pi^p}^M) =$
$T_{ell}^M(h_\infty^M h_p^M f_{\pi^p}^M)$. This proves the corollary. In the sum $T_{ell}^M(h_\infty^M h_p^M f_{\pi^p}^M)$ one
could also remove the central terms, which are zero owing to the presence of
the Archimedean orbital integrals of h_∞^M. $\square$

3.5 CAP Localization Revisited

At this point we apply the trace formula for M to write the sum in Corollary 3.2
in terms of automorphic representations of $M(\mathbb{A})$. Recall that the functions h_v^M for
$v \neq p, \infty$ are locally constant with compact support and independent of the twist n.

Concerning the Archimedean Place. Both the Frobenius trace formula of
Kottwitz and the topological trace formula of Goresky and MacPherson (in the
form of Chap. 2) were considered by us only under the simplifying assumption that
the derived group is simply connected. Since M_{der} is not simply connected, we may
nevertheless use the trace formula in the L^2 version given by Arthur [4]. This is
possible since the functions h_∞^M at the Archimedean place are linear combinations
of pseudocoefficients of discrete series representations of M_∞. Arthur's L^2-formula
can be brought into a CAP localized form in the same way as the topological trace
formula. This gives a spectral expression for the term in Corollary 3.2 similar to the
one in Corollary 2.2. In fact we have more than that. For $n \gg 0$ the parabolic terms
automatically vanish by our assumption that π is not CAP. The reason for this is the
same as the one in Sect. 2.8. We sketch the argument.

Concentration at Preferred Places. In Chap. 2 we saw that the trace formula is
simplified considerably if one discards CAP representations in the discrete spectrum. This can be achieved by inserting suitable test functions into the trace formula

called good projectors in Chap. 2, which concentrated the truncation at preferred places. In the construction of a good π-projector in Chap. 2 we specified one place v (not in $S \cup \{p\}$) where the support of the Satake transforms is "very" regular in addition to the Frobenius place p. This projector f_{π^p} was chosen for $G = GSp(4)$ such that for $n \gg 0$ truncation concentrates at two preferred places p, v. By transfer from G to M we now have to consider the local Hecke operators h_p^M and $f_{\pi^p}^M$ for the group M. These transferred functions do not define a good projector on M in the precise meaning of Chap. 2. Nevertheless, they act in the same way. For $n \gg 0$ truncation concentrates again at the preferred places p and v. Therefore, the argument of Sect. 2.8 carries over verbatim. In fact

The Bad Places. For the finite set S of ramified primes w, where $K_w \neq GSp(4, \mathbb{Z}_w)$, the existence of h_w^M is guaranteed by [36]. Except for their existence we need not care further about these fixed Hecke operators.

The Frobenius Prime p. Here, up to a constant $c(n)$ depending on n, the Satake transform of $h_p^{(n)M}$ turns out to be obtained by a rescaling (Lemma 10.2)

$$S(h_p^{(n)M})(\lambda) = c(n) \cdot S(h_p^{(1)M})(n \cdot \lambda)$$

for $\lambda \in \Lambda$.

The Unramified Places. For the places $v \neq p, \infty$, where $K_v = GSp(4, \mathbb{Z}_v)$, the fundamental lemma established in Chaps. 6–9 gives us precise information on the functions h_v^M. Recall h_v^G are in the spherical Hecke algebra $\mathcal{H}(G_v, \Omega_v)$ for the good maximal compact subgroup $\Omega_v = GSp(4, \mathbb{Z}_v)$. The homomorphism $b : \mathcal{H}(G_v, \Omega_v) \to \mathcal{H}(M_v, \Omega_v^M)$ between the local spherical Hecke algebras, defined by the commutative diagram

$$
\begin{array}{ccc}
\mathcal{H}(G_v, \Omega_v) & \xrightarrow{\ b\ } & \mathcal{H}(M_v, \Omega_v^M) \ , \\
{\scriptstyle S}\downarrow{\scriptstyle \sim} & & {\scriptstyle \sim}\downarrow{\scriptstyle S} \\
\mathbb{C}[\Lambda]^{W_G} & \longhookrightarrow & \mathbb{C}[\Lambda]^{W_M}
\end{array}
$$

describes for $v \neq p$ the spherical lift (Sect. 10.1)

$$h_v^M = b(h_v^G).$$

Hence, at the chosen preferred place $v \neq p$ the function h_v^G can be chosen such that $S(h_v^M)$ has sufficiently regular support in $\Lambda \subseteq \mathcal{X}_{M_v}$ with respect to the fixed Hecke operators h^M at the bad places and such that $S(h_v^G)$ has sufficiently regular support in $\Lambda \subseteq \mathcal{X}_{G_v}$ with respect to the fixed Hecke operators h^G at the bad places.

CAP Localization. The functions $f_{\pi^p}^M$ do not project onto a single irreducible representation of M. For instance, they cannot distinguish representations which are conjugate under the outer automorphism of M, which switches the two simple factors. However, they still allow us to eliminate the parabolic contributions to the trace formula. Without restriction of generality we can choose f_{π^p} in such a way that, by the endoscopic fundamental lemma, any irreducible automorphic representation of M which is a constituent of a globally induced representation from an automorphic representation of a proper parabolic subgroup of M is annihilated by $f_{\pi^p}^M$. This follows from the description of b given above, which describes the spherical endoscopic lift. We skip the details since we have to give the details of the argument later anyway (Lemma 4.27).

Concerning Finiteness. Irreducible representations of $M(\mathbb{A}_{fin})$ are given by pairs of irreducible automorphic representations π_1, π_2 of $Gl(2, \mathbb{A}_{fin})$, whose central characters $\omega_1 = \omega_2$ coincide. As $f_{\pi^{p \cup S}}^M$ is spherical, it is enough for us to consider the case where π_1 and π_2 are unramified representations of $Gl(2, \mathbb{A}_f)$, where $\mathbb{A}_{fin} = \mathbb{A}_f \times \prod_{v \in p \cup S} \mathbb{Q}_v$. If one of the two representations is of Eisenstein type including the one-dimensional case, then the endoscopic lift of the representation $\pi_1 \otimes \pi_2$ gives an automorphic irreducible spherical representation Π of $GSp(4, \mathbb{A}_f)$. From the endoscopic fundamental lemma it is clear that Π is then induced from an automorphic representation of the Levi subgroup of the Siegel parabolic subgroup (Lemma 4.27). As π was assumed to be not CAP, the π-projector f_{π^p} can be chosen such that it is orthogonal to all such lifts Π similar to the eliminations of induced representations coming from the halo. There are only finitely many possibilities for Π, because without restriction of generality the representations $\pi_1 \otimes \pi_2$ can be considered to have fixed discrete series components at the Archimedean place of some fixed level depending only on the fixed Hecke operators h_w^M for $w \in S$. Since the level is determined by the functions h_v^M for $v \in S$, and is independent from n, we get from the L^2-trace formula of Arthur for M after the process of CAP localization, which in the present case obviously also throws away all terms in the L^2-cohomology in cohomology degree different from 2, the following corollary

Corollary 3.3. *Suppose that the coefficient system is chosen subject to the unitary normalization. Then for $n \gg 0$ and a suitable choice of the good projector f_{π^p}, the right-hand side of the formula in Corollary 3.1 is equal to*

$$\sum_{\pi_1 \otimes \pi_2} tr(\pi_1 \otimes \pi_2)(h_\infty^M h_p^M f_{\pi^p}^M).$$

The sum runs over all irreducible cuspidal automorphic representations $\pi_1 \otimes \pi_2$ of $M(\mathbb{A})$ (described by a pair of cuspidal automorphic representations of $Gl(2, \mathbb{A})$ with the same central character $\omega_1 = \omega_2$) such that $\omega_{i,\infty}$ restricted to $A_M(\mathbb{R})^0$ is trivial.

Remark 3.4. That $\omega_{i,\infty}$ restricted to $A_M(\mathbb{R})^0$ is trivial comes from our *unitary* normalization considered on page 3, which makes E_λ pure of weight zero. In fact $E_\lambda \cong V_\lambda$ becomes self-dual. By this twist the central characters become $\omega_{\infty,i} = sign^{r_i}$. Notice $r_1 \equiv r_2$ modulo 2. In the L^2-trace formula π_1 and π_2, in general, need not be unitary without this normalization. They become unitary by a suitable twist with the idele norm character of $M(\mathbb{A})$.

Proof of Corollary 3.3. From the preceding discussion this is now a routine calculation. Recall

$$\pm h_\infty^M = h(\pi_{r_1} \otimes \pi_{r_2}) - h(\pi_{r_2} \otimes \pi_{r_1})$$

is the difference of two pseudocoefficients of discrete series representations of $M(\mathbb{R})$. Hence, the expression in Corollary 3.2 becomes the difference of two L^2-Lefschetz numbers for certain coefficient systems

$$\pm(\mathcal{L}_{\mu_1 \otimes \mu_2} - \mathcal{L}_{\mu_2 \otimes \mu_1}).$$

Since CAP localization eliminates the parabolic terms except for $P \neq M$ in Arthur's formula ([4], Theorem 6.1) for the L^2-Lefschetz number $\mathcal{L}_\mu = \mathcal{L}_\mu(h_p^M f_\pi^M)$, only the terms from the elliptic $M(\mathbb{Q})$-conjugacy trace remain:

$$\mathcal{L}_\mu(h_p^M f_\pi^M) = d(M) \sum_{\gamma \in M(\mathbb{Q})/\sim} \tau(M_\gamma^0)[M_\gamma : M_\gamma^0]^{-1} O_\gamma^{M(\mathbb{A})}(f_\mu h_p^M f_{\pi^p}^M) + \ldots,$$

where the sum is over all semisimple, $\mathbb{R}$-elliptic conjugacy classes in $M(\mathbb{Q})$, and where $f_\mu = (-1)^{q(M)} d(M)^{-1} h(\mu)$. Since $d(M) = 1$, the function $h(\mu)$ on $M(\mathbb{R})$ is a pseudocoefficient for the discrete series representation attached to μ. Since $q(M) = 2$, $f_\infty = h(\mu)$.

To understand the L^2-cohomology, observe that $f_{\pi^p}^M$ eliminates all spectral contributions except those coming from cuspidal cohomology in degree 2 (the underlying space is the product of two modular curves). From [4], Proposition 2.1, we get

$$\mathcal{L}_\mu(h_p^M f_{\pi^p}^M) = \sum_{\pi^M} m(\pi^M) \cdot \chi_\mu(\pi_\infty^M) \cdot tr\, \pi_{fin}^M(h_p^M f_{\pi^p}^M).$$

The sum runs over all irreducible representations π^M of $M(\mathbb{A})$ in the discrete spectrum of $L^2(M(\mathbb{Q}) \setminus M(\mathbb{A}), \xi_\mu)$ (see [4], p. 263; i.e., the space of measurable functions with the property $f(zg) = \xi_\mu(z)f(g)$ for $z \in A_G(\mathbb{R})^0$, where ξ_μ is the character of $A_M(\mathbb{R})^0$ defined by the central character $\lambda(zg) = \xi_\mu(z)^{-1}\lambda(g)$ of the representation λ, whose restriction to $M^1(\mathbb{A})$ also denoted $f(g)$ is square-integrable on $M^1(\mathbb{A})$; recall $M(\mathbb{A}) = M^1(\mathbb{A}) \times A_M(\mathbb{R})^0$). The multiplicity 1 theorem for $Gl(2)$ implies $m(\pi^M) = 1$. The trace of f_π^M eliminates all $\pi_1 \otimes \pi_2$ which are not cuspidal. The cohomological Euler characteristic $\chi_\mu(\pi_\infty^M)$ vanishes unless $\pi_\infty^M = \pi_{r_1} \otimes \pi_{r_2}$ is the unique discrete series representation of $M(\mathbb{R})$, determined by $\mu = \mu_1 \otimes \mu_2$. If it is nonvanishing, the Euler characteristic is $(-1)^{q(M)} = 1$ [4], Lemma 2.2. Hence, the claim follows owing to the presence of the pseudocoefficients which enter in the formula for h_∞^M, since the representation λ determines the Archimedean component of the central characters. □

3.6 The Fundamental Lemma

The function $f^M_{\pi^p}$ is chosen to be any locally constant function with compact support on $M(\mathbb{A}^p_{fin})$, which is attached to the function f_{π^p} on $GSp(4, \mathbb{A}^p_{fin})$ by a matching condition, which is given by an equality between orbital integrals:

$$\Delta(\eta, (t_1, t_2))O^{\kappa}_{\eta}(f_{\pi^p}) = SO^M_{(t_1, t_2)}(f^M_{\pi^p}).$$

The factor Δ is a transfer factor (see Chaps. 6, 7), and (t_1, t_2) is a representative in $Gl(2, \mathbb{A}^p_{fin})^2$ of a semisimple (G, M)-regular element in $M(\mathbb{A}_f)$, and η is a corresponding element in $GSp(4, \mathbb{A}^p_{fin})$. For the precise definition of η, we refer again to Chaps. 6 and 7.

Symmetry at $v \neq p, \infty$. Let us see what happens if we interchange t_1 and t_2? This interchange defines an outer automorphism of M, which is trivial on the center of M.

In case 2 (the situation considered in Chap. 7) the element η and the transfer factor depend only on the product $t_1 t_2$ and the left side is therefore a priori symmetric. In case 1 (considered in Chap. 6) this is not a priori clear. Interchanging t_1 and t_2 corresponds to a change of variables $x, x' \mapsto x, (x')^{\sigma}$, considered in Chap. 6. This changes η within its stable conjugacy class by the "factor" $\chi_{L/F}(-1)$. The transfer factor Δ, which depends only on x, x', thereby changes in the following way: $\Delta(x, (x')^{\sigma}) = \chi_{L/F}(-1)\Delta(x, x')$. This means that the right-hand side of the matching condition, stated above, is invariant under the interchange of the variables t_1 and t_2 in both cases.

In other words, if $f^M_{\pi^p}$ satisfies the matching identity above, then also the new function, which is obtained from the twist by the outer automorphism $(g_1, g_2) \mapsto (g_2, g_1)$ of M, satisfies the matching identity. Hence, we may replace $f^M_{\pi^p}$ by the arithmetic mean, and we can therefore assume that the function $f^M_{\pi^p}$ is symmetric under the considered automorphism of M.

Asymmetry at p, ∞. The functions h^H_p and h^H_∞, however, are not symmetric.

They are antisymmetric. For h^M_∞ this is evident from the formula given in the proof of Lemma 3.2. For h^M_p the antisymmetry follows from Kottwitz's description. If we simultaneously change s to $-s$ and η by precomposing with the outer automorphism, which switches the factors, then h^M_p does not change. So the effect of the outer automorphism is the same as that of the change of s to $-s$. However, changing s by $-1 \in Z(\hat{G})$ changes h^M_p by the factor $\mu_1(-1)^{-1} = -1$ from the remarks in [51] following formula (7.3).

Satake Parameters. The Satake transform of ϕ_n on $GSp(4, E)$ is $q_E^{\langle\delta,\mu\rangle} tr \, r_\mu(\varphi(\sigma_E))$ by Theorem 2.1.3 in [52]. Notice $q_E = p^n$. The spherical function $h^{(n)}_p$ is obtained from ϕ_n by cyclic base change, and hence corresponds to $tr \, \pi_\varphi(h_p) = p^{nd/2} tr \, r_\mu(\varphi(\sigma^n_F))$. See loc. cit. formula (2.2.1), where $d = 3$ is the dimension.

In terms of symplectic Satake parameters $\alpha_0, \alpha_1, \alpha_2$, which describe φ, the Satake transform of $h_p^{(n)}$ is

$$p^{3n/2}\left((\alpha_0)^n + (\alpha_0\alpha_1\alpha_2)^n + (\alpha_0\alpha_1)^n + (\alpha_0\alpha_2)^n\right).$$

From this we can also compute the Satake transform of $(h_p^{(n)})^M$, which is obtained from ϕ_n by a twisted cyclic base change. The twisted endoscopic lemma implies (Lemma 10.2)

$$tr(\pi_{1,p} \otimes \pi_{2,p})((h_p^{(n)})^M) = tr\ \pi_{1,p}(f^{(n)}) - tr\ \pi_{2,p}(f^{(n)}),$$

which up to a constant $f^{(n)}$ is the analogue for $Gl(2, \mathbb{Q}_p)$ of the spherical Hecke operator $h_p^{(n)}$ for $GSp(4, \mathbb{Q}_p)$. Again, this shows the asymmetry of this function. For spherical representations $\pi_{1,p}$ and $\pi_{2,p}$ of $Gl(2, \mathbb{Q}_p)$ with Satake parameters α, β and α', β', where $\alpha\beta = \alpha'\beta'$, the trace of $(h_p^{(n)})^M$ on $\pi_{1,p} \otimes \pi_{2,p}$ is

$$p^{3n/2}\left((\alpha)^n + (\beta)^n - (\alpha')^n - (\beta')^n\right).$$

This uniquely describes the spherical endoscopic lift π_p of the unramified representation $\pi_{1,p} \times \pi_{2,p}$. The symplectic Satake parameters of π_p are $\alpha_0 = \alpha$, $\alpha_0\alpha_1\alpha_2 = \beta$, $\alpha_0\alpha_1 = \alpha'$, and $\alpha_0\alpha_2 = \beta'$, which are well defined up to the action of the Weyl groups.

Conversely, two unramified representations $\pi_{1,p} \otimes \pi_{2,p}$ and $\pi'_{1,p} \otimes \pi'_{2,p}$ of M_p have the same symplectic unramified lift π_p if and only if they are conjugate under the outer automorphism, which switches the two factors of M.

Let us put things together. As both contributions from place p and place ∞ are antisymmetric and all the other contributions are symmetric, we can assume, by switching the two factors of M, that

$$\pi_{1,\infty} = \pi_{r_1}, \qquad \pi_{2,\infty} = \pi_{r_2}$$

normalized such that

$$r_1 > r_2.$$

Recall the weights $r_1 = k_1 + k_2 - 2$ and $r_2 = k_1 - k_2 + 2$ satisfy $r_1 - r_2 = 2(k_2 - 2) \geq 2$, and hence are different. From this normalization we now obtain every summand twice, which gives an additional factor 2. Furthermore, the trace of pseudocoefficients gives $tr(\pi_{r_1} \otimes \pi_{r_2})(h_\infty^M) = \pm 1$. The sign will be obtained later.[3]

Corollary 3.1 now implies

Corollary 3.4. *Under the assumptions of Lemma 3.1 and Corollary 3.3*

$$4 \cdot tr_s\left(Frob_p^n; H_{!,et}^\bullet(S_K(\overline{\mathbb{F}}_p), E_\lambda)(\pi^p)\right) - tr_s\left(h_p^{(n)}; H_!^\bullet(S_K(\mathbb{C}), E_\lambda)(\pi^p)\right)$$

---

[3] The reader might wonder why the sign $\pm$ is left undetermined. The reason is that at the end it most naturally comes out from the global multiplicity formula via Theorem 4.3 and Sect. 5.3.

is equal to

$$\pm 2 \cdot \sum_{\pi^M = \pi_1 \otimes \pi_2} tr \bigotimes_{v \neq p, \infty} \pi_v^M(f_{\pi^p}^M) \cdot \left(tr\, \pi_{1,p}(f_p^{(n)}) - tr\, \pi_{2,p}(f_p^{(n)}) \right).$$

The summation is over all cuspidal irreducible automorphic representations π_1 and π_2 of $Gl(2, \mathbb{A})$ with unitary central characters ω_i (for the unitary normalization of E_λ) such that:

1. $\omega_1 = \omega_2$.
2. $\pi_{1,\infty} = \pi_{r_1}$ and $\pi_{2,\infty} = \pi_{r_2}$ belong to the discrete series of $Gl(2, \mathbb{R})$ with fixed $r_1 > r_2$.

Remark 3.5. One may ask how many representations π^M contribute to the sum for the fixed choice of a good π-projector f_{π^p}. If there is a contribution at all, we call π a weak M-lift or a weak lift from $\pi_1 \otimes \pi_2$, otherwise we say that π is not an M-lift. To analyze this, check the orbits of Weyl groups. Suppose $\pi^M = \pi_1 \otimes \pi_2$ and $\pi'^M = \pi'_1 \otimes \pi'_2$ are two cuspidal automorphic representations that contribute to the sum. Then locally for all unramified places $v \neq p$ either $\pi_{1,v} \cong \pi'_{1,v}$ and $\pi_{2,v} \cong \pi'_{2,v}$ holds or in reversed order $\pi_{1,v} \cong \pi'_{2,v}$, $\pi_{2,v} \cong \pi'_{1,v}$. This implies the same for the global representations π_1, π_2 and π'_1, π'_2. By the strong multiplicity 1 theorem it is enough to show this outside a finite set of places. Suppose π'_1 is not isomorphic to π_1 or π_2. Then the quotient of the partial L-series with a suitable finite set S of places omitted

$$\frac{L^S \left((\pi'_1 \oplus \pi'_2) \otimes (\pi'_1)^*, s \right)}{L^S \left((\pi_1 \oplus \pi_2) \otimes (\pi'_1)^*, s \right)}$$

is by assumption (1). At $s = 1$ the numerator has a pole, but the denominator does not have one. This gives a contradiction. Hence, $\pi'_1 \cong \pi_1$ or $\pi'_1 \cong \pi_2$. But then $\pi'_1 \cong \pi_1$ by looking at the Archimedean place. Therefore, also $\pi'_2 \cong \pi_2$.

Let us return to the formula in Corollary 3.4. It holds for all $n \gg 0$. Assume π_f is automorphic but not CAP. Then we know from Chap. 1 that the π^p-isotypic component is concentrated in the middle cohomology of degree 3. By our "unitary" normalization, E_λ as a Weil sheaf is pure of weight 0; therefore, the module $H^3_{!,et}(S_K(\overline{\mathbb{F}}_p, E_\lambda)$ is pure of weight 3 by the Weil conjectures. Hence, by the Ramanujan conjecture for holomorphic discrete series representations in the case of $Gl(2)$, Corollary 3.4 implies that

$$p^{-3n/2} \cdot tr \left(h_p^{(n)}, H^\bullet_!(S_K(\mathbb{C}), E_\lambda)(\pi^p) \right)$$

is a finite linear combination of powers of complex numbers of absolute value 1. Therefore, every representation π_p of $GSp(4, \mathbb{Q}_p)$ in this module, which necessarily is spherical, has Satake parameters α_0, $\alpha_0 \alpha_1$, $\alpha_0 \alpha_2$, and $\alpha_0 \alpha_1 \alpha_2$ of absolute value 1. We thus obtain

Theorem 3.3. *Let $\pi = \otimes' \pi_v$ be an irreducible cuspidal automorphic representation of $GSp(4, \mathbb{A})$ with unitary central character and which is not CAP. Assume*

π_∞ belongs to the discrete series. Then at all places where π_v is spherical the generalized Ramanujan conjecture for π holds. In other words, the corresponding local
Satake parameters $\alpha_{0,v}$, $\alpha_{1,v}$, and $\alpha_{2,v}$ of the spherical representation π_v are complex numbers of absolute value 1.

3.6.1 Odds and Ends

In this section let π now be an irreducible cuspidal automorphic representation
which is neither CAP nor a weak M-lift in the sense above. Then the right side
of the formula in Corollary 3.1 or Corollary 3.4 vanishes. Write $\pi = \pi_\infty \otimes \pi_p \otimes \pi^p$,
where p denotes some chosen spherical place. Assume there is another irreducible
cuspidal automorphic representation $\pi' = \pi'_\infty \otimes \pi'_p \otimes \pi'^p$ of $GSp(4, \mathbb{A})$ such that
$\pi^p \cong \pi'^p$ are isomorphic. If π'_p is also spherical, then it also contributes to the
formula stated in Corollary 3.1, which now reads

$$(*) \qquad 4 \cdot tr_s\big(Frob_p^n; H^3_{!,et}(S_K(\overline{\mathbb{F}}_p), E_\lambda)(\pi^p)\big) = \sum_{\pi'} m(\pi') tr\, \pi'_p(h_p^{(n)}).$$

The sum runs over all irreducible automorphic representations π' counted with their
multiplicities $m(\pi')$ such that $\pi'_v \cong \pi_v$ for $v \neq p, \infty$ for unramified π'_p and π'_∞ belonging to the L-packet of discrete series representations determined by E_λ. Notice
$m(\pi')$ is the multiplicity of π'_p in $H^3_!(S_K(\mathbb{C}), E_\lambda)(\pi^p)$.

   If we now vary p, but keep π fixed, we see that for almost all p the previously
given representation π is uniquely determined by π^p. This follows from the finiteness of the number of isomorphism classes of automorphic representations with
fixed level and fixed representations at infinity. Hence, for almost all primes p the
right side of formula (*) consists of the single summand $m(\pi) \cdot tr\, \pi_p(h_p^{(n)})$.

   The semisimplification ρ_l of the $\overline{\mathbb{Q}}_l$-adic Galois representation of $\Gamma_{\mathbb{Q}}$ on the isotypic space $H^3_!(S_K \times_{\mathbb{Q}} \overline{\mathbb{Q}}, E_\lambda)(\pi^p)$ is determined by the Frobenius traces at almost
all places p, by the generalized Tchebotarev density theorem. So let us vary the
prime p for the moment. The finiteness theorems imply that for almost all p the sum
in (*) involves only one representation π', since π'_p is unramified. See also Sect. 4.1.
So we get for all p large enough

$$4 \cdot tr\, \rho_l(Frob_p^n) = m(\pi) \cdot tr\, \pi_p(h_p^{(n)}).$$

This holds for all large n, and therefore for all integers n. The formula implies
that $p^{-3/2} Frob_p$ has the eigenvalues $\alpha_{p,0}$, $\alpha_{p,0}\alpha_{p,1}$, $\alpha_0\alpha_{p,2}$, and $\alpha_{p,0}\alpha_{p,1}\alpha_{p,2}$ on
$H^3_!(S_K(\overline{\mathbb{F}}_p), E_\lambda)(\pi^p)$ each with multiplicity $m(\pi)$ for almost all primes p. Hence,
for almost all primes p

$$(**) \qquad det(1 - t \cdot \rho_l(p^{-3/2} Frob_p))^4 = L_p(\pi_p, t)^{-m(\pi)},$$

$$L_p(\pi_p, t)^{-1} = (1 - \alpha_{p,0}t)(1 - \alpha_{p,0}\alpha_{p,1}t)(1 - \alpha_{p,0}\alpha_{p,2}t)(1 - \alpha_{p,0}\alpha_{p,1}\alpha_{p,2}t).$$

In particular, the semisimplification ρ_l of the Galois representation only depends on the weak (or near) equivalence class of π. In [115] it is shown that the semisimplification is an isotypic multiple of a four-dimensional representation. This, of course, raises the question whether this Galois representation is irreducible and of dimension 4. The latter means that the multiplicity $m(\pi)$ is 1. This is expected and proofs have been announced by Arthur and Weselmann.

Chapter 4
Character Identities and Galois Representations Related to the Group $GSp(4)$

In this chapter we study the l-adic representations attached to cohomological Siegel modular forms. To understand their associated l-adic representations defined by the Galois action on the etale cohomology, one needs to understand the endoscopic lifts and vice versa, since for all automorphic representations in this lift the associated l-adic representations turn out to be smaller than what would be expected a priori.

The simple version of the Lefschetz trace formula used in Chap. 3 suffers from the particular restriction that for one specific prime – the chosen Frobenius prime p – the corresponding local Hecke operators at p always have to be deleted from consideration. As a consequence, this particular chosen Frobenius prime p together with the Archimedean place play a distinguished role. This situation in fact looks similar to the situation in Arthur's simple trace formula. Of course this is not a mere coincidence, but is definitely forced by the failure of the strong multiplicity 1 theorem for automorphic forms on $GSp(4)$. To bypass the technical difficulties that result from this, the l-adic representations of the absolute Galois group on the cohomology turn out to be very helpful. With their help we analyze the endoscopic lift. This allows us to understand the failure of the strong multiplicity 1 theorem caused by it.

In the second part of this chapter we describe the local endoscopic character lift in some more detail in terms of local representation theory. The local character identities in the trace formula define local L-packets. In some easy special cases they can be understood by local considerations alone. In the more complicated cases the local character identities of the endoscopic lift are obtained from global character identities through the global trace formula. Since this construction is unfortunately rather indirect, we explain at the end of this chapter how the underlying irreducible representations can be obtained alternatively from theta lifts as considered in [41].

R. Weissauer, *Endoscopy for GSp(4) and the Cohomology of Siegel Modular Threefolds*, 75
Lecture Notes in Mathematics 1968, DOI: 10.1007/978-3-540-89306-6_4,
© Springer-Verlag Berlin Heidelberg 2009

4.1 Technical Preliminaries

In the following we consider automorphic cuspidal irreducible representations π of the group $GSp(4, \mathbb{A})$, and we assume that π is not a cuspidal representation associated with a parabolic subgroup (CAP representation).

In this chapter we are interested in those representations which contribute to the cohomology of Siegel modular threefolds; hence, we assume for $\pi = \pi_\infty \pi_{fin}$ that π_∞ is contained in the L-packet of the Archimedean discrete series representations, say, of weight (k_1, k_2). Then, in particular, $k_1 \geq k_2 \geq 3$. Consider the corresponding local coefficient system $E = E_\lambda$ on the Shimura variety $S_K(G)$, where $\lambda_1 = k_1 - 3 \geq \lambda_2 = k_2 - 3 \geq 0$. For a suitable normalization depending only on λ or the weight (k_1, k_2), the representation π is unitary (as already explained in Chap. 1). Changing this normalization only amounts to twisting π by a suitable power of the idele norm character. By assumption, the representation π contributes to the cohomology of $H_!^3(S_K(\mathbb{C}), E_\lambda)$ for any principal congruence level subgroup $K \subseteq G(\mathbb{A}_{fin})$ such that $\pi^K \neq 0$.

Now let us fix λ and K for the moment.

We are interested in the classes of the irreducible cuspidal automorphic representations π' weakly equivalent to π, i.e., locally isomorphic to π at almost all places. Since π is not CAP, such π' are again cuspidal and not CAP.

The Set $\mathcal{R}$.   Among the classes weakly equivalent to π consider the subset $\mathcal{R}$ defined by those π' which nontrivially contribute to some cohomology group $H^\nu(S_K(G), E_\lambda)$ for our fixed K and λ. Then $(\pi')^K \neq 0$. Although only the group $G(\mathbb{A}_{fin})$ acts on the cohomology, all these representations are automorphic. Well-known results of continuous cohomology imply that π'_∞ belongs to the Archimedean local L-packet of the fixed representation π_∞ (for L-packets as defined in [91]). In particular, π'_∞ belongs to the discrete series. By the finite dimensionality of the cohomology groups, the set $\mathcal{R}$ is a finite set depending on π and K.

Frobenius Primes.   In the situation above we would now like to apply the trace formula to obtain information on the representations in $\mathcal{R}$. For instance, we would like to obtain information on all those representations $\pi' \in \mathcal{R}$ which are isomorphic to our fixed π outside some given finite set of primes S. Of course we would like to prescribe the set S. At the beginning it seems reasonable to exclude the set of primes dividing the level (notice at the moment the Archimedean place will be excluded a priori). Let S_0 temporarily denote this set of primes. But again the naive attempt to apply the trace formulas to control all representations π' in $\mathcal{R}$ isomorphic to π outside S_0 must fail. The reason is the trace formula in the form we consider it involves the Frobenius substitution at some unramified place p. By the choice of a prime $p \notin S_0$ we lose control over the representations π' at this specific place. This disturbing fact fortunately is not really a serious obstruction. Since $\mathcal{R}$ is finite there exists a prime p_0 such that $\pi'_p \cong \pi_p$ holds for all π' in R and all $p \geq p_0$. We can therefore enlarge the set S_0 of exceptional places in such a way that for every prime

p outside S the unramified local representation π'_p for $\pi' \in \mathcal{R}$ satisfies $\pi'_p \cong \pi_p$. Such a finite set S can be chosen, and will be called saturated.

Saturation of S. The use of saturated sets of exceptional places will be abandoned soon by Lemma 4.3. For the moment we choose S to be saturated and as small as possible. So

$$\pi'^S \cong \pi^S$$

for all π' in $\mathcal{R}$. In other words all $\pi' = \pi'_S \pi'^S$ in $\mathcal{R}$ have the form

$$\pi' \cong \pi'_\infty \otimes \pi'_S \otimes \pi^S,$$

where π'_∞ is in the discrete series L-packet determined by (k_1, k_2).

This being said, we now define a preliminary notion of the endoscopic lift. The following definition of a representation π attached to a pair $\sigma = (\sigma_1, \sigma_2)$ with respect to S in the sense below is a technical definition. It will be abandoned in two steps. It will soon be replaced by the notion of weak lift on page 87. However, in this chapter, where we study the cohomology of Siegel modular threefolds, we will always place a restriction on the Archimedean type $\sigma_{1,\infty}$ and $\sigma_{2,\infty}$. These restrictions will then also be abandoned in Chap. 5 by Definition 5.1.

4.1.1 Representations Attached to σ and S and p

Suppose we have a pair

$$\sigma = (\sigma_1, \sigma_2)$$

of irreducible automorphic cuspidal representations σ_1 and σ_2 of the group $Gl(2, \mathbb{A})$ with the same central characters $\omega_{\sigma_1} = \omega_{\sigma_2}$. Suppose $\sigma_{\infty,1}$ and $\sigma_{\infty,2}$ belong to the discrete series representations, which define holomorphic elliptic cusp forms of weight

$$r_1 = k_1 + k_2 - 2 \text{ and } r_2 = k_1 - k_2 + 2$$

(note $r > r_2 \geq 2$), respectively. Notice these are the conditions that also occur in Sect. 3.6.

The pair $\sigma = (\sigma_1, \sigma_2)$ defines an irreducible automorphic representation of the group $M(\mathbb{A})$ for

$$M = Gl(2)^2/Z_1$$

and the antidiagonal central subgroup $Z_1 \cong \mathbb{G}_m$.

Definition 4.1. Let S be a finite set of non-Archimedean places which contain the ramified primes of π. Then we say that the representation π, or more generally a representation π' with class in the set $\mathcal{R}$ defined by π, is *attached* to σ and S, if:

1. σ is unramified whenever π_v is unramified at v (v non-Archimedean).
2. The spinor L-series of π satisfies

$$L_v(\pi_v, s) = L_v(\sigma_{v,1}, s) L_v(\sigma_{v,2}, s)$$

for all non-Archimedean places $v \notin S$.

In the applications S will be saturated.

Notation. The Satake parameters of $\sigma_{v,1}$ and $\sigma_{v,2}$ at the unramified places $v \notin S \cup \{p\}$ are denoted α_v, β_v and α'_v, β'_v, respectively. Then for $v \notin S$

$$L_v(\pi_v, s) = (1 - \alpha_v p_v^{-s})^{-1}(1 - \beta_v p_v^{-s})^{-1}(1 - \alpha'_v p_v^{-s})^{-1}(1 - \beta'_v p_v^{-s})^{-1}.$$

Main Assumption. Assume π is attached to σ for some S. This assumption will be maintained in this chapter except at the end of Sect. 4.4. Notice that under this assumption π is a *weak lift* of σ in the sense of the more flexible Definition 5.1. See also page 87.

Remark 4.1. The auxiliary finite set of places S in this definition is introduced for technical reasons only and will be abandoned later. The Frobenius prime p will vary outside S. Hence, for representations π' in $\mathcal{R}$ which are attached to σ and $S \cup \{p\}$, the local representations π'_p might a priori depend on the prime p considered. However, if the set S is chosen to be saturated and π'_p is unramified, this will not be the case, as explained above.

Remark 4.2. If π is attached to σ and S, then by the strong multiplicity 1 theorem for the group $Gl(2, \mathbb{A})$ the representation σ is uniquely determined by the weak equivalence class of π. The converse is false. There exist different choices for π' attached to the same σ owing to a failure of the strong multiplicity 1 theorem for $GSp(4)$ as we will see. All such π' are contained in the same weak equivalence class, and if S is saturated the unramified representation outside S

$$\pi^S = \pi^S(\sigma) = \pi^S(\sigma^S)$$

is uniquely attached to σ. We will therefore say π^S is attached to σ^S.

We will also see soon that at places where σ_v is spherical the local representation π'_v for $\pi' \in \mathcal{R}$ is uniquely determined by σ_v (see Lemma 4.3) and does not depend on the π'_v chosen. This in fact implies that it was not really necessary to saturate the set of places S_0. This will not be true for the set of ramified places of π. Here the situation is more complicated. More precise information at these places will be obtained in Sect. 4.10.

4.2 Galois Representations

Consider a model S_K over $\overline{\mathbb{Q}}$ of the Shimura variety $S_K(\mathbb{C})$ as in Corollary 3.4. The absolute Galois group $\Gamma_{\mathbb{Q}} = Gal(\overline{\mathbb{Q}} : \mathbb{Q})$ acts on the direct limit

$$\lim_{\substack{K = K_S K^S \\ \to}} H_!^3(S_K, E)(\pi^S),$$

where this direct limit is over the cofinal set of principal congruence subgroups K_S of $GSp(4, \mathbb{A}_S)$ with $K^S = GSp(4, \mathbb{Z}^S)$ fixed. The Galois action commutes with the action of the Hecke operators in $GSp(4, \mathbb{A}_S)$; hence, it respects decomposition into irreducible representations π_S of $G(\mathbb{A}_S)$. Let $m(\pi_S)$ denote the multiplicity of the constituent $\pi_S \otimes \pi^S$ in $\lim_K H_!^3(S_K, E)$, which is a completely reducible $\overline{\mathbb{Q}}_l$-representation of $GSp(4, \mathbb{A}_S)$. Then π_S induces a representation

$$\rho_{\pi_S} : Gal(\overline{\mathbb{Q}} : \mathbb{Q}) \to Gl(m(\pi_S), \overline{\mathbb{Q}}_l).$$

By definition, it is obtained from a λ-adic representation by extension of scalars.

Fix K, E, and π. Fix a corresponding saturated set S for the subset $\mathcal{R}$ of the weak equivalence class of π. Then π^S is an unramified representation. Our main object of interest will be the finite-dimensional $\overline{\mathbb{Q}}_l$-vector space defined by the generalized eigenspace

$$V = H_!^3(S_K, E)(\pi^S), \qquad K = K_S K^S.$$

It is a module under the Hecke algebra $\mathcal{H}(G(\mathbb{A}_S)//K_S)$ of K_S-bi-invariant functions with compact support on $G(\mathbb{A}_S)$.

Remark 4.3. For irreducible π_S the space π_S^K is an irreducible $\mathcal{H}(G(\mathbb{A}_S)//K_S)$ module. If it is nontrivial, this module completely determines the representation π_S if K_S is a good small subgroup in the sense of [10], Corollary 3.9(ii). The groups K_S form a cofinal system. In the following we can assume K_S to be a good small congruence subgroup.

Finite Levels. Fixing such compact open level groups K, one can already determine the semisimplification of the finite-dimensional Galois representation ρ_{π_S} defined above by the direct limit of cohomology groups at some level K for a suitable good small level group K, provided $\pi_S^K \neq 0$ holds. We therefore obtain the decomposition

$$\boxed{V = H_!^3(S_K, E)(\pi^S) = \bigoplus_{\pi'_S} \rho_{\pi'_S} \otimes \pi'^K_S}.$$

The sum is over the finitely many representatives π'_S, for which the class of $\pi'_{fin} = \pi'_S \otimes \pi^S$ is in $\mathcal{R}$, i.e., $\pi' \in \mathcal{R}$ for $\pi' = \pi'_\infty \pi'_{fin}$, and for which π'_∞ is in the discrete series L-packet attached to (k_1, k_2).

4.2.1 The Classical λ-Adic Representations

Suppose π is attached to $\sigma = (\sigma_1, \sigma_2)$. We want to relate the λ-adic representation ρ_{π_S} defined for $\pi_S \pi^S$ to the λ-adic representations ρ_1, ρ_2 defined by Deligne for σ_1

and σ_2. The representations ρ_1, ρ_2 are irreducible $\overline{\mathbb{Q}}_l$-representations of dimension 2 and are obtained by scalar extension from two-dimensional λ-adic representations

$$\rho_{\lambda,1}, \rho_{\lambda,2}$$

attached to σ_1 and σ_2. The representations $\rho_{\lambda,i}$ are absolutely irreducible by a result obtained by Ribet. Since they are irreducible, ρ_1 and ρ_2 are characterized by their character, and hence by

$$tr\rho_{\lambda,1}(Frob_p) = p^{(r_1-1)/2}(\alpha_p + \beta_p), \qquad det\rho_{\lambda,1}(Frob_p) = p^{r_1-1}\alpha_p\beta_p,$$

$$tr\rho_{\lambda,2}(Frob_p) = p^{(r_2-1)/2}(\alpha'_p + \beta'_p), \qquad det\rho_{\lambda,2}(Frob_p) = p^{r_2-1}\alpha'_p\beta'_p.$$

This holds for all unramified primes of σ and σ', and so, in particular, for all $p \notin S$.

Remark: Here $Frob_p$ denotes the geometric Frobenius. In fact, we consider Galois representations dual to the ones usually considered in the literature.

Comparison with Hodge Structures (see Chap. 1). For the algebraic normalization $E_{\lambda_1,\lambda_2}^{alg}$ of the underlying coefficient system, the Hodge structure of the generalized eigenspace of π^S was computed in Chap. 1. It is of the Hodge type:

$$\left(V^{(r_1-1,0)} \oplus V^{(0,r_1-1)} \right) \oplus \left(V^{(r_2-1,0)} \oplus V^{(0,r_2-1)} \right)(2 - k_2).$$

Since $(r_1 - 1) - (r_2 - 1) = 2(k_2 - 2)$, this defines a Hodge structure on the generalized eigenspace which is pure of weight $k_1 + k_2 - 3 = r_1 - 1$.

4.2.2 Weak Reciprocity Law

We show that the representation ρ_{π_S} is related to the representations ρ_1 and $\rho_2 \otimes \nu_l^{k_2-2}$ in a rather easy way. Let ν_l denote the dual cyclotomic character $\nu_l : \Gamma_{\mathbb{Q}} \to \mathbb{Q}_l^*$ defined by

$$\nu_l(Frob_p) = p.$$

The semisimplification $(\rho_{\pi_S})^{ss}$ is determined by the characteristic polynomials of the Frobenius elements $Frob_p$ for $p \notin S$ chosen from a set of Dirichlet density 1. From the Frobenius relation proved by Chai and Faltings [19], p. 264, one knows that the characteristic polynomial $det(id - tFrob_p; \rho_{\pi_S})$ of the Frobenius on the generalized π^S-eigenspace divides a suitably high power of the local L-factor $L_p(\pi_p \otimes |.|_p^{-r-\frac{3}{2}}, t)$. Notice that π_p is unitary by normalization. This explains the twist with the character $\nu^{-r-\frac{3}{2}} = |.|^{\frac{k_1+k_2-3}{2}}$, where $\nu = |.|_p$ is the local character normalized by $\nu(p^{-1}) = p$.

Since σ^S is attached to π^S, the automorphic L-factor of π_p at the prime p is

$$L_p(\pi_p \otimes \nu^{-r-\frac{3}{2}}, t)^{-1} = det\left(id - Frob_p t; \rho_1 \bigoplus (\rho_2 \otimes \nu_l^{k_2-2})\right)$$

for all $p \notin S$. Up to permutations with elements from the Weyl group this implies for the Satake parameters $\alpha_{p,0}, \alpha_{p,1}, \alpha_{p,2}$ of the unramified representation π_p and for the Satake parameters α_p, β_p and α'_p, β'_p of $(\sigma_1)_p$ and $(\sigma_2)_p$ the identities

$$\alpha = \alpha_{p,0},$$

$$\beta = \alpha_{p,0} \cdot \alpha_{p,1} \cdot \alpha_{p,2},$$

$$\alpha' = \alpha_{p,0} \cdot \alpha_{p,1},$$

$$\beta' = \alpha_{p,0} \cdot \alpha_{p,2}.$$

Hence, the eigenvalues of the Frobenius elements $Frob_p$ for $p \notin S$ on the generalized π^S eigenspaces have to be among those four numbers. Since Ramanujan's conjecture holds for holomorphic elliptic modular forms, these numbers have absolute value 1. Therefore, Ramanujan's conjecture holds for π_p and all primes $p \notin S$. Hence, by purity, π only contributes to the cohomology of degree 3. See Chap. 1.

By the trace formula we show in the next sections the existence of integers $m_1(\pi_S), m_2(\pi_S)$ such that

$$\boxed{(\rho_{\pi_S})^{ss} = m_1(\pi_S)\rho_1 \bigoplus m_2(\pi_S)(\nu_l^{k_2-2} \otimes \rho_2)}.$$

To prove this, it suffices to have an equality of characteristic polynomials for the Frobenii of all primes $p \notin S$ in a set of Dirichlet density 1. Let us collect some facts on Complex Multiplication (CM) representations, which will be needed to show this.

4.2.3 CM Representations

By definition an irreducible cuspidal automorphic representation σ of $Gl(2, \mathbb{A})$ is a CM representation if there exists a character $\chi \neq 1$ such that $\sigma \otimes \chi \cong \sigma$. This implies $\chi^2 = 1$. Hence, χ defines a quadratic extension field F of $\mathbb{Q}$. If σ is a representation that comes from holomorphic elliptic modular forms of weight 2 or more, then F is an imaginary quadratic field extension of $\mathbb{Q}$.

Lemma 4.1. *There exists a set primes of positive Dirichlet density for which the Satake parameters $\alpha_p, \alpha'_p, \beta_p, \beta'_p$ are different from each other.*

Proof. For a number field L the absolute Galois group is denoted Γ_L. σ_1 and σ_2 are representations of $Gl(2, \mathbb{A})$ attached to holomorphic cusp forms of weights $r_1 > r_2 \geq 2$. Then $\sigma_1 \not\cong \kappa \otimes \sigma_2$ for all Dirichlet characters κ. Therefore, $\rho_{\lambda,1} \otimes \kappa \not\cong \rho_{\lambda,2} \otimes \nu_l^{k_2-2}$. The same holds for all Galois conjugates of these representations.

Hence, by a Goursat-type argument for the representation $\tau = \rho_{\lambda,1} \oplus (\rho_{\lambda,2} \otimes \nu_l^{k_2-2})$ (see, e.g., [110], p. 121 and Behauptung 5.3) either both representations σ_1, σ_2 are of CM type or the image of $\Gamma_{\mathbb{Q}}$ under the representation τ has a Lie group of dimension greater than the dimensions of the Lie groups of the images of $\Gamma_{\mathbb{Q}}$ for the two corresponding single representations $\rho_{\lambda,i}$. In this latter case we can apply Theorem 10 of [86]. If both σ_i are not of CM type, then the set of primes p for which $\sigma_{p,1} \cong \sigma_{p,2}$ holds has Dirichlet density 0. In fact, if none of the representations are CM, then conditions (a)–(c) of [86] are satisfied. Theorem 10 of [86] shows that the natural density is zero; hence the analytic Dirichlet density is zero (see [84], p. 76, for this notion).

In the case where both representations are CM with corresponding CM fields F_1 and F_2, both λ-adic representations are induced:

$$\rho_{\lambda,i} = Ind_{\Gamma_{F_i}}^{\Gamma_{\mathbb{Q}}}(\psi_i).$$

The ψ_i are grössen characters of $\mathbb{A}_{F_i}^* / F_i^*$ whose infinite type (reflex type) is determined by the weights r_i. The set of places Ω where both the quadratic extensions $F_1/\mathbb{Q}$ and $F_2/\mathbb{Q}$ split is a set of $\mathbb{Q}$-primes with Dirichlet density $\frac{1}{4}$ or greater. Now restrict the λ-adic characters ψ_i to the Galois group Γ_F of the composite field $F = F_1 \cdot F_2$. Since the infinite types of ψ_1 and ψ_2 are determined by $r_1 \neq r_2$, the quotient of the restricted λ-adic characters ψ_i is a one-dimensional λ-adic representation of Γ_F with infinite image. This implies $\sigma_{1,p} \not\cong \sigma_{2,p}$ for a subset of Ω of density $\frac{1}{4}$ or greater.

Finally, $\alpha_p = \beta_p$ for $p \notin S$ is equivalent to

$$\overline{\rho}_{\lambda,1}(Frob_p) \neq 1$$

for the associated "projective" representation $\overline{\rho}_{\lambda,1} : \Gamma_{\mathbb{Q}} \to PGl(2, E_{\lambda})$ attached to $\rho_{\lambda,i}$. Since $\overline{\rho}_{\lambda,1}$ does not have a finite image ([74], Theorem 4.3), the $\mathbb{Q}_l$-dimension of the Lie algebra defined by its image is positive. Hence, the set of primes $p \notin S$ where $\alpha_p = \beta_p$ holds has density 0. Similarly for α_p' and β_p'. $\square$

4.3 The Trace Formula Applied

We still suppose π is attached to σ and S for some cuspidal representation $\sigma = (\sigma_1, \sigma_2)$ and some saturated set S. Consider the generalized π^S-eigenspace V of the cohomology as in Corollary 3.1. The Frobenius $Frob_p$ acts on this space with eigenvalues (normalized by the factor $p^{r+\frac{3}{2}}$), say, γ_i, of absolute value 1 for all primes $p \notin S$. Let m_i denote the multiplicity of the eigenvalue γ_i. We may assume that the eigenvalues γ_i for the subsets $i \in I = \{1, 3\}$ and $i \in I^c = \{2, 4\}$ correspond to the Satake parameters of the representations $\sigma_{i,p}$:

$$(\gamma_1, \gamma_2, \gamma_3, \gamma_4) = (\alpha_p, \alpha_p', \beta_p, \beta_p').$$

Since S was chosen to be saturated, π^S is unramified outside S and only depends on σ^S and not on the particular representation in $\mathcal{R}$. This allows us to apply the trace formula in the following form.

The Trace Formula. Recall that $\pi = \pi_S \pi^S$ only contributes to cohomology in degree 3, since it satisfies the Ramanujan conjecture. From the trace formula in Corollary 3.4 we therefore obtain for $n \gg 0$ the *trace identity*

$$4 \cdot \sum_{\pi_S} tr \pi_S(f_S) \cdot \left(m_1(\pi_S)\alpha_p^n + m_2(\pi_S)(\alpha_p')^n + m_3(\pi_S)\beta_p^n + m_4(\pi_S)(\beta_p')^n \right)$$

$$- \sum_{\pi_S} tr(\pi_S)(f_S) \cdot \left(m_1(\pi_S) + m_2(\pi_S) + m_3(\pi_S) + m_4(\pi_S) \right) \cdot \left(\alpha_p^n + (\alpha_p')^n + \beta_p^n + (\beta_p')^n \right)$$

$$= \pm 2 \cdot tr\sigma_S(f_S^M) \cdot \left(\alpha_p^n + \beta_p^n - (\alpha_p')^n - (\beta_p')^n \right).$$

Here f_S is an arbitrary K_S-bi-invariant function on $GSp(4, \mathbb{A}_S)$ with compact support. The sum is over the finitely many π_S, which extend to an irreducible automorphic representation $\pi_\infty \pi_S \otimes \pi^S$ and represent the set of classes of representations $\mathcal{R}$ attached to σ. Notice π^S only depends on σ^S.

The Character Lift. The transfer $f_S \to f_S^M$ defines a pullback of distributions $\alpha \to \alpha^G$ for stable distributions α. See Rogawski [76], p. 183ff. Working with the group $Gl(2)^2$, we can drop the stability assumption. Thus, we obtain a distribution lift for invariant distributions. In particular for the distribution $\alpha = \chi_{\sigma_S}$ defined by the character of representation σ_S of $Gl(2, \mathbb{A}_S)^2$, this defines the distributions $\chi_{\sigma_S}^G$ as follows:

$$\chi_{\sigma_S}^G(f_S) = tr \, \sigma_S(f_S^M).$$

See also Sect. 4.4.

For a suitable choice of $p \notin S$ and a suitable choice of n, the factor $\alpha_p^n + \beta_p^n - \alpha'^n_p - \beta'^n_p$ is different from zero (Lemma 4.1). So, by the corresponding trace identity above, we can express $\chi_{\sigma_S}^G(f_S)$ as a linear combination of the characters $\chi_{\pi_S}(f_S)$ which appear on the left side of the trace identity. This gives an expression of the form

$$\chi_{\sigma_S}^G(f_S) = \sum_{\pi_S} n(\pi_S) \cdot \chi_{\pi_S}(f_S)$$

for all K_S-bi-invariant functions with compact support. Here $n(\pi_S)$ are certain complex numbers, and the sum is over all the finitely many π_S that occur in $\mathcal{R}$.

If we substitute this into the right side of the trace identity above, we obtain new character identities. They hold for all primes $p \notin S$ and all integers $n \gg 0$. Since the representations π_S in $\mathcal{R}$ satisfy $\pi_S^{K_S} \neq 0$, and since we haven chosen K_S to be good small subgroups, linear independency of characters implies that we obtain identities of the form

$$(*) \qquad \sum_{i=1}^{4} 4m_i \gamma_i^n - \left(\sum_{i=1}^{4} m_i\right)\left(\sum_i \gamma_i^n\right) = c \sum_{i \in I} \gamma_i^n - c \sum_{i \notin I} \gamma_i^n$$

for all $n \gg 0$, where

$$m_i = m_i(\pi_S) \quad \text{and} \quad c = \pm 2 \cdot n(\pi_S)$$

for fixed π_S and the normalized eigenvalues γ_i of the Frobenius $Frob_p$ on (V, ρ_{π_S}) for $p \notin S$. Notice, the $m_i \geq 0$ are integers.

Remark 4.4. Any identity like $(*)$ that holds for all $n \gg 0$ and $\gamma_i \neq 0$ is an identity that holds for all n. (To show this, multiply by t^n, take the sum from $n = 0$ to ∞, and consider its poles at $t \to \gamma_i$.)

Now we vary $p \notin S$ and discuss several cases.

First Assume $\sigma_{1,p} \not\cong \sigma_{2,p}$. Since the central characters of σ_1, σ_2 coincide, this is equivalent to $\{\alpha_p, \beta_p\} \cap \{\alpha'_p, \beta'_p\} = \emptyset$. Then, if $\alpha_p \neq \beta_p$, by the linear independence of characters

$$4m_1 = m_1 + m_2 + m_2 + m_4 + c,$$

$$4m_3 = m_1 + m_2 + m_2 + m_4 + c,$$

and similarly

$$4m_2 = m_1 + m_2 + m_2 + m_4 - c,$$

$$4m_4 = m_1 + m_2 + m_2 + m_4 - c.$$

Hence, $m_1 = m_3$ and $m_2 = m_4$ and

$$m_1 = \frac{dim(V) + c}{4} \in \mathbb{N},$$

$$m_2 = \frac{dim(V) - c}{4} \in \mathbb{N}.$$

Since

$$m_1 - m_2 = \frac{c}{2}, \qquad m := m_1 + m_2 = \frac{dim(V)}{2},$$

c and $dim(V)$ are even integers.

Remark 4.5. By Lemma 4.1 there exist $p \notin S$ for which $\sigma_{1,p} \not\cong \sigma_{2,p}$. From this we deduce $\frac{1}{4}(dim(V) \pm c) \in \mathbb{N}$, and that $dim(V) = 2m$ is of even dimension, by considering just one of these places. The numbers c and $dim(V)$ and also m_i are independent of the place $p \notin S$ considered; hence, in particular, the coefficients

$$\boxed{\pm n(\pi_S) = m_1(\pi_S) - m_2(\pi_S) \in \mathbb{Z}}$$

are integers.

Definition 4.2. Using the integers $m_1 = m_1(\pi_S)$ and $m_2 = m_2(\pi_S)$ just defined, we set

$$\tau = m_1 \rho_1 \oplus m_2(\nu_l^{k_2-2} \otimes \rho_2).$$

We claim that the normalized Frobenius has the same characteristic polynomial for the two representations τ and (V, ρ_{π_S}) for all $p \notin S$. For $\sigma_{1,p} \not\cong \sigma_{2,p}$ and $\alpha_p \neq \beta_p$ this has been shown already.

Now assume $\sigma_{1,p} \not\cong \sigma_{2,p}$, but $\alpha_p = \beta_p$. Then, by the arguments above, we only get

$$4(m_1 + m_3) = 2(m_1 + m_2 + m_3 + m_4) + 2c.$$

But still $m_1 + m_3 = \frac{1}{2} \cdot (dim(V) + c)$. Hence, τ and V again have the same characteristic polynomial for $Frob_p$.

Next Assume $\sigma_{1,p} \cong \sigma_{2,p}$. Then by reordering, we can achieve $\alpha_p = \alpha_p'$ and $\beta_p = \beta_p'$. Then the c terms cancel in the comparison, and for $\alpha_p \neq \beta_p$ we get

$$4(m_1 + m_2) = 2(m_1 + m_2 + m_3 + m_4) = 4m,$$

$$4(m_3 + m_4) = 2(m_1 + m_2 + m_3 + m_4) = 4m.$$

Hence, $m_1 + m_2 = m = m_3 + m_4$, and the characteristic polynomial of the normalized Frobenius on V yields

$$(1 - \alpha_p t)^m (1 - \beta_p t)^m = (1 - \alpha_p t)^{m_1}(1 - \beta_p t)^{m_3}(1 - \alpha_p' t)^{m_2}(1 - \beta_p' t)^{m_4},$$

which again is the characteristic polynomial of τ.

Needless to say, the same also holds in the final case where all eigenvalues coincide, since $m_1 + m_2 + m_3 + m_4 = dim(V) = dim(\tau)$. By the Cebotarev density theorem this implies

Lemma 4.2. *If π is attached to σ, then with our previous notation*

$$(\rho_{\pi_S})^{ss} = m_1(\pi_S)\rho_1 \bigoplus m_2(\pi_S)(\nu_l^{k_2-2} \otimes \rho_2).$$

Local Character Lifts. The local transfers $f_v \to f_v^{M_v}$ define local distributions $\chi_{\sigma_v}^G$ for each place $v \in S$, which are uniquely defined by σ_v once the transfer factor Δ_v has been specified. Let us assume this. Then the semilocal formulas $\prod_{v \in S} \chi_{\sigma_v}^G(f_v) = \chi_{\sigma_S}^G(f_S) = \sum_{\pi_S} n(\pi_S)\chi_{\pi_S}(f_S)$ obtained above imply local formulas for each $v \in S$:

$$\boxed{\chi_{\sigma_v}^G(f_v) = \sum_{\pi_v} n(\sigma_v, \pi_v) \cdot \chi_{\pi_v}(f_v)}$$

The sum is over the finitely many local irreducible admissible representations π_v at v of the representations π in $\mathcal{R}$. Here f_v is an arbitrary K_v-bi-invariant function with compact support on G_v.

The numbers $n(\sigma_v, \pi_v)$ are uniquely defined and depend only on σ_v and the local transfer factor Δ_v chosen. Furthermore, for any saturated finite set S of places

$$n(\pi_S) = \prod_{v \in S} n(\sigma_v, \pi_v)$$

holds by the linear independence of characters of the Hecke algebra $\mathcal{H}(G(\mathbb{A}_S)//K_S)$.

So far these character relations obtained are valid only for functions f_S which are K_S-bi-invariant. So they involve only those representations π_S which admit nontrivial K_S-invariant vectors. But at this stage we are now free to make K_S smaller. In the limit over a cofinal system we obtain a local character relation

$$\chi^G_{\sigma_v} = \sum_{\pi_v} n(\sigma_v, \pi_v) \cdot \chi_{\pi_v}, \qquad n(\sigma_v, \pi_v) \in \mathbb{C},$$

at the cost that this is only a countable sum over (unitary) characters. This is an identity of distributions on $G(\mathbb{A}_S)$. For each locally constant function on $G(\mathbb{A}_S)$ with compact support inserted in the identity, almost all summands are zero. Hence, this countable sum is absolutely convergent for such test functions. It is well known that linear independence of characters of unitary representations holds in this setting (see [42] and [76], Proposition 13.8.1). Thus we extract the local character relations from these semilocal ones by the linear independence of characters. Formally we may view the sum in the local character identity to be a sum over all unitary admissible irreducible representations of G_v. Since we are free to enlarge S provided $K_v = GSp(4, \mathbb{Z}_v)$ holds at the added places, we obtain

Corollary 4.1. *Let π be an irreducible cuspidal representation attached to some irreducible cuspidal automorphic representation σ. Suppose π_{fin} contributes nontrivially to some cohomology group $H^\nu(S_{K_S}, E_\lambda)$. Then there exists a sign $\varepsilon \in \{\pm 1\}$, which only depends on λ such that for any finite saturated set S_0 of non-Archimedean primes, and for an arbitrary set S_1 of unramified places $v \notin S$ of π and σ where $K_v = GSp(4, \mathbb{Z}_v)$, the following relations hold for $S = S_0 \sqcup S_1$:*

$$n(\pi_S) = \prod_{v \in S} n(\sigma_v, \pi_v),$$

$$m_1(\pi_S) - m_2(\pi_S) = \varepsilon \cdot n(\pi_S),$$

$$\rho^{ss}_{\pi_S} \otimes \nu_l^r \cong m_1(\pi_S)\rho_1 \bigoplus m_2(\pi_S)\rho_2 \otimes \nu_l^{k_2-2}.$$

Now we want to get rid of the technical assumptions which we still have to make regarding the set S (namely, to contain a saturated set S_0 as above). The next corollary shows that without restriction of generality, S can be chosen to be an arbitrary set of non-Archimedean places containing the ramified primes of the given representation π.

Lemma 4.3. *Suppose π and π' are in $\mathcal{R}$, then $\pi'_v \cong \pi_v$ if π_v and π'_v are spherical.*

Proof. For spherical Hecke operators f_v, i.e., $K_v = GSp(4, \mathbb{Z}_v)$-bi-invariant functions with compact support, we have

$$\chi_{\sigma_v}^G(f_v) = \sum n(\sigma_v, \pi_v')(\pi_v')^{K_v}(f_v),$$

considered as an identity on the spherical Hecke algebra. The left-hand side satisfies

$$\chi_{\sigma_v}^G(f_v) = \chi_{\sigma_v}(f_v^M) = tr\ \sigma_v(f_v^M) = tr\ \pi_v^+(f_v).$$

The first two equalities hold by definition. The last identity (Shintani identity) holds for the canonical endoscopic ring homomorphism of spherical Hecke algebras $f_v \rightarrow f_v^M$ (provided the local transfer factor Δ_v was chosen properly, which we tacitly assume) and for a canonical unramified representation

$$\pi_v^+ = \pi_+(\sigma_v),$$

whose Satake parameters are defined by the endoscopic L-embedding. (Recall that, in particular, this implies σ_v is unramified.) This is an immediate consequence of the fundamental lemma (see Chaps. 6, 7). Inserting this formula into the left-hand side of the character identity, linear independence of characters implies $\pi_v' \cong \pi_v^+$ for all spherical π_v' for which $n(\sigma_v, \pi'_v) \neq 0$ holds. $\square$

Remark 4.6. The last lemma allows us to choose the set S to be arbitrary. In fact, to obtain something nontrivial, we should take S to be the set of places where σ ramifies. It is now convenient to abandon the notion that "π is a attached to σ" and S, and replace it by the following.

Definition 4.3. Given a pair $\sigma = (\sigma_1, \sigma_2)$ of unitary irreducible cuspidal automorphic representations of $Gl(2, \mathbb{A})$ whose central characters coincide, we call an irreducible cuspidal automorphic representation π of $GSp(4, \mathbb{A})$ a weak lift of σ if for almost all places

$$L^S(\pi_S, s) = L^S(\sigma_1, s)L^S(\sigma_2, s)$$

holds. (These places can be assumed to be unramified without restriction of generality. Recall that in this chapter we still retain our assumptions regarding the Archimedean places of σ_1 and σ_2.)

Suppose $\pi = \pi_\infty \pi_{fin}$ is an irreducible cuspidal automorphic representation of the group $GSp(4, \mathbb{A})$, which is *cohomological* (with respect to E_λ). By this we mean that some generalized eigenspace of some cohomology group $\lim_K H_!^\nu(S_K, E_\lambda)(\pi^S)$ is nontrivial. Then, up to a twist by some power of the idele norm character, giving the unitary normalization, π is a preunitary representation.

Corollary 4.2. *Let π_{fin} be an irreducible constituent of the admissible preunitary representation*

$$\lim_K H_!^3(S_K, E(r))(\pi^S) = \bigoplus_{\pi_{fin}} \rho_{\pi_{fin}} \otimes \pi_{fin}$$

of $G(\mathbb{A}_{fin})$. All $\pi = \pi_\infty \pi_{fin}$ in this sum are cuspidal irreducible automorphic representations with infinite component π_∞ in the Archimedean L-packet of the holomorphic discrete series representations of weight $(k_1, k_2) = (\lambda_1 + 3, \lambda_2 + 3)$. Suppose π (with unitary normalization in the sense of Chap. 1) is a weak lift of some irreducible automorphic representation $\sigma = (\sigma_1, \sigma_2)$. Then σ_i are cuspidal automorphic representations of $Gl(2, \mathbb{A})$ attached to holomorphic elliptic modular forms of weight $r_1 > r_2 \geq 2$, where $r_1 = k_1 + k_2 - 2$ and $r_2 = k_1 - k_2 + 2$. In particular, π is not CAP. Furthermore, the semisimplification of the finite-dimensional $\overline{\mathbb{Q}_l}$-representation $\rho_{\pi_{fin}}$ (in the algebraic normalization) is isomorphic to

$$\rho_{\pi_{fin}}^{ss} \otimes \nu_l^r \cong m_2(\pi_{fin})(\rho_2 \otimes \nu_l^{k_2-2}) \bigoplus m_1(\pi_{fin})\rho_1.$$

The integers $m_1(\pi_{fin})$ and $m_2(\pi_{fin})$ are related by

$$m_1(\pi_{fin}) - m_2(\pi_{fin}) = \varepsilon \cdot \prod_{v \neq \infty} n(\sigma_v, \pi_v) \quad , \quad \pi_{fin} = \bigotimes_{v \neq \infty} \pi_v,$$

where $n(\sigma_v, \pi_v)$ are complex numbers uniquely determined (for fixed transfer factors) by the local character identities

$$\chi_{\sigma_v}^G = \sum n(\sigma_v, \pi_v)\pi_v.$$

The sign $\varepsilon = \pm 1$ is a fixed sign. (It is an Archimedean contribution. We show $\varepsilon = -1$ in Chap. 8.) The sum is a countable, absolutely converging sum over the unitary spectrum of $GSp(4, \mathbb{Q}_v)$. If σ_v is spherical, then there exists a unique spherical representation π_v such that $n(\sigma_v, \pi_v) \neq 0$. For this representation $n(\sigma_v, \pi_v) = 1$ holds.

Remark 4.7. In Chap. 5 we prove $m_1(\pi_{fin}) + m_2(\pi_{fin}) = 1$. Since Shelstad [90, 91] has shown

$$\chi_{\sigma_\infty}^G = \pi_+(\sigma_\infty) - \pi_-(\sigma_\infty)$$

for the nonholomorphic discrete series representation $\pi_+(\sigma_\infty)$ and the holomorphic discrete series representation $\pi_-(\sigma_\infty)$ of the Archimedean L-packet of weight (k_1, k_2), this gives the global multiplicity formula for weak lifts π:

$$m(\pi) = \frac{1}{2}\Big(1 + \prod_v n(\sigma_v, \pi_v)\Big) \quad \pi \text{ weak lift of } \sigma, \ \sigma \text{ generic}.$$

The product is now over all places including the Archimedean. In fact, in Corollary 4.4 the following multiplicity formulas will be shown:

$$m\big(\pi_+(\sigma_\infty) \otimes \pi_{fin}\big) = m_2(\pi_{fin}),$$

$$m\big(\pi_-(\sigma_\infty) \otimes \pi_{fin}\big) = m_1(\pi_{fin}).$$

Remark 4.8. If σ is ramified at a unique non-Archimedean place v, Lemma 4.3 and Corollary 4.2 force all $n(\sigma_v, \pi_v)$ to be integers. One can easily show that this implies $n(\sigma_v, \pi_v)$ is integral for all unitary generic representations σ_v. By character twists, one easily reduces some local calculations for induced representations σ_v to the case of discrete series, which are handled by embedding techniques. Details are given in Sect. 4.5. For a more uniform approach, see also Sect. 5.1.

4.4 Rationality

For a weak lift π of some representation σ for M, the Galois representation ρ_{π_S} is a direct sum of two-dimensional Galois representations ρ_1 and ρ_2 attached to certain elliptic modular forms. The characteristic polynomials of the Frobenius elements have their coefficients in some fixed number field E_σ. This follows from Corollary 4.1, since the same is true for the Galois representations ρ_1 and ρ_2 attached to σ_1 and σ_2.

On the other hand, the representation π is realized on a cohomology group with complex coefficients. Considering the $\mathbb{Q}$-structure defined by the Betti realizations of these representations, we see that the induced natural action of the group $Aut(\mathbb{C}/\mathbb{Q})$ on the coefficients of the cohomology group permutes the isotypic automorphic subspaces. Since $\pi^K \neq 0$ implies $(\pi^\tau)^K \neq 0$ for $\tau \in Aut(\mathbb{C}/\mathbb{Q})$, this action of $Aut(\mathbb{C}/\mathbb{Q})$ has finitely many orbits. Its fixed field E will be called the *field of rationality* for π_{fin}. The character of π has values in this field E since

$$\tau \in Aut(\mathbb{C}/E) \iff \chi_\pi \text{ is } E\text{-valued.}$$

However, it is not clear whether the representation π can be defined over E owing to the possible effect of Schur multipliers. We will show

$$E \subseteq E_\sigma$$

(an immediate consequence of Lemma 4.10 and Theorem 4.5).

Consider the set Σ_E of embeddings $\tau : E \to \overline{\mathbb{Q}}$. Choose some fixed embeddings of $\overline{\mathbb{Q}}$, into $\mathbb{C}$ and into $\overline{\mathbb{Q}_l}$ and the completion $\mathbb{C}_l$. Then one can consider the action of the absolute Galois group $\Gamma_{\mathbb{Q}}$ on the coefficients of the corresponding cohomology theories. This allows us to recover the λ-adic representation attached to π in terms of the λ-adic representations attached to σ_1, σ_2 using the isomorphism of $\overline{\mathbb{Q}_l}$-representations given in Lemma 4.2. Here we use the well-known fact that there is no problem with Schur multipliers for the two-dimensional Galois representations attached to elliptic modular forms.

Matching Conditions. The (not uniquely defined) assignments $f \mapsto f^M$ are defined by the matching conditions

$$SO_t^M(f^M) = \Delta(t, \eta)O_\eta^\kappa(f).$$

Here $t \in T_M(F) \subseteq M(F)$ and $\eta \in T_G(F) \subseteq G(F)$ must be related by an admissible homomorphism between tori T_M and T_G. The precise description of these matching conditions is quite involved, and therefore we do not want to give it here. See, e.g., [54], (5.5.1). However, let us make some general remarks:

(a) The transfer factors $\Delta(t, \eta)$ and the κ-integrals $O_\eta^\kappa(f)$ are defined for elliptic endoscopic data $(M, \mathcal{M}, s, \xi)$, where in our case except for the maximal datum $(G, {}^L G, 1, id)$ there is up to equivalence only one elliptic endoscopic datum given by $(M, {}^L M, s, \xi)$. See [36]. It is described in Sect. 6.3.

(b) For any maximal F-torus T_M in M this datum, in particular, defines a character κ and, roughly speaking, the κ-orbital integrals as a certain average of orbital integrals.

(c) Whenever the set $\mathcal{D}_G(T) = ker(H^1(F, T) \to H^1(F, G))$ is trivial, the character κ becomes trivial and $O_t^\kappa(f)$ becomes the orbital integral $O_t^G(f) = \int_{G_t \backslash G} f(g^{-1} t g) dg/dg_t$, and also the transfer factor $\Delta(t, \eta)$ becomes easy to describe.

(d) In our case the maximal F-tori T_M of M are determined by a pair (L_1, L_2) of quadratic etale algebras L_i over F. T_M is anisotropic if and only if both L_i are quadratic field extensions of F. Notice T_M is isomorphic to $\mathbb{G}_m^3$, or $\mathbb{G}_m \times Res_{L/F}(\mathbb{G}_m)$ or $[Res_{L_1/F}(\mathbb{G}_m) \times Res_{L_2/F}(\mathbb{G}_m)]/\mathbb{G}_m$ for quadratic extension fields L or L_1, L_2, respectively, of F. A glance at the list of maximal F tori in G (see page 94) implies that only the tori T of types (4)–(7) can be F-isomorphic to one of the maximal tori T_M of M. Furthermore, if this is the case, $H^1(F, T)$ is trivial (Hilbert 90) except for the anisotropic cases. In the anisotropic cases the exact sequence

$$1 \to H^1(F, T_M) \to H^2(F, \mathbb{G}_m) \to H^2(F, Res_{L_1/F}(\mathbb{G}_m) \times Res_{L_2/F}(\mathbb{G}_m))$$

forces the group $H^1(F, T_M) = Br(F)[2]$ to be cyclic of order 2 (for the anisotropic tori T_M). Therefore, since $H^1(F, GSp(4)) = 1$, we obtain for $\mathcal{D}_G(T) = ker(H^1(F, T) \to H^1(F, G))$ the following:

Lemma 4.4. *For maximal F-tori in G coming from M the group $\mathcal{D}_G(T)$ is*

$$\mathcal{D}_G(T) = \mathbb{Z}/2\mathbb{Z}$$

for F-anisotropic tori, and is trivial otherwise.

(e) For the anisotropic tori we give explicit formulas for the transfer factors $\Delta(t, \eta)$ for $t \in T_M$ and anisotropic tori T_M in Chaps. 6–8. The transfer factors for the remaining tori, which are not anisotropic, are very easy to determine, and this will be skipped.

4.4.1 Character Transfer

The matching conditions define a distribution lift $\alpha \mapsto \alpha^G$ (see also Hales [36]) via

$$\int_{G(F)/Z(F)} f(g)\alpha^G(g)dg = \int_{M(F)/Z(F)} f^M(h)\alpha(h)dh.$$

Here α is an invariant distribution on $M(F)$, which transforms with respect to a character ω under the center $Z(F)$, and f is a locally constant function, compact modulo the center, which transforms according to the character ω^{-1} under Z. We will replace M by its z-extension $Gl(2)^2$, also denoted M by abuse of notation, in the following. Similar assumptions will be made regarding f^M. Notice that in general the central characters of f and f^M need not coincide, owing to the transformation properties of the transfer factor Δ. However, in our case they are equal by an inspection of the transfer factor. See Theorems 6.1 and 7.1.

Following [76], p. 182, one can compute α^G.

Stability for M. Let F be a *non-Archimedean* local field of characteristic zero in the following: For a maximal F-torus T in a reductive F-group G one defines

$$\mathcal{D}(T) = \mathcal{D}_G(T) = Ker(H^1(F,T) \to H^1(F,G)).$$

Notice $\mathcal{D}_{GSp(4)}(T) = H^1(F,T)$, since $H^1(F,GSp(4)) = 1$. On the other hand $\mathcal{D}_M(T) = 1$.

Proof: For any maximal F-torus T in M, its preimage in $Gl(2)^2$ is of the form $Res_{L_1/F}(\mathbb{G}_m) \times Res_{L_2/F}(\mathbb{G}_m)$. Since the Galois cohomology of this torus vanishes and since $H^1(F,Gl(2)) = 1$, we obtain a diagram

$$
\begin{array}{ccccc}
1 & \longrightarrow & H^1(F,T) & \lhook\joinrel\longrightarrow & H^2(F,\mathbb{G}_m) \\
& & \downarrow & & \| \\
1 & \longrightarrow & H^1(F,M) & \lhook\joinrel\longrightarrow & H^2(F,\mathbb{G}_m)
\end{array}
$$

which proves the injectivity $H^1(F,T) \hookrightarrow H^1(F,M)$. Hence, for sufficiently regular semisimple elements t we have

$$SO_t^M(f^M) = O_t^M(f^M).$$

We only consider sufficiently regular elements t in the following, since this suffices for our purposes.

Notation.

1. Let $T \subseteq G$ be a maximal F-torus. Let $t \in T(F)$ be a sufficiently regular element. Then for $\delta \in \mathcal{D}_G(T) = Ker(H^1(F,T) \to H^1(F,G))$ one can define

$t^\delta \in T^\delta(F)$ (another maximal F-torus of G uniquely defined up to F-conjugacy in G) by $t^\delta = g^{-1}tg$ if δ is represented by the cocycle $a(\sigma) = g^\sigma g^{-1}$, $g \in G(\overline{F})$ (as in [76], p. 183). Recall $H^1(F, G) = 1$ for $G = GSp(4)$; hence, $\mathcal{D}_G = H^1(F, T)$. The result above implies

$$\mathcal{D}_M(T) = 1.$$

For the convenience of the reader we recall some of the definitions:

2. Fix an admissible embedding $T \to G$ in the sense of [60]. It comes together with a homomorphism ρ_T of the Galois group of the splitting field of T to the abstract Weyl group W_G. In this way the Galois group $\Gamma = Gal(\overline{F} : F)$ acts by conjugation with ρ_T on the Weyl group W_G (since the group G is split). Let $W_F(T, G)$ denote the subgroup of Γ-invariant elements in W_G. It contains the relative Weyl group $W(T, G) = N_G(T)(F)/T(F)$. By the non-Abelian Galois cohomology sequence

$$W(T, G) \to W_F(T, G) \xrightarrow{d} H^1(F, T) \to H^1(F, N_G(T)) \to H^1(F, W_G),$$

attached to the normalizer $N_G(T)$ of T in G, the coboundary map d maps $w \in W_G^\Gamma = W_F(T, G)$ with representative $n \in N_G(\overline{F})$ to the class of $d(w) = n^\sigma n^{-1}$. More generally $w \in W_F(T, G)$ with representative n acts on $\mathcal{D}(T)$, and sends $\delta \in \mathcal{D}(T)$, represented by the cocycle $\delta = a(\sigma) = g^{-1}g^\sigma$ (split by $g \in G(\overline{F})$), to the cohomology class of the cocycle $w(\delta) = n^\sigma g^\sigma g^{-1} n^{-1}$ in $\mathcal{D}(T)$. Notice $w(\delta) \in \mathcal{D}(T)$ and $image(\delta) = image(\delta^w)$ in $H^1(F, N_T)$. Furthermore, $w^\delta \in \mathcal{D}(T)$ is independent of the choice of representative n and the choice of splitting element g. In general this defines an action of $W_F(T, G)$ on the set $\mathcal{D}(T)$. But this action need not respect the group structure defined by $\mathcal{D}(T) \subseteq H^1(F, T)$ in general.

3. $Ker(H^1(F, N_G(T)) \to H^1(F, G))$ can be identified with the set $\mathcal{C}_G$ of $G(F)$-conjugacy classes of maximal F-tori T in G. The image $\mathcal{C}(T)$ of $\mathcal{D}(T)$ under $H^1(F, T) \to H^1(F, N(T))$ is the set of $G(F)$-conjugacy classes of F-tori which are stably equivalent to T. Furthermore, the fibers of the natural map induced by the Galois cohomology sequence

$$\mathcal{D}(T) \longrightarrow \left\{ \begin{matrix} G(F)\text{-conjugacy classes} \\ \text{of maximal } F\text{-tori in } G \end{matrix} \right\}$$

can be identified with the left cosets

$$W(T, G) \backslash W_F(T, G).$$

If we choose representatives $[T'] \in \mathcal{C}$, and representatives T in each stable equivalence class, then for a function f on $\mathcal{C}$

$$\sum_{[T'] \in \mathcal{C}} \frac{f(T')}{W(T', G)} = \sum_{\mathcal{C}(T) \in \mathcal{C}} \frac{1}{W_F(T, G)} \sum_{\delta \in \mathcal{D}(T)} f(T^\delta).$$

Similarly, we have formulas for M.

4. Let

$$\tau = \Delta \cdot D_M/D_G$$

be the product over the transfer factors $\Delta_I \Delta_{II} \Delta_{III}$ in the sense of Langlands and Shelstad [60].

The ∗-Invariance. We define the outer automorphism $t \mapsto t^*$ of M by

$$t = (t_1, t_2) \in Gl(2, F)^2/F^* \quad \mapsto \quad t^* = (t_2, t_1) \in Gl(2, F)^2/F^*.$$

Obviously $t^{**} = t$. For a distribution α on M put $\alpha^*(f(t)) = \alpha(f(t^*))$. Define $\varepsilon(T)$ to be 1 if T^* is $M(F)$-conjugate to T in M, and otherwise $\varepsilon(T) = 2$.

Lemma 4.5. $(\alpha^*)^G = \alpha^G$.

Remark on the Proof. The first guess would be to check whether the right side of

$$SO_t^M(f^M) = \Delta(t, \eta)O_\eta^\kappa(f)$$

is invariant under $t \mapsto t^*$. If this were the case, we could simply replace f^M by its average $\frac{1}{2}(f^M(t^*) + f^M(t))$, since in principal we are free to choose any matching pair of functions (f, f^M). This would immediately imply $\alpha^G = (\alpha^*)^G$.

However, this thought is somewhat too naive. The matching condition of Langlands and Shelstad specifies the relation between t and η: η is linked with t via some admissible embedding. This condition will be destroyed by simply replacing t by t^*. In general, the assignment $t^* \mapsto \eta$ will not be given by an admissible embedding any more. In fact, this phenomenon occurs for one type of torus (namely, in elliptic case I, if in addition $\chi_{L/F}(-1) = -1$ holds; see type (4) in the list of tori below). In this sense the tori of type (4) are rather curious. They will be examined in Chap. 8. For all other cases of tori the relevant admissibility condition is preserved under replacement of t by t^*, and also $\Delta(t, \eta) = \Delta(t^*, \eta)$ holds in these other cases. See Lemmas 6.5 and 7.3. In fact, the proof of Lemma 4.5 is given in Sect. 4.5 as a side result of the proof of Lemma 4.6.

An Assumption. The space of invariant distributions α on M splits into two subspaces, those anti-invariant $\alpha(t^*) = -\alpha(t)$ and those invariant $\alpha(t^*) = \alpha(t)$ under the involution ∗. By the last lemma $\alpha^G = 0$ holds for anti-invariant distributions α.

In the following we therefore always assume α to be ∗-invariant. Hence, suppose α is a ∗-invariant distribution on M. Then the argument for [76], Lemma 12.5.1, implies

Lemma 4.6. *For $\delta \in \mathcal{D}(T)$ and sufficiently regular $t \in T(F)$ and all maximal F-tori of G coming from M*

$$(D_G\alpha^G)(t^\delta) = \kappa(\delta) \cdot \tau(t)D_M(t) \cdot (\alpha + \alpha^*)(t)$$

$$= \varepsilon(T) \cdot \sum_{w \in W_F(T,M)\backslash W_F(T,G)} \tau(wtw^{-1})\kappa(w(\delta))(D_M\alpha)(wtw^{-1}).$$

In particular, $(D_G\alpha^G)(t^\delta) = \kappa(\delta) \cdot (D_G\alpha^G)(t)$ for $\delta \in \mathcal{D}(T)$.

Proof Postponed. See Sect. 4.5. □

Support. For further use let us discuss the support of the distribution $\alpha^G(t)$. For this we list the different $G(F)$-conjugacy classes of maximal F-tori of G. This list contains seven distinct types. Notice that the Galois group $G(L/F)$ of the splitting field L of a maximal torus T is a subgroup of the abstract Weyl group. Hence tori in $Gl(2)$ split over a quadratic extension field L and are isomorphic to $Res_{L/F}(\mathbb{G}_m)$. This easily gives a description of all maximal F-tori of the group $Gl(2)^2$, and hence for M. Notice the support of α^G (considered for sufficiently regular semisimple elements) is contained in the union of the F-tori of G that come from M by an admissible embedding.

4.4.2 The List of Maximal F-Tori of G

Maximal F-Tori of G that do not come from M by an admissible embedding:

1. $T(F)$ is isomorphic to $\{x \in E^* \mid Norm_{E/E^+}(x) \in F^*\}$, where E/F is an extension of F of degree 4, whose splitting field is either cyclic of degree 4 or dihedral of degree 8 over F. E contains a quadratic extension field E^+ of F.
2. $T(F) \cong \{x = (x_1, x_2) \in L_1^* \times L_2^* \mid Norm_{L_1/F}(x_1) = Norm_{L_2/F}(x_2)\}$ for two nonisomorphic quadratic field extensions L_1/F and L_2/F.
3. $T(F) \cong L^* \times F^* \subseteq Q(F)$, where $Q \subseteq G$ is the Klingen parabolic subgroup.

Elliptic maximal F-tori of G that come from M by an admissible embedding:

4. Elliptic case I: $T(F) \cong \{x = (x_1, x_2) \in L^* \times L^* \mid Norm_{L/F}(x_1) = Norm_{L/F}(x_2)\}$ for a quadratic field extension L/F. These tori (see Chap. 6) come from the tori $L^* \times L^*/F^*$ of M. We have $\mathcal{D}_G(T) = H^1(F,T) \cong \mathbb{Z}/2\mathbb{Z}$ and $\varepsilon(T) = 1$. The Galois action is via w_{max}, which is the longest element of W_G. The Weyl group is the semidirect product of $\langle w_1, w_2 \rangle$ and the reflection ι attached to the short root. We have $w_{max} = w_1 w_2$ and $\iota w_1 = w_2 \iota$. Furthermore, $W_F(T,G) = W_G$ and $W_F(T,M) = W_M$. The embedding $W_F(T,M) \hookrightarrow W_F(T,G)$ is an isomorphism onto the subgroup generated by the permutation ι and the element w_{max}. The outer automorphism $*$ acts on T like $w_2 \in W_F(T,G)$. We have $W(T,M) = W_F(T,M)$. Finally, we have $W(G,T) = W_F(G,T)$ iff $\chi_{L/F}(-1) = 1$, in which case T^δ is not F-conjugate to T for the nontrivial element $\delta \in \mathcal{D}_G(T)$. In the case $\chi_{L/F}(-1) = -1$ we have T^δ and T are $G(F)$-conjugate, and we can even assume $T = T^\delta$ by choosing a

suitable representative. In this case, if $\psi : T_M(F) \ni t \mapsto \eta \in T(F)$ is an admissible embedding, then $\psi(t^*) = t^\delta$ holds. Finally, $W(G,T) = W_F(T,M) \subseteq W_F(T,G)$ in this case.

5. Elliptic case II: These tori are $T(F) \cong (L_1^* \times L_2^*)/F^*$ for two nonisomorphic quadratic field extensions L_1/F and L_2/F (see Chap. 7). They come from M with $\varepsilon(T) = 2$. Then $W_F(T,G)$ is the subgroup of W_G generated by the two elements w_{max}, ι (image of Galois), which can be identified with the image of $W_F(T,M) = W_M$. Also $\mathcal{D}_G(T) \cong \mathbb{Z}/2\mathbb{Z}$.

Nonelliptic maximal F-tori of G that come from M by an admissible embedding:

6. These tori are $T(F) \cong L^* \times F^* \subseteq P(F)$ for quadratic extension fields L/F, contained in the Siegel parabolic $P \subseteq G$. They come from the tori $T_M = (L^* \times (F^*)^2)/F^*$ of M via the embedding

$$(x, diag(t_2, t_2'))/(t \cdot id, t^{-1} \cdot id) \mapsto blockdiag(t_2' det(x) x^{-t}, t_2 x).$$

Here $x \in L^* \subseteq Gl(2, F)$ and $t_2, t_2' \in F^*$. We have $\varepsilon(T) = 2$ and $\mathcal{D}_G(T) = 1$. Furthermore, $W_F(T,G)$ is generated by w_{max}, ι and $W_F(T,G) = W_F(T,M) = W_M$.

7. The split torus $T(F) \cong (F^*)^3$ obviously comes from M with $\varepsilon(T) = 1$. Furthermore, $W(T,G) = W_F(T,G) = W_G$ and $W(T,M) = W_F(T,M) = W_M$, which maps to the subgroup $\langle \iota, w_{max} \rangle$ of W_G. The admissible embedding is the one used in (6) for diagonal matrices $x = diag(t_1, t_1')$. The involution $*$ acts on the diagonal split torus T in M. This action corresponds to the action of the element $w_2 \in W_G$.

Lemma 4.7. *For all maximal F-tori T in M the following holds:*

$$\#W(T,M) = \#W_F(T,M)$$

and

$$\varepsilon(T) \cdot \#\big(W_F(T,M) \setminus W_F(T,G)\big) = 2.$$

Proof. This is done by a case-by-case inspection, and will be skipped. □

Remarks Concerning the Weyl Groups. Over the algebraic closure $\overline{F}$ of F all underlying admissible homomorphisms of tori $T_M \to T_G$ becomes conjugate to the homomorphism

$$\big(diag(t_1, t_1'), diag(t_2, t_2')\big) \mapsto diag\big(t_1' t_2', t_1 t_2', t_1 t_2, t_2 t_1'\big)$$

by Lemma 8.1. This isomorphism induces the isomorphism $\langle W_M, * \rangle \hookrightarrow W_G$ referred to in the list of tori given above.

Character Twists. Let us denote $\nu_M(g_1, g_2) = det(g_1 g_2)$ for $(g_1, g_2)/(t\, id, t^{-1}\, id) \in M$ and let ν_G denote the symplectic similitude factor $\nu_G : G(F) \to F^*$.

Then in cases (4)–(7) an admissible isomorphism of tori $\psi : T_M \cong T_G$ satisfies $\nu_G(\psi(t)) = \nu_M(t)$

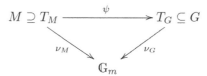

It is enough to show this over the algebraic closure $\overline{F}$, where this follows from Lemma 8.1. Furthermore, $\nu_G(t^\delta) = \nu_G(t)$ for $\delta \in \mathcal{D}_G(T)$. Both statements are easy to verify, and together they imply

$$\big(\chi(\nu_M(t)) \cdot \alpha(t)\big)^G = \chi(\nu(t)) \cdot \alpha^G(t).$$

Therefore, we have

Lemma 4.8. $\chi_{\sigma_v} = \sum n(\sigma_v, \pi_v)\chi_{\pi_v}$ *implies* $\chi_{\sigma_v \otimes \chi_v} = \sum n(\sigma_v, \pi_v)\chi_{\pi_v \otimes \chi_v}$.

Lemma 4.9. *Suppose* $\sigma_v = (\sigma_{v,1}, \sigma_{v,2})$. *Then* $n(\sigma_v, \pi_v) \neq 0$ *implies that the representations* π_v, $\sigma_{v,1}$, *and* $\sigma_{v,2}$ *have the same central character.*

Proof. The admissible isomorphism $T_M \to T_G$ maps $(t, t) \times (1, 1)$ and $(1, 1) \times (t, t)$ to (t, t, t, t). Furthermore, the transfer factors do not change the central character. Hence, Lemma 4.6 proves the claim. □

This being said, we come back to what we already formulated at the beginning of this section, the rationality statement $E \subseteq E_\sigma$.

Since τ has values ± 1, and κ is a quadratic character (see Corollaries 6.1 and 7.1) all coefficients of the formula for α^G stated in Lemma 4.6 are in $\mathbb{Q}$ except possibly the terms $\alpha = \chi_{\sigma_v}$. Therefore, the first statement of the next lemma follows.

Lemma 4.10. *Suppose* $\chi^G_{\sigma_v} = \sum n(\sigma_v, \pi_v)\chi_{\pi_v}$ *so that for compact open K there are only finitely many* $n(\sigma_v, \pi_v) \neq 0$ *for which* $\pi^K_v \neq 0$ *holds. Then*

$$\chi^G_{\sigma^\tau_v} = (\chi^G_{\sigma_v})^\tau$$

and $n(\sigma_v, \pi_v)^\tau = n(\sigma^\tau_v, \pi^\tau_v)$.

Proof and explanation. For an irreducible admissible representation π_v of G_v and $\tau \in Aut(\mathbb{C}/\mathbb{Q})$ or $Aut(\overline{\mathbb{Q}}_l/\mathbb{Q})$ the representations π^τ_v are defined as follows. Choose an admissible basis (extending the bases of the fixed spaces of a cofinal decreasing system of compact open groups) of the representation space W and consider the $\mathbb{Q}$-vector space generated by it. This choice of basis induces an action of τ on the $\mathbb{C}$-vector space W, preserving the fixed spaces of a cofinite system of compact open subgroups of the representation. Then $\pi^\tau_v(g) = \tau \circ \pi_v(g) \circ \tau^{-1}$ is again a $\mathbb{C}$-linear irreducible representation of G_v on W. For K in the cofinal system the subspace W^K is stable under τ. Hence, $W^{\pi^\tau_v(K)} = \tau(W^{\pi_v(K)}) = W^{\pi_v(K)}$, so π^τ_v is admissible. The character is $\chi_{\pi^\tau_v} = \chi^\tau_{\pi_v}$. The isomorphism class of π^τ

is independent of the choice of the admissible basis. Since $\chi^\tau_{\tilde{\sigma}_v} = \chi_{\sigma^\tau_v}$, we get $\sum n(\sigma_v, \pi_v)^\tau \pi^\tau_v = \sum n(\sigma^\tau_v, \pi^\tau_v)\pi^\tau_v$. For a fixed compact open subgroup $K \subseteq G_v$ assume that only finitely many representations satisfy $\pi^K_v \neq 0$ and $n(\sigma_v, \pi_v) \neq 0$ and $n(\sigma^\tau_v) \neq 0$. Then this implies

$$n(\sigma_v, \pi_v)^\tau = n(\sigma^\tau_v, \pi^\tau_v).$$

Since we do not assume that the representations π_v are preunitary (in fact for the representations that arise from the trace formula they are), we may not be able to apply linear independence of characters directly. But we can argue as follows. We can consider the spaces of fixed vectors under small compact open subgroups. If the coefficient relation held, this would contradict the linear independency of a finite number of irreducible characters of the Hecke algebra of K-bi-invariant functions with compact support. $\square$

Remark 4.9. We will show in Sect. 4.11 for tempered σ_v that $\chi^G_{\sigma_v} = \chi_{\pi_1} - \chi_{\pi_2}$ holds for two irreducible representations π_1 and π_2; hence, $\chi^G_{\sigma^\tau_v} = \chi_{\pi^\tau_1} - \chi_{\pi^\tau_2}$ by the last lemma. If $\sigma^\tau_v \cong \sigma_v$, then $\pi^\tau_i \cong \pi_i$ by linear independence of characters. This implies the rationality statement $E \subseteq E_\sigma$.

4.4.3 Concerning $\mathbb{Q}_l$-Rationality

Let us return to the representation on etale cohomology with coefficients $\overline{\mathbb{Q}}_l$ on $H^3_!(S_K, E)$. It decomposes into a direct sum

$$\bigoplus_{\pi_{fin}} \rho_{\pi_{fin}} \otimes \pi_{fin}$$

as a representation under $\Gamma_\mathbb{Q} \times G(\mathbb{A}_{fin})$. Let $V(\sigma)$ be the subspace corresponding to all weak lifts π_{fin} of σ. For constituents of $V(\sigma)$ the representation $\rho_{\pi_{fin}}$ is a finite-dimensional $\overline{\mathbb{Q}}_l$-representation of $\Gamma_\mathbb{Q}$ of the form

$$\rho^{ss}_{\pi_{fin}} = m_1(\pi_{fin})\rho_1 \oplus m_2(\pi_{fin})(\rho_2 \otimes \nu_l).$$

For $\tau \in Aut(\mathbb{C}/\mathbb{Q}) = Aut(\overline{\mathbb{Q}}_l/\mathbb{Q})$ obviously $V(\sigma)^\tau = V(\sigma^\tau)$.

Corollary 4.3. *In the situation of Corollary 4.2 the set of representations π_{fin} with the property $n(\sigma_{fin}, \pi_{fin}) = \prod_v n(\sigma_v, \pi_v) \neq 0$ is Galois stable with finite orbits in the sense of*

$$n(\sigma_{fin}, \pi_{fin}) \neq 0 \Leftrightarrow n(\sigma^\tau_{fin}, \pi^\tau_{fin}) \neq 0.$$

Furthermore, $m_1(\pi^\tau_{fin}) = m_1(\pi_{fin})$ and $m_2(\pi^\tau_{fin}) = m_2(\pi_{fin})$.

Proof. The first assertion follows from Lemma 4.10 and $n(\sigma_S, \pi_S) = n(\sigma_{fin}, \pi_{fin}) = \prod_v n(\sigma_v, \pi_v)$ and the integrality of $n(\sigma_{fin}, \pi_{fin}) = \varepsilon(m_1(\pi_{fin}) - m_2(\pi_{fin}))$ (Corollary 4.1). It implies

$$n(\sigma_{fin}, \pi_{fin}) = n(\sigma_{fin}^\tau, \pi_{fin}^\tau).$$

For the last assertion notice that conjugation by τ induces a τ-linear isomorphism

$$\tau : \pi_{fin} \otimes \rho_{\pi_{fin}} \rightarrow \pi_{fin}^\tau \otimes \rho_{\pi_{fin}}^\tau.$$

Hence, obviously $\rho_{\pi_{fin}}^\tau = \rho_{\pi_{fin}^\tau}$ and $dim(\rho_{\pi_{fin}}^\tau) = dim(\rho_{\pi_{fin}}^\tau) = dim(\rho_{\pi_{fin}^\tau})$. This implies

$$m_1(\pi_{fin}) + m_2(\pi_{fin}) = m_1(\pi_{fin}^\tau) + m_2(\pi_{fin}^\tau).$$

Together with $\varepsilon(m_1(\pi_{fin}) - m_2(\pi_{fin})) = \varepsilon(m_1(\pi_{fin}^\tau) - m_2(\pi_{fin}^\tau))$, since ε does not depend on π_{fin}, this implies $m_i(\pi_{fin}^\tau) = m_i(\pi_{fin})$ for $i = 1, 2$. $\square$

Corollary 4.4. *In the situation of Corollary 4.2 we have*

$$m_1(\pi_{fin}) = dim_{\mathbb{C}}\left(H^{k_1+k_2-3,0}(S(\mathbb{C}), E_{\mathbb{C}}(r))(\pi_{fin})\right),$$

$$m_2(\pi_{fin}) = dim_{\mathbb{C}}\left(H^{k_1-1,k_2-2}(S(\mathbb{C}), E_{\mathbb{C}}(r))(\pi_{fin})\right).$$

Proof. Notice the generalized π_{fin}-eigenspace is cuspidal by our assumptions. For weak endoscopic lifts $\pi = \pi_\infty \otimes \pi_{fin}$ we defined the field $E = E(\pi_{fin})$ of rationality, a finite extension field of $\mathbb{Q}$. The subspace

$$\tilde{H}_{\mathbb{C}}^{p,q} = \bigoplus_{\tau \in \Sigma_E} \left(H^{p,q}(S(\mathbb{C}), E_{\mathbb{C}}(r))(\pi_{fin}^\tau) \oplus H^{q,p}(S(\mathbb{C}), E_{\mathbb{C}}(r))(\pi_{fin}^\tau)\right)$$

is stable under $Aut(\mathbb{C}/\mathbb{Q})$. Hence, the $\tilde{H}^{p,q}$ define a $\mathbb{Q}$-rational sub-Hodge structure of the $\mathbb{Q}$-rational pure Hodge structure defined by cuspidality on the inner Betti cohomology group $H_!^3(S(\mathbb{C}), E(r))$. The dimension of each summand indexed by τ is twice the dimension

$$dim_{\mathbb{C}}\left(H^{p,q}(S(\mathbb{C}), E_{\mathbb{C}}(r))(\pi_{fin}^\tau)\right)$$

by well-known results of continuous cohomology. Hence, $dim_{\mathbb{C}}\left(H^{p,q}(S(\mathbb{C}), E_{\mathbb{C}}(r))(\pi_{fin}^\tau)\right)$ and $dim_{\mathbb{C}}\left(H^{q,p}(S(\mathbb{C}), E_{\mathbb{C}}(r))(\pi_{fin}^\tau)\right)$ are equal. Since $Aut(\mathbb{C}/\mathbb{Q})$ simply permutes the summands $(H^{p,q}(S(\mathbb{C}), E_{\mathbb{C}}(r))(\pi_{fin}^\tau) \oplus H^{q,p}(S(\mathbb{C}), E_{\mathbb{C}}(r))(\pi_{fin}^\tau))$ for the different π_{fin}^τ, this implies

$$dim_{\mathbb{C}}\left(H^{p,q}(S(\mathbb{C}), E_{\mathbb{C}}(r))(\pi_{fin})\right) = dim_{\mathbb{C}}\left(H^{q,p}(S(\mathbb{C}), E_{\mathbb{C}}(r))(\pi_{fin})\right)$$

for all p, q, and τ.

Similarly for the subspace $\tilde{H}_l = \bigoplus_{\tau \in \Sigma_E} H_{et,!}^3(S(\mathbb{C}), E_{\overline{\mathbb{Q}_l}}(r))(\pi_{fin}^\tau)$ of the $\overline{\mathbb{Q}_l}$-adic cohomology. This space is defined over $\mathbb{Q}_l$. It arises from a $\mathbb{Q}_l$-subspace of the $\mathbb{Q}_l$-adic cohomology group $H_{et,!}^3(S, E_l(r))$ by extension of scalars to $\overline{\mathbb{Q}_l}$. Hence, it is of Hodge–Tate type since it is a direct summand of a Hodge–Tate representation (using Chai and Faltings [19], Chapter VI, Theorem 6.2(i)). From loc. cit. further-more, after tensoring coefficients with $\mathbb{C}_l$, the Hodge–Tate weights can be read off from the de Rham cohomology using the comparison isomorphism and an identifi-cation $\mathbb{C} = \mathbb{C}_l$. The comparison isomorphism between the cohomology induces an isomorphism of the subspaces

$$(\tilde{H}_l \otimes_{\overline{\mathbb{Q}}_l} \mathbb{C}_l) \cong (\tilde{H}^{k_1+k_2-3,0} \otimes \mathbb{C}) \oplus (\tilde{H}^{k_1-1,k_2-2} \otimes \mathbb{C}).$$

Since the Hodge–Tate decomposition of ρ_1 is of type $(r_1 - 1, 0)$ and $(0, r_1 - 1)$ and the Hodge–Tate decomposition of ρ_2 is of type $(r_2 - 1, 0)$ and $(r_2 - 2)$, Corollary 4.2 therefore implies $\sum_\tau m_1(\pi_{fin}^\tau) = \sum_\tau dim_{\mathbb{C}} H^{k_1+k_2-3,0}(\pi_{fin}^\tau)$ and similarly for $\sum_\tau m_2(\pi_{fin}^\tau)$ and $\sum_\tau dim_{\mathbb{C}} H^{k_1-1,k_2-2,0}(\pi_{fin}^\tau)$. Since the multiplic-ities $m_i(\pi_{fin}^\tau)$ are independent of τ (Corollary 4.3), and since the corresponding statement was shown for the dimensions $dim_{\mathbb{C}}(H^{p,q}(S(\mathbb{C}), E_{\mathbb{C}}(r))(\pi_{fin}))$, this implies the stronger formulas of Corollary 4.4, which hold for each constituent π_{fin}^τ. $\square$

4.5 Orthogonality Relations

Let F denote the local non-Archimedean field $\mathbb{Q}_v$.

The multiplicity formulas for weak endoscopic lifts from M are encoded in char-acter identities

$$\chi_{\sigma_v}^G = \sum n(\sigma_v, \pi_v)\pi_v,$$

which are defined locally. In special cases we were able to prove the existence of such an identity. Indeed this holds whenever σ_v is the local representation of a global representation $\sigma = (\sigma_1, \sigma_2)$, where $\sigma_{i,\infty}$ are in the holomorphic discrete series of weight r_i. This proves the local identities for a large class of local representations, but not for all. Hence, this is not enough. To show that such a local character identity holds for an irreducible admissible representation σ_v of $M_v = M(F)$ one could replace $\sigma_v = (\sigma_{v,1}, \sigma_{v,2})$ by $\sigma_v \otimes \chi_v = (\sigma_{v,1} \otimes \chi_v, \sigma_{v,2} \otimes \chi_v)$ for some arbitrary character χ_v (see Lemma 4.8). Let us assume this for the moment.

Our first aim in Lemma 4.11 is to show that such a local character identity holds for all local irreducible representations σ_v in the discrete series. To show that it is enough to find a global representation σ for which σ_v is the corresponding local representation, at least up to a local character twist. This is the strategy of the proof. However, we show more. We also prove that the coefficients $n(\sigma_v, \pi_v)$ are all in-tegers. By the global multiplicity formulas already obtained, and by the way these coefficients were defined, this would immediately follow if we can find the global representation σ in such a way that for all non-Archimedean places $w \neq v$ the local character identities $\chi_{\sigma_w}^G = \sum n(\sigma_w, \pi_w)\pi_w$ involve only coefficients with $n(\sigma_w, \pi_w) \in \{-1, 0, 1\}$. This would suffice, since $\prod_{w \neq \infty} n(\sigma_w, \pi_w)$ are inte-gers by Corollary 4.1. However, so far the only representations where we know

$n(\sigma_w, \pi_w) \in \{0, 1\}$ by Lemma 4.3 are the spherical representations. So we could try to choose σ to be unramified outside v. This is a possible strategy. For those who do not like to waste energy on this, we prefer to invoke Lemmas 4.12 and 4.27 (an obvious extension of Lemma 4.3, whose proofs are elementary and local) to be more flexible with the construction of σ.

To prove the next lemma, let us therefore assume that σ_v is a representation of M_v in the discrete series.

Global Embeddings. Consider unitary irreducible cuspidal automorphic representations σ of the group $Gl(2, \mathbb{A})$. Since $\mathbb{A}^*/\mathbb{Q}^* = \mathbb{R}_{>0}^* \times \prod_v \mathbb{Z}_v^*$, the central character ω of σ can be viewed as a character of $\prod_v \mathbb{Z}_v^* \to \mathbb{C}^*$. Given a local character $\chi_v : F^* \to \mathbb{C}^*$ at a non-Archimedean place, one can extend $\chi_v|\mathbb{Z}_v^*$ to a global character χ ramified only at v. Hence, χ_v can be written as a product of a local unramified character χ_v^0 and a global character ramified only at v. Hence, for given σ_v, by replacing σ_v with the unramified character twist $\sigma_v \otimes (\omega_v^0)^{-1/2}$, one can find an extension ω of the central character of σ_v to a global Dirichlet character ω, which is ramified only at ∞ and v. So we can assume without restriction of generality that the central character of σ_v is the local component of an idele class character ω.

Now consider some representations σ_∞ and σ_v from the discrete series at the Archimedean place and at the fixed non-Archimedean place v whose central characters are given by the global character ω. Hence, σ_∞ is in the holomorphic discrete series of weight r. And r is even or odd, depending on whether the Dirichlet character ω is even or odd:

1. For cuspidal σ_v there always exist global cuspidal automorphic representations σ in $L^2_{cusp}(G(\mathbb{Q}) \setminus G(\mathbb{A}), \omega)$ which realize the local representation at the places v and at ∞ such that σ_w is unramified at all other places $w \neq \infty, v$. See [25], p. 110ff.

2. If σ_v is not cuspidal, then $\sigma_v = Sp(\chi_0) = Sp \otimes \chi_0$ is a special representation. By a local character twist we may assume $\chi_0 = 1$. Then the central character is trivial, so we could choose the trivial global character ω. One could ask whether there exists again a global cuspidal automorphic representation σ in $L^2_{cusp}(G(\mathbb{Q})Z_G(\mathbb{A}) \setminus G(\mathbb{A}))$ which realizes σ_v, and a fixed discrete series representation with trivial central character at ∞ such that σ_w is unramified at all other places $w \neq \infty, v$. This is not true in general, since otherwise the space of elliptic cusp forms of level 2 would be nonzero. In other words the conjecture formulated in [25], p. 110, does not hold. But for our purposes it is enough to have this result not for a fixed discrete series representation but rather for some discrete series representation σ_∞, say, of weight r sufficiently large, and up to a local character twist at the place v. In this weaker form the statement above holds. We will give a proof.

Proof. $\sigma_v = Sp$ has nontrivial fixed vectors for Iwahori subgroups of $Gl(2, F)$. Since the supercuspidal and ramified induced representations do not have a nontrivial I-fixed vector, the only admissible irreducible representations σ_v with trivial central character ω_v and nontrivial Iwahori fixed vector are unramified representations or the special discrete series representations $Sp \otimes \chi_{0,v}$ (where $\chi_{0,v}$ is an unramified quadratic character). Hence, the space of new forms in $[\Gamma_0(p), r]_0$ (classical elliptic cusp forms of even weight r for the group $\Gamma_0(p)$) is realized by automorphic forms, whose local representation space σ_p at the prime p is Sp or $Sp(\chi_{0,v})$. One could distinguish them by the Fricke involution w. For our purposes it is enough to realize one of the local representations Sp or $Sp(\chi_{0,v})$ as the local component of a global cuspidal representation. Hence, it suffices to have

$$dim([\Gamma_0(p), r]) > 2 \cdot dim([Sl(2, \mathbb{Z}), r]_0)$$

for $r \geq r_0(p)$ at a given prime $p = p_v$, which is a well-known fact. In fact, this is true for $r = 2$ except in the cases $p = 2, 3, 57, 13$ where the genus g of the corresponding modular curve $X_0(p)$ vanishes. If $g = 0$, then $vol(H/\Gamma_0(p)) = \sum_{i=1}^{r} \frac{e_i-1}{e_i} > 0$, since there are two inequivalent cusps [92], Theorem 2.20. Hence, $dim([\Gamma_0(p), 2k]_0) = -1 + \sum_{i=1}^{r}[k \cdot \frac{e_i-1}{e_i}] > 0$ for either $k = 3$ or $k = 4$ [92], Theorem 2.23.

It is obvious that the weight r can be chosen to be arbitrarily large.

The next step is to extend these results to the group $M(\mathbb{A})$ instead of $Gl(2, \mathbb{A})$. Of course a representation $\sigma_v = (\sigma_{1v}, \sigma_{2v})$ of M_v is in the discrete series if and only if both σ_{iv} are in the discrete series. Again by an unramified character twist we can assume that the central character $\omega_{\sigma_1, v} = \omega_{\sigma_2, v}$ comes from a global idele class character ω. The results above immediately carry over, except in the case where $\sigma_{1,v}$ and $\sigma_{2,v}$ are both special, but not isomorphic. Then $\sigma_{2,v} = \sigma_{1,v} \otimes \chi_{0,v}$ for some local character $\chi_{0,v}$ with the property $\chi_{0,v}^2 = 1$. Hence, $\chi_{0,v}$ defines a local quadratic field extension L_v/F_v. Choose a global $L/\mathbb{Q}$, which locally gives L_v. This defines a global character $\chi_0 = \chi_{L/F}$ which extends $\chi_{0,v}$. Without restriction of generality we can assume that $\sigma_{1,v}$ is the local component of a global cuspidal representation σ_1, which is unramified at all places outside v. Then $\sigma_2 = \sigma_1 \otimes \chi_0$ globally realizes $\sigma_{2,v}$. The local components $\sigma_{w,2} = \sigma_{w,1} \otimes \chi_{0,w}$ for $w \neq v$ are ramified if and only if $\chi_{0,w}$ ramifies. Fortunately, for $\sigma_w = (\sigma_{1,w}, \sigma_{1,w} \otimes \chi_{0,w})$ the character lift $\chi_{\sigma_w}^G$ is the character $\chi_{\pi_{+w}}(\sigma_w)$ of a single irreducible representation $\pi_+(\sigma_w)$ for all $w \neq v, \infty$. This follows from local computations, for which we refer to Lemma 4.12. See also case 2 in the list on page 154 for an even more precise result. For $\sigma_w = (\sigma_{1,w}, \sigma_{2,w})$ and $w \neq v$ this implies

$$n(\sigma_w, \pi_w) = 1 \quad \text{or} \quad n(\sigma_w, \pi_w) = 0$$

for all π_w.

Thus, up to a local character twist, any *discrete* series representation $\sigma_v, v \in S$ of M_v can be realized as the local component of a global cuspidal representation σ, which is an induced representation σ_w for all non-Archimedean $w \neq v$ and is a representation of the discrete series at $w = \infty$. Therefore, the local character expression stated in Corollary 4.2 exists for all discrete series representations σ_v of M_v in the following strong form. $\square$

Lemma 4.11. *If σ_v belongs to the discrete series of M_v, then $\chi^G_{\sigma_v} = \sum n(\sigma_v, \pi_v)\pi_v$ holds with integer coefficients $n(\sigma_v, \pi_v)$.*

4.5.1 Elliptic Scalar Products

Consider scalar products on the set of elliptic regular elements (=elliptic locus) as in Rogawski [76] or Kazhdan [44]. For the moment fix some central characters. Let $\alpha = \chi_{\sigma_v}$ be a $*$-invariant distribution on M.

Assume σ_v to be irreducible or to be of the form

$$\sigma_v = (\sigma_{v,1}, \sigma_{v,1})$$

or

$$\sigma_v = (\sigma_{v,1}, \sigma_{v,2}) \bigoplus (\sigma_{v,2}, \sigma_{v,1})$$

for irreducible discrete series representations $\sigma_{v,1}, \sigma_{v,2}$ of $Gl(2, F)$, where $\sigma_{v,1}$ and $\sigma_{v,2}$ are supposed to have the same central character. Then the L^2-norm on the elliptic locus is

$$\langle \chi_\sigma, \chi_\sigma \rangle_{M,e} = 1 \text{ or } 2, \text{ respectively,}$$

depending on the cases considered, by the well-known orthogonality relations for discrete series representations [44].

Similarly, for the character χ_π of irreducible discrete series representations π of G (see also [76], Proposition 12.5.2) we have

$$\langle \chi_\pi, \chi_\pi \rangle_{G,e} = 1.$$

Scalar Product Formulas. Let α be a $*$-invariant class function on M. Similarly to in [76], p. 183ff, for distributions α the endoscopic lift satisfies

$$\langle \alpha^G_1, \alpha^G_2 \rangle_{G,e} = 4\langle \alpha_1, \alpha_2 \rangle_{M,e}.$$

For this result (see Lemma 4.14) besides an explicit formula for α^G two further ingredients are needed:

1. $|\tau(t)|^2 = 1$, where $\tau = \Delta_I \Delta_{II} \Delta_{III}$, which can in this case be directly checked (without explicit computation) or by using the explicit formulas given later.
2. $[\mathcal{D}_G(T) : \mathcal{D}_M(T)] = 2$ for all elliptic tori in M.

4.5.2 The Formula for α^G

We first prove the formula for the distribution $\alpha^G(t)$, which was already stated in Lemma 4.6. Obviously the support of $\alpha^G(t)$ is contained in the union of

$G(F)$-conjugacy classes of tori T of types (4)–(7). For this terminology we refer to the list of maximal F-tori given after the statement of Lemma 4.6. For our purposes it is enough to consider elements $t \in T(F)$ which are sufficiently regular. Cases (6) and (7) are trivial and are left to the reader. The cases of elliptic tori are the relevant cases, and we do the calculations case by case.

Elliptic Case I for $\chi_{L/F}(-1) = 1$. This is type (4) in the list of tori. There is one $M(F)$-conjugacy class of type T in M. A representative of this class can be chosen with $T^* = T$. Notice $W(T, M) = W_M$ has order 4.

There are two stably conjugate, but not $G(F)$-conjugate classes of such tori in G. Fix representatives T and $T' = T^\delta$, where δ generates the group $\mathcal{D}_G(T)$ of order 2. The map $H^1(F, T) \to H^1(F, N(T))$ is injective. Its "kernel" is trivial $W_F(T, G)/W(T, G) = 1$, where $W_F(T, G)$ has order 8. The Weyl integration formula gives

$$\frac{1}{4} \int_{T/Z} D_M(t)^2 \alpha(t) O_t^M(f^M)dt = \frac{1}{8} \int_{T/Z} D_G(t)^2 \alpha^G(t) O_t^G(f)dt$$

$$+ \frac{1}{8} \int_{T'/Z} D_G(t')^2 \alpha^G(t') O_{t'}^G(f)dt'.$$

Notice $\int_{T/Z}$ here actually means integration over $T(F)/Z(F)$.

Now we insert the matching condition $O_t^M(f^M) = SO_t^M(f^M) = \Delta(t, \eta) O_\eta^\kappa(f)$, where η is obtained from t by an admissible embedding of tori. By abuse of notation we identify $T_M = T = T_G$ via such an embedding. Then this formula becomes

$$(\tau D_G/D_M)(t) \cdot (O_t^G(f) - O_{t'}^G(f)),$$

where $t' \in T'(F)$ is stably conjugate to t, but not $G(F)$-conjugate to t. This gives

$$\frac{1}{4} \int_{T/Z} D_M(t) D_G(t) \tau(t) \alpha(t) (O_t^G(f) - O_{t^\delta}^G(f))dt = \frac{1}{8} \int_{T/Z} D_G(t)^2 \alpha^G(t) O_t^G(f)dt$$

$$+ \frac{1}{8} \int_{T'/Z} D_G(t')^2 \alpha^G(t') O_{t'}^G(f)dt'.$$

Orbital integrals $O_t^G(f)$ of test functions $f \in C_c^\infty(G(F))$ only separate the $W(G, T) = W_G = \langle W_M, * \rangle$ orbits in $T(F)$. But they separate the tori T and T'. Hence, we can assume $O_{t'}^G(f) = 0$ near $t' = t^\delta$ for a suitable choice of f (t and t' are not $G(F)$-conjugate). Thus,

$$\frac{1}{4} \int_{T/Z} D_M(t) D_G(t) \tau(t) \alpha(t) O_t^G(f)dt = \frac{1}{8} \int_{T/Z} D_G(t)^2 \alpha^G(t) O_t^G(f)dt.$$

After making the left side $W(T, G) = W_G$ invariant, we can compare both sides at the (sufficiently regular) point t in $T(F)/W(T, G)$. Since W_G is generated by $W(T, M) = W_M$ and $*$, we obtain

$$D_G(t)\alpha^G(t) = \tau(t)D_M(t)\alpha(t) + \tau(t^*)D_M(t^*)\alpha(t^*)$$
$$= \tau(t)D_M(t)(\alpha(t) + \alpha(t^*)).$$

A similar argument for $t^\delta = t'$ gives

$$\alpha^G(t^\delta) = \kappa(\delta)\alpha^G(t) = -\alpha^G(t).$$

Notice $\tau(t^*) = \chi_{L/F}(-1)\tau(t) = \tau(t)$.

Elliptic case I for $\chi_{L/F}(-1) = 1$. Then there is a unique $M(F)$-conjugacy class of T in M, and the representative $T = T_M$ can be chosen to be invariant under $*$; furthermore, there is a unique $G(F)$-conjugacy class of such tori in G. The map $H^1(F,T) = \mathcal{D}_G(T) \to H^1(F,N(T))$ is trivial; hence, $\mathcal{D}_G(T) = W_F(T,G)/W(T,G)$ (via the exact non-Abelian Galois cohomology sequence attached to $W_G = N(T)/T$ over $\overline{F}$). Furthermore, $W(T,G) = W_F(M,T) = W_M$ and $W_F(T,G) = W_G = \langle W_M, * \rangle$. Since $\#W(T,G) = \#W(T,M) = 4$, the Weyl integration formula implies

$$\frac{1}{4}\int_{T/Z} D_M(t)^2\alpha(t)O_t^M(f^M)dt = \frac{1}{4}\int_{T/Z} D_G(t)^2\alpha^G(t)O_t^G(f)dt.$$

Hence, again by insertion of the matching condition for $O_t^M(f^M)$

$$\frac{1}{4}\int_{T/Z} D_M(t)D_G(t)\tau(t)\alpha(t)(O_t^G(f) - O_{t'}^G(f))dt = \frac{1}{4}\int_{T/Z} D_G(t)^2\alpha^G(t)O_t^G(f)dt.$$

Notice $t, t' \in T_G$ without restriction of generality in this case, and $t' = t^\delta$ for an element δ in the Weyl group $W_F(T,G)$. Recall $T_M = T_M^*$. If the underlying fixed admissible isomorphism maps $t \in T_M$ to $t \in T_G$ (notice the abuse of notation that results from our identifications $T_M = T = T_G$ using this admissible isomorphism), then $t^* \in T_M$ maps to t' under the same admissible embedding. See the example at the end of Sect. 6.2.

Hence, we obtain $t' = t^*$. By $(t^*)^* = t$, after making a substitution $t \to t^*$, we obtain

$$\frac{1}{4}\int_{T/Z} D_G(t)D_M(t)\tau(t)\alpha(t)O_t^G(f)dt - \frac{1}{4}\int_{T/Z} D_G(t^*)D_M(t^*)\tau(t^*)\alpha(t^*))O_t^G(f)dt$$

$$= \frac{1}{4}\int_{T/Z} D_G(t)^2\alpha^G(t)O_t^G(f)dt.$$

Thus, since the $O_t^G(f)$ separate the $W(T,G)$-orbits in $T(F)$,

$$D_G(t)\alpha^G(t) = \tau(t)D_M(t)\alpha(t) - \tau(t^*)D_M(t^*)\alpha(t^*) = \tau(t)D_M(t)(\alpha(t) + \alpha(t^*)).$$

Notice $D_G(t) = D_G(t^*)$ and $D_M(t^*) = D_M(t)$ (see Sect. 6.3), and (by Lemma 6.5)

$$\tau(t^*) = \chi_{L/F}(-1)\tau(t) = -\tau(t).$$

Of course, this also implies $\alpha^G(t^\delta) = \alpha^G(t^*) = \kappa(\delta)\alpha(t)$.

Elliptic tori II. These are the tori of type (5) in the list of maximal F-tori of G. In this case there are two F-conjugacy classes of such tori in M, with representatives T_M and T_M^* (the image of T under the involution $*$ of M). An admissible embedding ψ of T_M into G defines an admissible embedding $\psi \circ *$ of $T^* = T_M^*$ into G. Note that these define the same Δ-factor (see Lemma 7.3). Thus, the argument in [76], pp. 183–184, gives (with an additional contribution by combining the contribution for T_M and T_M^* in M!) the desired formula. In fact, $W(T, M) = W(T^*, M) = W_M$ and $W_F(T, G) = W(T, G) = W_M \subseteq W_G$ now all have order 4. Since $\Delta_{T^*, T_G}(t^*, t) = \Delta_{T, T_G}(t, t) = \Delta(t, t)$ for $t^* \in T_M^*$ (Lemma 7.3)

$$\frac{1}{4} \int_{T/Z} D_M(t)(\alpha + \alpha^*)(t)\tau(t)D_G(t)O_t^\kappa(f)dt$$

$$= \frac{1}{4} \int_{T/Z} D_M^2(t)\alpha(t)\Delta(t, t)O_t^\kappa(f)dt + \frac{1}{4} \int_{T^*/Z} D_M^2(t)\alpha(t)\Delta(t^*, t)O_t^\kappa(f)dt.$$

By the Weyl formula this is

$$= \frac{1}{4} \int_{T/Z} D_G(t)^2\alpha^G(t)O_t^G(f)dt + \frac{1}{4} \int_{T'/Z} D_G(t')^2\alpha^G(t')O_{t'}^G(f)dt.$$

Since the test functions f separate the tori T and T' in G, we can compare the terms for $t \in T(F)$ and $t' \in T'(F)$ separately. This gives

$$D_G(t)\alpha^G(t^\delta) = \kappa(\delta)\tau(t)D_M(t)(\alpha(t) + \alpha(t^*)).$$

Since $W_F(T, M) = W_F(T, G)$, this could also be written in the form

$$D_G(t)\alpha^G(t^\delta) = \sum_{w \in W_F(T,M) \backslash W_F(T,G)} \tau(t^w)D_M(t^w)\kappa(w(\delta))(\alpha + \alpha^*)(t^w).$$

Notice $\alpha + \alpha^* = 2\alpha = \varepsilon(T) \cdot \alpha$ if $\alpha = \alpha^*$ in this case.

Split Tori. Here κ is trivial and $W(T, G) = \langle W(T, M), * \rangle$ is of order 8 and $\varepsilon(T) = 1$. Notice $\frac{1}{8} \int_{T/Z} D_M(t)(\alpha + \alpha^*)(t)\tau(t)D_G(t)O_t^G(f)dt = \frac{1}{4} \int_{T/Z} D_M^2(t)\alpha(t)\Delta(t, t)O_t^\kappa(f)dt$. By the Weyl integration formula this is $\frac{1}{8} \int_{T/Z} D_G^2(t)\alpha^G(t)O_t^G(f)dt$. Hence,

$$(D_G\alpha^G)(t) = D_M(t)(\alpha + \alpha^*)(t),$$

since evidently $\tau(t) = 1$ in this case, as follows from the definition of $\tau = \Delta_I\Delta_{II}\Delta_{III}$ given by Langlands and Shelstad [60].

Semisplit Tori. Notice $T(F) \cong L^* \times F^*$ and $\varepsilon(T) = 2$ and $W(T, M) = W_F(T, M) = W_F(T, G) = W(T, G)$; hence, T_M and T_M^* give contributions. Now κ is trivial, and $\tau(t) = \Delta_I \Delta_{II} \Delta_{III} = 1$, since $\Delta_I = 1$ (Hilbert 90), $\Delta_{II} = 1$ (the Galois group stabilizes the set of positive roots outside M and the corresponding χ-subdata can be chosen trivially), and henceforth $\Delta_{III} = 1$ is the trivial character. Again

$$(D_G \alpha^G)(t) = D_M(t)(\alpha + \alpha^*)(t)$$

in this case.

So now we come to the end. Adding the terms from all types gives the desired identity, which proves Lemma 4.6. Since all the terms only depend on $\alpha + \alpha^*$, this also proves Lemma 4.5.

4.5.3 Induced Representations

Assume $\sigma_v \cong Ind_{B_M}^M(\rho_v)$ is induced from a character ρ_v of the Borel subgroup B_M of M. The character χ_{σ_v} of σ_v has the regular semisimple support contained in the conjugates of the split torus $T_s^M \subseteq B_M$. On the split torus $T_s = T_s^M$ itself the character of σ_v is given by the sum of four conjugates ρ_v^w of the character ρ_v of $T_s(F)$

$$D_M(t) \chi_{\sigma_v}(t) = \sum_{w \in W_M} \rho_v(t^w), \qquad t \in T_s^M(F).$$

From Lemma 4.6 and its proof we obtain for $\alpha = \chi_{\sigma_v}$

$$D_G(t) \chi_{\sigma_v}^G(t) = D_M(t)(\alpha + \alpha^*) = \sum_{w \in W_M \backslash W_G} D_M(t^w) \chi_{\sigma_v}(t^w),$$

where $t \in T_s^G(F)$ is identified with $t \in T_s^M(F)$ by an admissible isomorphism. Observe that in this case κ is trivial and $W_F(T, M) = W_M$ and $W_F(T, G) = W_G$ and $\tau = 1$.

Fix an admissible isomorphism of split tori of M and G, respectively. This also includes a choice of Borel subgroups B_M and B containing the split torus. View ρ_v as a character of both split tori via the chosen admissible isomorphism. Because $\chi_{\sigma_v}^G$ vanishes on the nonsplit conjugacy classes of tori in G, the formulas above imply

Lemma 4.12.

$$\boxed{\left(\chi_{Ind_{B_M}^M(\rho_v)}\right)^G = \chi_{Ind_B^G(\rho_v)}}.$$

A similar formula holds for representations $\sigma_v = Ind_{P_M}^M(\rho_v)$ induced from a Levi subgroup of a proper maximal parabolic subgroup P_M of M, where

$$\rho_v\Big((x, diag(t_2, t_2'))/(t\,id, t^{-1}id)\Big) = \tau_v(x) \chi_{1,v}(t_2) \chi_{2,v}(t_2').$$

Here P_M is a parabolic subgroup in M with Levi quotient $Gl(2, F) \times (F^*)^2/F^*$, and of course $(x, diag(t_2, t_2')) \in Gl(2, F) \times (F^*)^2$ denotes representatives in P_M modulo the unipotent radical. Assume the central character ω_{τ_v} of τ_v satisfies

$$\omega_{\tau_v} = \chi_{1,v}\chi_{2,v}.$$

In that case, we will see that the lifted character $\chi^G_{\sigma_v} = \frac{1}{2}\chi^G_{\sigma_v + \sigma^*}$ is the character of a corresponding representation of $GSp(4, F)$ induced from the Levi subgroup of the Siegel parabolic P. As above by Lemma 4.6 $D_G(t)\chi^G_{\sigma_v}(t) = D_M(t)(\alpha + \alpha^*) = D_M(t)\chi_{\sigma_v + \sigma_v^*}(t)$ for all t in the nonanisotropic tori. For $t \in T(F) \cong L^* \times F^* \subseteq P_M(F)$ one of the two distributions $\chi_{\sigma_v}(t)$ and $\chi_{\sigma_v}(t^*)$ vanishes by the choice of σ_v. Furthermore, the factor $\Delta_{IV} = D_G/D_M = D_P/D_{P_M}$ compensates for the factors appearing in the character formula for induced representations. Hence,

$$\boxed{(\chi_{Ind^M_{P_M}(\rho_v)})^G = \chi_{Ind^G_P(\tilde{\rho}_v)}},$$

where $\tilde{\rho}_v$ is the representation of the Siegel parabolic subgroup $P(F) \subseteq G(F)$ given by the representation

$$\tilde{\rho}_v(blockdiag(\lambda g^{-t}, g)) = \tau_v(g)\chi_{2,v}(\lambda/det(g)), \qquad \lambda \in F^*, g \in Gl(2, F),$$

of its Levi subgroup of block diagonal matrices in $G(F)$. It remains to check semisimple regular elements in the split torus T_s. For that observe that the character $\chi_{Ind^G_P(\tilde{\rho}_v)}$ is invariant under $w_2 \in W_G$, which corresponds to the involution $*$ on the split torus of M. This compensates for the factor $1/2$.

Corollary 4.5. *For any irreducible admissible representation σ_v of the local group M_v the lift $\chi^G_{\sigma_v}$ can be expressed as a (locally finite) sum $\chi^G_{\sigma_v} = \sum n(\pi_v)\chi_{\pi_v}$ of irreducible characters of the group G_v with integral coefficients $n(\pi_v) \in \mathbb{Z}$.*

Proof. It is enough to consider generators of the Grothendieck group up to character twists. For the group M elliptic tempered representations necessarily belong to the discrete series. Hence, every irreducible character χ_{σ_v} can be written as a finite integral linear combination of discrete series characters and a finite number of induced representations. For discrete series representations σ_v the statement has already been given (Lemma 4.11). For induced representations the claim follows from the remarks above. Hence, the coefficients $n(\pi_v)$ are integers. That moreover the sum is locally finite (see the definition on page 110) is obvious. This proves the claim. $\square$

4.5.4 Scalar Product Formula

For invariant distributions β_1, β_2 on $G(F)$ one formally defines

$$\langle \beta_1, \beta_2 \rangle_{G,e} = \int_{G^e} \beta_1(g)\overline{\beta_2(g)}dg$$

whenever the integral on the right side is well defined. Integration is over the regular elliptic subset G^e of $G(F)$. Hence, by the Weyl integration formula

$$\langle \beta_1, \beta_2 \rangle_{G,e} = \sum_{\substack{T \text{ elliptic tori } \subseteq G \text{ from } M \\ \text{modulo } G(F)\text{-conjugacy}}} \frac{1}{\#W(T,G)} \int_{T/Z} D_G(t)\beta_1(t)\overline{D_G(t)\beta_2(t)}dt.$$

Suppose $D_G\beta_i(t^\delta) = \kappa(\delta)D_G\beta_i(t)$ for all $\delta \in \mathcal{D}(T)$ and all elliptic maximal tori T in G. Then the integrand is invariant under $\mathcal{D}_G(T)$, meaning that

$$\langle \beta_1, \beta_2 \rangle_{G,e} = \sum_{\substack{T \text{ elliptic tori} \subseteq G \text{ from } M \\ \text{modulo stable conjugacy}}} \frac{\#\mathcal{D}_G(T)}{\#W_F(T,G)} \int_{T/Z} D_G(t)\beta_1(t)\overline{D_G(t)\beta_2(t)}dt.$$

See Sect. 4.4.2 for an explanation.

The Scalar Product Formula. Assume α_i are invariant distributions on $M(F)$ such that $\alpha_i = \alpha_i^*$ for $i = 1, 2$. Then both distribution lifts $(D_G\alpha_i^G)(t) = 2 \cdot \tau(t)(D_M\alpha_i)(t)$ satisfy $D_G\alpha_i(t^\delta) = \kappa(\delta)D_G\alpha_i(t)$ for all $\delta \in \mathcal{D}(T)$ and all maximal elliptic F-tori T in G (Lemma 4.6). Hence,

$$\langle \alpha_1^G, \alpha_2^G \rangle_{G,e} = 4 \cdot \sum_{\substack{T \text{ elliptic tori} \subseteq G \text{ from } M \\ \text{modulo stable conjugacy}}} \frac{\#\mathcal{D}_G(T)}{\#W_F(T,G)} \int_{T/Z} |\tau(t)|^2 D_M(t)^2 \alpha_1(t)\overline{\alpha_2(t)}dt.$$

The sum is over all stable conjugacy classes of elliptic maximal F-tori T in G coming from M by an admissible embedding. Since $|\tau(t)|^2 = 1$ and

$$\frac{\varepsilon(T)}{\#W(T,M)} = \frac{\#\mathcal{D}_G(T)}{\#W_F(T,G)},$$

by Lemmas 4.4 and 4.7, this becomes

$$\sum_{\substack{T \text{ elliptic tori} \subseteq M \\ \text{modulo } M(F)\text{- conjugacy}}} \frac{4}{\#W(T,M)} \int_{T/Z} D_M(t)^2 \alpha_1(t)\overline{\alpha_2(t)}dt = 4 \cdot \langle \alpha_1, \alpha_2 \rangle_{M,e}.$$

Lemma 4.13. *Let $\alpha_i = \alpha_i^*$ be $*$-invariant class functions on $M(F)$. Then*

$$\langle \alpha_1^G, \alpha_2^G \rangle_{G,e} = 4\langle \alpha_1, \alpha_2 \rangle_{M,e}.$$

4.5.5 Character Formula

Let $\sigma_v = (\sigma_{v,1}, \sigma_{v,2})$ be an irreducible admissible representation of $M(F)$, which belongs to the discrete series. We distinguish two cases: (1) $\sigma_v^* \ncong \sigma_v$ and (2) $\sigma_v^* \cong \sigma_v$. Since σ_v^* is defined as $\sigma_v \circ *$ for the involution $*$ of the group M, which flips the two copies of $Gl(2)$, the second case is equivalent to $\sigma_{v,1} \cong \sigma_{v,2}$.

Lemma 4.14. *We have*

$$\langle \chi_{\sigma_v}^G, \chi_{\sigma_v}^G \rangle_{G,e} = 2$$

in the first case (σ is not $*$-invariant) and

$$\langle \chi_{\sigma}^G, \chi_{\sigma}^G \rangle_{G,e} = 4$$

in the second case (σ_v is $*$-invariant).

Proof. Suppose $\sigma_v^* \not\cong \sigma_v$. Put $\alpha = \chi_{\sigma_v^*} + \chi_{\sigma_v}$. Then $\langle \alpha, \alpha \rangle_{M,e} = 2$; hence $\langle \alpha^G, \alpha^G \rangle_{G,e} = 8$. But $\alpha^G = 2\chi_{\sigma_v}^G$ by Lemma 4.5; hence, $\langle \chi_{\sigma_v}^G, \chi_{\sigma_v}^G \rangle_{G,e} = 2$. If $\sigma_v \cong \sigma_v^*$, put $\alpha = \chi_{\sigma_v}$. Then $\langle \alpha, \alpha \rangle_{M,e} = 1$; hence $\langle \alpha^G, \alpha^G \rangle_{G,e} = 4$ and $\langle \chi_{\sigma_v}^G, \chi_{\sigma_v}^G \rangle_{G,e} = 1$. This completes the proof. $\square$

4.6 Exponents and Tempered Representations

In the following we use the notation and results from [44]. Let $R(G) = R_{\mathbb{Z}}(G) \otimes_{\mathbb{Z}} \mathbb{C}$ denote the Grothendieck group as in [44], p. 3. Let $R_I(G) \subseteq R(G)$ denote the subspace spanned by the classes of representations fully induced from proper parabolic subgroups. $R_I(G)$ is generated by the images of the maps $i_{G,L} : R(L) \to R(G)$ defined by induction, from proper standard parabolic subgroups P of G with Levi group L. "Parabolic" in the following should always be understood to mean standard parabolic, containing a fixed minimal parabolic.

It is possible to arrange the character expression $\chi_{\sigma_v}^G = \sum n(\sigma_v, \pi_v)\pi_v$ in the form

$$\chi_{\sigma_v}^G = \sum \tilde{n}(\pi_v) \cdot \chi_{\pi_v} + \sum_{P \neq G} \sum_{\rho_v} m(\rho_v) \cdot Ind_P^G(\rho_v)$$

with integer coefficients $\tilde{n}(\pi_v)$ and $m(\rho_v)$ such that only tempered representations π_v occur in the first summation. This uses the property that the coefficients $n(\sigma_v, \pi_v)$ are integral (Corollary 4.5), and a result from [15], Sects. XI.2.11 and XI.2.13 (see also [44], p. 6). Using the notation in [15], we sketch the argument for the convenience of the reader. Every nontempered representation π_v occurs as a Langlands quotient $J(P, \rho_v, \chi_v)$ of a standard induced representation $I(P, \rho_v, \chi_v)$. The multiplicity $d(\pi_v)$ of the Langlands quotient $\pi_v \cong J(P, \rho_v, \chi_v)$ in the composition series of $I(P, \rho_v, \chi_v)$ is 1 [15], Sect. XI.2.13. Hence, in the integral Grothendieck group, π_v can be expressed by the class of the induced representation $class(I(P, \rho_v, \chi_v)) \in R_I(G)$, and some integral linear combination of nontempered or tempered representations with smaller Langlands parameter $< \lambda_\pi$ [15], Sect. XI.2.13. The nontempered representations in this expression can again be written as Langlands quotients. This process terminates after finitely many steps by the finiteness of exponents of a given (induced) representation. This gives an expression for nontempered representations as a $\mathbb{Z}$-linear combination $\sum \tilde{n}(\pi_v)\pi_v + R_I(G)$ of tempered representations modulo $R_I(G)$. Inserting this expression into the summation gives the desired expansion $\chi^G_{\sigma_v} = \sum \tilde{n}(\pi_v)\chi_{\pi_v}$ with integral coefficients $\tilde{n}(\pi_v)$ modulo a sum whose summands are in $R_I(G)$.

However, one remark is in order. Since we have not simply reordered the summation, it remains for us to show that the new sum $\sum_{\pi_v} \tilde{n}(\sigma_v, \pi_v)\pi_v$ is still absolutely convergent. For this we write the character sums so that they are indexed by infinitesimal characters χ.

Infinitesimal Characters [10, 11]. An infinitesimal character χ is a $G(F)$-conjugacy class of pairs (L, ρ_v), where L is a standard Levi subgroup and ρ_v is a cuspidal irreducible representation of L. See [11], Sect. 2.1. For a representative (L, ρ_v) of χ, fix some parabolic $P = LN$ and consider all representations that occur as irreducible constituents of representations induced from cuspidal representations of the Levi component of P. Call the finite set of inequivalent representatives of these classes R_χ. Then one can expand $\chi^G_{\sigma_v}$, by reordering, in the form

$$\chi^G_{\sigma_v} = \sum_\chi \pi_\chi,$$

$$\pi_\chi = \sum n_{\chi,i}\chi_{\pi_i}, \qquad \pi_i \in R_\chi.$$

Locally Finite Sums. Suppose we have a cofinal system of good small open compact subgroups K_ν of G_v in the sense of [10] such that all K_ν are normal subgroups of K_0

$$K = K_\nu \lhd K_0$$

and such that $G(F) = P(F) \cdot K_0$ holds for standard parabolic subgroups P of G. In our cases $GSp(2n)$ and $Gl(n)$ the principal congruence subgroup forms a cofinal system K_ν of normal subgroups of K_0 such that $G(F) = P(F) \cdot K_0$ holds for every standard parabolic subgroup P containing the group of diagonal matrices. An infinite sum $\sum n_\chi \pi_\chi$, $\pi_\chi = \sum n_{\chi,i}\chi_{\pi_i}$ for $\pi_i \in R_\chi$ is called *locally finite*, if for every $K = K_\nu$ in the cofinal system there exists only a finite number of infinitesimal characters χ represented by (L, ρ_v) such that $\rho_v^{K \cap L} \neq 0$ and $\pi_\chi \neq 0$ holds:

1. By our assumptions regarding K the vanishing $\rho_v^{K \cap L} = 0$ implies $Ind_P^G(\rho_v)^K = 0$. Hence, $\pi_\chi(f) = 0$ for a K-bi-invariant function f unless $\rho_v^{K \cap L} \neq 0$. Thus, $\rho_v^{K \cap L} = 0$ implies $tr\, \pi_v(f) = 0$ for all $\pi_v \in R_\chi$. Therefore, locally finite sums are absolutely convergent. Furthermore, since R_χ is finite, there are only finitely many $n_{\chi,i}$ for which π_i^K is nonzero.

Conversely:

2. The summation $\sum n(\pi_v)\chi_{\pi_v}$, so that for every compact open $K \subseteq G$ there exist only finitely many π_v for which $\pi_v^K \neq 0$ and $n(\pi_v) \neq 0$ holds, is necessarily locally finite. Obviously it is absolutely convergent, so we can write it as a sum $\sum_\chi \pi_\chi$ indexed by infinitesimal characters χ by reordering. But $\pi_v \in R_\chi$ and

$\pi_v^K = 0$ implies $\rho_v^{L \cap K} = 0$ [10], Proposition 3.8. Hence, $\rho_v^{L \cap K} \neq 0$ and $\pi_\chi \neq 0$ implies $\pi_v^K \neq 0$ and $n(\pi_v) \neq 0$ for some $\pi_v \in R_\chi$. But there are only finitely such π_v, and hence there are only finitely such χ.

3. The Jacquet functors are known to be compatible with infinitesimal characters (see [11], Sect. 2.4; see also the argument preceding Lemma 4.17); therefore, they "map" locally finite sums to locally finite sums in an algebraic sense. See the proof of Lemma 4.17. Under certain local L^1-conditions, properties of the distributions are inherited by this process as shown later in the crucial Lemma 4.18.

From this it is clear that $\chi_{\sigma_v}^G = \sum_{\pi_v} n(\sigma_v, \pi_v)\pi_v$ is a locally finite sum. If we reorder the summation by first grouping into combinations π_χ of irreducible representations with the same infinitesimal character χ, the further splitting up into two sums, as defined above, obviously gives two locally finite sums, since in fact we have to split up the inner summands $\pi_\chi = \tilde{\pi}_\chi + \pi_{I,\chi}$ only when they are different from zero $\pi_\chi \neq 0$. So both new sums, in particular the sum $\sum_{\pi_v} \tilde{n}(\pi_v)\pi_v = \sum_\chi \tilde{\pi}_\chi$, are locally finite.

Remark 4.10. It follows from [44], Theorems A and F, that the pairings $\langle .,. \rangle_{G,e}$ and $\langle .,. \rangle_{M,e}$ have the spaces $R_I(G)$ and $R_I(M)$, respectively, in their radical. Hence, to compute $\langle \chi_{\sigma_v}^G, . \rangle_{G,e}$ one can replace $\chi_{\sigma_v}^G$ by the sum $\sum_{\pi_v} \tilde{n}_v(\pi_v)\chi_{\pi_v}$ defined above:

$$\langle \chi_{\sigma_v}^G, \beta \rangle_{G,e} = \langle \sum_{\pi_v} \tilde{n}_v(\pi_v)\chi_{\pi_v}, \beta \rangle_{G,e}.$$

Pseudocoefficients. For irreducible local representations $\pi = \pi_v$ in the discrete series there exist pseudocoefficients f_π [44], Theorem K, i.e., locally constant functions with compact support such that $tr\ \pi'(f_\pi) = \delta_{\pi,\pi'}$ holds for all tempered irreducible representations π', so that furthermore $O_t^G(f_\pi) = \chi_\pi(t)$ holds for all regular semisimple elliptic elements $t \in G(F)$ and $O_t^G(f_\pi) = 0$ holds for all regular semisimple nonelliptic elements. In loc. cit. this is stated for groups with anisotropic center. But these statements hold in general. For cuspidal π_v the conditions imply $tr\ \pi'(f_\pi) = \delta_{\pi,\pi'}$ for all irreducible representations π'. Finally, $\langle f_\pi, f_{\pi'} \rangle_G = \langle f_\pi, f_{\pi'} \rangle_{G,e} = \sum \frac{1}{W(T,G)} \int_{T/Z} D_G(t)^2 O_t^G(f_{\pi'}) O_t^G(f_{\pi'}) dt = tr\ \pi'(f_\pi) = \delta_{\pi,\pi'}$.

An Application. Let $f = \sum_{i=1}^r \tilde{n}(\pi_i) f_{\pi_i}$ be a finite linear combination of pseudocoefficients f_{π_i} for discrete series representations π_i. Then f is locally constant with compact support, has vanishing orbital integrals outside the elliptic locus, and the orbital integrals are given by characters on the elliptic locus. Hence,

$$\int_G f(g)\chi_{\sigma_v}^G d\mu = \int_{G^e} f(g)\chi_{\sigma_v}^G d\mu.$$

Now $\chi_{\sigma_v}^G = \sum_{\pi_v} \tilde{n}(\pi_v)\chi_{\pi_v} + \sum_{P \neq G, \rho_v} m(\rho_v) \cdot Ind_P^G(\rho_v)$, where both sums are locally finite sums. Since characters of induced representations vanish on the elliptic locus, we get

$$\int_{G^e} f(g)\chi^G_{\sigma_v}d\mu = \int_{G^e} f(g)\sum_{\pi_v}\tilde{n}(\pi_v)\chi_{\pi_v}(g)d\mu.$$

Since the sum is locally finite, and f is K-invariant for some compact open subgroup K, we obtain

$$\int_{G^e} f(g)\sum_{\pi_v}\tilde{n}(\pi_v)\chi_{\pi_v}(g)d\mu = \sum_{\pi_v}\tilde{n}(\pi_v)\chi_{\pi_v}(f).$$

Since f is a finite linear combination of pseudocoefficients, and since the sum involves only tempered representations, this becomes $\sum_{i=1}^r \tilde{n}(\pi_v)^2$. Therefore,

$$\int_{G^e} f(g)\chi^G_{\sigma_v}d\mu = \sum_{i=1}^r \tilde{n}(\pi_v)^2.$$

By the Schwartz inequality

$$|\int_{G^e} f(g)\chi^G_{\sigma_v}d\mu|^2 \le \langle\chi^G_{\sigma_v},\chi^G_{\sigma_v}\rangle_{G,e}\langle f,f\rangle_{G,e} = \langle\chi^G_{\sigma_v},\chi^G_{\sigma_v}\rangle_{G,e}\sum_{i=1}^r \tilde{n}(\pi_i)^2.$$

Equality holds if and only if $\chi^G_{\sigma_v} = O^G(f)$ holds on the elliptic locus G^e. By the formula above, the Schwartz inequality implies $\sum_{i=1}^r \tilde{n}(\pi_{v_i})^2 \le \langle\chi^G_{\sigma_v},\chi^G_{\sigma_v}\rangle_e$. Hence, in the limit

$$\sum_{\pi_v\in\Pi^2(G_v)} \tilde{n}(\pi_v)^2 \le \langle\chi^G_{\sigma_v},\chi^G_{\sigma_v}\rangle_e,$$

where the summation is over all isomorphism classes $\Pi^2(G)$ of discrete series representations of $G(F)$. Again, equality holds if and only if $\chi^G_{\sigma_v} = O^G(f)$ holds on the elliptic locus G^e.

Now we use:

- The integrality $\tilde{n}(\pi_v) \in \mathbb{Z}$ (Corollary 4.5).
- $\langle\chi^G_{\sigma_v},\chi^G_{\sigma_v}\rangle_e \in \{2,4\}$ (Lemma 4.14 for irreducible σ_v).

Therefore, the inequality from above implies

Lemma 4.15. *For irreducible representations σ_v of M_v the character lift $\chi^G_{\sigma_v}$ can be written as a locally finite sum*

$$\sum \tilde{n}(\pi_v)\pi_v$$

of tempered representations π_v, and a locally finite sum of characters of representations induced from proper parabolic subgroups P of G. The multiplicities of the induced representations in these sums are integral. Furthermore, the coefficients $\tilde{n}(\pi_v)$ are integral. $\tilde{n}(\pi_v) \in \{-1,0,1\}$ for π_v in the discrete series.

Proof. If some $|\tilde{n}(\pi_v)| \ne 1$ for some discrete series representations π_v, then $|\tilde{n}(\pi_v)| = 2$ by the inequality. But then, as explained above, $O_t(f_{\pi_v}) = \chi^G_{\sigma_v}(t)$ holds almost everywhere on G^e. Hence,

$$\chi^G_{\sigma_v}(t) = \chi_{\pi_v}(t), \qquad t \in G^e.$$

But this gives a contradiction looking at formal degrees (the case $r = 1$ below). $\square$

Formal Degrees. Assume $\sum_{i=1}^r \tilde{n}(\pi_{v_i})^2 = \langle \chi^G_{\sigma_v}, \chi^G_{\sigma_v} \rangle_e$ holds. Then

$$\chi^G_{\sigma_v} = \sum_{i=1}^r \tilde{n}(\pi_i)\chi_{\pi_i}$$

holds as a character identity on the regular elliptic locus G^e. $\chi^G_{\sigma_v}$ vanishes identically on elliptic tori of G corresponding to a cyclic field extension of $\mathbb{Q}_v$ of degree 4 (type 1 in the list of maximal F-tori of G). From Rogawski [76], pp. 194–195, any finite character relation $\sum_{i=1}^r \tilde{n}(\pi_i)\chi_{\pi_i} = 0$ between characters of discrete series representations on some elliptic maximal torus implies $\sum d(\pi_v)\tilde{n}(\pi_v) = 0$. Since the formal degrees $d(\pi_v)$ of the discrete series representations π_v are positive, not all of the $\tilde{n}(\pi_i)$ can be positive or negative, respectively. In particular, this excludes $r = 1$. This was used in the proof of Lemma 4.15.

Corollary 4.6. $\tilde{n}(\pi_v) \neq 0$ *can occur for at most two discrete series representations if σ_v is not $*$-invariant, and for at most four if σ_v is $*$-invariant.*

In fact, if there are two or four discrete series representations, respectively, with coefficient ± 1, then this determines $\chi^G_{\sigma_v}$ on the elliptic locus in terms of these discrete series representations. However, we will see that the situation is slightly more complicated. In general, there exist limits of discrete series which compensate for the difference on the elliptic set. Recall that tempered irreducible representations are called elliptic or limits of discrete series if their character is nontrivial on the regular elliptic locus G^e. For instance, irreducible tempered representations, which are fully induced from a proper parabolic subgroup, are not elliptic.

Remark 4.11. Suppose σ_v is not $*$-invariant. Suppose $\tilde{n}(\pi_1), \tilde{n}(\pi_2) = \pm 1$ holds for two nonisomorphic discrete series representations π_v. Denote them by $\pi_+(\sigma_v)$ and $\pi_-(\sigma_v)$. Then the argument based on formal degrees implies $\tilde{n}(\pi_+(\sigma_v)) + \tilde{n}(\pi_-(\sigma_v)) = 0$. Without restriction of generality, we can therefore assume $\tilde{n}(\pi_\pm(\sigma_v)) = \pm 1$. But then, as explained above,

$$\chi^G_{\sigma_v}(t) = \chi_{\pi^+_v}(t) - \chi_{\pi^-_v}(t) \quad , \quad t \in G^e$$

holds on the elliptic locus.

To understand what happens if the number of discrete series representations π_v with $\tilde{n}(\pi_v) \neq 0$ is less than or equal to one, we have to study the asymptotic behavior of the distribution $\chi^G_{\sigma_v}$ at infinity.

4.6.1 The Asymptotic Behavior

For a representation (V, π) and a parabolic subgroup $P = LN$, define the unnormalized Jacquet module $V(P)$ to be the space of $N(F)$-coinvariants, considered as a module under $L(F)$. As in [10], Sect. 2.5, define

$$r_P^G(V) = V(P) \otimes \delta_P^{-1/2}.$$

r_P^G induces a homomorphism between the Grothendieck groups $r_P^G : R(G) \rightarrow R(L)$. For fixed *standard* parabolic subgroups P with Levi group L we also write r_{LG} instead of r_P^G.

Asymptotic Independence. Fix an infinitesimal character χ of the Bernstein center represented by (L, ρ_v), $P_L = LN$. Consider the finitely generated $\mathbb{Z}$-submodule $R_\chi(G)$ of $R(G)$ generated by the classes $\pi_i \in R_\chi$ of all irreducible representations with infinitesimal character χ. Let $AD_\chi \subseteq R_\chi(G)$ be the kernel of the homomorphism $r_P^G : R_\chi(G) \rightarrow R(L)$. We say asymptotic independence holds for χ if $AD_\chi = 0$.

Suppose f is a locally finite sum $f = \sum_\chi \pi_\chi$, and suppose each π_χ is a finite linear combination $\pi_\chi = \sum n_{\chi,i} \chi_{\pi_{\chi,i}} \in R_\chi$. For a parabolic subgroup Q we associate the locally finite sum

$$f(Q) = \sum_\chi \pi_\chi(Q),$$

where $\pi_\chi(Q) = \sum_i n_{\chi,i} \pi_{\chi,i}(Q) \in R(L_Q)$ is the sum of (the characters of) the Jacquet module of the representations $\pi_{\chi,i}$ with respect to Q.

Claim 4.1. $f(Q) = 0$ implies $\pi_\chi(Q) = 0$ for each infinitesimal character $\chi = (L, \rho_v)$ of G.

Proof. (see [11], Sect. 2.4) Let M be the Levi component of Q. Then from [11], p. 189,

$$r_{MG} \circ i_{GL} = \sum_{w \in W(M) \backslash W / W(L)} \left(i_{M, wLw^{-1} \cap M} \right) \circ w \circ \left(r_{L \cap w^{-1} Mw, L} \right)$$

for minimal length representatives w. Here $W = W(T_s, G)$ is the Weyl group of G, and similarly $W(M)$, etc. are the corresponding Weyl groups of the Levi subgroups, considered as subgroups of W. For irreducible admissible cuspidal representations ρ_v therefore,

$$r_{MG} \circ i_{GL}(\rho_v) = \sum_{w \in W(M) \backslash W / W(L), wLw^{-1} \subseteq M} i_{M, wLw^{-1}} \left(w(\rho_v) \right).$$

So for irreducible admissible representations π_v with infinitesimal character (L, ρ_v) the class of $r_{MG}(\pi_v)$ in $R(M)$ is a linear combination of representations of M with infinitesimal characters $w(L, \rho_v)w^{-1}$ where $wLw^{-1} \subseteq M$, since ρ_v is cuspidal. $\square$

This Implies. Suppose π_v and π'_v are irreducible admissible representations of $G(F)$ with different infinitesimal characters (L, ρ_v) and (L', ρ'_v). Then the infinitesimal characters of the constituents (representations of M) of $r_{MG}(\pi_v)$ and $r_{MG}(\pi'_v)$ are different. Otherwise there would exist $w, w' \in W$ and $w'' \in W(M)$ from the last statement above such that $w''w(L, \rho_v)w^{-1}w''^{-1} = w'(L', \rho'_v)(w')^{-1}$. Hence, (L, ρ_v) and (L', ρ'_v) would be conjugate in $G(F)$, i.e., π_v and π'_v would have the same infinitesimal character, contradicting the assumption.

Furthermore, for $\chi = (L, \rho_v)$ we get $r_{LG}i_{GL}(\rho_v) = \sum_{w \in W(L) \backslash W/W(L),\ wLw^{-1}=L} w(\rho_v)$ or

$$r_{LG}i_{GL}(\rho_v) = \sum_{w \in W_L/W(L)} w(\rho_v),$$

where W_L is the group of all $w \in W$ such that $wLw^{-1} = L$. It contains $W(L)$ as a normal subgroup since $w \in W_L$ permutes the roots of L.

Lemma 4.16. *Let χ be represented by (L, ρ_v). Then $AD_\chi = 0$ holds unless χ is "ramified" or "irregular," i.e., there exists a nontrivial element $w \in W_L/W(L)$ such that $w(\rho_v) \cong \rho_v$.*

Proof. Suppose χ is not ramified. Then $r_{GL}i_{GL}(\rho_v)$ is a multiplicity-free linear combination of the irreducible representations $w(\rho_v)$ for $w \in W_L/W(L)$. For non-isomorphic representations π_v and π'_v with infinitesimal character χ, $r_{LG}(\pi_v)$ and $r_{LG}(\pi'_v)$ therefore have disjoint support. Therefore, $AD_\chi = 0$. $\square$

In particular, this implies the next lemma, and since r_{MG} is up to a normalization just the virtual Jacquet module, it also implies the claim.

Lemma 4.17. *$r_{MG}(\sum_\chi \pi_\chi) = 0$ implies $r_{MG}(\pi_\chi) = 0$ for all χ.*

Corollary 4.7. *$\pi_\chi(Q) = 0$ implies, by inductivity of the Jacquet functor, $\pi_\chi \in AD_\chi$ for all infinitesimal characters χ, which are represented by (L, ρ_v) so that Q contains a parabolic group $P_L = LN$ with Levi group L.*

Hence, $\pi_\chi(Q) = 0$ for all proper standard parabolic subgroups Q of G implies that either χ corresponds to a cuspidal representation (G, ρ) of $G(F)$ itself or $\pi_\chi \in AD_\chi$.

4.6.2 The Theorem of Deligne and Casselman

Suppose T_1, T_2 are distributions on $G(F)$ represented by locally finite sums of irreducible representations $T_i(f) = \sum n_i(\pi_v)\chi_\pi(f)$. If both distributions are represented locally by an L^1-function, we show in this section how to derive from the distribution identity $T_1 = T_2$ the corresponding distribution identity $T_1(Q) = T_2(Q)$.

Let G be a reductive group over a p-adic local field. Choose a minimal parabolic subgroup $P_\emptyset$ with simple roots Δ. For a subset $\theta \subseteq \Delta$ one defines the torus $A_\theta = \cap_{\alpha \in \theta} kern(\alpha)$ and $P_\theta = Zentr(A_\theta)R_u(P_\emptyset)$. The unipotent radical of P_θ is generated by the unipotent subgroups, which belong to those positive roots, which are not linear combinations of roots in θ. There exists a reductive group L_θ such that $Zentr(A_\theta) = L_\theta$. The roots in $\Delta \setminus \theta$ define characters on $A_\theta \subseteq L_\theta$. An element $a \in A_\theta$ is called P_θ-contractive provided

$$|\alpha(a)| < 1 \qquad \text{holds for all } \alpha \notin \theta.$$

Let A_θ^- denote the set of P_θ-contractive elements.

Let g be a regular semisimple element of G. Let A be the maximal split torus in $Z_G(g)$. Some power g^k of g has the form $g^k = a \cdot s, a \in A, s \in S$, where S is the maximal anisotropic torus in $Z_G(g)$. After replacement of g by a suitable conjugate, there exists a standard parabolic group P_θ such that $a \in A_\theta = \cap_{\alpha \in \theta} kern(\alpha)$ holds and

$$|\alpha(a)| \leq 1 \quad \text{for all } \alpha \in \Delta,$$

where Δ is the set of simple roots associated with the standard minimal parabolic group $P_\emptyset$. Let $\theta(g)$ be the set of simple roots with the property $|\alpha(a)| = 1$. Notice $\theta \subseteq \theta(g) \subseteq \Delta$. Let P_θ (contained in $P \subseteq G$) denote the standard parabolic subgroup attached to $\theta(g)$. We say, the conjugacy class of g belongs to the P-stratum. By definition, g is a semisimple regular element contained in the Levi component of P; hence, g acts on the unnormalized Jacquet module $\pi_v(P)$. The Deligne–Casselman theorem gives the following formula for the character χ_{π_v} of a finitely generated admissible representation π_v of G

$$\boxed{\chi_{\pi_v}(g) = \chi_{\pi_v(P)}(g)}.$$

See [17], Theorem 5.2. Hence, for example, the support of the character of an irreducible cuspidal representation is contained in the set of topologically quasiunipotent elements.

Similarly to this statement, we prove for $Q = P_\theta, A_Q = A_\theta$, and $L_Q = L_\theta$ (for some θ in Δ) the following.

Lemma 4.18. *Let G be a reductive p-adic group. Suppose f is an invariant distribution on $G(F)$ defined by a locally finite sum of characters which is represented by a locally integrable function (also denoted f) on the set G^{rs} of semisimple, regular elements of $G(F)$. So $f(\phi) = \int_{G^{rs}} f(g)\phi(g)dg$ holds for all locally constant functions ϕ on G with compact support. Let $Q = L_Q N_Q$ be a parabolic subgroup of G with Levi component L_Q and the split component A_Q. Then the following conditions are equivalent:*

(i) Let $\Omega \subseteq L_Q$ be a compact subset. For all semisimple regular elements $g \in \Omega$ $f(ag) = 0$ holds for all sufficiently Q-contractive elements $a \in A_Q^-$, i.e., $|\alpha(a)| < \epsilon(\Omega) \ll 1$ for all roots $\alpha \notin \theta$.

(i)' For all semisimple regular elements $g \in \Omega$ $f(ag) = 0$ holds for all elements a of A_Q^- with $|\alpha(a)| < \epsilon(\Omega) \ll 1$ for $\alpha \notin \theta$, for which $(\log(|\alpha(a)|))_{\alpha \notin \theta}$ is contained in a fixed open cone of the Weyl chamber attached to A_Q^-.
(ii) $f(Q) = 0$ holds (in the sense of page 114).

Proof. Let $Q^- = L_Q N_Q^-$ be the parabolic opposite to $Q = L_Q N_Q$. Choose some small good compact open subgroup $K \subseteq G$ such that

$$K = (K \cap N_Q^-)(K \cap L_Q)(K \cap N_Q^+).$$

Put $\tilde{K} = K \cap L_Q$.

Let $g \in L_Q$ be regular semisimple and let $a \in A_Q$ be Q-contractive. By replacing a by a suitable power and g by $a^n g$, one can assume that a and g are contained in

$$L_Q^- = \{g \in L_Q \mid g(K \cap N_Q)g^{-1} \subseteq K \cap N_Q \text{ and } g(K \cap N_Q^-)g^{-1} \subseteq K \cap N_Q^-\}.$$

Let 1_C denote the characteristic function of a set C. Then we obtain

$$(*) \qquad f(meas(K_i agK_i)^{-1} 1_{K_i agK_i}) = f(Q)(meas(\tilde{K}_i ag\tilde{K}_i)^{-1} 1_{\tilde{K}_i ag\tilde{K}_i}).$$

Observe that only finitely many irreducible admissible characters in the summation defining f are relevant on both sides of the equality; thus we get this from [11], Sect. 5.3, term by term.

Remark: If the distributions f and $f(Q)$ are both smooth at the point ag, then – by taking the limit $K \to 0$ – we obtain $f(ag) = f(Q)(ag)$.

Suppose (ii) holds. Then the identity (*) above shows that $f(meas(K_i agK_i)^{-1} 1_{K_i agK_i})$ vanishes for all $a \in A_Q^-$ which are highly contractive. This is independent of K. By a covering argument this implies $\int_C f(g)dg = 0$ for all compact subsets C contained in this range of points. This gives $f(ag) = 0$ (almost everywhere) for $a \in A_Q^-$ which satisfy $|\alpha(a)| < \epsilon(g), \alpha \notin \theta$. It is clear that this argument works locally uniformly.

Conversely assume $f(ag) = 0$ for all highly Q-contractive $a \in A_Q^-$. Assume K is chosen to be small enough such that every element in $KagK$ is conjugate to an element $a'g'$ with $a', g' \in L_Q^-$ and $g' \in \Omega(g)$ and a' satisfying $|a'| \leq \epsilon(\Omega)$. Then $f(meas(KagK)^{-1} 1_{KagK})$ vanishes and therefore

$$f(Q)(meas(\tilde{K}_i ag\tilde{K}_i)^{-1} 1_{\tilde{K}_i ag\tilde{K}_i}) = \sum n_i \omega_{\sigma_i}(a) \chi_{\sigma_i}(meas(\tilde{K}_i ag\tilde{K}_i)^{-1} 1_{\tilde{K}_i ag\tilde{K}_i})$$

vanishes. Summation is over the irreducible representations σ_i of L_Q, which appear in $f(Q)$. This sum actually is a finite sum. Almost all summands are zero. Using linear independency of a finite number of different central characters ω_{σ_i} on cones as in Sect. 2.7, we get the following. If this finite sum vanishes for all a in a open subcone of A_Q^-, then the sum vanishes for all $a \in A_Q$. This shows for every $g \in L_Q$ and every $a \in A_Q$ that $f(Q)(1_{\tilde{K}ag\tilde{K}}) = 0$ holds.

How do the $\tilde{K}$ depend on the particular elements $a, g \in L_Q$ we started from? I claim that the good small subgroup $K = K(a, g)$ can be chosen such that $\tilde{K} = L_Q \cap K(a, g)$ is a suitably fixed small subgroup $\tilde{K}$ of L_Q independent of a and g. This would imply $f(Q)(1_{\tilde{K}ag\tilde{K}}) = 0$ for all $ag \in L_Q$, because the regular semisimple elements are dense. Then $f(Q) = 0$ follows by letting $\tilde{K}$ get arbitrary small.

To construct K we use the remarks in [10], p. 16. Choose a lattice Λ to construct K as in [10]. If we replace Λ by $p^i\Lambda$ we get a cofinal system of good small subgroups K_i of G with respect to Q and a^m and $a^n g$ for suitable n, m. Now choose a small compact open K_0 in L_Q, which stabilizes $\Lambda \cap Lie(N_Q)$ and $\Lambda \cap Lie(N_Q^-)$ and $\Lambda \cap Lie(L_Q)$. Then K_0 normalizes all K_i or equivalently the intersections of K_i with N_Q, N_Q^-, L_Q. Therefore, we have $(K_0 K_i) \cap N_Q = K_i \cap N_Q$ and $(K_0 K_i) \cap N_Q^- = K_i \cap N_Q^-$, as the decompositions $N_Q^- L_Q N_Q$ are unique. Then we also have

$$(K_0 K_i) = (K_0 K_i \cap N_Q^-)(K_0 K_i \cap L_Q)(K_0 K_i \cap N_Q^+).$$

Therefore, not only K_i but also $K_0 K_i = K_i K_0$ satisfies the assumptions necessary for the argument in [11], Sect. 5.3. But now $\bigcup_i (K_0 K_i) = K_0$. Therefore, for i sufficiently large, $K(a, g)$ can be chosen to be of the form $K_0 K_i$. Then $\tilde{K} = K_0$ for i sufficiently large. This completes the proof of the lemma. $\square$

Corollary 4.8. *If σ_v is a cuspidal irreducible representation of M, then the distribution $f = \chi_{\sigma_v}^G$ defined by the endoscopic character lift is a locally finite sum of cuspidal characters and characters with infinitesimal character χ, for which $AD_\chi \neq 0$ holds.*

Proof. If σ_v is a cuspidal irreducible representation of M, then the distribution χ_{σ_v} is a local L^1-function. Hence, it is easy to see from the explicit formulas for the transfer factor that $f = \chi_{\sigma_v}^G$ is also a local L^1-distribution. The distribution $f = \chi_{\sigma_v}^G$ also satisfies condition (i) of Lemma 4.18 for all three conjugacy classes of proper parabolic subgroups. This follows immediately from Lemma 4.6 and the cuspidality of σ_v. Therefore, by the implication (i)$\Longrightarrow$(ii) of Lemma 4.18 we have $f(Q) = 0$ for all proper parabolic subgroups. Hence, χ_{σ_v} is a locally finite sum of cuspidal characters and characters with infinitesimal character χ, for which $AD_\chi \neq 0$ holds. See Lemma 4.17 and Corollary 4.7. $\square$

4.7 The Classification of Local Representations

Let F be a local field. The symplectic group of similitudes $GSp(4, F)$ is defined as the subgroup of $Gl(4, F)$ of all matrices g which satisfy $g^t J g = \lambda(g) J, \lambda(g) \in F^*$ for

$$J = \begin{pmatrix} 0 & -E \\ E & 0 \end{pmatrix}.$$

The standard parabolic subgroups are supposed to contain the Borel group B consisting of all matrices $g = \left(\begin{smallmatrix} A & B \\ 0 & D \end{smallmatrix}\right) \in GSp(4, F)$ for which D is upper triangular. Then A is a lower triangular matrix.

In the following ω denotes central characters, i.e., ω_π denotes the central character of π, etc. In the first two sections we consider representations induced from cuspidal representations of maximal parabolic subgroups. According to Shahidi [87], Theorem 6.1 and Waldspurger [106], Sect. 5, such representations have a multiplicity 1 decomposition series. There are two maximal proper standard parabolic subgroups, the Siegel parabolic P and the Klingen parabolic Q.

4.7.1 The Siegel Parabolic P (Results Obtained by Shahidi)

Representations induced from P sometimes contain discrete series representations [88], Proposition 6.1 and Theorem 5.1. The result in [88] is the following. Let σ_1 and χ be irreducible admissible representations of $Gl(2, \mathbb{Q}_v)$ and $Gl(1, \mathbb{Q}_v)$, respectively, and let

$$\sigma = \sigma_1 \boxtimes \chi$$

be the irreducible representation of the Levi component of the Siegel parabolic $M \cong Gl(2) \times Gl(1)$ defined by $\sigma(x, \tau) = \sigma_1(x) \cdot \chi(\tau)$. For the representation $I(\sigma) = Ind_P^G(\sigma)$ of G induced from P (using unitary normalization)

$$P \ni \begin{pmatrix} x & * \\ 0 & \tau \cdot x^{-t} \end{pmatrix} \mapsto \sigma_1(x)\chi(\tau)$$

we write

Short Notation. $I(\sigma) = \sigma_1 \lhd \chi$.
Notice

$$(\sigma_1 \lhd \chi) \otimes \chi' \cong \sigma_1 \lhd \chi\chi',$$

$$\omega_{\sigma_1 \lhd \chi} = \omega_{\sigma_1} \chi^2.$$

For cuspidal unitary σ_1 the induced representation $\sigma_1 \lhd \chi$ is irreducible. More generally, consider the representation $\sigma_1 \nu^s \lhd \chi$ for $\sigma_1 \nu^s(x) := \sigma_1(x)|det(x)|^s$, where σ_1 is *unitary cuspidal*. In the notation in [114] this is $\Pi(\sigma, s) \otimes \nu^s$.

For unitary cuspidal σ_1 the representation $\sigma_1 \nu^s \lhd \chi$ is reducible if and only if the following two conditions hold: (1) $s = \pm 1/2$ and (2) the central character ω_{σ_1} of σ_1 is trivial $\omega_{\sigma_1} = 1$:

– In the reducible case $s = 1/2$ the Langlands quotient $J_P(\nu^{1/2}, \sigma_1 \boxtimes \chi)$ is irreducible unitary nontempered and not generic. The corresponding kernel is a generic *discrete series* representation

$$D = \delta(\sigma_1 \nu^{\frac{1}{2}} \lhd \chi)$$

as shown by Silberger (we refer to [87, 88]).

The representation D is irreducible. Bruhat theory ([93], Corollary 2.12.2(1)) implies that the induced representation is multiplicity free, since $\sigma_1 \nu^{1/2}$ is not stable under the nontrivial element of the Weyl group. By the same reasoning there are at most two exponents; therefore, there are at most two irreducible constituents. Hence, D is irreducible.

4.7.2 The Klingen Parabolic Q (Results Obtained by Waldspurger)

Representations induced from Q were studied by Waldspurger [106]. For irreducible admissible cuspidal representations χ and ρ of $Gl(1)$ and $Gl(2)$, respectively, let σ be $\chi \boxtimes \rho$ considered as a representation of the Levi subgroup $Gl(1) \times Gl(2)$ of Q. Then the representation $I(\sigma) = Ind_Q^G(\sigma)$ induced from the representation σ (using unitary normalization)

$$Q \ni \begin{pmatrix} a & 0 & b & * \\ * & x & * & * \\ c & 0 & d & * \\ 0 & 0 & 0 & y \end{pmatrix} \mapsto \chi(x)\rho\left(\begin{pmatrix} a & b \\ c & d \end{pmatrix}\right)$$

is multiplicity free and has at most two composition factors.

Short notation. $I(\chi \boxtimes \rho) = \chi \times \rho$.
Notice

$$(\chi \times \rho) \otimes \chi' = \chi \times (\rho \otimes \chi'),$$

$$\omega_{\chi \times \rho} = \chi \omega_\rho.$$

For cuspidal ρ the representation $\chi \times \rho$ is reducible if and only if one of the following two conditions holds:

1. $\chi = 1$.
2. $\chi = \nu^{\pm 1} \chi_0$ and ρ has CM multiplication by χ_0. (By this we mean $\chi_0^2 = 1, \chi_0 \neq 1$ such that $\rho \otimes \chi_0 \cong \rho$.)

Reducibility for cuspidal ρ in cases (2) yields an irreducible nontempered Langlands quotient and an irreducible generic *discrete series* representation

$$\delta(\nu \chi_0 \times \rho).$$

This follows again from Silberger's result (see [87] and [88], p. 285, Theorem 5.1).

For cuspidal ρ reducibility in cases (1) produces two tempered representations of G, which are relatives in the sense of Kazhdan [44] called $Wp_+(\rho), Wp_-(\rho)$. Hence, the unitary representation $1 \times \rho$ decomposes

$$\boxed{Wp_+(\rho) \oplus Wp_-(\rho) = 1 \times \rho}.$$

Notably $r_{GP}(Wp_\pm(\rho)) = 0$ and $r_{QG}(Wp_\pm(\rho)) = 1 \times \rho$ implies

$$Wp_+(\rho) - Wp_-(\rho) \in AD_\chi$$

for the infinitesimal character represented by the representation $1 \boxtimes \rho$ of the Levi group of Q. Exactly one of the two representations $Wp_\pm(\rho)$ admits a Whittaker model (see [105], p. 64, the remark in loc. cit. following Lemma 8.2; we will give a detailed discussion of this later). Fix $Wp_+(\rho)$ to be the representation with the Whittaker model. The definition of representations $Wp_+(\rho)$ and $Wp_-(\rho)$ will be extended later to include also the case where ρ is an arbitrary discrete series representation of $Gl(2, F)$. For this we refer to the discussion of case (b) on page 128.

4.7.3 The Borel Group B (Results Obtained by Tadic and Rodier)

In this section we give a review of results obtained by Tadic [100] on the (nonunitary) principal series representations. These are induced from representations $\chi_1 \boxtimes \chi_2 \boxtimes \chi$ of the Borel group B attached to three (not necessarily unitary) characters χ_1, χ_2, χ, which are defined by

$$B \ni \begin{pmatrix} x & 0 & * & * \\ * & y & * & * \\ 0 & 0 & t/x & * \\ 0 & 0 & 0 & t/y \end{pmatrix} \mapsto \chi_2(x)\chi_1(y)\chi(t).$$

We denote this induced representation (using unitary normalization) by

$$\chi_1 \times \chi_2 \lhd \chi,$$

using the short notation used in [100]. For the reader who wants to compare with the results in [100], it should be remarked that Tadic used a slightly different realization of the symplectic group of similitudes, which differs by conjugation in $Gl(4, F)$. If we take this into account and use corresponding choices of Borel groups, then the definition above matches the definition in [100].

For characters χ_1, χ_2 we write $\chi_1 \times \chi_2 = Ind_{B_{Gl_2}}^{Gl_2}(\chi_1, \chi_2) = \frac{\chi_1}{\chi_2} \lhd \chi_2$.

We have the obvious isomorphism

$$\chi_1 \times \chi_2 \lhd \chi \cong \sigma_1 \lhd \chi,$$

where

$$\sigma_1 = \chi_1 \times \chi_2 = Ind_{B_{Gl(2)}}^{Gl(2)}(\chi_1, \chi_2) \cong Ind_{\overline{B}_{Gl(2)}}^{Gl(2)}(\chi_2, \chi_1) \cong \frac{\chi_1}{\chi_2} \lhd \chi_2,$$

and where $B_{Gl(2)}$ is the group of upper triangular matrices in $Gl(2, F)$. This is consistent with the short notation. Also observe $\chi_0 \lhd \chi \cong \chi \chi_0 \times \chi \cong \chi \times \chi_0 \chi \cong \chi_0 \lhd \chi_0 \chi$ for $\chi_0^2 = 1, \chi_0 \neq 1$.

Furthermore, there is an obvious isomorphism

$$\chi_1 \times \chi_2 \lhd \chi \cong Ind_Q^G(\chi_1 \boxtimes \rho),$$

where

$$\rho = \chi_2 \lhd \chi = Ind_{B_{Gl(2)}}^{Gl(2)}(\chi_2\chi, \chi).$$

This is again consistent with the short notation introduced above.

Reducibility. The nonunitary principal series representations $\chi_1 \times \chi_2 \lhd \chi$ of the group $GSp(4, \mathbb{Q}_v)$ are irreducible unless $\chi_1\chi_2 = \nu^{\pm 1}$ or $\chi_1/\chi_2 = \nu^{\pm 1}$ or $\chi_1 = \nu^{\pm 1}$ or $\chi_2 = \nu^{\pm 1}$ holds, where $\nu = |.|$. See [100], Theorem 7.9 and Remark 7.10. Observe

$$\left(\chi_1 \times \chi_2 \lhd \chi\right) \otimes \chi' = \chi_1 \times \chi_2 \lhd \chi\chi',$$

$$\omega_{\chi_1 \times \chi_2 \lhd \chi} = \chi_1\chi_2\chi^2.$$

Furthermore, one has the following equalities

$$\chi_1 \times \chi_2 \lhd \chi \equiv \chi_2 \times \chi_1 \lhd \chi,$$

$$\chi_1 \times \chi_2 \lhd \chi \equiv \chi_1 \times \chi_2^{-1} \lhd \chi_2\chi$$

valid in the Grothendieck group $R_{\mathbb{Z}}(G)$ of representations of $G = GSp(4, \mathbb{Q}_v)$. In the *reducible* cases this allows us to do the determination of the irreducible composition factors in the two basic cases $\chi_2 = \nu$ or $\chi_1/\chi_2 = \nu$. Hence, it is enough to consider the two types of representation.

Basic Reducible Cases.

$$\boxed{\chi_1 \times \nu \lhd \chi} \quad \text{or} \quad \boxed{\nu^{1/2}\chi_0 \times \nu^{-1/2}\chi_0 \lhd \chi}.$$

Regular Cases. According to results obtained by Rodier (see [100], Remark 8.3) the induced representation $\chi_1 \times \chi_2 \lhd \chi$ is multiplicity free if the characters (χ_1, χ_2, χ) are *regular*. Regularity means $\chi_1 \neq \chi_2^{\pm 1}$ and $\chi_1 \neq 1$ and $\chi_2 \neq 1$ (i.e., the character is unramified for the action of the Weyl group). Condition (iii) of Tadic [100], proposition 8.1(b) is void for $n = 2$. So in the regular case there are one, two, or four constituents, and the length of $\chi_1 \times \chi_2 \lhd \chi$ is 2^s ([100], Remark 8.3), where $s = s(\chi_1 \times \chi_2 \lhd \chi)$ is the cardinality of a subset

$$S = S(\chi_1 \times \chi_2 \lhd \chi)$$

of the coroots of $GSp(4)$ defined as follows: $\chi_1 \boxtimes \chi_2 \boxtimes \chi$ defines a character φ of the maximal torus $T \subseteq B$ of diagonal matrices. Each coroot $\alpha^\vee$ is a cocharacter, and hence defines a homomorphism $\alpha^\vee : \mathbb{G}_m \to T$. Define $S(\chi_1 \times \chi_2 \lhd \chi)$ to be the set of all coroots $\alpha^\vee$ for which $\varphi \circ \alpha^\vee = \nu$ holds. For the two simple positive roots the homomorphisms are $\alpha_1^\vee(t) = diag(t^{-1}, t, t, t^{-1})$ and $\alpha_2^\vee(t) = diag(t, 1, t^{-1}, 1)$. See [100], p. 36ff.

Remark 4.12. There are eight coroots, but the cardinality of S is at most 3. The orbits of $1 \boxtimes \nu \boxtimes \chi$ under the Weyl group become the only cases of characters where the cardinality of S is 3 with eight irreducible constituents. This is the important case (b), which is *not regular* and will be discussed later.

In the reducible regular cases there are either two or four irreducible constituents. To determine them, we look at the basic cases.

First Basic Case. Applied to the exact sequence

$$0 \to Sp(\chi_0) \to \nu^{1/2}\chi_0 \times \nu^{-1/2}\chi_0 \to \chi_0 \circ det \to 0$$

of representations of the group $Gl(2)$, induction from Q gives

$$0 \to \chi_1 \times Sp(\chi_0) \to \chi_1 \times \nu \triangleleft \nu^{-1/2}\chi_0 \to \chi_1 \times (\chi_0 \circ det) \to 0.$$

The representations L and R that occur on the left and right are irreducible for $\chi_1 \neq 1, \nu^{\pm 1}, \nu^{\pm 2}$.

Proof: $\chi_1 \times \nu \triangleleft \nu^{-1/2}\chi_0$ is regular and $\#S = 2$ in these cases. Similarly, the cases $\chi_1 = \nu^{\pm 2}$ are equivalent in the Grothendieck group

$$\nu^{\pm 2} \times \nu \triangleleft \nu^{-1/2}\chi_0 \equiv \nu^2 \times \nu \triangleleft \nu^{-1/2}\chi_0,$$

and for the first basic case this is the only regular case with four constituents, whereas the other remaining cases $\chi_1 = 1, \nu^{\pm 1}$ essentially give the *nonregular* cases (b) and (c), which will be discussed later (see page 128ff).

The Second Basic Case.

$$0 \to Sp(\chi) \triangleleft \chi' \to \nu^{1/2}\chi \times \nu^{-1/2}\chi \triangleleft \chi' \to (\chi \circ det) \triangleleft \chi' \to 0$$

is obtained by induction from P. The representations on the left and right are irreducible if $\chi \neq \chi_0\nu^{\pm 1/2}$ for all $\chi_0, \chi_0^2 = 1$ and $\chi \neq \nu^{\pm 3/2}$ and $\chi^2 \neq 1$.

Proof: $\#S = 2$ and $\nu^{1/2}\chi \times \nu^{-1/2}\chi \triangleleft \chi'$ is regular in these cases. Similarly the remaining cases $\chi = \chi_0\nu^{\pm 1/2}$, $\chi_0^2 = 1, \chi_0 \neq 1$, and $\chi = \nu^{\pm 3/2}$ are regular with four constituents, and have discrete series constituents. For $\chi = \nu^{3/2}$ – the corresponding principal series representation

$$\nu^2 \times \nu \triangleleft \chi'$$

already came up in the first basic case – two of its constituents are preunitary: the Steinberg representation and the one-dimensional representation. The other two constituents are nonunitary by a theorem of Casselman. So all regular induced representations with more than two constituents already occur in the second basic case, and they were just described, since the remaining cases are nonregular: Notice $\chi_0 = 1, \chi = \nu^{\pm 1/2}$ is *nonregular* of type (b). The *nonregular* case $\chi^2 = 1$ is case (a) below (see page 127).

Discrete Series.   Up to the obvious modifications with respect to the action of the Weyl group described above, which gives equivalent elements in the Grothendieck group, they occur in (nonunitary) regular principal series representations possibly only for the cases $\nu^2 \times \nu \lhd \chi$ or $\nu\chi_0 \times \chi_0 \lhd \chi$ for $\chi_0^2 = 1, \chi_0 \neq 1$. In these cases there exists a unique irreducible *discrete series* subquotient of the principal series. This subquotient is denoted

$$\delta(\nu^2 \times \nu \lhd \chi)$$

(Steinberg representations) or

$$\delta(\nu\chi_0 \times \chi_0 \lhd \chi)$$

as a constituent of $Sp(\chi_0)\nu^{1/2} \lhd \chi\nu^{-1/2}$ together with the Langlands quotient $J_P(\rho_1\nu^{1/2} \lhd \chi\nu^{-1/2})$ for the tempered representation $\rho_1 = Sp(\chi_0)$. Both are constituents of $\nu\chi_0 \times \chi_0 \lhd \chi$. For a proof see [100], Theorem 8.5, which uses the different notation $\delta(k, f, \chi)$, or the remark on page 132. These discrete series representations are generic representations [100], p. 42. In the nonregular cases it turns out that the nonunitary principal series do not have constituents in the discrete series. This will also become clear from the subsequent discussion of the three relevant nonregular cases (a)–(c).

Concerning Tempered Representations. No relatives are contained in unitary principal series since they are always irreducible (e.g., [100], Corollary 7.6).

4.7.4 The Borel Group B (The Nonregular Reducible Principal Series)

It remains for us to examine the nonregular cases of the reducible nonunitary principal series representations. See also [75, 79, 100]. For later applications it is crucial for us to determine the exponents of the irreducible constituents of these representations. Up to equivalence in the Grothendieck group, only the three following nonregular cases remain to be considered:

(a) $\nu^{1/2}\chi_0 \times \nu^{-1/2}\chi_0 \lhd \chi,$ $\chi_0^2 = 1$
(b) $1 \times \nu \lhd \chi$
(c) $\nu \times \nu \lhd \chi$

We Start with a Short Summary. Situation (a) turns out not to be interesting since there are only the two obvious irreducible constituents $Sp(\chi_0) \lhd \chi$ (tempered) and $J = \chi_0 \circ det \lhd \chi$ (nontempered), completely analogous to the regular second basic case. In case (c) there are again two irreducible constituents; both are nontempered. In the interesting case (b) there are four irreducible constituents: J, J' (nontempered) and T_+, T_- (tempered). Warning: Temperedness for a representation of $G(F)$ – here and for the rest of this chapter – will usually mean temperedness up to a suitable character twist of the representation.

Finally, none of the irreducible constituents in cases (a)–(c) belong to the discrete series.

Exponents. For an irreducible admissible representation (π, V) of a reductive group G over F and a standard parabolic group P one can attach characters $E(P, \pi)$ as in [15], Chapters XI.1.8. The elements of $E(P, \pi)$ are the characters of $T(F)$ for the maximal split torus T in the standard Borel group B that appear in $r_{\overline{P}G}(V)$ as generalized eigenvalues, where $\overline{P}$ is the parabolic group opposite to P [15], Chapters XI.3.3. $T(F)$ is contained in $P(F)$, and hence acts on the normalized Jacquet module $r_{PG}(V) = \delta_P^{-1/2} \otimes V(P)$. The longest element w_{max} maps $\overline{P}$ to P: hence, $w_{max}(E(P, \pi))$ are the generalized eigenvalues of $T(F)$ on $r_{PG}(V)$. The absolute values of the characters in $E(P, \pi)$ define the set $^0E(P, \pi)$ as in [15], Chapters XI.1.13. Let $-w_{max}$ denote the opposite involution.[1] We call the elements $-w_{max}{}^0(E(P, \pi))$ the *exponents* of π with respect to P. Let $\alpha_1, \ldots \alpha_l$ be the simple roots attached to the standard Borel group. Let $\beta_1, \ldots, \beta_l$ be defined by $(\alpha_i, \beta_j) = \delta_{ij}$. Then the Lie algebra of the maximal split torus of the Levi group M of P is spanned by $\beta_i, i \in F$, modulo the center of $Lie(M)$, for a unique subset $F \subseteq \{1, \ldots, l\}$ which characterizes $P = P_F$. Recall the following facts:

1. π belongs to the discrete series of G (up to a central character twist) if for a minimal standard parabolic P_F for which $E(P_F, \pi) \neq \emptyset$ we have $(\nu, \beta_i) < 0$ for all $\nu \in {}^0E(P, \pi)$ and all $i \notin F$ [15], Chapter XI.1.14.
2. π is tempered (up to a central character twist) if for a minimal standard parabolic P_F for which $E(P_F, \pi) \neq \emptyset$ we have $(\nu, \beta_i) \leq 0$ for all $\nu \in {}^0E(P, \pi)$ and all i [15], Chapter XI.2.3.
3. For P_F with Levi group M_F and a unitary tempered representation of M_F let χ be a character of $M_F(F)$. Suppose $|\chi|$, considered as a character of the points $T(F)$ of the maximal split torus T of G, has exponent ν such that $(\nu, \alpha_i) > 0$ holds for all $i \notin F$. Then the induced representation $Ind_{P_F(F)}^{G(F)}(\chi \otimes \sigma)$ has a unique irreducible quotient, the Langlands quotient $J_{P_F}(\chi, \sigma)$ (up to a slight change in the notation, this is [15], Chapters XI.2.6 and XI.2.7).
4. Every irreducible admissible representation is equivalent to some (essentially) unique Langlands quotient [15], Chapters XI.2.10 and XI.2.11.

Example Gl(2). For convenience, since this is used later, consider the special and the trivial representations Sp and $1_{Gl(2)}$ of $Gl(2)$. Let $Sp(\chi)$ be the discrete series representation constituent (submodule) of $\nu^{1/2}\chi \times \nu^{-1/2}\chi = \nu \lhd \chi\nu^{-1/2}$. Then $Sp(\chi) \cong Sp \otimes (\chi \circ det)$. The diagonal matrices act on $r_{Gl(2)B_{Gl(2)}}(Sp)$ by

$$\delta_{B_{Gl(2)}}^{1/2} = \nu^{1/2} \boxtimes \nu^{-1/2}.$$

Recall $B_{Gl(2)}$ is the group of upper triangular matrices. Hence,

---

[1] For the groups considered $-w_{max} = id.$

$$E(B_{Gl(2)}, Sp) = \{w_{max}(\delta_{B_{Gl(2)}^{1/2}})\} = \{\delta_{B_{Gl(2)}^{-1/2}}\}$$

and

$$\{E(B_{Gl(2)}, 1)\} = \{w_{max}(\delta_{B_{Gl(2)}^{-1/2}})\} = \{\delta_{B_{Gl(2)}^{1/2}}\}.$$

So the exponents are $-\frac{1}{2}\alpha_1$ and $\frac{1}{2}\alpha_1$.

The Case *GSp(4)*. Now consider the group $G = GSp(4)$ and its principal series representations for the split torus $T \subseteq B$ of G. For an admissible representation V the irreducible constituents of $r_{BG}(V)$ are characters $\mu : T(F) \to \mathbb{C}^*$. Their absolute values $|\mu|(t), t \in Sp(4, F) \cap T(F)$, will be listed in root coordinates as a point in the Euclidean space $\mathbb{R}^2$ spanned by the roots such that for $t = diag(x, y, x^{-1}, y^{-1})$ the vector $(v, w) \in \mathbb{R}^2$ corresponds to the character

$$(v, w) \in \mathbb{R}^2 \longleftrightarrow |\mu| = \nu^v \boxtimes \nu^w \boxtimes \chi$$

(for simplicity of notation we ignore similitudes, hence χ may be arbitrary). The vector $(v, w) \in \mathbb{R}^2$ will be called the *exponent* of the character μ. The multiplicity of an exponent will be the dimension of the corresponding generalized eigenspace.

The positive simple roots α_1 and α_2 of B correspond to the characters

$$\mu(t) = |\mu(t)| = (\nu \boxtimes \nu^{-1} \boxtimes 1)(t) = |x|^{-1}|y|,$$

$$\mu(t) = |\mu(t)| = (1 \boxtimes \nu^2 \boxtimes \nu^{-1})(t) = |x|^2,$$

with the coordinates $\alpha_1 = (1, -1)$ and $\alpha_2 = (0, 2)$. Hence, $\beta_1 = (1, 1)$ and $\beta_2 = (2, 0)$ up to some positive scaling factor, which we can ignore. An exponent (v, w) is tempered if $\nu = w_{max}(v, w) = (-v, -w)$ satisfies $(\nu, \beta_1) = -v - w \leq 0$ and $(\nu, \beta_2) = -2v \leq 0$. In other words if $v \geq 0$ and $v + w \geq 0$ holds, or equivalently if (v, w) is in the closed cone spanned by α_1 and α_2. An exponent is called discrete if it is in the open cone

$$\mathbb{R}_{>0} \cdot \alpha_1 + \mathbb{R}_{>0} \cdot \alpha_2.$$

α_1 is the short root and α_2 is the long root of $Sp(4)$. The element w_{max} of maximal length is in the center of W_G. Let $s_{\alpha_1}, s_{\alpha_2}$ denote the simple reflections attached to α_1 and α_2. Then s_{α_i} generates $W(M)$ for the Levi component M of the parabolic groups P (for $i = 1$) and Q (for $i = 2$).

Shuffle Formulas. Define maps from $R_{\mathbb{Z}}(B)$ to $R_{\mathbb{Z}}(B)$ by

$$s_{\alpha_1}(\chi_1 \boxtimes \chi_2 \boxtimes \chi) = \chi_2 \boxtimes \chi_1 \boxtimes \chi,$$

$$s_{\alpha_2}(\chi_1 \boxtimes \chi_2 \boxtimes \chi) = \chi_1 \boxtimes \chi_2^{-1} \boxtimes \chi_2\chi.$$

We will use the formulas ([11], Lemma 5.4) to compute the Jacquet modules of induced representations in cases (a)–(c). For the group $GSp(4)$ these formulas are:

$$r_{PG} \circ i_{GP}(V) = V + w_{max}s_{\alpha_1}(V) + i_{PB} \circ s_{\alpha_2} \circ r_{BP}(V). \tag{a}$$

Notice $w_{max}s_{\alpha_1}(\sigma_1 \boxtimes \chi) = \sigma_1^* \boxtimes \omega_{\sigma_1}\chi$ and $W_P \backslash W_G/W_P = \{id, w_{max}s_{\alpha_1}, s_{\alpha_2}\}$ (representatives of minimal length), where in the following W_P and W_Q stand for the Weyl group $W(L)$ of the standard Levi subgroup L of P and Q, respectively,.

$$r_{QG} \circ i_{GP}(V) = i_{QB} \circ r_{BP}(V) + i_{QB} \circ s_{\alpha_1}s_{\alpha_2} \circ r_{BP}(V) \qquad \text{(b)}$$

$$r_{QG} \circ i_{GQ}(V) = V + w_{max}s_{\alpha_2}(V) + i_{QB} \circ s_{\alpha_1} \circ r_{BQ}(V). \qquad \text{(c)}$$

Notice $w_{max}s_{\alpha_2}(\chi \boxtimes \rho) = \chi^{-1} \boxtimes \rho\chi$.

$$r_{PG} \circ i_{GQ}(V) = i_{PB} \circ r_{BQ}(V) + i_{PB} \circ s_{\alpha_2}s_{\alpha_1} \circ r_{BQ}(V). \qquad \text{(d)}$$

It is easy to see that the substitutions listed are minimal representatives of $W_Q \backslash W_G/W_P$ $W_Q \backslash W_G/W_Q$, or $W_P \backslash W_G/W_Q$.

Nonregular Case (a) (Discussion of $\pi = \nu^{1/2}\chi_0 \times \nu^{-1/2}\chi_0 \triangleleft \chi$ Where $\chi_0^2 = 1$)

Representations of type (a) have two irreducible constituents Π and J, where Π is

$$\Pi = Sp(\chi_0) \triangleleft \chi.$$

Π is tempered (up to the χ-twist) but not elliptic, whereas J is not tempered.

Proof. By a twist we may assume χ to be unitary. Partial induction with respect to P yields two composition factors of π: the tempered $\Pi = Sp(\chi_0) \triangleleft \chi$ and $J = \chi_0 \circ det \triangleleft \chi$. $\square$

The Representation Π. By formula (a) the P-Jacquet module of Π has the following three constituents:

$$r_{PG}(\Pi) = 2 \cdot Sp(\chi_0) \boxtimes \chi + (\nu^{1/2}\chi_0 \times \nu^{1/2}\chi_0) \boxtimes \chi\chi_0\nu^{-1/2},$$

two special ones with real exponents $\frac{1}{2}\alpha_1$ and one with the real exponent $\frac{1}{2}(\alpha_1 + \alpha_2)$ (by applying r_{BP}). The latter occurs with multiplicity 2 in the $Gl(2, F)$ representation $(\nu^{1/2}\chi_0 \times \nu^{1/2}\chi_0)$. The tempered unitary representation Π is completely reducible. It cannot contain a discrete series representation δ ([44], Sect. 1, Lemma 4) since otherwise δ could be embedded into two different representations both induced from unitary discrete series, either induced from (G, δ) or induced from $(P, Sp(\chi_0))$. This is impossible! Therefore, every constituent of Π must contain one of the two "nondiscrete" tempered exponents $\frac{1}{2}\alpha_1$. Thus, there are at most two irreducible constituents, one of them must contain $(\nu^{1/2}\chi_0 \times \nu^{1/2}\chi_0) \boxtimes \chi\chi_0\nu^{-1/2}$, and hence contains both the exponents $\frac{1}{2}(\alpha_1 + \alpha_2)$ (since these occur with multiplicity 2 in $(\nu^{1/2}\chi_0 \times \nu^{1/2}\chi_0) \boxtimes \chi\chi_0\nu^{-1/2}$). Formula (b) with respect to the other maximal parabolic group Q implies that

$$r_{QG}(\Pi) = \nu^{1/2}\chi_0 \boxtimes (\nu^{-1/2}\chi_0 \lhd \chi) + \nu^{1/2}\chi_0 \boxtimes (\nu^{1/2}\chi_0 \lhd \chi\chi_0\nu^{-1/2})$$

has two irreducible constituents. After applying r_{QB}, each of them has the exponents $\frac{1}{2}\alpha_1, \frac{1}{2}(\alpha_1 + \alpha_2)$ with multiplicities 1. Using both results together, we conclude that any constituent of Π with an exponent $\frac{1}{2}(\alpha_1 + \alpha_2)$ must contain this exponent with multiplicity 2, and then by the last observation it must contain all four exponents together. Hence Π is irreducible with exponents $\frac{1}{2}\alpha_1, \frac{1}{2}(\alpha_1 + \alpha_2)$, each with multiplicity 2.

Now $J = \chi_0 \circ \boldsymbol{det} \lhd \chi$. The restriction of J to the symplectic group $Sp(4)$ gives the representation $I(s, \chi_0)$ of Kudla and Rallis [55] at the special point $s = 0$. According to [55], Proposition 5.1, this is an irreducible representation of $Sp(4, F)$. Hence, J is irreducible. According to [55], p. 235, its exponents are $-1/2\alpha_1 = (-1/2, 1/2)$ and $-1/2(\alpha_1 + \alpha_2) = (-1/2, -1/2)$, both with multiplicity 2 (be aware that Kudla and Rallis [55] used a different choice of the minimal parabolic).

Notice that $\chi_0\nu^{1/2} \times \chi_0\nu^{-1/2} \lhd \chi$ has the same irreducible constituents as $\chi_0\nu^{1/2} \times \chi_0\nu^{1/2} \lhd \chi\chi_0\nu^{-1/2}$. We already know that these are the constituents of $J \oplus \Pi$. Since Π is tempered but J is not, J must necessarily be (up to a character twist by $\chi\chi_0\nu^{-1/2}$) the Langlands quotient $J_P(\nu^{1/2}, (\chi_0 \times \chi_0) \boxtimes 1)$ attached to the tempered representation $(\chi_0 \times \chi_0) \boxtimes 1$ of the Levi component of P.

Nonregular Case (b) (Discussion of $\pi = 1 \times \nu \lhd \chi$)

This case is the most interesting case among the irregular cases. By a character twist – without restriction of generality – we may assume $\chi\nu^{\frac{1}{2}}$ to be unitary. In fact, one could make it trivial and invoke Rodier's result [75], Sects. 6 and 6.2. We will not do this. Since we need more information, we give an alternative proof.

Claim 4.2. There are four irreducible nonisomorphic constituents: two irreducible nontempered representations J and J' and two tempered nondiscrete representations T_+ and T_-, where T_+ and T_- are relatives in the tempered induced representation

$$T = T_+ \oplus T_- = 1 \times Sp(\chi\nu^{\frac{1}{2}}).$$

Definition 4.4. The two representations T_+, T_- will be denoted

$$\boxed{Wp_+(\rho) , \; Wp_-(\rho), \qquad \rho = Sp(\chi\nu^{\frac{1}{2}})}.$$

Remember that $Sp(\chi\nu^{\frac{1}{2}})$ is the special representation twisted by $\chi\nu^{\frac{1}{2}}$.

Remark 4.13. Notice, with this additional definition, the representations $Wp_\pm(\rho)$ are now defined for all irreducible discrete series representations of $Gl(2, F)$.

Proof. Starting from

$$0 \to Sp \to \nu^{1/2} \times \nu^{-1/2} \to 1 \circ det \to 0,$$

a twist by $\chi\nu^{\frac{1}{2}}$ and induction from Q gives an exact sequence for $\pi = 1 \times \nu \triangleleft \chi$

$$0 \to 1 \times Sp(\chi\nu^{\frac{1}{2}}) \to \pi \to 1 \times (\chi\nu^{\frac{1}{2}} \circ det) \to 0.$$

This gives two constituents of π: the tempered subrepresentation

$$T = 1 \times Sp(\chi\nu^{\frac{1}{2}})$$

and the quotient $1 \times (\chi\nu^{\frac{1}{2}} \circ det)$. Formulas (b) and (c) on page 127 imply

$$r_{QG}(T) = 1 \boxtimes Sp(\chi\nu^{1/2}) + 1 \boxtimes Sp(\chi\nu^{1/2}) + \nu \boxtimes (1 \triangleleft \chi),$$

$$r_{PG}(T) = 2 \cdot (Sp(\nu^{1/2} \boxtimes \chi)) + 2 \cdot (\nu^{1/2} \circ det \boxtimes \chi)$$

with the characters $2 \cdot (1 \boxtimes \nu \boxtimes \chi) + 2 \cdot (\nu \boxtimes 1 \boxtimes \chi)$ in $r_{BG}(T)$. So the exponents of the tempered constituent T are $\frac{1}{2}\alpha_2$ and $\alpha_1 + \frac{1}{2}\alpha_2$, each with multiplicity 2. The exponents of $1 \times (\chi\nu^{\frac{1}{2}} \circ det)$ are their negatives; hence, all are nontempered. $\square$

Hence:

1. Every irreducible constituent of $1 \times \nu \triangleleft \chi$ which has one of the exponents $\frac{1}{2}\alpha_2$ or $\alpha_1 + \frac{1}{2}\alpha_2$ is contained in T.
2. T cannot contain a discrete series representation since it is induced from a unitary discrete series representation of the Levi group of Q [44], Sect. 1, Lemma 4b. Therefore, every constituent of T has to contain one of the tempered but nondiscrete exponents $\frac{1}{2}\alpha_1$.
3. The representations $\pi = 1 \times \nu \triangleleft \chi$ and $\nu \times 1 \triangleleft \chi$ have the same irreducible constituents. For the latter representation one has an exact sequence using induction from P

$$0 \to L \to \nu \times 1 \triangleleft \chi \to R \to 0,$$

where $R = (\nu^{\frac{1}{2}} \circ det) \triangleleft \chi$ and $L = Sp(\nu^{\frac{1}{2}}) \triangleleft \chi$. R has a unique irreducible (nontempered) quotient, the Langlands quotient $J = J_Q(\nu, 1 \boxtimes 1 \triangleleft \chi\nu^{\frac{1}{2}}) \otimes \nu^{-\frac{1}{2}}$ of $\nu \times 1 \triangleleft \chi$. Let T_- denote the kernel of the projection from R to J

$$0 \to T_- \to R \to J \to 0.$$

The module $L = Sp(\nu^{\frac{1}{2}}) \triangleleft \chi$ also has a unique irreducible nontempered quotient, the Langlands quotient $J' = J_P(\nu^{\frac{1}{2}}, Sp \boxtimes \chi)$. Let T_+ be the kernel

$$0 \to T_+ \to L \to J' \to 0.$$

Remark 4.14. Later we also use the notation $T_+ = \theta_+(Sp)$, $T_- = \theta_-(1)$, and $J = \theta_+(1)$.

Exponents of L. Formula (b) on page 127 gives three irreducible constituents for $r_{QG}(L)$,

$$r_{QG}(L) = \nu \boxtimes (1 \lhd \chi) + 1 \boxtimes Sp(\nu^{1/2}\chi) + 1 \boxtimes (\nu^{1/2}\chi \circ det),$$

and after applying r_{QB}

$$(\nu \boxtimes 1 \boxtimes \chi + \nu \boxtimes 1 \boxtimes \chi) + 1 \boxtimes \nu \boxtimes \chi + 1 \boxtimes \nu^{-1} \boxtimes \nu\chi.$$

The exponents of the three summands are $\alpha_1 + \frac{1}{2}\alpha_2$, $\frac{1}{2}\alpha_2$, and $-\frac{1}{2}\alpha_2$, respectively. The last two have multiplicity 1, the first one has multiplicity 2. Thus, the constituent T' of L with exponent $\alpha_1 + \frac{1}{2}\alpha_2$ must contain this exponent with multiplicity 2, because these belong to the irreducible representation $\nu \boxtimes (1 \lhd \chi)$ of the Levi group of Q. If we compare this with the discussion of T, we see that T' is contained in T, and therefore cannot have exponent $-\frac{1}{2}\alpha_2$. It cannot be a discrete series constituent either. Hence, the tempered constituent T' contains in addition the exponent $\frac{1}{2}\alpha_2$.

As a consequence L has two irreducible constituents. One of them is T_+ with the real exponents $2(\alpha_1 + \frac{1}{2}\alpha_2)$ and $\frac{1}{2}\alpha_2$. Hence, T_+ is tempered. The other one is the Langlands quotient J' with the real nontempered exponent $-\frac{1}{2}\alpha_2$.

Finally,

$$r_{PG}(L) = 2 \cdot Sp(\nu^{1/2}) \boxtimes \chi + Sp(\nu^{-1/2}) \boxtimes \nu\chi + (\nu^{1/2} \circ det) \boxtimes \chi$$

has three constituents. Applying r_{PB} gives

$$\nu \boxtimes 1 \boxtimes \chi + 1 \boxtimes \nu^{-1} \boxtimes \nu\chi + (\nu \boxtimes 1 \boxtimes \chi + 1 \boxtimes \nu \boxtimes \chi)$$

with exponents $\alpha_1 + \frac{1}{2}\alpha_2$, $-\frac{1}{2}\alpha_2$, and $\frac{1}{2}\alpha_2, \alpha_1 + \frac{1}{2}\alpha_2$.

4. Since we already know that the tempered subrepresentation $T \subseteq \pi$ contains all irreducible tempered constituents of π, it must contain the tempered constituent T_+. By a comparison of exponents of R and T_+, we see that the tempered representation T/T_+ has exponent $\frac{1}{2}\alpha_2$ with multiplicity 1. Hence, T/T_+ must be irreducible. Since J and J' are irreducible and nontempered, the tempered representation T/T_+ must therefore be isomorphic to $T_- \neq 0$. Hence, J has the remaining exponents $-\alpha_1 - \frac{1}{2}\alpha_2$ with multiplicity 2 and $-\frac{1}{2}\alpha_2$.

5. For the module $R = (\nu^{\frac{1}{2}} \circ det) \lhd \chi$ one has an alternative description by the following.

4.7.5 Results Obtained by Kudla and Rallis

Let us temporarily work with the group $Sp(4, F)$ instead of $GSp(4, F)$. This is enough, since the exponents in the irregular case (b) can already be distinguished from each other on that level.

The $Sp(4, F)$ analogue of the module R can be identified with the module

$$R = R_2(2, 2) = R_2(V_1)$$

in the notation in [55]. V_1 is the space underlying a split rank 4 quadratic form. By [55], pp. 211–212, Theorem 2(v), the module R is isomorphic to the induced representation $I_2(1/2, 1)$. The intertwining operator $M_2^*(1/2, 1) : I_2(1/2, 1) \to I_2(-1/2, 1)$ (see [55], p. 212) maps the maximal quotient J of R to the irreducible maximal submodule $R_2(1, 1)$ of $I_2(-1/2, 1)$. See [55] p. 211, Theorem 2(i). It therefore has a composition series of length 2 with unique irreducible quotient $R_2(1, 1)$

$$J = R_2(1, 1)$$

and a unique irreducible subrepresentation $R_2(4)$; therefore,

$$T_- = R_2(4) = R_2(V_2).$$

V_2 is the quadratic space of rank 4 for the quaternion norm form.

For the convenience of the reader we give some more details. One has to consider the theorem in [55] for the trivial character χ_V and $n = 2$, $s_0 = \pm 1/2$ in the cases:

(i) $m = 2$, $O(1, 1)$

and

(v) $m = 4$, where $V_1 = V(2, 2)$ is split and $V_2 = V(4)$ is quaternionic anisotropic, both having trivial character χ_V. Finally, choose $V_{1,0} = V(1, 1)$ to be split of dimension 2, with trivial character χ_V. Then $R_2(4)$ is irreducible ([55], Proposition 6.10), contained in $R_2(2, 2) = I(1/2, 1) = Ind_P^G(\nu^{1/2} \circ det \boxtimes \chi) = R$ [55], Proposition 3.4(iii) and also p. 254, bottom. The quotient is irreducible and isomorphic to $R_2(1, 1)$. For this see pp. 211, case (i) and 255 in [55]. Hence,

$$0 \to R_2(4) \to R_2(2, 2) \to R_2(1, 1) \to 0.$$

From [55], Propositions 4.5 and 4.6, we get the real exponents of the two constituents: $(0, 1)$ for $R_2(4)$ and $(-1, 0), (-1, 0), (0, -1)$ for $J = R_2(1, 1)$. This is the notation in [55] and corresponds to the real exponents after the identifications $\alpha_1 = (1, -1)$ and $\alpha_2 = (0, 2)$. Hence, the real exponents are $\frac{1}{2}\alpha_2$ and $-\alpha_1 - \frac{1}{2}\alpha_2$, $-\alpha_1 - \frac{1}{2}\alpha_2$, $-1/2\alpha_2$ in accordance with what we have already obtained. This completes our sketchy argument concerning the group $Sp(4, F)$.

6. We return to consider representations of the group $GSp(4, F)$ instead of $Sp(4, F)$. Let us come back to our starting point and summarize. We have shown

$$T = T_- \oplus T_+.$$

Secondly, we proved

$$r_{QG}(T_+) = \nu \boxtimes (1 \lhd \chi) + 1 \boxtimes Sp(\chi \nu^{1/2})$$

and

$$r_{QG}(T_-) = 1 \boxtimes Sp(\chi \nu^{1/2}).$$

Hence, the difference is

$$\boxed{r_{QG}(T_+ - T_-) = \nu \boxtimes (1 \triangleleft \chi)},$$

which gives $r_{BG}(T_+ - T_-) = 2 \cdot (\nu \boxtimes 1 \boxtimes \chi)$ by applying r_{QB}. Comparing with the formula obtained for $r_{PG}(T)$, this gives

$$\boxed{r_{PG}(T_+ - T_-) = 2(Sp(\nu^{1/2}) \boxtimes \chi)}.$$

Remark 4.15. Later it will be necessary to know the exponents of the *discrete* series representations, which are contained in the regular principal series representations and which were determined by Tadic. In fact, they are contained in $\chi_0 \times \chi_0 \nu \triangleleft \chi$. The arguments used for case (b) can also be repeated for the representations $\chi_0 \times \chi_0 \nu \triangleleft \chi$, with $\chi_0^2 = 1, \chi_0 \neq 1$ (regular case). Especially analogues of the modules L, T_+, and J can be defined. The discrete series representations $\delta(\nu \chi_0 \times \chi_0 \triangleleft \chi \nu^{-1/2})$ are then contained in the correspondingly defined module T_+ with exactly one exponent $\alpha_1 + \frac{1}{2}\alpha_2$ (it has multiplicity 2). It is then clear from the discussion above that

$$r_{QG}(\delta(\nu \chi_0 \times \chi_0 \triangleleft \chi \nu^{-1/2})) = \nu \chi_0 \boxtimes (\chi_0 \triangleleft \chi \nu^{-1/2}),$$

$$r_{PG}(\delta(\nu \chi_0 \times \chi_0 \triangleleft \chi \nu^{-1/2})) = Sp(\nu^{1/2}\chi_0) \boxtimes \chi \nu^{-1/2} + Sp(\nu^{1/2}\chi_0) \boxtimes \chi \chi_0 \nu^{-1/2},$$

$$r_{BG}(\delta(\nu \chi_0 \times \chi_0 \triangleleft \chi \nu^{-1/2})) = \nu \chi_0 \boxtimes \chi_0 \boxtimes \chi_0 \chi \nu^{-1/2} + \nu \chi_0 \boxtimes \chi_0 \boxtimes \chi \nu^{-1/2}.$$

One can simply repeat the arguments used above for the analysis of the module L. We skip the details.

Nonregular Case (c) (the Representation $\pi = \nu \times \nu \triangleleft \chi$)

In this case π has two irreducible nonisomorphic nontempered constituents denoted J, J'. Furthermore, the infinitesimal character $(B, \nu \times \nu \triangleleft \chi)$ satisfies $AD_{(B,\nu \times \nu \triangleleft \chi)} = 0$.

Proof. π has two irreducible constituents (see [75], Sect. 6.3, in the case $\chi = 1$). By a character twist, the general case $\chi \neq 1$ follows as well. So it remains for us to show the remaining assertions:

1. The representation $\nu \times \nu \triangleleft \chi$ has a unique irreducible Langlands quotient

$$J = J_P(\nu, (1 \times 1) \boxtimes \chi).$$

Thus,

$$\pi = \nu \times \nu \lhd \chi \to J \to 0$$

for the nontempered Langlands quotient J.

2. There exists an exact sequence

$$0 \to L \to \pi \to R \to 0,$$

where the module $L = \nu \times Sp(\chi\nu^{\frac{1}{2}})$ admits a unique irreducible Langlands quotient $J' = J_Q(\nu, 1 \boxtimes Sp) \otimes \chi\nu^{\frac{1}{2}}$

$$L \to J' \to 0,$$

which is nontempered and not isomorphic to J by the Langlands classification theory. Of course the quotient R of $\pi = \nu \times \nu \lhd \chi$ admits a nontrivial map

$$R \to J \to 0.$$

So there are at least two irreducible nonisomorphic nontempered constituents of π. Since π has only two irreducible constituents, we are done. □

4.8 Summary

All the reducible induced representations of $GSp(4, F)$ were listed in the last section. The Langlands quotients were described in those cases where the induced representation was reducible and "ramified" (with respect to the action of the Weyl group).

Recall an irreducible tempered representation is elliptic if its character is nontrivial on the regular elliptic locus of the group G. An irreducible tempered elliptic admissible representation is called a limit of a discrete series if it is not in the discrete series.

Lemma 4.19. *The limits of discrete series of $GSp(4, F)$ are the irreducible admissible representations $Wp_+(\rho)$ and $Wp_-(\rho)$*

$$1 \times \rho = Wp_+(\rho) \oplus Wp_-(\rho),$$

where ρ is an irreducible discrete series representation of the group $Gl(2, F)$.

Proof. Suppose π is a limit of a discrete series, but is not in the discrete series itself. Then π is tempered and is a constituent of a proper induced representation. Induction is from a unitary discrete series representation of a Levi subgroup of a proper parabolic. If π is fully induced, it cannot be elliptic. Hence, we have to look for reducible representations of this type. The unitary principal series are always irreducible. Therefore, it is enough to consider representations of G induced from one of the two maximal parabolic subgroups. If induction is from a cuspidal representation ρ, then the only reducible case occurs for induction from Q, where the possible candidates for elliptic families are the representations $Wp_+(\rho), Wp_-(\rho)$ found by Waldspurger [106]. It remains for us to see that they actually are elliptic. According to [44], Proposition 1, it is enough to show that $W_\pm(\rho)$ cannot be linearly expressed in the Grothendieck group by properly induced representations. See also [21], p. 116. The only induced representations containing $W_\pm(\rho)$ are of the form $W_+(\rho) + W_-(\rho)$ (in the Grothendieck group). Hence, $W_\pm(\rho)$ are elliptic for cuspidal ρ.

To complete our discussion it remains for us to consider the cases where induction is from a unitary special representation of the Levi component of a maximal parabolic subgroup. The special representation can be embedded into an induced representation. By induction in steps this leaves us to consider the corresponding principal series representations

$$\chi_u \times \nu \lhd \chi$$

(in the case of the parabolic Q) and

$$\chi_u \nu^{1/2} \times \chi_u \nu^{-1/2} \lhd \chi$$

(in the case of the parabolic P) where the unitary condition for the special representations implies that χ_u is unitary. So let us look at the composition factors of such principal series representations. If there are only two irreducible constituents, then these constituents themselves are fully induced from a maximal parabolic; hence, they cannot be elliptic. In fact they are the tempered representations $\chi_u \times Sp(\chi \nu^{1/2})$ in the first case (induction from Q),and $Sp(\chi_u) \lhd \chi$ in the second case (induction from P). So it remains for us to consider constituents of nonunitary principal series of the type above with more than two irreducible constituents. The only cases to consider are the nonregular cases (a)–(c). The reason is that for the basic reducible regular cases $\chi_1 \times \nu \lhd \chi$ and $\nu^{1/2'} \chi \times \nu^{-1/2} \chi' \lhd \chi$ the only two possible cases where there are more than two constituents are $\nu^2 \times \nu \lhd \chi$ and $\nu \chi_0 \times \chi_0 \lhd \chi, \chi_0^2 = 1, \chi_0 \neq 1$ (up to changes from the action of the Weyl group). See page 123. Hence, to have a common constituent with $\chi_u \times \nu \lhd \chi$ or $\chi_u \nu^{1/2} \times \chi_u \nu^{-1/2} \lhd \chi$, the character χ_u cannot be nonunitary. However, case (c) did not have tempered constituents. In case (a) we found only one irreducible tempered constituent Π, which itself was induced $\Pi = Sp(\chi_0) \lhd \chi$.

However, in case (b) we found that the tempered representation $1 \times Sp(\chi \nu^{\frac{1}{2}})$ splits into two irreducible tempered constituents T_+ and T_-. They are relatives in the sense of Kazhdan and, in fact, it only remains for us to show that they are elliptic. By Proposition 1 in [44], it is enough to show that $T_\pm$ cannot be expressed as a linear combination of induced representations in the Grothendieck group. But this follows from our discussion of the nonregular case 1(b), in particular our claim (4.2) and its proof. What can be combined by induced representations must have the form $\alpha(T_- + J) + \beta(T_+ + J') + \gamma(T_- + T_+) + \delta(J + J')$ for $\alpha, \beta, \gamma, \delta \in \mathbb{Z}$. This shows that T_+ and T_- are tempered elliptic. They define the representations $W_\pm(\rho)$ for discrete series representations ρ, which are not cuspidal. This completes the proof of Lemma 4.19. □

A similar discussion determines all infinitesimal characters for which $AD_\chi \neq 0$. For an infinitesimal character $\chi = (L, \rho)$ one has $AD_\chi = 0$ unless ρ is "ramified" with respect to the Weyl group. So only irregular principal series and the representations induced from cuspidal representations of the Levi components of the two maximal parabolic subgroups P and Q are relevant. Our discussions of exponents in cases (a)–(c) and the results mentioned by Shahidi and Waldspurger show

Lemma 4.20. *Asymptotic independence only fails for infinitesimal characters*

$$\chi = (L_Q, 1 \boxtimes \rho),$$

where ρ is a cuspidal representation of $Gl(2, F)$. In this case χ corresponds to the two relatives $Wp_+(\rho), Wp_-(\rho)$ with $AD_\chi = \mathbb{Z}(Wp_+(\rho) - Wp_-(\rho))$.

Proof. The case where ρ is special does not occur. In that case $T_+ = Wp_+(\rho)$ and $T_- = Wp_-(\rho)$ have different exponents as shown in the discussion of case (b). □

4.8.1 List of Irreducible Unitary Discrete Series Representations of G

1. The unitary irreducible cuspidal representations of G.

The second types arise as constituents of representations induced from the maximal parabolic P (Siegel parabolic).

2. The series $\delta(\rho_1 \nu^{1/2} \lhd \chi \nu^{-1/2})$ for unitary χ and for irreducible unitary discrete series representations $\rho_1 \not\cong Sp$ of $Gl(2, F)$, whose central character is trivial $\omega_{\rho_1} = 1$. The central character of $\delta(\rho_1 \nu^{1/2} \lhd \chi \nu^{-1/2})$ is χ^2.

Notice that either ρ_1 is cuspidal or ρ_1 is a special representation with trivial central character. In the second case $\rho_1 = Sp(\chi_0)$ and $\chi_0^2 = 1$, but $\chi_0 \neq 1$. This second type of discrete series representation, contained in the regular principal series representations, was found by Tadic: Recall $\delta(Sp(\chi_0)\nu^{1/2} \lhd \chi \nu^{-1/2}) = \delta(\nu \chi_0 \times \chi_0 \lhd \chi \nu^{-1/2})$ for $\chi_0 \neq 1, \chi_0^2 = 1$ and unitary χ is one of the four irreducible constituents of the regular principal series $\nu \chi_0 \times \chi_0 \lhd \chi \nu^{-1/2}$. See the beginning of Section 4.7.4 and our Remark 4.15 on page 132.

$$0 \to \delta(Sp(\chi_0)\nu^{1/2} \lhd \chi \nu^{-1/2}) \to Sp(\chi_0)\nu^{1/2} \lhd \chi \nu^{-1/2}$$
$$\to J_P(Sp(\chi_0)\nu^{1/2} \lhd \chi \nu^{-1/2}) \to 0.$$

Finally, notice that the "missing" discrete series, corresponding to the forbidden choice $\rho_1 = Sp$ and the forbidden choice $\chi_0 = 1$, comes from the nonregular $1 \times \nu \lhd \chi \nu^{1/2}$ (case (b)). In this case one only has the "virtual" discrete series counterpart $T_+ - T_-$ instead of ρ_1, which, however, is only defined as an element of the Grothendieck group $R_{\mathbb{Z}}(G)$.

The third type of discrete series representation comes from constituents of representations induced from the maximal parabolic Q.

3. The series $\delta(\nu\chi_0 \times \sigma\nu^{-1/2})$ for unitary (automatically irreducible cuspidal) discrete series representations σ of $Gl(2)$ with CM by χ_0, i.e., $\chi_0 \neq 1$, $\chi_0^2 = 1$ and $\sigma \otimes \chi_0 \cong \sigma$. Notice that the central character of $\delta(\nu\chi_0 \times \sigma\nu^{-1/2})$ is $\chi_0\omega_\sigma$.

Notice that special representations are never CM. Finally there are:

4. The Steinberg representations $\delta(\nu^2 \times \nu \lhd \chi\nu^{-3/2})$ for unitary χ. The central character of the Steinberg representation is χ^2.

4.9 Asymptotics

The Parabolic Subgroups of M. For the group $M(F) = Gl(2, F) \times Gl(2, F)/$ $\{(t, t^{-1}) \mid t \in F^*\}$ we fix the parabolic subgroups

$$B_M(F) = \{\begin{pmatrix} * & * \\ 0 & * \end{pmatrix} \times \begin{pmatrix} * & 0 \\ * & * \end{pmatrix} \ mod \ F^*\},$$

and

$$P_M(F) = \{\begin{pmatrix} * & * \\ * & * \end{pmatrix} \times \begin{pmatrix} * & 0 \\ * & * \end{pmatrix} \ mod \ F^*\}.$$

With respect to the action of the automorphism group of M these are representatives of the proper parabolic subgroups of M. The simple roots defined by the Borel subgroup B_M are the characters of the maximal torus of diagonal matrices which map

$$(diag(t_1, t_1'), diag(t_2, t_2')) \ mod \ F^*$$

to t_1/t_1' and t_2'/t_2.

Parabolic Subgroups of $G = GSp(4)$. We fix the Borel group $B \subseteq G$

$$B(F) = \{\begin{pmatrix} * & 0 & * & * \\ * & * & * & * \\ 0 & 0 & * & * \\ 0 & 0 & 0 & * \end{pmatrix} \in G(F)\},$$

the Siegel parabolic P, and the Klingen parabolic subgroup Q, containing B. The simple positive roots are the characters which map the diagonal matrices

$$diag(x_1, x_2, x_1', x_2') \in B(F)$$

to x_2/x_1 (short root) and to x_1/x_1' (long root).

Levi Components. We identify the Levi components of the parabolic groups $P_M \subseteq M$ and $P \subseteq G$ via the following map: $\psi : Levi(P_M) \to Levi(P)$

$$\psi\left((x, diag(t_2, t_2')) \ mod \ F^*\right) = \begin{pmatrix} t_2'det(x)x^{-t} & 0 \\ 0 & t_2x \end{pmatrix}.$$

This induces an isomorphism between the maximal split tori $Levi(B_M) \rightarrow Levi(B)$

$$(diag(t_1, t_1'), diag(t_2, t_2')) \bmod F^* \mapsto diag(t_2't_1', t_2't_1, t_2t_1, t_2t_1'),$$

whose inverse is

$$diag(x_1, x_2, x_1', x_2') \mapsto (diag(x_2, x_1), diag(x_1'/x_2, 1)) \bmod F^*.$$

In the list of maximal F-tori of G (see page 94), those which come from M via an admissible embedding and which are also contained in a proper parabolic subgroup of G, are the tori of type (6) or (7). These cases were:

6. $T(F) \cong L^* \times F^* \subseteq P(F)$, such that $\epsilon(T) = 2$, $W_F(T, G) = W_F(T, M)$, and $\mathcal{D}_G(T) = 1$.
7. $T(F) \cong (F^*)^3 \subseteq B(F)$, such that $\epsilon(T) = 1$, $W_F(T, M) = W_M$ is a subgroup of index 2 in $W_F(T, M)$.

Among the maximal tori in $Levi(Q)$ only the split tori come from M, but they are already contained in B. For tori T contained in parabolic subgroups of G, which come from M (cases 6 and 7) the formula for the character lift α^G for a $*$-invariant distribution α on $M(F)$ simplifies to

$$D_G(t)\alpha^G(t) = 2 \cdot D_M(t)\alpha(t), \qquad t \in T(F),$$

where $D_G(t) = \prod_\alpha |\alpha(t) - 1|^{1/2}$ for $t \in T(F)$ semisimple regular is the product over all roots α of T.

4.9.1 The Character Lift

For $\sigma_v = (\sigma_{1,v}, \sigma_{2,v})$, where $\sigma_{i,v}$ are irreducible admissible representations of $Gl(2, F)$ with the same central character, we now construct distributions $f = \chi_{\sigma_v}^G - D$, where D is a finite sum of characters of irreducible admissible representations of $GSp(4, F)$ so that the Jacquet coefficients $f(\mathcal{P})$ vanish for the parabolic subgroups $\mathcal{P}$ of G. For cuspidal representations we set $D = 0$ (using Corollary 4.8). For induced representations σ_v of $M(F)$ the character lift is completely understood. So it remains for us to consider the following cases:

$$\sigma_v = Sp(\chi) \otimes \rho, \quad \omega_\rho = \chi^2.$$

Since special representations are character twists $Sp(\chi) = Sp \otimes \chi$ of the Steinberg representation Sp, the central character is a square. The endoscopic character lift commutes with character twists (Lemma 4.8). Hence, we can assume $\chi = 1$ or $\omega_\rho = 1$. We distinguish the cases $\sigma_v^* \not\cong \sigma_v$ and $\sigma_v^* \cong \sigma_v$. Let α be the character of $\sigma_v \oplus \sigma_v^*$ in the first case, and the character of σ_v in the second. So there are the cases:

1. $\sigma_v \oplus \sigma_v^* = (Sp \otimes \rho) \bigoplus (\rho \otimes Sp)$,

   where ρ is an irreducible representation of $Gl(2, F)$ with trivial central character, which is either ρ cuspidal or $Sp(\chi_0)$ for a quadratic character $\chi_0 \neq 1$, $\chi_0^2 = 1$.

2. $\sigma_v = Sp \otimes Sp$.

Let α denote the character of $\sigma_v \oplus \sigma_v^*$ or the character of σ_v in these cases.

The Distribution f. In the *first case* $\sigma_v = (Sp, \rho)$ put

$$f = \alpha^G - 2 \cdot \chi_{\delta(\rho\nu^{1/2} \triangleleft \nu^{-1/2})}.$$

Then $\omega_\rho = 1$ by assumption. For a general central character write $\sigma_v = Sp(\chi) \otimes \rho_1\chi$, where $\omega_{\rho_1} = 1$ and replace $\chi_{\delta(\rho\nu^{1/2} \triangleleft \nu^{-1/2})}$ by the character of the discrete series representation $\delta(\rho_1\nu^{\frac{1}{2}} \triangleleft \chi\nu^{-1/2})$ of $G(F)$.

In the *second case* $\sigma_v = (Sp, Sp)$ put

$$f = \alpha^G - \chi_{(T_+ - T_-)},$$

where $T_+ \oplus T_- = 1 \times Sp$ defines the tempered irreducible constituents of $1 \times \nu \triangleleft \nu^{-1/2}$. For general central characters, i.e., for $\sigma_v = (Sp(\chi), Sp(\chi))$, choose the correction term $D = (T_+ - T_-) \otimes \chi$. Notice $T_\pm \otimes \chi = Wp_\pm(Sp(\chi))$ are the two irreducible summands of $1 \times Sp(\chi) = 1 \times \nu \triangleleft \nu^{-1/2}\chi$ (Lemma 4.19).

Lemma 4.21. *For the distribution f defined above $f(P) = f(B) = f(Q) = 0$ holds for the Jacquet functor of the proper parabolic subgroups P, B, and Q.*

Remark 4.16. In the first case the lemma implies $(\chi^G_{\rho \otimes Sp} - \chi_{\delta(\rho\nu^{1/2} \triangleleft \nu^{-1/2})})(\mathcal{P}) = 0$ for all proper parabolic standard subgroups $\mathcal{P}$ of G (Lemma 4.6).

Proof of Lemma 4.21. It is enough to show $r_{\mathcal{P}G}(f) = 0$ for the proper parabolic subgroups $\mathcal{P}$ of G for the normalized Jacquet functor $r_{\mathcal{P}G} : R_{\mathbb{Z}}(G) \to R_{\mathbb{Z}}(Levi(\mathcal{P}))$.

Claim 4.3. If $ag \in \mathcal{P}$ is semisimple regular, so that $a \in A_{\mathcal{P}}$ is sufficiently $\mathcal{P}$-contractive and g is in a fixed compact set Ω, then

$$r_{\mathcal{P}G}(\alpha^G)(ag) = (D_G\alpha^G)(ag).$$

To prove the claim it is enough to show

$$\chi_{r_{\mathcal{P}G}(\pi)}(ag) = (\delta_{\mathcal{P}}^{-1/2}\chi_{\pi(\mathcal{P})})(ag) = (D_G\chi_\pi)(ag) = (\delta_{\mathcal{P}}^{-1/2}\chi_\pi)(ag).$$

For that let g be semisimple regular with $g \in T$. Then there exists an integer n such that $g^n = a \cdot s$, where $a \in A_\theta$ and $s \in S$, where $S \subseteq T$ is anisotropic and $T = A_\theta \cdot S$. Remember Δ was a fixed basis of simple roots, $\mathcal{P} = P_\theta$ for $\theta \subseteq \Delta$ a standard parabolic with split component A_θ. Then for every root α of T we get $|\alpha(g)| = |a \cdot s|^{1/n} = |a|^{1/n}$. Now suppose $a \in A_{\mathcal{P}}$ is $\mathcal{P}$-contractive and g varies in some fixed compact set $\Omega \subseteq \mathcal{P}$. Then we get for a sufficiently $\mathcal{P}$-contractive

$$D_G(ag) = D_{Levi(\mathcal{P})}(g)\delta_{\mathcal{P}}^{-1/2}(a).$$

Together with Lemma 4.18 this proves the claim.

By a similar formula for the group M the proof of Lemma 4.21 is therefore reduced to the calculation of the normalized Jacquet functors of certain representations. Notice that to keep track of asymptotic conditions, one has to consider elements $ag \in T$ as elements of the two different groups M and G. To keep track of this we deal with the two cases $\sigma_v^* \not\cong \sigma_v$ and $\sigma_v^* \cong \sigma_v$ separately.

Discussion of the First Case, ρ Cuspidal. For the *Siegel parabolic* $P \subseteq G$ we identified the Levi components of $P_M \subseteq M$ and $P \subseteq G$ via an isomorphism ψ. This identifies elements $t = (x, diag(t_2, t_2')) \mod F^*$ in $P_M(F)$ with elements of $P(F)$, also denoted t. With this notation one obtains $\delta_{P_M}^{\frac{1}{2}}(t) = |t_2'/t_2| = \delta_P(t)^{1/6}$ for $t = (x, diag(t_2, t_2')) \mod F^*$. Thus, $\delta(\rho\nu^{1/2} \triangleleft \nu^{-1/2}) = \delta(\rho\delta_P^{1/6} \triangleleft \nu^{-1/2})$.

The Siegel parabolic $P \subseteq G$ is the standard parabolic group attached to the short root α_1. For $t = ga$ with $g \in \Omega \cap P$ and $a \in A_P$ and $a = diag(t_2'c, t_2'c, t_2c, t_2c)$ with $|\alpha_2(a)| = |t_2'/t_2| < \epsilon(\Omega) \ll 1$ Lemma 4.18 implies

$$r_{PG}\left(\chi_{\delta(\rho\nu^{1/2} \triangleleft \nu^{-1/2})}\right)(t) = \chi_{\rho\nu^{\frac{1}{2}} \boxtimes \nu^{-1/2}}(ag), \qquad |\alpha_2(a)| \ll 1,$$

since $\delta(\rho\nu^{1/2} \triangleleft \nu^{-1/2})$ is one of two irreducible constituents of the induced representation $\rho\nu^{1/2} \triangleleft \nu^{-1/2}$. The normalized Jacquet module of the induced representation has the two irreducible constituents $\rho\nu^{1/2} \boxtimes \nu^{-1/2}$ and $\rho\nu^{-1/2} \boxtimes \nu^{1/2}$. The discrete series representation corresponds to the tempered exponent. Hence, $r_{PG}(\delta(\rho\nu^{1/2} \triangleleft \nu^{-1/2})) = \rho\nu^{1/2} \boxtimes \nu^{-1/2}$. So for sufficiently $\mathcal{P}$-contractive elements a, $g \in \Omega$ and $t = ag$

$$r_{PG}(\delta(\rho\nu^{1/2} \triangleleft \nu^{-1/2}))(t) = \chi_{\rho\nu^{1/2}}(t_2 det(x)x^{-t})\nu^{-1/2}(t_2t_2' det(x)) = \chi_\rho(x)|t_2'/t_2|^{1/2}$$

using $\omega_\rho = 1$. On the other hand, $r_{PG}(\alpha^G)(ag) = (D_G\alpha_G)(ag)$ is equal to

$$2 \cdot D_M(ag)\left(\chi_{\rho \otimes Sp} + \chi_{Sp \otimes \rho}\right)(ag) = 2 \cdot r_{P_M M}\left(\chi_{\rho \otimes Sp} + \chi_{Sp \otimes \rho}\right)(ag)$$

for $|\alpha_2(a)| = |t_2'/t_2| \ll 1$. This is equivalent to the condition $|\beta(\psi^{-1}(a))| = |t_2'/t_2| \ll 1$, where β is the positive root of P_M. Hence,

$$r_{PG}(\alpha^G)(ag) = 2 \cdot r_{P_M M}\left(\chi_{\rho \otimes Sp} + \chi_{Sp \otimes \rho}\right)(ag)$$

$$= 2 \cdot r_{P_M M}\left(\chi_{\rho \otimes Sp}\right)(ag)$$

$$= 2 \cdot \chi_{\rho \otimes \delta_{P_M}^{1/2}}(ag) = 2\chi_\rho(x)|t_2'/t_2|^{1/2}$$

for $|\beta(a)| \ll 1$ and g staying in some fixed compact subset Ω. Therefore, for all such a

$$(r_{GP}(f))(ag) = 0.$$

This implies $r_{GP}(f) = 0$, or equivalently $f(P) = 0$. In particular, $f(B) = 0$ (Lemma 4.18).

First Case, $\rho = Sp(\chi_0)$ Noncuspidal. We briefly indicate which modifications have to be made in the noncuspidal case $\rho = Sp(\chi_0)$. For $\delta = \delta(\rho\nu^{1/2} \triangleleft \nu^{-1/2})$

$$r_{PG}(\delta)(t) = (1 + \chi_0(t_2 t_2' det(x)))\chi_\rho(x)|t_2'/t_2|^{1/2},$$

since $r_{PG}(\delta) = \rho\nu^{\frac{1}{2}} \triangleleft \nu^{-1/2} + \rho\nu^{\frac{1}{2}} \triangleleft \chi_0\nu^{-1/2}$. Since, furthermore,

$$r_{PG}(\alpha^G)(ag) = 2 \cdot r_{P_M M}\left(\chi_{\rho\otimes Sp} + \chi_{Sp\otimes\rho}\right)(ag)$$

$$= 2\chi_\rho(x)|t_2'/t_2|^{1/2} + 2\chi_{Sp}(x)\chi_0(t_2 t_2')|t_2'/t_2|^{1/2},$$

which coincides with the expression for $r_{PG}(\delta)(ag)$, we get the same conclusion as for cuspidal ρ.

First Case, but Second Maximal Parabolic Q. Now we show $f(Q) = 0$ or equivalently $r_{QG}(f) = 0$. By definition the difference $f(t)$ of the two distributions α^G and $\delta(\rho\nu^{1/2} \triangleleft \nu^{-1/2})$ vanishes for elliptic semisimple regular elements t in the Levi group of Q. So it suffices to consider regular elements t in the split tori

$$t = (diag(t_1, t_1'), diag(t_2, t_2')) \, mod \, F^* \in B_M(F) \subseteq M(F)$$

and the image (under the isomorphism ψ)

$$t = diag(t_1' t_2', t_1 t_2', t_1 t_2, t_1' t_2) \in B(F) \subseteq G(F).$$

Then

$$(r_{QG}f)(t) = (r_{QG}\alpha^G)(t) - 2 \cdot \chi_{r_{QG}(\delta(\rho\nu^{1/2}\triangleleft\nu^{-1/2}))}(t) = (r_{QG}(\alpha^G))(t),$$

since $r_{QG}(\delta(\rho\nu^{1/2} \triangleleft \nu^{-1/2})) = 0$ as a consequence of the shuffle formula (page 126).

For $r_{GQ}(f)(t) = 0$ it is enough to consider elements $t = ag$ with the property $|\alpha_1(a)| \ll 1$, g in some fixed compact subset of Q (Lemma 4.18). These conditions imply $|\alpha_1(t)| = |t_1/t_1'| \ll 1$ and $0 < const_1 < |t_1' t_2'/t_1 t_2| \leq const_2$ for t. In other words $|t_2'/t_2| < const_2|t_1/t_1'| \ll 1$. For such elements $t = ag$ with the property $|\alpha_1(a)| \ll 1$, g in some fixed compact subset of Q, we have

$$(r_{QG}f)(t) = (r_{QG}\alpha^G)(t) = D_G(t)\alpha^G(t)$$

$$= 2 \cdot D_M(t)(\chi_{\rho\otimes Sp} + \chi_{Sp\otimes\rho})(t).$$

Notice, $t = ag$ is also B_M-contractive under the assumptions made. Hence,

$$(r_{QG}f)(t) = 2 \cdot D_M(t)(\chi_{\rho \otimes Sp} + \chi_{Sp \otimes \rho})(t) = 0$$

for elements t chosen as above, since ρ is cuspidal. This implies $f(Q) = 0$ for cuspidal ρ.

If $\rho = Sp(\chi_0)$ and $\chi_0^2 = 1, \chi_0 \neq 1$, it is again enough to consider regular elements $t \in Q$ in a split torus. A similar computation gives

$$r_{QG}(f)(t) = 2 \cdot (\chi_{r_{M_P M}(\rho \otimes Sp + Sp \otimes \rho)}) - 2 \cdot \chi_{r_{BG}(\delta(\rho \nu^{1/2} \triangleleft \nu^{-1/2})}.$$

This difference vanishes since the first term in parentheses is $\chi_0 \nu \boxtimes \chi_0 \boxtimes \nu^{-1/2}(t) + \chi_0 \nu \boxtimes \chi_0 \boxtimes \chi_0 \nu^{-1/2}(t)$, whereas the second term is $|t_1 t_2'/t_1' t_2|^{1/2}(\chi_0(t_1 t_1') + \chi_0(t_2 t_2'))$. See the remark on page 132.

The Second Case ($\rho = Sp$ **and** $\alpha = \chi_{Sp \otimes Sp}$)**.** The vanishing $f(P) = 0$. We already got (see page 132)

$$r_{PG}(T_+ - T_-) = 2 \cdot \left(Sp(\nu^{\frac{1}{2}}) \boxtimes \nu^{-1/2}\right).$$

We know $D_G(t)\alpha^G(t) = 2 \cdot D_M(t)\alpha(t)$ holds for all regular semisimple $t \in G$ coming from M. For $(x, diag(t_2, t_2')) \in P_M$ with $|t_2'/t_2| \ll 1$ and x in a fixed compact set Ω therefore

$$r_{PG}(\alpha^G) = 2 \cdot r_{P_M M}(\alpha).$$

Hence,

$$r_{PG}(\alpha^G - \chi_{(T_+ - T_-)}) = 0,$$

as a consequence of Lemma 4.18 and the formula

$$r_{P_M M}(\alpha)(x, diag(t_2, t_2')) = r_{P_M M}(\chi_{Sp \otimes Sp})(x, diag(t_2, t_2')) = \chi_{Sp}(x)|t_2'/t_2|^{\frac{1}{2}}$$

together with (already used in the first case)

$$\chi_{Sp(\nu^{1/2}) \times \nu^{-1/2}}(diag(t_2' det(x)x^{-t}, t_2 x)) = \chi_{Sp}(x)|t_2'/t_2|^{1/2},$$

where both formulas hold in the above-mentioned range for x, t_2, t_2'.

The vanishing $f(Q) = 0$ is shown as in the first case. We only used the following two facts: $f(B) = 0$ as a consequence of $f(P_Q) = 0$, and $r_{QG}(f)$ as a character of $Levi(Q)(F)$ vanishes on the elliptic regular elements. The analogue of the first statement has been shown already. The second assertion holds for $f = \alpha^G - \chi_{T_+ - T_-}$ since it holds for the distribution α^G by the defining formula (Lemma 4.6) and since it holds for the character of $r_{QG}(T_+ - T_-)$. Recall $T_+ \oplus T_- \cong 1 \times Sp$ and (by the formula on page 132)

$$r_{QG}\left(T_+ - T_-\right) = \nu \boxtimes (1 \triangleleft \nu^{-1/2}).$$

This formula implies that $r_{QG}(T_+ - T_-)$ vanishes on the elliptic locus of $Levi(Q)(F)$ since it is represented by the character of a representation induced from its Borel subgroup. This completes the proof of Lemma 4.21. $\square$

Lemma 4.22. *For all irreducible discrete series representations ρ of $Gl(2, F)$*

$$\langle Wp_\pm(\rho), Wp_\pm(\rho)\rangle_{G,e} = 1;$$

hence,

$$\langle Wp_+(\rho) - Wp_-(\rho), Wp_+(\rho) - Wp_-(\rho)\rangle_{G,e} = 4.$$

Proof. $W_\pm = W_\pm(\rho)$ are elliptic irreducible representations (Lemma 4.19). Therefore, $\langle W_+, W_+\rangle_{G,e} = a > 0$. $W_+ \oplus W_-$ is induced; hence, its character vanishes on G^e. Therefore, $\langle W_+, W_-\rangle_{G,e} = -a$. Then also $\langle W_-, W_-\rangle_{G,e} = a$; thus, $\langle W_+ - W_-, W_+ - W_-\rangle_{G,e} = 4a$. The scalar products $\langle \pi, \pi'\rangle_{G,e}$ are integers (see Theorem 21 in [80] in the essential case of groups with an anisotropic center). Suppose W_+ occurs nontrivially in the endoscopic character lift (see Theorem 4.4 in Sect. 4.10), then $a \in \mathbb{Z}$ and $4a \le 4$ by Lemma 4.14. This implies the claim $a = 1$. Alternatively this follows from [2]: since the R-group in this situation is $R_\rho = W_\rho = \mathbb{Z}/2\mathbb{Z}$, its central extension $\tilde{R}_\sigma$ (in the notation in [2]) can be assumed to be $\tilde{R}_\sigma = R_\sigma$, as the Schur multiplier of cyclic groups is trivial. Then, by formula (1*) in [2], we get

$$a = |R_\rho|^{-1} \sum_{r \in R_{\rho,reg}} |d(r)| = 1,$$

since $d(r) = 2$ holds for the nontrivial element $r \in R_{\rho,reg} = R_\rho \setminus \{1\}$. $\square$

Corollary 4.9. *(a) Suppose $\sigma_v = \sigma_{1,v} \otimes \sigma_{2,v}$ is an irreducible cuspidal series representation of $M(F)$ which is not invariant under the outer automorphism $*$. Then*

$$\chi^G_{\sigma_v} = \chi_{\pi_+(\sigma_v)} - \chi_{\pi_-(\sigma_v)}$$

for two irreducible cuspidal representations $\pi_+(\sigma_v), \pi_-(\sigma_v)$.
(b) If σ_v is irreducible cuspidal and invariant under $*$, either $\chi^G_{\sigma_v} = \tilde{n}(\pi_{v,1})\chi_{\pi_{v,1}} + \tilde{n}(\pi_{v,2})\chi_{\pi_{v,2}} + \tilde{n}(\pi_{v,3})\chi_{\pi_{v,3}} + \tilde{n}(\pi_{v,4})\chi_{\pi_{v,4}}$ where the representations $\pi_{v,i}$ are irreducible admissible nonisomorphic cuspidal representations with coefficients $\tilde{n}(\pi_{v,i}) \in \{1, -1\}$, or alternatively

$$\pm\chi^G_{\sigma_v} = \chi_{Wp_+(\rho_v)} - \chi_{Wp_-(\rho_v)},$$

where ρ_v is a irreducible cuspidal representations of $Gl(2, F)$.

Proof. Once we have determined the infinitesimal characters which contribute to the character expression for $\chi^G_{\sigma_v}$, this follows from the integrality of the coefficients $\tilde{n}(\pi_v)$ (Corollary 4.5) and the elliptic scalar product formula (Lemma 4.14). In fact, for cuspidal σ_v only cuspidal representations π_v contribute or infinitesimal characters χ for which $AD_\chi \neq 0$ (Corollary 4.8). These noncuspidal contributions are integral combinations of the form $Wp_+(\rho_v) - Wp_-(\rho_v)$ with cuspidal ρ_v (Lemmas 4.19 and 4.20). This gives an expression $\chi^G_{\sigma_v}$ of the form

$$\chi^G_{\sigma_v} = \sum_{\pi_v \text{ cuspidal}} \tilde{n}(\pi_v)\chi_{\pi_v} + \sum_{\rho_v \text{ cuspidal}} n(\rho_v)\left(\chi_{Wp_+(\rho_v)} - \chi_{Wp_-(\rho_v)}\right)$$

on regular semisimple elements, where π_v are irreducible cuspidal and $\tilde{n}(\pi_v) \in \{0, \pm1\}$ and where $n(\rho_v) \in \mathbb{Z}$ and ρ_v runs over cuspidal irreducible admissible representations of $Gl(2, F)$ (Lemma 4.15 and Corollary 4.8). Since $\chi^G_{\sigma_v}$ is an integral linear combination of characters of cuspidal discrete series representations and an integral linear combination of terms $Wp_+(\rho_v) - Wp_-(\rho_v)$ for cuspidal ρ_v's, the remaining statements follow from the formula $\langle \chi^G_{\sigma_v}, \chi^G_{\sigma_v} \rangle_{G,e} = 2$ or 4 in cases (a) and (b), respectively (Lemma 4.13). This together with Lemma 4.22 implies $n(\rho_v) = 0$ in case (a). In case (b) $\sigma^*_v \cong \sigma_v$ it implies that $n(\rho_v) \neq 0$ holds for at most one representation ρ_v. But then again by Lemmas 4.13 and 4.22 we obtain $\chi^G_{\sigma_v} = \pm(Wp_+(\rho_v) - Wp_-(\rho_v))$ on the elliptic locus. For the remaining assertion see Corollary 4.6 and the remark thereafter. This completes the proof of Corollary 4.9. $\square$

If σ_v is not cuspidal but belongs to the discrete series, then use Lemma 4.21. If we replace $\chi^G_{\sigma_v}$ by the distribution f of Lemma 4.21, Lemma 4.21 plays a role similar to that of Corollary 4.8 for the character lift for cuspidal σ_v. Again we get an expansion

$$f = \sum_{\pi_v \text{ cuspidal}} \tilde{n}(\pi_v)\chi_{\pi_v} + \sum_{\rho_v \text{ cuspidal}} n(\rho_v)\left(\chi_{Wp_+(\rho_v)} - \chi_{Wp_-(\rho_v)}\right),$$

where π_v are irreducible cuspidal and $\tilde{n}(\pi_v) \in \{0, \pm1\}$ and where $n(\rho_v) \in \mathbb{Z}$ and ρ_v runs over irreducible cuspidal admissible representations of $Gl(2, F)$ (Lemma 4.15 and Corollary 4.8). Using this, we can express $\chi^G_{\sigma_v}$ as a character sum in such a way that the arguments from the cuspidal case carry over verbatim. This proves

Corollary 4.10. *A similar statement holds for σ_v in the discrete series, but for σ_v not cuspidal. In this situation the statements of Corollary 4.9 carry over with the following replacements:*

(a) *For $\sigma_v = (\rho_{1,v} \otimes \chi_v) \otimes Sp(\chi_v) = (\rho_{1,v} \otimes \chi_v, Sp(\chi_v))$, where $\omega_{\rho_{1,v}} = 1$ and $\rho_v = \rho_{1,v} \otimes \chi_v \not\cong Sp(\chi_v)$ is in the discrete series, put $\pi_+(\sigma_v) = \delta(\rho_{1,v}\nu^{1/2} \triangleleft \chi_v\nu^{-1/2})$ (see the list of discrete series on page 135). Then*

$$\chi^G_{\sigma_v} = \chi_{\pi_+(\sigma_v)} - \chi_{\pi_-(\sigma_v)}$$

for some cuspidal irreducible representation $\pi_-(\sigma_v)$.

(b) *For $\sigma_v = Sp(\chi_v) \otimes Sp(\chi_v) = (Sp(\chi_v), Sp(\chi_v))$ put $\rho_v = Sp(\chi_v)$. Then*

$$\chi^G_{\sigma_v} = (\chi_{T_+} - \chi_{T_-}) \otimes \chi_v = \chi_{Wp_+(\rho_v)} - \chi_{Wp_-(\rho_v)}.$$

4.10 Whittaker Models

In this section global arguments are used, and the relation between the endoscopic lift and theta lifts plays a role.

Let $\pi = \otimes \pi_v$ be a global cuspidal irreducible automorphic representation of $GSp(4, \mathbb{A})$. Then π is ψ-generic if it has a global Whittaker model with respect to the character ψ. Since any two nondegenerate characters ψ are conjugate for the group $GSp(4)$, we fix ψ and do not mention it further. Generic will mean ψ-generic. Under the assumption that $\pi = \otimes \pi_v$ is globally generic, all π_v are locally generic, i.e., have a Whittaker model. Conversely, assume all local components π_v are generic. If the degree 5 standard L-function $\zeta(\pi, s)$ of π does not have zeros on the line $Re(s) = 1$, then π is globally generic [9], p. 49.

Suppose π is a weak lift from M attached to an irreducible cuspidal automorphic form σ with holomorphic infinite component $\sigma_\infty = \sigma_{1,\infty} \times \sigma_{2,\infty}$ of weights $k_i \geq 2$, then the σ_i are all tempered by the Ramanujan conjecture. Furthermore, the degree 5 standard zeta function is $\zeta(\pi, s) = \zeta(s) L(\sigma_1 \otimes \sigma_2^*, s)$. It does not have a zero on the line $Re(s) = 1$. See [8], p. 200, and [89]. Since the zeta function of a weak lift does not vanish on $Re(s) = 1$, we get

Theorem 4.1. *A weak endoscopic lift from M is generic if and only if it is locally generic for all places v.*

Theorem 4.2 ([95], p. 295). *Two cuspidal, irreducible generic representations of $GSp(4, \mathbb{A})$ which are locally isomorphic for almost all places v are isomorphic.*

By a theorem of Waldspurger, Moeglin, and Rodier for irreducible admissible representations σ_v of a quasisplit reductive group over a local field F_v the following assertions are equivalent [87], p. 325:

1. σ_v is generic.
2. The germ expansion of the character of σ_v has a nontrivial (positive) contribution from at least one of the maximal (regular) nilpotent orbits.

By the character formulas this implies that for generic irreducible representations on a (quasisplit) Levi component, the corresponding induced representation has at least one generic irreducible constituent. Notice that this does not imply that the induced representation itself has a Whittaker model. However, in the case of the completely reducible induced representation $1 \times \rho = Wp_+(\rho) \oplus Wp_-(\rho)$ it implies that one of the representations $Wp_\pm(\rho)$ has a Whittaker model. On the other hand, considering the double cosets $B(F) \backslash G(F) / Q(F)$, one can easily show $dim(Hom_{B(F)}(1 \times \rho, \psi)) \leq 1$ (an independent and stronger result will be proved later; see Corollary 4.16). Hence,

Lemma 4.23. *For fixed ρ exactly one of the two representations $Wp_+(\rho), Wp_-(\rho)$ is generic.*

Another consequence is

Lemma 4.24 ([87], Proposition 9.6). *For tempered irreducible generic represen-tations σ_v of M_v at least one of the coefficients $\tilde{n}(\pi_{v,i})$ of $\chi^G_{\sigma_v} = \sum \tilde{n}(\pi_i)\pi_{v,i}$ in Corollaries 4.9 and 4.10 is positive $n(\pi_{v,i}) > 0$, and the corresponding $\pi_{v,i}$ is locally generic for $GSp(4, F_v)$.*

Notation. Fix one such generic $\pi_{v,i}$ as in Lemma 4.24, and denote it by

$$\pi_+(\sigma_v) \quad \sigma_v \text{ generic, irreducible }.$$

Inspecting Corollaries 4.9 and 4.10, we see that $\pi_+(\sigma_v)$ is obviously *unique* ex-cept possibly in the situation of Corollary 4.9(b). However, in this case we will show in the proof of Theorem 4.4 via a global argument, using Lemma 4.26 and Theorem 4.3, that the first alternative in the statement of Corollary 4.9(b) cannot occur. Hence, the generic representations $\pi_+(\sigma_v)$ turn out to be uniquely defined by the tempered representation σ_v, which justifies the notation.

4.10.1 Theta Lifts

The connected component of the group of similitudes of a split quadratic form in four variables

$$GSO(2, 2) = (Gl(2) \times Gl(2))/\{(t, t^{-1}), t \in F^*\}$$

is isomorphic to the group M. To determine the generic representation $\pi_+(\sigma_v)$ we consider the local theta lift from $M = GSO(2, 2)$ to $G = GSp(4)$ stud-ied by Soudry [94], [96], p. 363ff, and [70] (loc. cit. p. 514 local theory). For the corresponding global lift see [71], p. 416, and [94]. From general properties of theta lifts one knows that for a global *generic* cuspidal automorphic representa-tion $\sigma = \sigma_1 \times \sigma_2$ of $M(\mathbb{A})$ the theta lift, denoted $\Theta_+(\sigma)$, contains an automorphic representation $\theta_+(\sigma)$ of $GSp(4)$, which does not vanish and has a global Whit-taker model. Notice $\sigma = (\sigma_1, \sigma_2)$ is generic if it is nondegenerate in the sense $dim(\sigma_i) \neq 1$ for $i = 1, 2$ (locally or globally).

Of course $(GSO(2, 2), GSp(4))$ is not an honest dual reductive pair in the sense of Howe. The underlying dual reductive pair $(O(2, 2), Sp(4))$ was studied in [41]. In fact, the restriction of the representation $\Theta_+(\sigma)$ to $Sp(4, \mathbb{A})$ is a sum of repre-sentations of the type stated in Theorem 8.1 in [41], up to character twists μ in the notation in [41].

The global representations $\Theta_+(\sigma)$ attached to a generic cuspidal automorphic representations σ are cuspidal representations, provided the corresponding theta lift of σ from $GSO(2, 2)$ to $Gl(2)$ vanishes. This can be proved for $GSp(4)$ similarly as for $Sp(4)$ [41], Theorem 8.1 and p. 78ff. The lift of σ to $Gl(2)$ vanishes for $\sigma = \sigma_1 \times \sigma_2$ if $\sigma_1 \not\cong \sigma_2$. Therefore, for $\sigma_1 \not\cong \sigma_2$ the lift of σ to $Gl(2)$ vanishes and $\Theta_+(\sigma)$ is cuspidal. Then every irreducible constituent π of $\Theta_+(\sigma)$ is a weak cuspidal lift (in our sense) attached to the automorphic cuspidal representation σ from M. This is a consequence of Theorem 2.4 in [70] or formula (*) in [71], p. 417. In fact formula (*) in [71] computes the degree 4 standard L-series of π at almost all places v:

$$L_v(\theta_+(\sigma), s) = L_v(\sigma_1, s)L_v(\sigma_2, s).$$

Soudry [97] implied that π is not CAP if $\sigma = (\sigma_1, \sigma_2)$ is generic (i.e., if none of the σ_i are one-dimensional). Recall that there is at least one irreducible cuspidal automorphic constituent π of $\Theta_+(\sigma)$ which has a global Whittaker model. Denote this constituent

$$\theta_+(\sigma).$$

$\theta_+(\sigma)$ is irreducible and furthermore uniquely defined by σ by the generic strong multiplicity 1 Theorems 4.1 and 4.2 stated above.

Theorem 4.3. *Suppose* $\sigma = (\sigma_1, \sigma_2)$ *is a (generic) irreducible cuspidal automorphic representation of* $M(\mathbb{A})$ *with* $\sigma_1 \not\cong \sigma_2$. *Then the global theta lift* $\Pi_+(\sigma)$ *is cuspidal and contains a unique irreducible generic representation*

$$\theta_+(\sigma)$$

called the global theta lift. *This cuspidal irreducible representation is not CAP, and is a weak endoscopic lift attached to* σ. *Its local components*

$$\theta_+(\sigma)_v$$

have Whittaker models.

The local representations $\theta_+(\sigma)_v$ in fact will be shown to depend only on σ_v, and they will turn out to be uniquely determined by the local theta correspondence and their property to have a local Whittaker model. See Lemma 4.25(b). Hence, we write $\theta_+(\sigma)_v = \theta_+(\sigma_v)$.

4.10.2 The Local Theta Lift

According to the Howe conjecture, there should exist a well-defined local correspondence between irreducible representations

$$\sigma_v \mapsto \theta_+(\sigma_v),$$

where $\theta_+(\sigma_v)$ denotes the unique irreducible quotient of the local theta lift $\Pi_+(\sigma_v)$ considered in [96]. However Howe's conjecture is only proved for dual reductive pairs when the residue characteristic is different from 2. But we do not consider an honest dual reductive pair and we furthermore need to include the case of odd residue characteristic 2. Therefore, we remark that fortunately there are two important cases where the local theta lift $\theta_+(\sigma_v)$ is locally defined in a unique way.

Lemma 4.25. *Let* σ_v *be an irreducible admissible representation of* $M = GSO(2,2)_v$. *Then up to isomorphism there is at most one irreducible quotient* π_v *of the local theta lift* $\Theta_+(\sigma_v)$ *if we impose the following restrictions on* σ_v *and* π_v. *Either*

(a) if $\sigma_v \cong Ind(\tilde{\chi}_1, \tilde{\chi}_2) \times Ind(\chi_1, \chi_2)$ *is irreducible unramified and the quotient* π_v *of* $\Theta(\sigma_v)$ *is assumed to be unramified. Then*

$$\pi_v \cong \tilde{\chi}_1\chi_1^{-1} \times \tilde{\chi}_2\chi_1^{-1} \lhd \chi_1.$$

(b) or if σ_v *is generic and the quotient* π_v *is also assumed to be generic.*

Proof. See [71], p. 417ff, and [96], Theorem 3.1. The second part of Lemma 4.25 allows us to define a *local theta lift*

$$\sigma_v \text{ irreducible, generic} \mapsto \theta_+(\sigma_v) \text{ irreducible, generic}$$

for all generic σ_v for which the theta representation $\Pi_+(\sigma_v)$ admits an admissible irreducible generic quotient. This local assignment then obviously commutes with character twists. It remains for us to examine under which conditions the local theta representation $\Pi_+(\sigma_v)$ admits an admissible irreducible generic quotient. $\square$

Existence of Generic Quotients. By Theorem 4.3 $\Pi_+(\sigma_v)$ admits an admissible irreducible generic quotient if σ_v can be realized as the local component of a global generic cuspidal automorphic representation σ of $M(F)$. Up to a local character twist this is the case for all representations of $M(F)$ in the discrete series (see the arguments preceding Lemma 4.11). Hence, using the compatibility with local character twists, this allows us to define the local theta lift $\theta_+(\sigma_v)$ for all discrete series representations σ_v of M_v.

The local theta lift $\theta_+(\sigma_v)$ is also defined for all irreducible induced representations

$$\sigma_v = Ind(\tilde{\chi}_{1,v}, \tilde{\chi}_{2,v}) \times Ind(\chi_{1,v}, \chi_{2,v})$$

for *unitary* characters $\tilde{\chi}_{1,v}, \tilde{\chi}_{2,v}, \chi_{1,v}, \chi_{2,v}$. Notice unitary principal series of $GSp(4, F_v)$ are irreducible (e.g., [100], Corollary 7.6). The assertion made is a generalization of the assertion of Lemma 4.25(a) in the spherical case. The argument for the proof of the latter assertion given in [71], p. 417, can be extended to hold also in a more general case, where one allows one of the characters to ramify. In this case the map ν constructed in [70], p. 418, still properly exists (not just as the value of a meromorphic function). By Tate theory the integration over b and t in loc. cit. remains well defined in the sense of analytic continuation. In fact, we only have to assume $\tilde{\chi}_{2,v}/\chi_{1,v}(a) \neq 1$ to make the integral in loc. cit. well defined. Without restriction of generality, this can be assumed unless $\tilde{\chi}_{1,v} = \tilde{\chi}_{2,v} = \chi_{1,v} = \chi_{2,v}$ holds. This exceptional case is, however, a character twist of the spherical case, which is already known. The map ν induces a nontrivial intertwining operator

$$T_v : \text{(big Weil representation)} \otimes \sigma_v \to \tilde{\chi}_{1,v}\chi_{1,v}^{-1} \times \tilde{\chi}_{2,v}\chi_{1,v}^{-1} \lhd \chi_{1,v},$$

where we refer to [71], p. 417, for details. If the image of T_v has a generic quotient, we are done. Since the unitary principal series of $GSp(4, F_v)$ are irreducible, and since representations induced from generic representations contain a generic constituent, this finishes the argument.

There is a further case of particular importance.

Lemma 4.26. *Suppose* $\sigma_v = (\rho_v, \rho_v)$, *where* ρ_v *is an irreducible cuspidal representation of* $Gl(2, F_v)$. *Then* $\Theta_+(\sigma_v)$, *and its quotient* $1 \times \rho_v$, *admit a unique generic irreducible quotient* $\theta_+(\sigma_v)$

$$\theta_+(\sigma_v) = Wp_+(\rho_v).$$

Proof. See [106], Lemma 8.2 and p. 64. Recall Y' in loc. cit. pp. 62–63 is $Y' = D_v^2$, where $D_v = M_{2,2}(F_v)$. The space D_v can be identified with the quadratic space of a split quadratic form, realized by the quadratic form det (the determinant on the space D_v of 2×2-matrices). The group $M = GSO(2,2)$ acts by $(h_1, h_2) \circ x = h_1 x h_2'$ for $(h_1, h_2) \in M$ and $x \in D_v$.

It is shown in [106] that the Weil representation on the Schwartz space $S(D_v^2 \times F^*)$ admits a nontrivial map

$$\zeta : S(D_v^2 \times F_v^*) \rightarrow Ind_Q^G(1 \boxtimes \rho_v) \otimes \sigma_v, \qquad \sigma_v = \rho_v \times \rho_v,$$

equivariant with respect to the action of the group $GSp(4, F)$ and the group $M(F) = Gl(2, F)^2 / F^*$, where $M(F)$ acts through the Weil representation on the left and by the representation $\sigma_v = \rho_v \times \rho_v$ on the right. The image of ζ is one of the two irreducible constituents $Wp_\pm(\rho_v)$ of the induced representation $Ind_Q^G(1 \boxtimes \rho_v)$ [106], Lemma 8.2(2). We know already that exactly one of them – namely $Wp_-(\rho_v)$ by definition – is generic.

We claim that the image of ζ is generic; hence, it is the generic constituent $Wp_+(\rho_v)$. This was already expected by Waldspurger (remark after Lemma 8.2 in [106]). The uniqueness statement then follows from Lemma 4.25(b) above.

Proof of the claim. Suppose the image of ζ is not the generic representation $Wp_+(\sigma_v)$. From a result shown in Corollary 4.15 in Sect. 4.12, $Wp_-(\sigma_v)$ is in the image of the anisotropic theta lift, also defined in Sect. 4.12. More precisely

$$\theta_-(\sigma_{2,v} \times \sigma_{2,v}) \cong Wp_-(\sigma_{2,v}).$$

In fact, this statement, proved later, excludes the possibility that $\theta_+(\sigma_v)$ is isomorphic to $Wp_-(\sigma_{2,v})$, and therefore proves the claim, since otherwise the restrictions of $\theta_+(\sigma_v)$ and $\theta_-(\sigma_v)$ to the subgroup $Sp(4, F_v)$ would have a common irreducible component. But by Theorem 9.4 in [41], the (big) anisotropic and isotropic Weil representations attached to the dual reductive pairs $(O_4, Sp(4))$ and $(O(2,2), Sp(4))$ have exactly one common irreducible representation of $Sp(4, F_v)$. The proof of Theorem 9.4 in [41] implies more: This common representation is related to the trivial representation of $SO(4, F_v)$ under the Howe correspondence. For this notice that the proof in [41] was obtained by studying $Sp(4, F_v)$-invariant distributions on the tensor product of the big Weil representations. Such distributions factorize over the Dirac distribution of the point zero in the compact Schrödinger model. The group $SO(4, F_v)$ therefore acts trivially on this distribution. This being said, we only have to exclude the property that the anisotropic Weil representation $\theta_-(\sigma_v)$ with respect to a cuspidal representation $\sigma_v = (\rho_v, \rho_v)$ under restriction to $Sp(4, F_v)$ contains the trivial representation of $SO(4, F_v)$. This could only happen if ρ were a special representation. But this is excluded since ρ_v is cuspidal by assumption. This completes the proof of the claim and the proof of Lemma 4.26. □

4.10.3 The Local L-Packets

For an irreducible admissible representation σ_v of $M(F)$ we define the local packet attached to σ_v to be the set of classes of irreducible admissible representations π_v of $GSp(4, F)$ for which $n(\sigma_v, \pi_v) \neq 0$ holds in the expansion $\chi_{\sigma_v}^G = \sum_{\pi_v} n(\sigma_v, \pi_v) \chi_{\pi_v}$. With this definition the local packets are finite (Lemma 4.12 and Corollaries 4.9 and 4.10). If σ is generic and preunitary, then we call this packet the local L-packet.

Remember the previously used notation (introduced after Lemma 4.24) to denote $\pi_+(\sigma_v)$ as any *generic* irreducible representation π_i in the local packet, defined by σ_v, which occurs in the character expansion $\chi(\sigma_v) = \sum n(\sigma_v, \pi_v) \pi_v$ with a *positive* coefficient $n(\sigma_v, \pi_v) > 0$.

Theorem 4.4. *Let σ_v belong to the discrete series representation of $M(F)$. Suppose π_v is in the local packet of σ_v. Suppose π_v is generic with the property $n(\sigma_v, \pi_v) > 0$. Then*

$$\pi_+(\sigma_v) \cong \theta_+(\sigma_v).$$

In particular, $\pi_+(\sigma_v)$ is uniquely defined by these properties.

Hence, by Lemma 4.26 we have

Corollary 4.11. *For cuspidal representations ρ_v of $Gl(2, F_v)$ and $\sigma_v = (\rho_v, \rho_v)$*

$$\pi_+(\sigma_v) \cong Wp_+(\rho_v).$$

Proof of Theorem 4.4. We may assume v is non-Archimedean (for the Archimedean case, see [4]). We realize $\sigma_v = (\rho_v, \rho_v)$ as the local component of some cuspidal irreducible automorphic representation of $M(\mathbb{A})$ such that $\sigma_\infty = (\sigma_{\infty,1}, \sigma_{\infty,2})$, where $\sigma_{\infty,i}$ are holomorphic of weight $r_1 \neq r_2$ sufficiently large (see page 100). Then σ is automatically generic, since none of the two local factors $\sigma_{v,i}$ can be one-dimensional. It is tempered by results obtained by Deligne and Carayol. Furthermore, $\sigma_1 \not\cong \sigma_2$ globally. Therefore, the global theta lift $\theta_+(\sigma)$ is irreducible, cuspidal, nonvanishing, and generic. It is not a CAP, and a weak lift attached to σ (Theorem 4.3). $\pi = \theta_+(\sigma)$ contributes to cohomology (Corollary 4.2); hence, it is detected by the multiplicity formula (Corollaries 4.2 and 4.4).

On the other hand consider the local packets attached to the local representations σ_w of the global representation σ at all other places $w \neq v$. For each w there exists at least one representation $\pi_w = \pi_+(\sigma_w)$ in the local packet of σ_w for which π_w is generic with $n(\sigma_w, \pi_w) > 0$ (Lemma 4.24). Fix any such local choice so that π_w is unramified at almost all places and define $\pi = \otimes_w \pi_w$. Temporarily suppose this representation π of $GSp(4, \mathbb{A})$ is automorphic. Notice σ_1, σ_2 are cuspidal. If this is the case, π cannot be weakly equivalent to an automorphic Eisenstein representation since at all unramified places (Lemma 4.3) its L-factor is

$$L_v(\tilde{\chi}_1 \tilde{\chi}_1^{-1} \times \tilde{\chi}_2 \chi_1^{-1} \lhd \chi_1, s) = L_v(\sigma_1, s) L_v(\sigma_2, s).$$

Then for a suitable character twist the left side has poles where the right side does not have poles. This is a contradiction. Hence, π is cuspidal, but not CAP. Since π_∞ is in the local Archimedean L-packet of σ_∞, it is in the discrete series. Hence, the representation π contributes to the cohomology group $H_!^3$ (see Chap. 1). Since all local components π_w are chosen to be generic, the representation π is globally generic (Theorem 4.1). By the generic strong multiplicity 1 theorem (Theorem 4.2) therefore

$$\theta_+(\sigma) \cong \pi_+(\sigma).$$

This proves Theorem 4.4, but of course it remains for us to show that $\pi_+(\sigma)$ is automorphic.

For this use the property that π_∞ is generic. Hence, π_∞ does not belong to the holomorphic discrete series. Thus, by Corollary 4.4, the automorphic multiplicity is

$$m(\pi) = m(\pi_\infty \otimes \pi_{fin}) = m_2(\pi_{fin}).$$

Hence, by the multiplicity formula of Corollary 4.4

$$m_2(\pi_{fin}) = m_1(\pi_{fin}) - \varepsilon \prod_{w \neq \infty} n(\sigma_w, \pi_w) \geq -\varepsilon = 1.$$

This uses the fact that all π_w were chosen to satisfy $n(\sigma_w, \pi_w) > 0$, and it furthermore uses the fact that the constant ε is -1, which amounts to showing $m_1(\pi_-(\sigma_\infty) \otimes \prod_{w \neq \infty} \theta_+(\sigma_w)) = 0$ (which comes out easily in Chap. 8). Therefore, π is automorphic. This completes the proof of Theorem 4.4. The argument furthermore implies $m_1(\pi_+(\sigma)_{fin}) = 0$ by the strong generic multiplicity 1 theorem. $\square$

This improves our previous results and gives

Theorem 4.5. *For all irreducible unitary discrete series representations $\sigma_v = (\sigma_{1,v}, \sigma_{2,v})$ of $M(F)$ the non-Archimedean character lift is given by*

$$\boxed{\chi_{\sigma_v}^G = \chi_{\pi_+(\sigma_v)} - \chi_{\pi_-(\sigma_v)}}$$

with irreducible representations $\pi_+(\sigma_v)$ and $\pi_-(\sigma_v)$ as follows:

(a) Suppose $\sigma_{v,1} \not\cong \sigma_{v,2}$. Then $\pi_-(\sigma_v)$ is cuspidal. $\pi_+(\sigma_v)$ is cuspidal if and only if σ_v is cuspidal. If σ_v is not cuspidal but is in the discrete series, then $\sigma_v = (\rho_{1,v} \otimes \chi_v, Sp(\chi_v))$, where $\omega_{\rho_{1,v}} = 1$ and $\rho_v = \rho_{1,v} \otimes \chi_v \not\cong Sp(\chi_v)$ is in the discrete series of $Gl(2, F)$. In this case the representation $\pi_+(\sigma_v)$ is $\delta(\rho_{1,v}\nu^{1/2} \lhd \chi_v\nu^{-1/2})$. Hence, $\pi_+(\sigma_v)$ is in the discrete series of $GSp(4, F)$, but is not cuspidal.

(b) Suppose $\sigma_{1,v} \cong \sigma_{2,v}$. Then $\sigma_{i,v} \cong \rho_v$ belongs to the discrete series of $Gl(2, F)$ and $\pi_\pm(\sigma_v) \cong Wp_\pm(\rho_v)$.

Furthermore, $\pi_+(\sigma_v) \cong \theta_+(\sigma_v)$ has a Whittaker model.[2]

Proof. This improves Corollaries 4.9 and 4.10 in two ways. First improvement: For $\sigma_v = (\rho_v, \rho_v)$ and cuspidal ρ_v

$$\chi^G_{\sigma_v} = \chi_{Wp_+(\rho_v)} - \chi_{Wp_-(\rho_v)},$$

since by Corollary 4.11 $Wp_+(\rho_v)$ must appear in the character expression. The correct signs are determined by Theorem 4.3 and Corollary 4.11: $Wp_+(\rho_v)$ is a generic representation, whereas $Wp_-(\rho_v)$ is not generic (see page 121). This argument carries over to the case $\sigma = (\rho_v, \rho_v)$ where $\rho_v = Sp(\chi_v)$ is a special representation. Without restriction of generality we can assume $\chi_v = 1$ by a character twist. Then $T_- = Wp_-(Sp)$ is tempered elliptic (Lemma 4.19) but does not admit a Whittaker model. Finally $T_+ = \theta_+(Sp \times Sp) = Wp_+(Sp)$ by Theorem 4.4. Second improvement: $\pi_+(\sigma_v) \cong \theta_+(\sigma_v)$, which follows from Theorem 4.4. $\square$

Theorem 4.6. *Let π_v be a discrete series representation of $GSp(4, F)$ over a non-Archimedean local field of characteristic zero. If π_v is not cuspidal, then π_v is generic.*

Proof. The representations $\delta(k, f, \chi)$ in the notation in [100], p. 42, have Whittaker models. Furthermore, a discrete series constituent in $\chi_1 \times \chi_2 \lhd \chi$ in the regular case is of the form $\delta(k, f, \chi)$ [100], Theorem 8.5. In the nonregular cases (a)–(c) of the nonunitary principal series, there exists no discrete series constituent. This covers all cases that occur in the principal series. The discrete series representations contained in representations induced from cuspidal representations of a Levi subgroup of the Klingen parabolic are generic [88], p. 285 and Theorem 5.1. Finally Silberger [93] proved that discrete series constituents of representations which are induced from cuspidal representations of a Levi subgroup of the Siegel parabolic P are generic. $\square$

4.11 The Endoscopic Character Lift

In this section we summarize the facts that have been collected so far.

---

[2] Notice that $\pi_-(\sigma_v)$ does not admit a Whittaker model. See Corollary 4.16 and Chap. 5.

Let F be a p-adic local field. Let M be the group $Gl(2) \times Gl(2)$ and let G be $GSp(4)$. Let $*$ denote the involution of M, which switches the two factors. Let P denote the Siegel parabolic in G. Let P_M denote the parabolic in M, which is the product of $Gl(2)$ in the first factor and the group of lower triangular matrices in the second factor. The map $(x, diag(t_2, t_2)) \mapsto blockdiag(t_2' det(x) x^{-t}, t_2 x)$ identifies the Levi components of P_M and P, and hence relates representations

$$\rho(x, diag(t_2, t_2')) = \sigma_1(x) \chi_1(t_2) \chi_2(t_2')$$

of the Levi component of $P_M(F)$ with representations

$$\tilde{\rho}(blockdiag(\lambda \cdot g^{-t}, g)) = \sigma_1(g) \chi_2(\lambda / det(g))$$

of the Levi group of $P(F)$. With these notations and identifications we construct a homomorphism $r(\alpha) = \alpha^G$

$$r = r_M^G : R_{\mathbb{Z}}[M(F)] \to R_{\mathbb{Z}}[G(F)]$$

between the Grothendieck groups of finitely generated admissible representations (the endoscopic character lift) with the following properties:

1. It describes the endoscopic lift locally

$$\alpha(f^M) = r(\alpha)(f), \quad f \in C_c^{\infty}(G).$$

2. It commutes with Galois twists (Lemma 4.10)

$$r(\sigma^\tau) = r(\sigma)^\tau.$$

3. It commutes with character twists (Lemma 4.8)

$$r(\sigma \otimes \chi) = r(\sigma) \otimes \chi.$$

4. It commutes with involution $*$

$$r(\sigma^*) = r(\sigma).$$

5. It commutes with induction

$$r \circ r_{P_M}^M(\rho) = r_P^G(\tilde{\rho}), \qquad \rho \in R_{\mathbb{Z}}[Gl(2, F) \times F^*]$$

($\tilde{\rho}$ is defined by ρ via the identification of Levi components of P_M and P above).
6. It preserves central characters, i.e., the central characters of the irreducible representations which occurs nontrivially in $r(\sigma)$ and the central character of σ (defined by the central character of σ_1 or σ_2) coincide.

For the last statement notice that irreducible admissible representations σ of G appear in the form

$$\sigma = (\sigma_1, \sigma_2),$$

where σ_i are irreducible admissible representations of $Gl(2, F)$ with equal central character ω_{σ_i}.

Here and in the following we use the convention to write $r(\sigma) = r(\chi_\sigma)$, etc. In other words, we do not distinguish between irreducible representations and their characters.

Recall. The group $\tilde{M}$ generated by $M = GSO(2,2)$ together with the outer automorphism $*$ is isomorphic to the group $GO(2,2)$, the group of orthogonal similitudes of a split quadratic form of rank 4. In this disguise, the endoscopic lift r is related to the local theta correspondence $\sigma_v \longleftrightarrow \theta_+(\sigma_v)$ between $\tilde{M} = GO(2,2)$ and $G = GSp(4)$. For $\sigma = (\sigma_1, \sigma_2)$ we have $\sigma^* \cong (\sigma_2, \sigma_1)$.

Let us explicitly describe this local lift r on the irreducible admissible tempered representations of M. The relevant cases are:

1. Description of r in the discrete series case. Suppose σ is an irreducible, admissible representation of $M(F)$ which belongs to the discrete series. Under this assumption there exists a set $\{\pi_+(\sigma), \pi_-(\sigma)\}$ of two isomorphism classes of irreducible, admissible representations of $G(F)$ attached to σ such that

$$r(\sigma) = \pi_+(\sigma) - \pi_-(\sigma).$$

Furthermore, the representations $\pi_+(\sigma)$ have Whittaker models, and:

(I) Either $\sigma^* \cong \sigma$. Then, although σ belongs to the discrete series, both $\pi_\pm(\sigma)$ do not belong to the discrete series. They are tempered elliptic representations of $G(F)$ (limits of discrete series).

(II) Or $\sigma^* \not\cong \sigma$. Then for σ in the discrete series, the representation $\pi_-(\sigma)$ is always cuspidal and $\pi_+(\sigma_v)$ belongs to the discrete series. Furthermore, $\pi_+(\sigma)$ is cuspidal if and only if σ is cuspidal.

More precisely, the representations $\pi_\pm(\sigma)$ are described as follows:

1a. *If σ_i are both cuspidal and not isomorphic, then $\pi_\pm(\sigma)$ are both cuspidal.*

1b. *The case where $\sigma_1 \cong \sigma_2$ is cuspidal.* In this case $\pi_\pm(\sigma) = Wp_\pm(\sigma_1)$. Remember these representations are elliptic relatives in the sense of Kazhdan. In particular, they are not cuspidal. Only $Wp_+(\sigma_1)$ has a Whittaker model. In Sect. 4.12 we will also show

$$\pi_-(\sigma) = Wp_-(\sigma_1) = \theta_-(\sigma_1).$$

Note that $\theta_-(\sigma_1)$ is an anisotropic theta lift, which in general is well defined only up to some additional hypotheses (e.g., Howe conjecture, residue characteristic not equal to 2). But by Corollary 4.17 of the next section it is defined unconditionally:

$$r(\sigma) = Wp_+(\sigma_1) - Wp_-(\sigma_1) = \theta_+(\sigma_1) - \theta_-(\sigma_1).$$

1c. *If $\sigma_1 = \rho_1 \otimes \chi$ is in the discrete series, and $\sigma_2 = Sp(\chi) \not\cong \sigma_1$ is a special representation, then $\pi_+(\sigma) = \delta(\rho_1 \nu^{1/2} \triangleleft \chi\nu^{-1/2})$ is in the discrete series, but is not cuspidal, and $\pi_-(\sigma)$ is cuspidal*

$$r\big((\sigma_1, Sp(\chi))\big) = \delta(\rho_1 \nu^{1/2} \lhd \chi \nu^{-1/2}) - \pi_-(\sigma).$$

1d. *If* $\sigma_i = Sp(\chi)$ *are both the same special representation, then* $\pi_\pm(\sigma) = Wp_\pm(\sigma_1) = (T_+ - T_-) \otimes \chi$ with tempered $T_\pm$. By the discussion of the nonregular case (b) (in Sect. 4.7) $T_- = R_2(4)$ is the anisotropic theta lift $\theta_-\big((1, 1)\big)$. Therefore, T_- has no Whittaker model. $T_+ = \theta_+(\sigma)$ is the indefinite theta lift (see Sect. 4.12)

$$r\big((Sp(\chi), Sp(\chi))\big) = (T_+ - T_-) \otimes \chi = \theta_+\big((Sp(\chi), Sp(\chi))\big) - \theta_-\big((\chi, \chi)\big).$$

This completes the description of the lift $r(\sigma)$ for discrete series representations. Now consider the tempered case and two further cases, which are important for the Saito–Kurokawa lift. We state the results and then we give the proof:

2. *Tempered, but not discrete series.* Suppose $\sigma = (\sigma_1, \sigma_2)$, where σ_1 is irreducible tempered and $\sigma_2 = Ind(\chi_1, \chi_2)$ is induced from a pair of unitary characters. Then $r(\sigma)$ is represented by the class of the irreducible tempered representation

$$r\big((\sigma_1, Ind(\chi_1, \chi_2))\big) = \sigma_1 \chi_1^{-1} \lhd \chi_1.$$

2'. *Suppose both representations* σ_i *are irreducible unitary generic. Then* $r(\sigma) = \pi_+(\sigma)$ *is irreducible (for* $\sigma = (\sigma_1, \sigma_2)$*), except in those cases which were listed in* (1).

Definition 4.5. Suppose σ is a *preunitary* irreducible *generic* admissible representation of $M(F)$. Then the irreducible representations π, which occur nontrivially in the linear combination $r(\sigma)$ – viewed as a subset of the equivalence classes of irreducible admissible representations of $GSp(4, F)$ – will be called the *local L-packet* attached to σ.

Hence, by (1) and (2') these local L-packets have cardinality either 2 or 1.

If σ is preunitary but not generic, at least one of the two representations σ_i of $Gl(2, F)$ defining σ is one-dimensional. In these cases the image packets defined by the support of the lift r do not define "L-packets," but rather describe the "Arthur packets." For the Archimedean analogue see [3], case 1.4.1, with $f_M(\psi) = r\big((\sigma_1, 1)\big)$ in the notation of loc. cit. These are relevant for the Saito–Kurokawa lift.

Hence, suppose $dim(\sigma_2) = 1$. A character twist allows us to assume – without restriction of generality – that $\sigma_2 = 1$ is the trivial representation. Then the central character is trivial, and hence $\omega_{\sigma_1} = 1$:

3. *Degenerate case.* If σ_1 is cuspidal ($\omega_{\sigma_1} = 1$) or of the form $\sigma_1 = Sp(\chi_0)$ for $\chi_0 \neq 1$ but $\chi_0^2 = 1$, then

$$r((\sigma_1, 1)) = J_P(\nu^{1/2}, \sigma_1 \boxtimes 1) + \pi_-(\sigma_1)$$

is the sum of two irreducible representations. $J_P(\nu^{1/2}, \sigma_1 \boxtimes 1)$ is nontempered, and $\pi_-(\sigma_1)$ is cuspidal. The second representation $\pi_-(\sigma_1) := \pi_-((\sigma_1, Sp))$ is by definition the one that appears also in the lift $r((\sigma_1, Sp)) = \pi_+((\sigma_1, Sp)) - \pi_-((\sigma_1, Sp))$.

Furthermore,

$$r((Sp, 1)) = J' + T_-$$

is a sum of two irreducible representations, where T_- is the tempered representation that also appears in the lift $r((Sp, Sp))$. Finally,

$$r((\sigma_1, 1)) = \chi_1 \circ det \triangleleft \chi_2$$

is irreducible nontempered if $\sigma_1 = Ind(\chi_1, \chi_2)$ is tempered.

Last but not least:

4. *The totally degenerate case.* $r((1, 1)) = J - J'$ and the remarkable case 5.

$$r((\chi_0, 1)) = A + B - C - D$$

for $\chi_0 \neq 1, \chi_0^2 = 1$ with four irreducible representations A, B, C, D described in the proof below.

Concerning the proof. By Theorem 4.5 it is enough to consider the assertions made for cases (2)–(4). We use compatibility of r with induction, which gives the following explicit formulas

$$r\Big((Ind(\tilde{\chi}_1, \tilde{\chi}_2), Ind(\chi_1, \chi_2))\Big) \equiv \tilde{\chi}_1 \chi_1^{-1} \times \tilde{\chi}_2 \chi_1^{-1} \triangleleft \chi_1,$$

and more generally the *induction formula*, which makes the result given on page 106 more explicit

Lemma 4.27.

$$\boxed{r\Big((\sigma_1, Ind(\chi_1, \chi_2))\Big) \equiv \sigma_1 \chi_1^{-1} \triangleleft \chi_1}.$$

Proof. The Levi components of P_M and P (see pages 106ff and 136) were identified by the comparison map ψ

$$(x, diag(t_2, t_2')) \in P_M \mapsto blockdiag(\lambda g^{-t}, g) = blockdiag(t_2' det(x) x^{-t}, t_2 x) \in P,$$

where in this formula t_2' is assumed normalized to be $t_2' = 1$, which identifies the image of $(x, diag(t_2, t_2'))$ in $M(F)$ with a unique representative in $Gl(2, F) \times Gl(2, F)$. Via this isomorphism ψ the representation

$$\rho(x, diag(t_2, t_2')) = \sigma_1(x) \chi_1(t_2) \chi_2(t_2')$$

of the standard Levi group of P_M is related to the representation

$$\tilde{\rho}\big(blockdiag(\lambda g^{-t}, g)\big) = \sigma_1(g)\chi_2(\lambda/det(g))$$

of the standard Levi group of P. To make this compatible with the notation for induction from P to G used in Sect. 4.7, we now write $blockdiag(\lambda \cdot g^{-t}, g) = blockdiag(x, \tau \cdot x^{-t})$. Hence, $g = \tau \cdot x^{-t}$ and $\lambda = \tau$. So $\tilde{\rho}$ maps $blockdiag(x, \tau x^{-t})$ to $\sigma_1(g)\chi_2(\lambda/det(g)) = (\sigma_1\chi_2^{-1})(g)\chi_2(\lambda) = (\sigma_1\chi_2^{-1})(\tau x^{-t})\chi_2(\tau)$. Using the contragredient representation, this gives $(\sigma_1\chi_2^{-1})^*(x)(\omega_{\sigma_1}\chi_2^{-2}\chi_2)(\tau) = (\sigma_1\chi_1^{-1})(x)\chi_1(\tau)$. In the symbolic short notation used in Sect. 4.7 the lift $r\big((\sigma_1, Ind(\chi_1, \chi_2))\big)$ becomes the induced representation

$$\sigma_1\chi_1^{-1} \lhd \chi_1.$$

This proves the induction formula. □

Proof of case (2). By the last lemma and from the previous results it remains for us to show that

$$r(\sigma_1, Ind(\chi_1, \chi_2)) \equiv \sigma_1\chi_1^{-1} \lhd \chi_1$$

is irreducible and tempered for all irreducible tempered representations σ_1 of $Gl(2, F)$ and unitary characters χ_1 and is at least irreducible for all unitary σ_1 and all $\sigma_2 = Ind(\chi_1, \chi_2)$ in the unitary complementary series.

We distinguish certain cases. First, for special or cuspidal σ_1 and unitary χ_1 the representation $\sigma_1\chi_1^{-1} \lhd \chi_1$ is irreducible tempered by the results obtained by Shahidi (σ_1 cuspidal) and by Tadic (σ_1 special): In fact for $\chi_1^2 = 1$, the last case is implicitly contained in the discussion of case (a) on page 127. If $\chi_1^2 \neq 1$ only for the nonunitary characters $\chi_1 = \nu^{\pm 3/2}$ and $\chi_1 = \chi_0\nu^{\pm 1/2}$ the representation $Sp(\chi_1^{-1}) \lhd \chi_1$ becomes reducible. See the second basic reducible case on page 123. We leave the details as an exercise. Hence, in fact, the representation $r(\sigma_1, Ind(\chi_1, \chi_2)) \equiv \sigma_1\chi_1^{-1} \lhd \chi_1$ is still irreducible for $Ind(\chi_1, \chi_2)$ in the unitary complementary series, since then $\chi_1 = \chi\nu^{-s}$ and $\chi_2 = \chi\nu^s$ holds for a unitary character χ and $-\frac{1}{2} < s < \frac{1}{2}$.

Second, $r(Ind(\tilde{\chi}_1, \tilde{\chi}_2) \times Ind(\chi_1, \chi_2))$ is irreducible tempered for unitary characters $\chi_i, \tilde{\chi}_i$, since the unitary principal series of G are irreducible (Rodier–Tadic). But the representation remains to be irreducible if either $\sigma_1 = Ind(\tilde{\chi}_1, \tilde{\chi}_2)$ or $\sigma_2 = Ind(\chi_1, \chi_2)$, or both of them belong to the unitary complementary series. For this notice that for characters $\chi_1, \chi_2, \tilde{\chi}_1, \tilde{\chi}_2$ such that $\chi_1\chi_2 = \tilde{\chi}_1\tilde{\chi}_2$ the representation $r(I(\tilde{\chi}_1, \tilde{\chi}_2), I(\chi_1, \chi_2)) \equiv \frac{\tilde{\chi}_1}{\tilde{\chi}_2} \times \frac{\chi_2}{\chi_1} \lhd \chi_1$ is irreducible except for $\tilde{\chi}_1/\tilde{\chi}_2 = \nu^{\pm 1} \iff \tilde{\chi}_1\tilde{\chi}_2/\chi_1^2 = \nu^{\pm 1}$ (that means $I(\tilde{\chi}_1, \tilde{\chi}_2)$ or $I(\chi_1, \chi_2)$ is not irreducible) or $\tilde{\chi}_2/\chi_1 = \nu^{\pm 1}$ (or $\tilde{\chi}_1/\chi_1 = \nu^{\pm 1}$, which amounts to an interchange of $\tilde{\chi}_1$ and $\tilde{\chi}_2$). The first cases need not be considered. In the latter case we can assume $\tilde{\chi}_2/\chi_1 = \nu$ by an interchange of (χ_1, χ_2) and $(\tilde{\chi}_1, \tilde{\chi}_2)$. Hence, $(\tilde{\chi}_1, \tilde{\chi}_2) = (\nu^{-1}\chi_2, \nu\chi_1)$ by $\chi_1\chi_2 = \tilde{\chi}_1\tilde{\chi}_2$, which gives $\tilde{\chi}_1/\tilde{\chi}_2 = \nu^{-2}\chi_2/\chi_1$. This cannot happen if the σ_i belong to the complementary series and/or the tempered principal series. This proves the assertions made in case (2). □

Let us turn to the degenerate case $r\big((\sigma_1, \sigma_2)\big)$ where $\sigma_2 = 1$ is the trivial representation.

Proof of Case (3). First assume σ_1 is cuspidal or $\sigma_1 = Sp(\chi_0), \chi_0 \neq 1$ and without restriction of generality with central character $\omega_{\sigma_1} = 1$. Then $r(\sigma)$ is

$$r\big((\sigma_1, Ind(\nu^{-1/2}, \nu^{1/2}))\big) - r\big((\sigma_1, Sp)\big) = (\sigma_1\nu^{1/2}) \triangleleft \nu^{-1/2} - r\big((\sigma_1, Sp)\big)$$

$$= \big(\delta(\sigma_1\nu^{1/2} \triangleleft \nu^{-1/2}) + J_P(\sigma_1\nu^{1/2} \triangleleft \nu^{-1/2})\big) - \big(\delta(\sigma_1\nu^{1/2} \triangleleft \nu^{-1/2}) - \pi_-(\sigma_1)\big)$$

$$= J_P(\sigma_1\nu^{1/2} \triangleleft \nu^{-1/2}) + \pi_-(\sigma_1).$$

Here we used $\sigma_1\nu^{1/2} \triangleleft \nu^{-1/2} \equiv \delta(\sigma_1\nu^{1/2} \triangleleft \nu^{-1/2}) + J_P(\sigma_1\nu^{1/2} \triangleleft \nu^{-1/2})$ in the Grothendieck group. $J_P(\sigma_1\nu^{1/2} \triangleleft \nu^{-1/2})$ is the nontempered Langlands quotient, which arises together with the discrete series representation $D = \delta(\sigma_1\nu^{1/2} \triangleleft \nu^{-1/2})$ in $(\sigma_1\nu^{1/2}) \triangleleft \nu^{-1/2}$. See page 135 for the special representations, and page 119 for cuspidal σ_1. Recall that $\pi_-(\sigma_1)$ is a cuspidal representation. Hence,

$$r\big((Sp(\chi_0), 1)\big) = J_P(Sp(\chi_0)\nu^{1/2} \triangleleft \chi_0\nu^{-1/2}) + \pi_-(Sp(\chi_0)).$$

Next consider the case $\sigma_1 = Sp$, where $r\big((Sp, 1)\big) = r\big((Sp, Ind(\nu^{-1/2}, \nu^{1/2}))\big) - r\big((Sp, Sp)\big)$. $r\big((Sp, Ind(\nu^{-1/2}, \nu^{1/2}))\big) = Sp(\nu^{1/2}) \triangleleft \nu^{-1/2}$ is the representation denoted $L \equiv T_+ + J'$ in the discussion of the nonregular case (a) on page 127. Recall

$$J' = J_P(\nu^{1/2}, Sp \boxtimes \nu^{-1/2}) = J_P(Sp(\nu^{1/2}) \triangleleft \nu^{-1/2})$$

is an irreducible nontempered Langlands quotient. On the other hand, $r\big((Sp, Sp)\big) = T_+ - T_-$. Hence, $r\big((Sp, 1)\big) = J' + T_-$.

The case of induced representations $\sigma_1 = Ind(\tilde{\chi}_1, \tilde{\chi}_2)$ for unitary characters $\tilde{\chi}_i$ remains. Then $\tilde{\chi}_1\tilde{\chi}_2 = 1$, since $\omega_{\sigma_1} = 1$. In the Grothendieck group $Sp + 1 \equiv Ind(\nu^{-1/2}, \nu^{1/2})$. Thus, $r\big((Ind(\tilde{\chi}_1, \tilde{\chi}_2), 1)\big) = \tilde{\chi}_1\nu^{1/2} \times \tilde{\chi}_2\nu^{1/2} \triangleleft \nu^{-1/2} - r\big((Ind(\tilde{\chi}_1, \tilde{\chi}_2), Sp)\big)$ by the induction formula. Also $r\big((Ind(\tilde{\chi}_1, \tilde{\chi}_2), Sp)\big) = r\big((Sp, Ind(\tilde{\chi}_1, \tilde{\chi}_2))\big) = Sp(\tilde{\chi}_1^{-1}) \triangleleft \tilde{\chi}_1$, again by the induction formula. In fact $Ind(\tilde{\chi}_1, \tilde{\chi}_2)$, when considered as an element of the Grothendieck group, does not depend on the choice of the Borel group. $\tilde{\chi}_1\nu^{1/2} \times \tilde{\chi}_2\nu^{1/2} \triangleleft \nu^{-1/2} \equiv \tilde{\chi}_1\nu^{1/2} \times \tilde{\chi}_1\nu^{-1/2} \triangleleft \tilde{\chi}_2$ (on the level of Grothendieck groups) has two irreducible constituents. One is $Sp(\tilde{\chi}_1) \triangleleft \tilde{\chi}_2$. The other is $\tilde{\chi}_1 \circ det \triangleleft \tilde{\chi}_2$, which is nontempered. See page 134. Altogether this shows that $r\big((Ind(\tilde{\chi}_1, \tilde{\chi}_2), 1)\big) = \tilde{\chi}_1 \circ det \triangleleft \tilde{\chi}_2$ is irreducible nontempered. $\square$

Proof of cases (4) and (5). To compute $r\big((1, 1)\big)$ consider $r\big((Ind(\nu^{1/2}, \nu^{-1/2}), Ind(\nu^{-1/2}, \nu^{1/2}))\big)$. By the induction formula this is $\nu \times 1 \triangleleft \nu^{-1/2} \equiv 1 \times \nu \triangleleft \nu^{-1/2}$ in the Grothendieck group. According to the discussion of the nonregular case (b) in Sect. 4.7 (see page 128), this is $J + J' + T_+ + T_-$. Since $r\big((Sp, Sp)\big) = T_+ - T_-$, this implies $2 \cdot r\big((Sp, 1)\big) + r\big((1, 1)\big) = J + J' + 2T_-$. Subtracting $2 \cdot r\big((Sp, 1)\big) = 2J' + 2T_-$, this gives $r\big((1, 1)\big) = J - J'$.

Now the final case $\chi_0 \neq 1, \chi_0^2 = 1$:

$$r\big((\chi_0, 1)\big) = r\big((Ind(\chi_0\nu^{-1/2}, \chi_0\nu^{1/2}), 1)\big) - r\big((Sp(\chi_0), 1)\big)$$
$$= r\big((1, Ind(\chi_0\nu^{-1/2}, \chi_0\nu^{1/2}))\big) - r\big((Sp(\chi_0), 1)\big).$$

By the induction formula $r\big((1, Ind(\chi_0\nu^{-1/2}, \chi_0\nu^{1/2}))\big) = \chi_0\nu^{1/2} \circ det \triangleleft \chi_0\nu^{-1/2}$.
It is a quotient of $\chi_0\nu \times \chi_0 \triangleleft \chi_0\nu^{-1/2}$ (second basic case on page 123 for $\chi = \nu^{1/2}\chi_0$, regular case with four irreducible constituents) and

$$0 \to Sp(\nu^{1/2}\chi_0) \triangleleft \chi_0\nu^{-1/2} \to \nu\chi_0 \times \chi_0 \triangleleft \chi_0\nu^{-1/2} \to (\nu^{1/2}\chi_0 \circ det) \triangleleft \chi_0\nu^{-1/2} \to 0.$$

Therefore, $r\big((\chi_0, 1)\big) = A + B - J_P(Sp(\chi_0)\nu^{1/2} \triangleleft \chi_0\nu^{-1/2}) - \pi_-(Sp(\chi_0))$ using the earlier discussion of case (3). The first two terms denote the two irreducible constituents A, B of $\chi_0\nu^{1/2} \circ det \triangleleft \chi_0\nu^{-1/2}$. $\square$

Final Remark.   The local description of the endoscopic character lift carries over to arbitrary local fields of characteristic zero. See Sect. 5.1.

4.12 The Anisotropic Endoscopic Theta Lift

Consider the inner form M_c of M, which is characterized by

$$M_c(F) = D^* \times D^*/\{(x, x^{-1}) \mid x \in F^*\} \hookrightarrow GO(4, q)(F),$$

where D is a quaternion skew field over the local field F, and where

$$q = Norm_{D/F} : D \to F$$

is the norm form $v\bar{v} = q(v)$. The group $M_c(F)$ acts on D by

$$(d_1, d_2) \cdot v = d_1 v \bar{d}_2, \qquad d_1, d_2 \in D^*, v \in D.$$

The kernel SO of the map $M_c(F) \to F^*$, which maps (d_1, d_2) to $Norm_{D/F}(d_1 d_2)$, is compact. Irreducible admissible representations of $M_c(F)$ are of the form $\check{\sigma}(d_1, d_2) = \check{\sigma}(d_1) \otimes \check{\sigma}_2(d_2)$, where $\check{\sigma}_i$ are irreducible admissible representations of D^*, whose central characters coincide. We write $\check{\sigma} = (\check{\sigma}_1, \check{\sigma}_2)$.

The bilinear form $2 \cdot B(v, w) = q(v+w) - q(v) - q(w)$ is $2 \cdot B(v, w) = v\bar{w} + w\bar{v}$; hence, $B(v, w) = \frac{1}{2} tr_{D/F}(v\bar{w})$ for the reduced trace $tr_{D/F}$. Extend the bilinear form B from D to the form $B \oplus B$ on D^2. Fix some nontrivial additive character ψ of F. Every nontrivial additive character $\psi(x)$ is of the form $\psi_t(x) = \psi_1(tx)$ for some $t \in F^*$. Let $S(D)$ and $S(D^2)$ denote the space of Schwartz–Bruhat functions on the vector spaces D and D^2, respectively. $M_c(F)$ naturally acts on $S(D)$ and $S(D^2)$, and hence also on its subgroup SO. For $X, Y \in D^2$ and symmetric 2×2-matrices with coefficients in F put

$$Q(T, X, Y) = Trace \left(T \cdot \begin{pmatrix} B(x_1, y_1) & B(x_1, y_2) \\ B(x_2, y_1) & B(x_2, y_2) \end{pmatrix} \right), \qquad X = (x_1, x_2),$$

$$Y = (y_1, y_2).$$

The Weil Representation π. This representation of $GSp(4, F)$ is defined on the Schwartz space $S(D^2 \times F^*)$, and is given for generators of the group. Since the norm form q of D represents every $t \in F^*$, the Weil constant $\varepsilon(q, \psi_t) = \int \psi_t(q(x))dx/|\int \psi_t(q(x))dx|$ does not depend on t (the integral is the limit of integrals over $\wp_F^{-n}$, since $\psi_t(q(x))$ is not in L^1). Hence, the general formulas of the literature are simplified considerably in this case. They are

$$\pi\left(\begin{pmatrix} E & T \\ 0 & E \end{pmatrix}\right)\Phi(X, t) = \psi_t(Q(T, X, X)) \cdot \Phi(X, t),$$

where $X = (x_1, x_2) \in D^2$. For $X \circ A = (x_1 a_{11} + x_2 a_{21}, x_1 a_{12} + x_2 a_{22})$ and $A = \begin{pmatrix} a_{11} & a_{12} \\ a_{21} & a_{22} \end{pmatrix}$

$$\pi\left(\begin{pmatrix} A & 0 \\ 0 & A^{-t} \end{pmatrix}\right)\Phi(X, t) = |det(A)|^2\Phi(X \circ A, t).$$

Then

$$\pi\left(\begin{pmatrix} 0 & E \\ -E & 0 \end{pmatrix}\right)\Phi(X, t) = \varepsilon_2(q, \psi_t)\hat{\Phi}(X, t),$$

with the Fourier transform defined by

$$\hat{\Phi}(X, t) = \int_D \Phi(Y, t)\psi_t\left(2Q(E, X, Y)\right)dY_{\psi_t}.$$

Here $\varepsilon_2(q, \psi_t)$ denotes the Weil constant, and dY_{ψ_t} denotes the self-dual Haar measure on D with respect to this Fourier transform. Finally,

$$\pi\left(\begin{pmatrix} E & 0 \\ 0 & \lambda E \end{pmatrix}\right)\Phi(X, t) = \Phi(X, t\lambda^{-1}).$$

The representation π is a preunitary left representation on $S(D^2 \times F^*)$, and defines a unitary representation on the Hilbert space $L^2(D^2 \times F^*)$. The Weil representation commutes with the unitary right action of the group $M_c(F)$ defined by

$$\pi((d_1, d_2))\Phi(X, t) = |q(d_1 d_2)|^2 \cdot \Phi(d_1 X \overline{d}_2, q(d_1 d_2)^{-1}t).$$

Notice an element $\lambda \in F^*$ can be viewed as an element of the center $\{(\lambda, 1)\}$ of $M_c(F)$ or alternatively as an element $diag(\lambda E, \lambda E)$ in the center of $GSp(4, F)$. In both senses it acts by $\Phi(X, t) \mapsto |\lambda|^4\Phi(\lambda X, t/\lambda^2)$. For the isotropic case, see [78].

Central Characters. For unitary characters ω of F^* one defines a surjective projection map

$$S(D^2 \times F^*) \to S(D^2 \times F^*, \omega)$$

by

$$\Phi(X, t) \mapsto \int_{F^*} |\lambda|^4\Phi(\lambda X, t\lambda^{-2})\omega(\lambda)^{-1}d^\bullet\lambda.$$

Notice that every equivariant map from $S(D^2 \times F^*)$ to a vector space, on which the center of $G(F)$ or $M_c(F)$ acts by ω, factorizes over this projection map. Hence, there is an induced action of $G(F) \times M_c(F)$ on the space $S(D^2 \times F^*, \omega)$, which makes the projection map equivariant with respect to $G(F) \times M_c(F)$. The representation thus obtained is a ω-representation, i.e., the center of $G(F)$ acts by the character ω.

Let G^0 and M_c^0 denote the subgroups of $G(F)$ and $M_c(F)$, respectively, of finite index, defined by all elements whose similitude factor respectively whose norm is in $(F^*)^2$. Let $S_{t_0}(D^2 \times F^*, \omega)$ be the subspace of $S(D^2 \times F^*, \omega)$ of all functions whose support with respect to the second variable $t \in F^*$ is contained in the coset $t_0(F^*)^2 \subseteq F^*$. By choosing representatives t_0 of $F^*/(F^*)^2$, we obtain a decomposition

$$S(D^2 \times F^*, \omega) = \bigoplus_{t_0 \in F^*/(F^*)^2} S_{t_0}(D^2 \times F^*, \omega),$$

where the subspaces $S_{t_0}(D^2 \times F^*, \omega)$ are stable under the action of the subgroup $G^0 \times M_c^0$. Let ω_0 be the restriction of the character ω to the subgroup $\{\pm 1\} \subseteq F^*$. Then, by specializing the value of t, we obtain identifications

$$ev_{t_0} : S_{t_0}(D^2 \times F^*, \omega) \cong S(D^2, \omega_0),$$

$$\Phi(X, t) \mapsto \Phi(X, t_0).$$

Using this isomorphisms, the action of $G^0 \times M_c^0$ transports to an action on the space $S(D^2)$ denoted π_{t_0}. Hence, as a $G^0 \times M_c^0$ module,

$$S(D^2 \times F^*, \omega) \cong \bigoplus_{t_0 \in F^*/(F^*)^2} (S(D^2), \pi_{t_0}).$$

Lemma 4.28. *The Weil representation π of $GSp(4, F)$ on $S(D^2 \times F^*, \omega)$ is admissible. Its restriction to the subgroup $Sp(4, F)$ is preunitary.*

Proof. For both statements it is enough to consider the action of the subgroup $Sp(4, F)$. One is thus reduced to the finitely many representations $S(D^2), \pi_{t_0})$, for which the claim is easy to verify. $\square$

The Weil Representation $\tilde{\pi}$. Similarly one defines a unitary Weil representation $\tilde{\pi}$ of the group $Gl(2, F)$ on $S(D \times F^*)$ by the formulas

$$\tilde{\pi}(\begin{pmatrix} 1 & \tau \\ 0 & 1 \end{pmatrix})\Phi(x, t) = \psi_t(\tau B(x, x)) \cdot \Phi(x, t),$$

$$\tilde{\pi}(\begin{pmatrix} a & 0 \\ 0 & a^{-1} \end{pmatrix})\Phi(x, t) = |a|^2 \Phi(xa, t),$$

$$\tilde{\pi}(\begin{pmatrix} 0 & 1 \\ -1 & 0 \end{pmatrix})\Phi(x, t) = \varepsilon_1(q, \psi_t)\hat{\Phi}(x, t),$$

where now $\varepsilon_1(q, \psi_1) = -1$ for the standard additive character $\psi = \psi_1$ of F, and where dy_{ψ_t} denotes the self-dual Haar measure on D for Fourier transform with respect to ψ_t

$$\hat{\Phi}(x, t) = \int_D \Phi(y, t) \psi_t (2 \cdot B(x, y)) dy_{\psi_t}.$$

Again

$$\tilde{\pi}\left(\begin{pmatrix} 1 & 0 \\ 0 & \lambda \end{pmatrix}\right) \Phi(x, t) = \Phi(x, t\lambda^{-1}).$$

This defines a unitary left representation of $Gl(2, F)$. It commutes with the unitary right action of the group $M_c(F)$ on $S(D \times F^*)$ defined by

$$\tilde{\pi}((d_1, d_2)) \Phi(x, t) = |q(d_1 d_2)| \Phi(d_1 x \bar{d}_2, q(d_1 d_2)^{-1} t).$$

Similarly, as above one can restrict oneself to a subgroup of elements whose determinant or norm is contained in $(F^*)^2$ and describe the resulting representations within $S(D)$.

4.12.1 The Global Situation

Now we switch the notation. Let F be a number field with adele ring $\mathbb{A}$ and D a global quaternion algebra over F. There are global analogues π and $\tilde{\pi}$ of the local Weil representations defined above on the global Schwartz spaces $S(D_{\mathbb{A}}^2 \times \mathbb{A}^*)$ and $S(D_{\mathbb{A}} \times \mathbb{A}^*)$, respectively. The theta distribution

$$\vartheta : S(D_{\mathbb{A}}^i \times \mathbb{A}^*) \to \mathbb{C}, \qquad (i = 1, 2)$$

$$\vartheta_\Phi = \sum_{(X, t) \in D^i \times F^*} \Phi(X, t)$$

defines functions

$$\vartheta_\Phi(m, g) = \vartheta_{\pi(m,g)\Phi} \text{ and } \vartheta_{\tilde{\pi}(m,g)\Phi}, \qquad (m, g) \in M_c(\mathbb{A}) \times G(\mathbb{A})$$

on $M_c(\mathbb{A})/M_c(F) \times G(F) \backslash G(\mathbb{A})$ of moderate growth for $G = GSp(4)$ and $Gl(2)$, respectively. For an irreducible subspace τ of the space of cusp forms on $M_c(F) \backslash M_c(\mathbb{A})$ the theta representation $\Theta_-(\tau)$ is defined to be the $G(\mathbb{A})$-invariant space of all automorphic forms on $GSp(4, F) \backslash GSp(4, \mathbb{A})$ spanned by the functions

$$\int_{M_c(F) \backslash M_c(\mathbb{A})} \vartheta_\Phi(m \times g) \overline{f}(\overline{m}) dm,$$

Φ in $S(D_{\mathbb{A}}^2 \times \mathbb{A}^*)$, and where f runs over all functions in the representation space of τ. Let ω denote the central character of τ. Suppose the representation space $\Theta_-(\tau)$

is contained in the space of cusp forms. Let $\pi(\tau)$ be some irreducible constituent of it. By evaluation with $\Phi = \Phi_p \Phi^p$ for fixed Φ^p outside p one obtains local maps $\vartheta_{\Phi^p} : S(D^2 \times F^*) \to \pi_p$ at p. They factorize over the projection map

$$
\begin{array}{ccc}
S(D^2 \times F^*) & \xrightarrow{\quad \vartheta_{\Phi^p} \quad} & \pi_p \\
\downarrow & \nearrow & \\
S(D^2 \times F^*, \omega_p) & &
\end{array}
$$

4.12.2 The Anisotropic Theta Lift (Local Theory)

Let F be local again. Irreducible left representations $\breve{\sigma}$ of the group $M_c(F)$ are of the form $\breve{\sigma} = (\breve{\sigma}_1, \breve{\sigma}_2)$ for irreducible representations $\breve{\sigma}_1, \breve{\sigma}_2$ of the group D^*, whose central characters, denoted ω, coincide. Since D^* is compact modulo the center, the irreducible representations $\breve{\sigma}_i$, and hence also $\breve{\sigma}$, are finite-dimensional. Let $\breve{\sigma}^*$ denote the contragredient representation. The Weil representation $\tilde{\pi}$ defines a right action of $M_c(F)$. For the left action $\tilde{\pi}_l(h) = \tilde{\pi}(h^{-1})$, let W be the maximal quotient space of $S(D \times F^*)$, on which $M_c(F)$ acts from left via $\tilde{\pi}_l$ by the irreducible dual representation $\breve{\sigma}^*$. Then W is a quotient of $S(D \times F^*, \omega)$ for the central character ω of $\breve{\sigma}_i$. Since the action $\tilde{\pi}$ of $Gl(2, F)$ commutes with the action of $M_c(F)$, W is a module under $Gl(2, F)$. In fact it is isomorphic to the following $Gl(2, F)$-submodule of the tensor product

$$
S(D \times F^*, \breve{\sigma}) = \{\Phi \in S(D \times F^*, \omega) \otimes V_{\breve{\sigma}} \mid \tilde{\pi}(h)\Phi = \breve{\sigma}(h)\Phi , \; \forall h \in M_c(F)\}.
$$

In other words, the functions in this space satisfy

$$
\breve{\sigma}(d_1, d_2)\Phi(x, t) = |q(d_1, d_2)| \cdot \Phi(d_1 x \bar{d}_2, t/q(d_1 d_2)).
$$

In fact, $M_c(F)$ is compact modulo the center, and W is the maximal $\breve{\sigma}^*$-quotient of $S(D \times F^*, \omega)$, where the center acts by a scalar. So the last claim is a consequence of the theorem of Peter and Weyl and Schur's lemma.

Similarly, one defines $GSp(4, F)$-modules $S(D^2 \times F^*, \breve{\sigma})$.

Lemma 4.29. *The $Gl(2, F)$-representation space $S(D \times F^*, \breve{\sigma})$ is zero unless $\breve{\sigma} \cong (\breve{\sigma}_1, \breve{\sigma}_2)$ with $\breve{\sigma}_1 \cong \breve{\sigma}_2$.*

Proof. $\tilde{\pi}(M_c(F))$ commutes with the natural action of the center F^* on $S(D \times F^*)$. If $\Phi(x, t) \in S(D \times F^*, \breve{\sigma})$ is not zero, then there exists a quasicharacter χ of F^* such that for a preimage Φ' of Φ in $S(D \times F^*) \otimes V_{\breve{\sigma}}$ we have

$$
\Phi_\chi(x) = \int_{z \in F^*} \int_{F^*} \Phi'(zx, tz^{-2})|z|^2 \omega(z)^{-1} \chi(t)^{-1} d^\bullet t d^\bullet z \neq 0.
$$

Then $\Phi_\chi \in S(D)$ and $\Phi_\chi(d_1 x \bar{d}_2) = (\chi|.|^{-1})(q(d_1 d_2))\check{\sigma}(d_1, d_2)\Phi_\chi(x)$. For $(d_2)^+ := (\bar{d}_2)^{-1} = d_2/q(d_2)$ without restriction of generality we can assume

$$\check{\sigma}_2^*(d_2) = \check{\sigma}_2(d_2^+)$$

to be defined on the same space. Then $\Phi_\chi(d_1 x d_2^{-1}) = (\chi|.|^{-1})(q(d_1)/q(d_2)) \cdot \check{\sigma}_1(d_1) \boxtimes \check{\sigma}_2^*(d_2) \Phi_\chi(x)$. The exact sequence $0 \to S(D^*) \to S(D) \to \mathbb{C} \to 0$ and the Peter–Weyl theorem for the compact group D^*/F^* and the nonvanishing of Φ_χ forces $\check{\sigma}_1 \cong \check{\sigma}_2$ as in [41], Theorem 9.1a. $\square$

The situation for the Weil representation of the group $GSp(4, F)$ is different.

Lemma 4.30. *For all irreducible unitary representations $\check{\sigma}$ the representation of $GSp(4, F)$ on $S(D^2 \times F^*, \check{\sigma})$ is nontrivial.*

Proof. Follows from [41], Theorem 9.1b, and [105]. $\square$

Definition 4.6. Let $\Theta_-(\check{\sigma})$ denote the admissible quotient representation $(S(D^2 \times F^*, \check{\sigma}), \pi(\check{\sigma}))$ of the Weil representation π of $GSp(4, F)$ defined on $S(D^2 \times F^*)$.

4.12.3 The Jacquet–Langlands Lift

Jacquet and Langlands defined a lift which assigns to an irreducible representation $\check{\rho}$ of the multiplicative group D^* of a quaternion algebra over the local field F an irreducible representation $\rho = JL(\check{\rho})$ of $Gl(2, F)$ in the discrete series. This defines a bijection between the classes of irreducible representations of D^* and the classes of admissible irreducible representations of $Gl(2, F)$ in the discrete series.

The lift $JL(\check{\rho})$ is defined as follows. Let ω be the central character of $\check{\rho}$. Let D^1 be the subgroup of D^* of elements of norm 1. Define a representation ρ of $Gl(2, F)$ (see [42], Propositions 1.3 and 1.5 for K a quaternion algebra) on the space of functions $\Phi(y)$ in

$$S(D, \check{\rho}) = \{\Phi \in S(D) \otimes V_{\check{\rho}} \mid \Phi(xh) = \check{\rho}(h)^{-1}\Phi(x) , h \in D^1\}$$

by

$$\rho(\begin{pmatrix} 1 & \tau \\ 0 & 1 \end{pmatrix})\Phi(y) = \psi(\tau B(x, x))\Phi(y),$$

$$\rho(\begin{pmatrix} a & 0 \\ 0 & a^{-1} \end{pmatrix})\Phi(y) = |a|^2 \Phi(ya),$$

$$\rho(\begin{pmatrix} 0 & 1 \\ -1 & 0 \end{pmatrix})\Phi(y) = \varepsilon_1(q, \psi)\hat{\Phi}(y),$$

for $\tau \in F, a \in F^*$, where $\varepsilon_1(q, \psi) = -1$ for our fixed standard additive character ψ of F, and where dy_ψ is the self-dual Haar measure on D for Fourier transform with respect to ψ

$$\hat{\Phi}(y) = \int_D \Phi(z)\psi(2 \cdot B(z,y))dz_\psi.$$

Finally, for an arbitrary $x_\lambda \in D^1 \setminus D^*$ with norm $q(x_\lambda) = \lambda$

$$\rho\left(\begin{matrix} \lambda & 0 \\ 0 & 1 \end{matrix}\right)\Phi(y) = |q(x_\lambda)| \cdot \check{\rho}(x_\lambda)\Phi(yx_\lambda).$$

Then $\rho\left(\begin{matrix} \lambda & 0 \\ 0 & \lambda \end{matrix}\right)\Phi(y) = \omega(\lambda)\Phi(y)$. According to [42], Theorem 4.2(ii), the representation ρ is isotypic, i.e., isomorphic to $deg(\check{\rho})$ copies of the irreducible representation $JL(\check{\rho})$.

4.12.4 The Intertwining Map b

Let $\check{\sigma} = (\check{\sigma}_1, \check{\sigma}_2)$ be an irreducible representation of $M_c(F)$ on $V_{\check{\sigma}} = V_{\check{\sigma}_1} \otimes V_{\check{\sigma}_2}$. Let $V^*_{\check{\sigma}_i}$ denote the dual space of $V_{\check{\sigma}_i}$ for $i = 1, 2$. Then we define a map

$$b: \quad V^*_{\check{\sigma}_1} \otimes S(D \times F^*, \check{\sigma}) \quad \longrightarrow \quad S(D, \check{\sigma}_2)$$

by

$$v_1^* \otimes \Phi(y, t) \mapsto v_1^*(\Phi(y, 1)).$$

b is well defined. For $h \in D^1$ we have $h^+ = h$ and $q(h) = 1$ by definition. Hence,

$$b(v_1^* \otimes \Phi)(yh) = v_1^*(\Phi(yh, 1)) = v_1^*(|q(h^{-1})|\Phi(y\overline{h}^{-1}, q(h)))$$

$$= v_1^*\left(\tilde{\pi}(1, h^{-1})\Phi(y, 1)\right) = v_1^*\left(\check{\sigma}_2(h^{-1})\Phi(y, 1)\right) = \check{\sigma}_2(h^{-1})v_1^*(\Phi(y, 1))$$

$$= \check{\sigma}_2(h)^{-1}b(v_1^* \otimes \Phi)(y).$$

This implies $b(v_1^* \otimes \Phi) \in S(D, \check{\sigma}_2)$.

Equivariance. b is equivariant with respect to the representations $triv_{V^*_{\check{\sigma}_1}} \otimes \tilde{\pi}$ of $Gl(2, F)$ on $V^*_{\check{\sigma}_1} \otimes S(D \times F^*, \check{\sigma}_1 \times \check{\sigma}_2)$ and $\rho = JL(\check{\sigma}_2)$ on the image space. Since b is defined by the specialization $t = 1$, this assertion is clear for the action of the subgroup $Sl(2, F)$ by a direct comparison of the generators of $Sl(2, F)$. Equivariance for the elements $diag(\lambda, 1)$ follows from

$$b(v_1^* \otimes \tilde{\pi}(diag(\lambda, 1))\Phi(y, t)) = v_1^*(\omega(\lambda)\Phi(y, \lambda))$$

and

$$\rho(diag(\lambda, 1))b(v_1^* \otimes \Phi(y, t)) = |\lambda|\check{\sigma}_2(x_\lambda)v_1^*(\Phi(yx_\lambda, 1)), \qquad q(x_\lambda) = \lambda$$

$$= |\lambda| \check{\sigma}_2(x_\lambda) v_1^*(\Phi(yx_\lambda, \lambda/q(x_\lambda)))$$

$$= \check{\sigma}_2(x_\lambda) v_1^*(\check{\sigma}_2(\overline{x}_\lambda)\Phi(y, \lambda))$$

$$= \check{\sigma}_2(x_\lambda \overline{x}_\lambda) v_1^*(\Phi(y, \lambda))$$

$$= v_1^*(\omega(\lambda)\Phi(y, \lambda)).$$

The Kirillov Model. Assume $\check{\sigma} = (\check{\sigma}_1, \check{\sigma}_2)$ for $\check{\sigma}_1 \cong \check{\sigma}_2$ (Lemma 4.29). Then without restriction of generality $V_{\check{\sigma}_1} = \mathbb{C}^d$ and $\check{\sigma}_2(d_2) = \check{\sigma}_1(d_2^+)^{-t}$ (both acting on $\mathbb{C}^d$). This identifies $V_{\check{\sigma}}$ with the space of matrices $M_{d,d}(\mathbb{C})$ so that the action is defined by

$$\check{\sigma}(d_1, d_2)X = \check{\sigma}_1(d_1)X\check{\sigma}_1(\overline{d}_2), \qquad X \in V_{\check{\sigma}} = M_{d,d}(\mathbb{C}).$$

For $\Phi(x,t) \in S(D \times F^*, \check{\sigma})$ and $y \neq 0$ then

$$\check{\sigma}(y)\Phi(1,t) = \check{\sigma}(y,1)\Phi(1,t) = |q(y)|\Phi(y, t/q(y)),$$

$$\Phi(1,t)\check{\sigma}(y) = \check{\sigma}(1,\overline{y})\Phi(1,t) = |q(y)|\Phi(y, t/q(y)).$$

Hence,

$$\check{\sigma}_1(d)\Phi(1,t) = \Phi(1,t)\check{\sigma}_1(d)$$

for all $d \in D^*$. Since $\check{\sigma}_1$ is irreducible, this implies

$$\Phi(1,t) = \varphi(1,t) \cdot id, \qquad \varphi(1,t) \in \mathbb{C}.$$

Hence, for all $y \neq 0$

$$\Phi(y,t) = \varphi(1, tq(y))|q(y)|^{-1} \cdot \check{\sigma}_1(y) \in M_{d,d}(\mathbb{C}).$$

Notice $dim(\check{\sigma}_2) > 1$ implies $\Phi(0,t) = 0$. Hence, the last formula immediately implies

Lemma 4.31. *For $d = dim(\sigma_2) > 1$ and $\check{\sigma} = (\check{\sigma}_2, \check{\sigma}_2)$, acting on $M_{d,d}(\mathbb{C})$ as above, the map $\Phi(x,t) \mapsto \varphi(t) = \Phi(1,t)$ defines an isomorphism*

$$S(D \times F^*, \check{\sigma}) = \left\{ \Phi(x,t) = |q(x)|^{-1}\check{\sigma}_1(x)\varphi(tq(x)) \;\middle|\; \varphi \in S(F^*) \right\} \cong S(F^*),$$

in the sense that $\Phi(0,t) = 0$. For $dim(\check{\sigma}_2) = 1$ the right side has codimension 1 in $S(D \times F^*, \check{\sigma})$.

b **is injective.** Let e_i^* be a basis of $V_{\check{\sigma}_1}^*$ and suppose

$$0 = b\left(\sum_{i=1}^d e_i^* \otimes \Phi_i(x,t)\right) = |q(x)|^{-1}\sum_{i=1}^d \varphi_i(tq(x))e_i^*(\check{\sigma}_1(x)) \in \mathbb{C}^d.$$

For $d = 1$ then obviously $\Phi(x, t) = 0$. For $d > 1$ we put $x = 1$, then $\sum_i e_i^* \varphi_i(t) = 0$. Hence, $\varphi_i(t) = 0$ for all i, and therefore $\Phi_i(x, t) = 0$ by the last lemma. Hence, b is injective.

By Theorem 4.2(ii) in [42], the representation ρ is isomorphic to $deg(\check{\rho}) \cdot JL(\check{\rho})$, where $JL(\check{\rho})$ is the irreducible Jacquet–Langlands lift of $\check{\sigma}$. Hence, the representation $(\check{\pi}, S(D \times F^*, \check{\sigma}))$ for $\check{\sigma} = (\check{\sigma}_2, \check{\sigma}_2)$ is irreducible and isomorphic to $JL(\check{\sigma}_2)$. Furthermore,

Corollary 4.12. *b is an isomorphism.*

Lemma 4.32. *Suppose $\check{\sigma} = (\check{\sigma}_1, \check{\sigma}_2)$ satisfies $\check{\sigma}_1 \cong \check{\sigma}_2$ for irreducible $\check{\sigma}_2$. Then the representation $\check{\pi}(\check{\sigma})$ of $Gl(2, F)$ on $S(D \times F^*, \check{\sigma})$ is irreducible and isomorphic to the Jacquet–Langlands lift $\sigma = JL(\check{\sigma}_2)$.*

4.12.5 The Q-Jacquet Module of the Anisotropic Weil Representation $\Theta_-(\sigma)$

The Weil representations π of $GSp(4, F)$ and $\check{\pi}$ of $Gl(2, F)$, defined above, are related as follows. For the Klingen parabolic subgroup Q of $GSp(4, F)$ the representation $\check{\pi}$, considered as a representation of the Levi group, is isomorphic to the normalized Jacquet module of π with respect to the parabolic Q up to a character twist

$$\delta_Q^{-1/2} \pi(Q) \cong \check{\pi} \otimes \nu.$$

The subgroup $N \subseteq Q(F)$ defined by all matrices

$$n(\tau) = \begin{pmatrix} 1 & 0 & 0 & 0 \\ 0 & 1 & 0 & \tau \\ 0 & 0 & 1 & 0 \\ 0 & 0 & 0 & 1 \end{pmatrix}, \qquad \tau \in F,$$

is the center of the unipotent radical of Q. For $\Phi \in S(D^2 \times F^*)$

$$\left(\pi(n(\tau)) - 1 \right) \Phi(x_1, x_2, t) = \left(\psi_t(\tau B(x_2, x_2)) - 1 \right) \Phi(x_1, x_2, t).$$

Since $\psi_t(\tau B(0, 0)) = 1$, the map $p : S(D^2 \times F^*) \to S(D \times F^*)$ defined by $p : \Phi(x_1, x_2, t) \mapsto \Phi(x_1, 0, t)$ factorizes over a quotient map

$$\bar{p} : S(D^2 \times F^*) / \left(\pi(N) - 1 \right) S(D^2 \times F^*) \xrightarrow{\simeq} S(D \times F^*)$$

such that

Claim 4.4. The induced map $\bar{p}$ is an isomorphism.

Proof. Any Schwartz–Bruhat function with $\Phi(x_1, 0, t) = 0$ is a finite linear combination of products $\phi_1(x_1, t)\phi_2(x_2)$ with $\phi_1 \in S(D \times F^*)$ and $\phi_2 \in S(D)$ and $\phi_2(0) = 0$. So one is reduced to showing that $\phi_2(0) = 0$ implies that there exists $\phi_3(x_2, t) \in S(D \times F^*)$ and $\tau \in F^*$ such that $\phi_2(x_2) = (\psi_t(\tau B(x_2, x_2)) - 1)\phi_3(x_2, t)$ for all t in the support of $Supp(\phi_1)$. Since this support condition restricts t to be bounded $|t| \geq |t_0|$, this can easily be achieved by choosing τ with $|\tau|$ sufficiently large, so $\psi(t\tau q(x_2)) = 1$ implies $x_2 \in V$ for some fixed open subset V of the point 0, on which ϕ_2 vanishes. Define $\phi_3(x_2, t)$ to be zero on V and to be $\phi_2(x_2)/(\psi_t(\tau B(x_2, x_2)) - 1)$ on the complement.

Notice that

$$p \in Hom_{Q(F)}\left(\delta_Q^{-1/2} \otimes S(D^2 \times F^*), S(D \times F^*) \otimes \nu\right)$$

is $Q(F)$-equivariant. For $\eta \in Q(F)$

$$\eta = \begin{pmatrix} \alpha & 0 & \beta & * \\ * & a & * & * \\ \gamma & 0 & \delta & * \\ 0 & 0 & 0 & \lambda a^{-1} \end{pmatrix},$$

we get after a slightly lengthy computation, for which one has to express the embedded involution in the Weyl group of Q as a product of elementary matrices in $GSp(4, F)$, the formula

$$\left(\pi(\eta)\Phi\right)_{(X,t)=(x,0,t)} = |a|^2 \cdot \tilde{\pi}(\begin{pmatrix} \alpha & \beta \\ \gamma & \delta \end{pmatrix})\Phi(x, 0, t).$$

The unipotent radical of $Q(F)$ acts trivially on the right side. This implies

$$S(D \times F^*) \cong S(D^2 \times F^*)/(\pi(N) - 1) \cong S(D^2 \times F^*)/(\pi(Rad(Q)) - 1).$$

Since $\delta_Q^{1/2}(\eta) = |a|^2|\lambda|^{-1}$ and $\lambda = \alpha\delta - \beta\gamma$, this computes the normalized Jacquet module of π with respect to Q. $\quad\square$

Lemma 4.33. *$\bar{p}$ induces a $Q(F)$-equivariant isomorphism $\delta_Q^{-1/2} \otimes \pi(Q) \cong 1 \boxtimes (\tilde{\pi} \otimes \nu)$.*

Concerning the Notation. On the right side of the isomorphism in the last lemma we used the notation given on page 120. In other words the standard Levi component of Q is identified with the group $F^* \times Gl(2, F)$ using the coordinates a and $\alpha, \beta, \gamma, \delta$ of η from above. In this sense, the representation is trivial on F^* and isomorphic to $\tilde{\pi} \otimes \nu$ as a representation of the second factor $Gl(2, F)$.

Furthermore, p is $M_c(F)$-equivariant for trivial reasons, up to the factor $|q(d_1 d_2)|$. Hence,

Lemma 4.34. $\overline{p} : \delta_Q^{-1/2} \otimes \pi(Q) \cong \tilde{\pi} \otimes \nu$ is $M_c(F)$-equivariant, where $M_c(F)$ acts on $\delta_Q^{-1/2} \otimes \pi(Q)$ by the representation induced from π, and by the action induced from $\tilde{\pi} \otimes \nu$ as a right module and $\tilde{\pi}_l \otimes \nu^{-1}$ as a left module.

For a given irreducible representation $\check{\sigma}$ of $M_c(F)$ we already defined the maximal quotient representation $\Theta_-(\check{\sigma})$ of $S(D^2 \times F^*, \pi)$, on which $M_c(F)$ acts from the left by $\check{\sigma}^*$. Therefore, by last two lemmas $\delta_Q^{-1/2}\Theta_-(\check{\sigma})(Q) \cong 1 \boxtimes S(D \times F^*, \check{\sigma} \otimes \nu^{-1})$ is the maximal quotient of $S(D \times F^*, \tilde{\pi})$ on which $\tilde{\pi} \otimes \nu$ acts by $\check{\sigma}^*$

$$
\begin{array}{ccc}
\pi & \longrightarrow & \Theta_-(\check{\sigma}) = S(D^2 \times F^*, \check{\sigma}) \\
\downarrow & & \downarrow \\
\delta_Q^{1/2}\pi(Q) & \longrightarrow & \Theta_-(\check{\sigma})(Q) \\
\| & & \| \\
1 \boxtimes (\tilde{\pi} \otimes \nu) & \longrightarrow & JL(\check{\sigma} \otimes \nu^{-1}) \otimes \nu
\end{array}
$$

Using $JL(\check{\sigma} \otimes \nu^{-1}) \otimes \nu = JL(\check{\sigma})$ (the Jacquet–Langlands correspondence commutes with character twists), we obtain from Lemmas 4.29 and 4.32

Corollary 4.13. *For an irreducible representation $\check{\sigma} = (\check{\sigma}_1, \check{\sigma}_2)$ of $M_c(F)$ the normalized Jacquet module $\delta_Q^{-1/2}\Theta_-(\check{\sigma})(Q)$ vanishes for $\check{\sigma}_1 \not\cong \check{\sigma}_2$ and satisfies*

$$
\boxed{\delta_Q^{-1/2} \otimes \Theta_-(\check{\sigma})(Q) \cong 1 \boxtimes JL(\check{\sigma})}
$$

for $\check{\sigma}_1 \cong \check{\sigma}_2$.

Therefore, $\overline{p}$ defines a nontrivial intertwining adjunction map

$$
\overline{p}(\check{\sigma}) : \left(S(D^2 \times F^*, \check{\sigma}), \pi(\check{\sigma}) \right) \to Ind_Q^G \left(1 \boxtimes (S(D \times F^*, \check{\sigma})), \tilde{\pi}(\check{\sigma}) \right)
$$

by the adjunction formula (e.g., [93], Theorem 2.4.3)

$$
Hom_G((S(D^2 \times F^*), \pi(\check{\sigma})), Ind_Q^G(\tau)) \cong Hom_Q(\delta_Q^{-1/2}\pi(Q), \tau).
$$

4.12.6 Whittaker Models

We claim that $(S(D^2 \times F^*), \pi)$ does not admit a nontrivial Whittaker functional. Let ψ_1, ψ_2 be nontrivial additive characters and suppose L is a Whittaker functional, i.e.,

$$
L : S(D^2 \times F^*) \to \mathbb{C}, \qquad L\left(\pi(n(x, y, u, v, t)\Phi\right) = \psi_1(x)\psi_2(y)L(\Phi)
$$

for all $\Phi \in S(D^2 \times F^*)$ and all

$$n(x,y,u,v,t) = \begin{pmatrix} 1 & 0 & x & u \\ y & 1 & v & \tau \\ 0 & 0 & 1 & -y \\ 0 & 0 & 0 & 1 \end{pmatrix} \in Q(F)$$

in the unipotent radical $U(F)$ of the Borel subgroup. The commutator group consists of the matrices with $x = y = 0$.

Since $L\big(\pi(n(\tau))\Phi\big) = L(\Phi)$ for all $\tau \in F$, the functional L factorizes $L = \overline{L} \circ \overline{p}$ over a functional $\overline{L} : S(D \times F^*) \to \mathbb{C}$. Abbreviate $p(\Phi) = \overline{\Phi} \in S(D \times F^*)$. Then by the $U(F)$-equivariance of L and the $Q(F)$-equivariance of $\overline{p}$

$$\psi_2(y) \cdot \overline{L}(\overline{\Phi}) = \psi_2(y) \cdot L(\Phi) = L(\pi(n(0,y,*,*,*)\Phi) = \overline{L}(\overline{\Phi})$$

for all $y \in F$. Therefore, $\overline{L}(\overline{\Phi}) = 0$ for all $\overline{\Phi}$; hence, $L = 0$.

Lemma 4.35. *The spaces* $\Theta_-(\check{\sigma}) = S(D^2 \times F^*, \check{\sigma})$ *do not admit Whittaker models.*

Proof. If one of these spaces admitted a nontrivial Whittaker functional, then by composition with $S(D^2 \times F^*) \twoheadrightarrow S(D^2 \times F^*, \check{\sigma})$ $S(D^2 \times F^*)$ would also admit a nontrivial Whittaker functional, contradicting the computation above. This proves the lemma. □

4.12.7 The Siegel Parabolic

The normalized Jacquet module of the Weil representation $(S(D^2 \times F^*), \pi)$ for the Siegel parabolic P in $GSp(4)$ is isomorphic to the representation of the Levi component $Gl(2, F) \times Gl(1, F)$ of P

$$\boxed{\delta_P^{-1/2} \otimes \pi(P) = \left(\nu^{1/2} \circ det\right) \boxtimes \left(\nu^{3/2} \otimes S(F^*)\right)},$$

in the notation given on page 119.

Proof. $Q(T, X, X) = 0$ for all symmetric 2×2-matrices T implies $X = 0$. Therefore,

$$S(D^2 \times F^*)/\left(\pi(N) - 1\right) \cong S(F^*),$$

induced by

$$\Phi(x_1, x_2, t) \mapsto \Phi(0, 0, t) \in S(F^*),$$

where N denotes the unipotent radical of P. The argument is similar to the one used in the case of the Klingen parabolic. □

Since $blockdiag(A, \lambda A^{-t})$ has modulus $\delta_P = |det(A)/\lambda|^{3/2}$ and induces an action on $\Phi(0, 0, t)$ in $S(F^*)$ defined by $\Phi(0, 0, t) \mapsto |det(A)|^2 \Phi(0, 0, t/\lambda)$, we get $\Phi(0,0,t) \mapsto |det(A)|^{1/2}|\lambda|^{3/2} \cdot \Phi(0,0,t/\lambda)$ for the action on the normalized Jacquet module $S(F^*)$.

By the adjunction formula this induces a nontrivial adjunction map from $(S(D^2 \times F^*), \pi)$ to the induced representation $Ind_P^G(\nu^{1/2} \circ det \boxtimes (\nu^{3/2} \otimes S(F^*)))$. In short notation

$$(S(D^2 \times F^*), \pi) \longrightarrow \nu^{1/2} \circ det \triangleleft (\nu^{3/2} \otimes S(F^*)).$$

There are analogous maps for the quotient representations on $\Theta_-(\check{\sigma}) = S(D^2 \times F^*, \check{\sigma})$.

Corollary 4.14. *For irreducible representations $\check{\sigma}_1 \not\cong \check{\sigma}_2$ and $\check{\sigma} = (\check{\sigma}_1, \check{\sigma}_2)$ the admissible representation $\pi(\check{\sigma})$ of $GSp(4, F)$ on $\Theta_-(\check{\sigma}) = S(D^2 \times F^*, \check{\sigma})$ is a cuspidal representation.*

Proof. For all parabolic subgroups $\mathcal{P}$ the Jacquet module $S(D^2 \times F^*, \check{\sigma})(\mathcal{P})$ is a quotient of the Jacquet module for the Klingen parabolic group, as the discussion above shows. Under the assumption made for $\check{\sigma}$, the Jacquet module with respect to the Klingen parabolic vanishes (Corollary 4.13). Hence, the Jacquet module with respect to all standard parabolic subgroups is trivial. This proves the claim.   □

Corollary 4.15. *Suppose $\check{\sigma}_1 \cong \check{\sigma}_2$ is irreducible and $dim(\check{\sigma}_2) > 1$. Then the image of the adjunction map $\overline{p}(\check{\sigma}) : \Theta_-(\check{\sigma}) \to 1 \times JL(\check{\sigma}_2)$ is*

$$Image(\overline{p}(\check{\sigma})) = Wp_-(\sigma) \text{ for } \sigma = JL(\check{\sigma}_2).$$

The kernel of the map $\overline{p}(\check{\sigma})$ is an admissible cuspidal ω-representation, i.e., a cuspidal representation on which the center acts by the central character ω of $\check{\sigma}$.

Proof. Under the assumptions made, $\check{\sigma}$ cannot be a quotient representation of $\delta_P^{-1/2} \otimes \pi(P) = (\nu^{1/2}\circ det) \boxtimes (\nu^{3/2} \otimes S(F^*))$; hence, $\Theta_-(\check{\sigma})(P) = 0$. Therefore, the kernel of the adjunction map for the Q-Jacquet module is a cuspidal representation. On the other hand, the image of

$$\overline{p}(\check{\sigma}) : S(D^2 \times F^*, \check{\sigma}) \to Ind_{Q(F)}^{G(F)}(1 \boxtimes S(D \times F^*, \check{\sigma})) = 1 \times JL(\check{\sigma}_2)$$

is a subrepresentation of $1 \times JL(\check{\sigma}_2)$ (Lemma 4.32) such that $JL(\check{\sigma}_2)$ is cuspidal by our assumptions, and which decomposes

$$1 \times JL(\check{\sigma}_2) = Wp_+(JL(\check{\sigma}_2)) \oplus Wp_-(JL(\check{\sigma}_2))$$

into the direct sum of two irreducible representations (Lemma 4.19). Since the image of $\overline{p}(\check{\sigma})$ does not admit a Whittaker model (Lemma 4.35) and $Wp_+(\check{\sigma})$ does admit a Whittaker model (Theorem 4.4 and Corollary 4.11), the image of $\overline{p}(\check{\sigma})$ must be $Wp_-(\sigma)$ where $\sigma = JL(\check{\sigma}_2)$.   □

4.12.8 The Anisotropic Theta Lift $\theta_-(\sigma)$ (Global Theory)

Again consider a quaternion algebra D over some totally real number field F. Suppose S_0 is a finite number of places including all Archimedean places. Suppose D ramifies at all places v in S_0. The global group M_c is anisotropic modulo the center over the global field F such that $M_c(F) = (D^* \times D^*)/F^*$. The same holds for all local fields at places $v \in S_0$.

Suppose we have admissible irreducible representations

$$\check{\sigma}_v = \check{\sigma}_{1,v} \times \check{\sigma}_{2,v}, \qquad (v \in S_0)$$

of $M_c(F_v)$ for all places v in S_0. Assume there exists a global idele class character ω such that ω_v is the central character of $\check{\sigma}_{1,v}$ and $\check{\sigma}_{2,v}$ at all places $v \in S_0$. Let us make the following additional assumptions:

Assumption 1. There are places $v', v'' \in S_0$ such that $dim(\check{\sigma}_{1,v'}) > 1$ and $dim(\check{\sigma}_{2,v''}) > 1$.

Assumption 2. For some place $v''' \in S_0$ we have $\check{\sigma}_{v''',1} \not\cong \check{\sigma}_{v''',2}$.

Since $M_c(F_v)$ is anisotropic modulo the center for $v \in S_0$, the unitary representation of $M_c(F_v)$ on $S(D_v^2 \times F_v^*, \omega_v)$ is completely reducible:

$$S(D_v^2 \times F_v^*, \omega_v) = \bigoplus_{\omega(\check{\sigma}_{2,v}) = \omega_v} S(D_v^2 \times F_v^*, \check{\sigma}_v), \quad v \in S_0.$$

The functions Φ_v with compact support are dense in the Schwartz space at the Archimedean places.

Suppose we have functions

$$\Phi_v \neq 0 \in S(D_v^2 \times F_v^*, \check{\sigma}_v) \quad (\forall v \in S_0)$$

such that $\Phi_v \in R_v$, where R_v denotes some fixed admissible subrepresentation of $S(D_v^2 \times F_v^*, \check{\sigma}_v)$. We can view Φ_v as elements in $S(D_v^2 \times F_v^*, \omega_v)$. Choose lifts

$$\tilde{\Phi}_v \in S(D_v^2 \times F_v^*) \quad (v \in S_0),$$

under the central projection $S(D_v^2 \times F_v^*) \to S(D_v^2 \times F_v^*)$ and some sufficiently close approximations Φ_v' so that Φ_v' are functions with compact support also for all Archimedean $v \in S_0$. There exist points $(X_v, t_v) \in D_v^2 \times F_v^*$ such that $\Phi_v'(X_v, t_v) \neq 0$. Approximate these finitely many local points by a global point $(X, t) \in D^2 \times F^*$ such that

$$\Phi_v'(X, t) \neq 0 \quad (\forall v \in S_0).$$

We may assume $X \neq 0$. Then choose a function $\prod_{v \notin S} \Phi_v'$ at the remaining places not in S_0 with a sufficiently small support containing the global point (X, t) such that $\Phi' = \prod_v \Phi_v'$ has the property

$$supp(\Phi') \cap (D^2 \times F^*) = \{(X, t)\}.$$

Then

$$\vartheta_{\Phi'}(1, 1) = \Phi'(X, t) \neq 0.$$

In particular, the theta distribution is

$$\vartheta_{\Phi'}(m, 1) \neq 0$$

as a function of the variable $m \in M_c(\mathbb{A})/M_c(F)$. Then it is easy to see that there also exists a choice of $\prod_{v \notin S_0} \tilde{\Phi}_v$ such that $\vartheta_{\tilde{\Phi}}(m, 1) \neq 0$. Hence, by assumption (1) there exists a cuspidal automorphic representation $\check{\sigma} = \check{\sigma}_1 \times \check{\sigma}_2$ of $M_c(\mathbb{A})$ of $dim(\check{\sigma}_i) \neq 1$ and a function $f = f_{S_0} f^{S_0}$ in the representation space of $\check{\sigma}$ such that

$$L_f(\tilde{\Phi})(g) = \int_{M_c(\mathbb{A})/M_c(F)} \vartheta_{\tilde{\Phi}}(m, 1)\overline{f}(m)dm \neq 0.$$

The corresponding theta lift maps to the space of cusp forms of $G(\mathbb{A})$ by assumption (2). It defines a subspace $\Theta_-(\check{\sigma})$ of the space of cusp forms of $GSp(4, \mathbb{A}_F)$. We can therefore use orthogonal projection $pr_\pi : \Theta_-(\check{\sigma}) \to \pi$ to (any) irreducible automorphic quotient π of the completely reducible representation space $\Theta_-(\check{\sigma})$. The composed map

$$pr_\pi \circ L_f : S(D^2(\mathbb{A}) \times \mathbb{A}^*) \to \pi$$

factorizes over a map

$$pr_\pi \circ \overline{L}_f : S(D_{\mathbb{A}}^2 \times A_F^*, \omega) \to \pi,$$

where ω denotes the central character of π (or of $\check{\sigma}$). The restriction of this map $pr_\pi \circ \overline{L}_f$ to elements in $S(D_v^2 \times F_v^*, \check{\sigma}_v) \otimes \prod_{w \neq v} \tilde{\Phi}_w \subseteq S(D_{\mathbb{A}}^2 \times A_F^*, \omega)$ defines for a suitable choice of $\prod_{w \neq v} \tilde{\Phi}_w$ a *nontrivial* intertwining map

$$R_v \to \pi_v.$$

In particular, π_v is a quotient of R_v.

Since we are interested here in the non-Archimedean places v for the case $F = \mathbb{Q}$, assume for the rest of this section.

Assumption 3. $F = \mathbb{Q}$ and $v \neq \infty$ and $dim(\check{\sigma}_{1,\infty}) = 1$ and $dim(\check{\sigma}_{2,\infty}) = k-2 > 1$.

Assumptions (1)–(3) together imply the following. By assumption (3) the cuspidal representation π constructed above contributes to cohomology. By Theorem 2.4(1) in [70], it is a weak lift of $\sigma = \sigma_1 \times \sigma_2$, where $\sigma_i = JL(\check{\sigma}_i)$ are cuspidal (holomorphic) automorphic representations of $Gl(2, F)$ obtained from $\check{\sigma}$ by the global Jacquet–Langlands lift. Since $dim(\check{\sigma}) \neq 1$, the Ramanujan conjectures holds for the representations σ_i by the theorem of Deligne. Hence, Theorem 2.4(1) in [70] also implies that π is not a CAP representation.

Now use Remark 4.4 in Sect. 4.3 (or Lemma 5.6 in Chap. 5, which follows from the Hasse–Brauer–Noether theorem). It implies $m_1(\pi) \neq m_2(\pi)$. Hence, $\pi = \otimes_v \pi_v$ is "detected" by the multiplicity formula: $n(\sigma_v, \pi_v) \neq 0$ holds for all places v. Therefore, π_v is isomorphic to either $\pi_+(\sigma_v)$ or $\pi_-(\sigma_v)$. Since π_v does not have a Whittaker model

$$\pi_v \cong \pi_-(\sigma_v).$$

Definition 4.7. We have now defined a unique irreducible quotient representation π_v of the anisotropic theta lift $\Theta_-(\check{\sigma}_v)$. We call this representation $\theta_-(\check{\sigma}_v)$, and we also write for $\sigma_v = JL(\check{\sigma}_v)$

$$\pi_v = \theta_-(\sigma_v).$$

With this notation we obtain

Corollary 4.16. *Assume that σ_v is an irreducible discrete series representation of $M(F_v)$ at a non-Archimedean place. Then*

$$r(\sigma_v) = \theta_+(\sigma_v) - \theta_-(\sigma_v).$$

Furthermore, $\theta_-(\sigma_v) = \pi_-(\sigma_v)$ does not have a Whittaker model.

The Archimedean case of this is well known.

Corollary 4.17. *Suppose $dim(\check{\sigma}_v) > 1$ or $\check{\sigma}_{1,v} \not\cong \check{\sigma}_{2,v}$. Then under the assumptions of Corollary 4.16, the Weil representation on $\Theta_-(\check{\sigma}_v) = S(D_v^2 \times F_v^*, \check{\sigma}_v)$ is isomorphic to a finite multiple of the cuspidal irreducible representation $\pi_-(\sigma_v)$.*

Proof of Corollary 4.17. Either $\check{\sigma}_{1,v} \not\cong \check{\sigma}_{2,v}$. Then $\Theta_-(\check{\sigma}_v)$ is cuspidal (Lemma 4.14). In the category of smooth ω_v-representations the cuspidal representations are projective. Since $\Theta_-(\check{\sigma}_v)$ is cuspidal admissible, and since the center acts by the character ω_v, this representation is completely reducible. Hence, every irreducible summand $R_v \subseteq \Theta_-(\check{\sigma}_v)$ must be isomorphic to $\pi_-(\sigma_v)$ by the arguments preceding Corollary 4.16.

Or $\check{\sigma}_{v,1} \cong \check{\sigma}_{2,v}$. Then there exists an exact sequence

$$0 \to K \to \Theta_-(\check{\sigma}_v) \to Wp_-(\sigma_v) \to 0$$

with ω_v-cuspidal kernel (Corollary 4.15). Hence, again K is completely reducible and every irreducible summand $R_v \subseteq K \subseteq \Theta_-(\check{\sigma}_v)$ must be isomorphic to $\pi_-(\sigma_v)$ by the arguments above. On the other hand, $\pi_-(\sigma_v) \cong Wp_-(\sigma_v)$ by Theorem 4.5. But $Wp_-(\sigma_v)$ is not cuspidal. Hence, $K = 0$ and $\Theta_-(\check{\sigma}_v) = \theta_-(\sigma_v)$ is irreducible. $\square$

Remark 4.17. For the omitted case $dim\check{\sigma}_v = 1$ and $\check{\sigma}_{1,v} \cong \check{\sigma}_{2,v}$ the representation $\pi_-(\sigma_v)$ is not cuspidal. Nevertheless an analogous result was obtained by Kudla and Rallis (see Sect. 4.12, page 130).

Chapter 5
Local and Global Endoscopy for *GSp*(4)

In this chapter we refine the global description of the endoscopic lift obtained in Corollary 4.2. The main result obtained in this chapter is Theorem 5.2, which is a special case of a global multiplicity formula conjectured by Arthur. We consider the symplectic group $GSp(4)$ over an arbitrary totally real number field F. In the special case $F = \mathbb{Q}$ and for irreducible automorphic representations π with π_∞ in the discrete series the proof of this theorem is given in Sects. 5.2 and 5.3. In Sect. 5.1 we explain how the local endoscopic character lift is constructed for arbitrary local base fields of characteristic zero. In Sects. 5.4 and 5.5 we explain how the Arthur–Selberg trace formula can be used to extend the results obtained in Sects. 5.2 and 5.3 to the case of arbitrary totally real number fields and arbitrary irreducible automorphic not necessarily cohomological representations π. The proof of Theorem 5.2 is based on two ingredients: the principle of exchange and the key formula (stated in Sect. 5.3). The latter can be directly deduced from weak versions of the trace formula (see Chap. 4, Corollary 4.2, or Sect. 5.5). It deals with simultaneous changes of global representations at two places. The principle of exchange, on the other hand, deals with an exchange of the representation at one place exclusively, and its proof boils down to a special case of the Hasse–Brauer–Noether theorem (Lemma 5.4). This is based on an explicit theta lift. At this point we use [56], which unfortunately forces us to make the restrictive assumption that F is a totally real number field. Since we apply results obtained in [56] in a rather specific case, it is very likely that this restriction on F can be removed.

5.1 The Local Endoscopic Lift

In this section we extend the results obtained in Chap. 4 to the case of arbitrary local fields F_v of characteristic zero. We restrict ourselves to the non-Archimedean case. For the Archimedean case see [90, 91].

Let $R_{\mathbb{Z}}[G_v]$ and $R_{\mathbb{Z}}[M_v]$ denote the Grothendieck group of finitely generated, admissible representations of G_v and M_v, respectively. For a finite sum

R. Weissauer, *Endoscopy for GSp(4) and the Cohomology of Siegel Modular Threefolds*, 175
Lecture Notes in Mathematics 1968, DOI: 10.1007/978-3-540-89306-6_5,
© Springer-Verlag Berlin Heidelberg 2009

$\alpha = \sum n_i \cdot \pi_{v,i} \in R_{\mathbb{Z}}(G_v)$ with integer coefficients n_i and irreducible, admissible representations $\pi_{v,i}$, and a locally constant function f_v with compact support on G_v, put $\alpha(f_v) = \sum n_i \cdot tr\, \pi_{v,i}(f_v)$. Similarly for the group M_v. Irreducible admissible representations σ_v of M_v are considered as a pair $(\sigma_{1v}, \sigma_{2v})$ of irreducible representations of $Gl(2, F_v)$ with equal central characters $\omega_{1v} = \omega_{2v}$. This common central character ω_v will be called the central character of σ_v. For a quasicharacter χ_v, the character twist for G_v is defined via the similitude homomorphisms $G_v \to F_v^*$ (Lemma 4.8), and similarly $\sigma_v \otimes \chi_v = (\sigma_{1,v} \otimes \chi_v, \sigma_{2,v} \otimes \chi_v)$. Recall $\sigma \mapsto \sigma^*$ was defined by $(\sigma_{1,v}, \sigma_{2,v})^* := (\sigma_{2,v}, \sigma_{1,v})$.

Theorem 5.1 (Nonarchimedean Lift). *There exists a unique homomorphism*

$$r = r_{M_v}^{G_v} : R_{\mathbb{Z}}[M_v] \to R_{\mathbb{Z}}[G_v]$$

between the Grothendieck groups with the following properties:

(i) The lift r is endoscopic: For $\alpha \in R_{\mathbb{Z}}[M_v]$

$$\alpha(f_v^{M_v}) = r(\alpha)(f_v)$$

holds for locally constant functions $f_v, f_v^{M_v}$ with compact support on G_v, and M_v, respectively, and matching orbital integrals.

(ii) The lift r preserves the central character, and commutes with character twists

$$r(\sigma_v \otimes \chi_v) = r(\sigma_v) \otimes \chi_v.$$

(iii) The lift r commutes with parabolic induction

$$r \circ r_{P_{M_v}}^{M_v}(\rho_v) = r_{P_v}^{G_v}(\rho_v),$$

where ρ_v is in $R_{\mathbb{Z}}[Gl(2, F_v) \times F_v^*]$, and where P_{M_v} is a maximal proper parabolic of M_v and P_v is the Siegel parabolic of G_v. Similarly for induction from the Borel groups B_{M_v}, B_v of M_v and G_v, we have

$$r \circ r_{B_{M_v}}^{M_v}(\rho_v) = r_{B_v}^{G_v}(\rho_v).$$

(iv) $r(\sigma^*) = r(\sigma)$.
(v) r commutes with Galois twists

$$r(\sigma_v^\tau) = r(\sigma_v)^\tau, \qquad \tau \in Aut(\mathbb{C}/\mathbb{Q}).$$

In addition the refined properties of Sect. 4.11 hold.

Concerning the proof. The above-mentioned properties of the endoscopic lift have already been shown for p-adic fields $\mathbb{Q}_p$. To extend this to the case of a general non-Archimedean local field F_v of characteristic zero, again one reduces the case to that of of irreducible representations σ_v in the discrete series by local arguments. For the global arguments we now replace the topological trace formula by the Arthur–Selberg trace formula. Choose a totally real number field F of degree at least 2 over $\mathbb{Q}$ with completion F_v at the non-Archimedean place v. Suppose one can find a global cuspidal automorphic representation $\sigma \not\cong \sigma^*$ extending the given discrete series σ_v, where σ can be chosen to be cuspidal at some additional non-Archimedean place $w \neq v$, belonging to the discrete series at all Archimedean places. Now one stabilizes the Arthur trace formula for the group $GSp(4, A_F)$, using the fundamental lemma for $GSp(4)$ (see Chaps. 6–8). For this it is enough to consider a simple version of the trace formula with a test function of the form $f = \prod_w f_w$, where f_w is a difference of pseudocoefficients of the discrete series representations $\pi_+(\sigma_\infty), \pi_-(\sigma_\infty)$ at all Archimedean places of F. Furthermore, we may assume f_w to be a matrix coefficient of a cuspidal representation $\pi_-(\sigma_w)$ at some auxiliary place w, say, where $F_w = \mathbb{Q}_w$ splits and where σ_w is cuspidal. Then Arthur's trace formula can be stabilized with arguments analogous to those in Chap. 4 starting from [6], Corollaries 7.2 and 7.4. For the details and further technical assumptions, we refer to Sect. 5.5. Once the Arthur trace formula has been stabilized, one can deduce from it the following statement: The endoscopic lift $r(\sigma_v)$ of the character of a discrete series representation σ_v of $M(F_v)$, a priori only defined as a distribution satisfying properties (i) and (iii) from above, can be expressed as a finite integral linear combination $r(\sigma_v) \in R_{\mathbb{Z}}[G_v]$ of characters of irreducible, admissible representations of $G(F_v)$. The construction of this "integral" lift is the crucial step (generalizing Lemma 4.11 or its Corollary 4.5), and it will be explained in more detail later. Once this is known, all properties of the lift are deduced completely analogously to the p-adic case. So, for the proof of the theorem, it remains for us to construct the integral lift, since up to a character twist σ_v can be embedded globally. See the comment on page 182. □

5.1.1 The Integral Character Lift

We want to show the existence of the integral character lift for irreducible, admissible representations σ_v of M_v. It is enough to assume σ_v is in the discrete series. The endoscopic transfer of distributions applied to the character of σ_v defines distributions on G_v, which will be called character lifts of the characters σ_v. We have to show that the character lift is a linear combination of the characters of irreducible representations

$$r(\sigma_v) = \sum_{\pi_v} n(\sigma_v, \pi_v) \cdot \pi_v.$$

This uniquely determines the transfer coefficients $n(\sigma_v, \pi_v)$, once they are known to exist, by linear independence of characters (see [76], Proposition 13.8.1). So it remains for us to show the existence of such an expansion, and integrality of the coefficients.

Archimedean Case. According to Shelstad [90, 91] this holds in the Archimedean case such that, furthermore, the Archimedean transfer coefficients $n(\sigma_\infty, \pi_\infty)$ belong to $\{\pm 1, 0\}$, and only finitely many are nonzero [91], Theorem 4.1.1(i).

Non-Archimedean Case. To obtain a corresponding local character lift for an arbitrary local field of characteristic zero, we copy the arguments used so far. We have to replace the global arguments. Since these arguments involve several steps, we first summarize for the convenience of the reader the different steps involved in the proof of the p-adic case. Afterwards we discuss how this carries over to a non-p-adic local field.

5.1.2 Steps (0)–(xi) of the Proof (in the p-Adic Case)

(A) Global Inputs. The stabilization of the global trace formula in Chaps. 3 and 4 was the source for the desired weak character identities $r(\sigma_v) = \sum n(\sigma_v, \pi_v) \cdot \pi_v$, with coefficients $n(\sigma_v, \pi_v)$ in $\mathbb{C}$. A priori these sums are not finite, but are locally finite and hence absolutely convergent (see Sect. 4.6 [76]). The coefficients $n(\sigma_v, \pi_v)$ are directly related to global multiplicities $m(\pi)$ of automorphic representations π. In fact, the basic result was the identity (Corollary 4.2)

$$(0) \qquad \prod_{v \neq \infty} n(\sigma_v, \pi_v) = m(\pi_\infty^W \otimes \pi_f) - m(\pi_\infty^H \otimes \pi_f),$$

for a global weak endoscopic automorphic lift $\pi_f = \otimes_{v \neq \infty} \pi_v$ of a global irreducible cuspidal automorphic representation $\sigma_f = \otimes_{v \neq \infty} \sigma_v$, for which $\sigma^* \not\cong \sigma$ holds, where $m(\pi)$ denote the multiplicity of π in the discrete spectrum, and $\pi_\infty^H, \pi_\infty^W$ denote the irreducible Archimedean representations defined by the Archimedean character lift $r(\sigma_\infty) = \pi_\infty^W - \pi_\infty^H$ defined by Shelstad [91].

Claim 5.1. Formula (0) implies for the local component σ_v of σ:

(i) $r(\sigma_v)$ is an absolutely convergent sum

$$r(\sigma_v) = n(\sigma_v) \cdot \sum_{\pi_v} \varepsilon(\sigma_v, \pi_v) \cdot \pi_v$$

for certain $n(\sigma_v) \in \mathbb{C}^*$ such that the following hold:
(ii) (integrality) $\varepsilon(\sigma_v, \pi_v) \in \mathbb{Z}$ and $\varepsilon(\sigma_v, \pi_v) = 1$ for $\pi_v = \pi_+(\sigma_v)$ (as defined below).
(iii) (product formula) $\prod_v n(\sigma_v) = 1$.
(iv) (Shelstad) $n(\sigma_\infty) = 1$.

Proof of the Claim. Theorem 4.3 provides the existence of a globally generic cuspidal representation $\pi_+ = \pi_+(\sigma)$ such that $(\pi_+)_v = \pi_+(\sigma_v)$ for all v, and $m(\pi_+) = 1$. Being generic implies $(\pi_+)_\infty = \pi_+(\sigma_\infty) = \pi_\infty^W(\sigma_\infty)$. Since

$m(\pi_+) = m(\pi_\infty^W \otimes (\pi_+)_f) = 1$, the principle of exchange (see Lemmas 5.4 and 5.5) implies $m(\pi_\infty^H \otimes (\pi_+)_f) = 0$. Hence, $\prod_{v\neq\infty} n(\sigma_v, \pi_+(\sigma_v)) = 1$ by formula (0). For

$$n(\sigma_v) := n(\sigma_v, \pi_+(\sigma_v))$$

we get $n(\sigma_\infty) = 1$ from [91], Theorem 4.1.1(i), and hence (iv). (iv) together with the product formula just obtained gives (iii). Assertion (i), on the other hand, is a consequence of the global stabilized trace formula. For $\varepsilon(\sigma_v, \pi_v) = n(\sigma_v, \pi_v)/n(\sigma_v, \pi_+(\sigma_v))$ obviously $\varepsilon(\sigma_v, \pi_v) = 1$ for $\pi_v = \pi_+(\sigma_v)$, by definition. So for (ii) only $\varepsilon(\sigma_v, \pi_v) \in \mathbb{Z}$ has to be shown. From [91] we can assume v is non-Archimedean. Put $\tilde{\pi}_f = \pi_v \otimes_{w\neq v,\infty} \pi_+(\sigma_w)$ and $(\pi_+)_f = \otimes_{w\neq\infty} \pi_+(\sigma_w)$. By (iii) and (iv) we have $\prod_{w\neq\infty} n(\sigma_w, \pi_+(\sigma_w)) = 1$; hence,

$$\varepsilon(\sigma_v, \pi_v) = \varepsilon(\sigma_v, \pi_v) \prod_{w\neq\infty} n(\sigma_w, \pi_+(\sigma_w)) =$$

$$\frac{n(\sigma_v, \pi_v)}{n(\sigma_v, \pi_+(\sigma_v))} \prod_{w\neq\infty} n(\sigma_w, \pi_+(\sigma_w)) = \prod_{w\neq\infty} n(\sigma_w, \tilde{\pi}_w).$$

Together with formula (0) applied for the representation $\tilde{\pi}_f$, instead of π_f, this gives

$$\varepsilon(\sigma_v, \pi_v) = m(\pi_\infty^W \otimes \tilde{\pi}_f) - m(\pi_\infty^H \otimes \tilde{\pi}_f),$$

and hence (ii), since the right side is an integer. This proves the claim. □

(B) **Local Inputs.** A purely local investigation of the character lift, using compatibility with parabolic induction, gives (Lemmas 4.12 and 4.27):

(v) $n(\sigma_v) = 1$ for generic σ_v not in the discrete series.

For the elliptic scalar product $< \eta_v, \omega_v >_e$ on the elliptic regular locus of G_v write $\|\eta_v\|_e^2 = < \eta_v, \eta_v >_e$. Then (see in particular Sect. 4.5 and Lemma 4.14) we have

(vi) (Weyl integration formula) $\|r(\sigma_v)\|_e^2 = A_v \cdot \|\sigma_v\|_e^2 = A_v$

for irreducible σ_v in the discrete series of M_v, where $A_v = 2$ or 4 depending on whether $\sigma_v^* \not\cong \sigma_v$ or $\sigma_v^* \cong \sigma_v$.

Furthermore (Sects. 4.6, 4.9, Lemma 4.20, and Corollary 4.8), for σ_v in the discrete series the classification of irreducible, admissible representations of G_v implies that there exists a second irreducible representation $\pi_-(\sigma_v)$ such that:

(vii) $\sum_{\pi_v} \varepsilon(\sigma_v, \pi_+(\sigma_v)) \cdot \pi_v = L(\sigma_v) + R(\sigma_v)$ for $L(\sigma_v) = \pi_+(\sigma_v) - \pi_-(\sigma_v)$

and

(viii) $\|L(\sigma_v)\|_e^2 = A_v$ and $< L(\sigma_v), R(\sigma_v) >= 0$

with the same coefficients $A_v = 2, 4$ as above.

Combining (A) and (B). For the finite set S of places, where the (suitable chosen global cuspidal irreducible) representation $\sigma \not\cong \sigma^*$ belongs locally to the discrete series, statements (vi) and (vii) above imply

$$\prod_{v \in S} \|r(\sigma_v)\|_e^2 = \prod_{v \in S} n(\sigma_v)^2 \cdot \prod_{v \in S} \|\sum_{\pi_v} \varepsilon(\sigma_v, \pi_+(\sigma_v)) \cdot \pi_v\|_e^2 \leq 1 \cdot \prod_{v \in S} \|L(\sigma_v)\|_e^2,$$

since $\prod_{v \in S} n(\sigma_v)^2 = 1$ by (iii)–(v). The inequality obtained is in fact an equality, since the left and the right sides are equal to $\prod_{v \in S} A_v$ by (vi) and (viii), respectively. That an equality holds forces $\|R(\sigma_v)\|_e^2 = 0$ from (vii) and (viii). Therefore,

(ix) $\sum_{\pi_v} \varepsilon(\sigma_v, \pi_v) \cdot \pi_v = \pi_+(\sigma_v) - \pi_-(\sigma_v)$

for σ_v in the discrete series (for σ) and $\|r(\sigma_v)\|_e^2 = \|\sum \varepsilon(\sigma_v, \pi_v)\pi_v\|_e^2$. Hence, $\|r(\sigma_v)\|_e^2 = |n(\sigma_v)|^2 \cdot \|\sum \varepsilon(\sigma_v, \pi_v)\pi_v\|_e^2$ implies $A_v = |n(\sigma_v)|^2 \cdot A_v$, or

(x) $|n(\sigma_v)| = 1$.

On the other hand, the local non-Archimedean theory of Whittaker models implies

(xi) $n(\sigma_v)$ is a positive real number,

since $\pi_-(\sigma_v)$ does not have local Whittaker models (Lemmas 4.35 and 5.6 and Corollary 4.16). This implies $n(\sigma_v) = 1$; hence, we obtain from (x) and (xi) the final formula

$$r(\sigma_v) = \pi_+(\sigma_v) - \pi_-(\sigma_v)$$

for all σ_v in the discrete series realized in a global representation σ. Since up to a character twist every irreducible representation σ_v in the discrete series of M_v can be realized as the local component of a global cuspidal irreducible representation σ of $M(\mathbb{A}_F)$, which follows from the existence of strong pseudocoefficients for $Gl(2, F_v)$, this completes the proof for σ_v in the discrete series. The general case is reduced to this case by purely local methods (as in Sects. 4.5–4.11).

How this Generalizes. To see how this carries over to the case of an arbitrary non-Archimedean local field of characteristic zero, observe that all the arguments above were of general nature, and hence carry over to a non-Archimedean field of characteristic zero immediately, except that the trace identity (0) needs some equivalent in the general case, which now will be provided by the strong multiplicity 1 theorem for $M(\mathbb{A}_F)$ (or $Gl(2, \mathbb{A}_F)$) and a version of the Arthur trace formula (Lemma 5.8 in Sect. 5.5). Formula (0) then again implies (i)–(iv), from which one deduces (v)–(xi) as above. Finally (0)–(xi) imply all properties of the local endoscopic character lift $r(\sigma_v)$ as in the p-adic case. This includes the statements of Theorem 5.1.

Concerning the Proof of Formula (0). To be accurate, one has to use an avatar of Lemma 5.8 for two reasons. On one hand, in the formulation and proof of Lemma 5.8, the existence and properties of the local character lift will already be used in special cases. Secondly, using an avatar gets rid of twisting by multipliers in the proof of Lemma 5.8. By examining the different steps in the proof of Lemma 5.8, one sees that formula (0) can be proved by making slight changes in the argument. First, it is enough to use the existence of the local character lift only in situations where the formula is already known, say, for the Archimedean places [91] and for places where the number field F splits, i.e., where $F_v \cong \mathbb{Q}_p$ is a p-adic field. Second, for the proof of (0) multipliers are not needed, since in this case it suffices to use a strongly simplified version of the trace formula of Deligne–Kazhdan type. We sketch the main steps of the argument which gives formula (0) at the end of the proof of Lemma 5.8 in Sect. 5.5. For the moment the reader is advised to skip this proof at first, since arguments from Sects. 5.2–5.4 are also used. □

This being said, we return to the representations $\pi_\pm(\sigma_v)$ defining the local L-packets. They can be further described in terms of the general classification of irreducible admissible representations of $GSp(4, F_v)$ as in Sect. 4.11. But they can also be described in terms of a theta lift (Weil representation) in Sect. 4.12.

5.1.3 Complements on the Local Theta Lift

Extended Jacquet–Langlands Correspondence. Let D_v be the nonsplit quaternion algebra over F_v, and $\check{M}_v = (D_v^* \times D_v^*)/F_v^*$. The quotient is defined by the embedding $t \mapsto (t, t^{-1})$ of F_v^* in $D_v^* \times D_v^*$.

The Jacquet–Langlands lift describes the irreducible discrete series representations of the group $Gl(2, F_v)$ in terms of the irreducible representations of the multiplicative group D_v^* of the quaternion algebra. We can use this to define an extended lift, denoted JL, from the irreducible admissible representations of the group $\check{M}_v$ to the irreducible, admissible, discrete series representations of the group M_v. In fact the description of $\check{M}_v$ and that of M_v as a quotient of two copies of D_v^* and $Gl(2, F_v)$, respectively, allows us to extend the Jacquet–Langlands lift JL in the obvious way. By this the irreducible, admissible discrete series representations σ_v of M_v correspond uniquely to irreducible, admissible representations $\check{\sigma}_v$ of $\check{M}_v$ and vice versa via the correspondence $\sigma_v = JL(\check{\sigma}_v)$.

Quaternary Theta Lifts. M_v can be considered as the group of special orthogonal similitudes $GSO(V)$ of a split quaternary quadratic form over F_v. Similarly $\check{M}_v$ can be considered as the group of special orthogonal similitudes $GSO(\check{V})$ of the quaternary anisotropic quadratic form over F_v. This defines the theta correspondence for the groups $GO(V)$ and $GSp(4)$ as in Chap. 4 for arbitrary number fields F. It relates representations σ_v^{ext} and π_v of $GO(V)$ and $GSp(4)$ if $\sigma_v^{ext} \otimes \pi_v$ is a quotient of the big theta representation $\tilde{w}$ defined in [104], Sect. 4, or in Chap. 4. Although the big theta representation in [104] is defined in a slightly different way, we will use results obtained in [104] whenever they carry over: The groups $GSO(V)$ are subgroups of index 2 in $GO(V)$, so in general irreducible admissible representations σ_v of $GSO(V)$ cannot be extended uniquely to irreducible representations of the group $GO(V)$. If $\sigma_v \cong \sigma_v^*$, there exist possibly two nonisomorphic extensions σ_v^{ext} of the representation σ_v to $GO(V)$. One can consider the representations on $GSp(4, F_v)$ attached to σ_v^{ext} by the theta correspondence.

Notice there is a subtlety present in the definition of the theta correspondence: Either one deals with it on the level of representations of the groups $O(V), Sp(2n)$ or one deals with it on the extended level of the groups $GO(V), GSp(2n)$. Both points of view are of relevance. The representations obtained from the theta lift for $GSp(4, F_v)$, say, with a fixed central character of the center Z_v of $GSp(4, F_v)$, can be obtained – via compact induction from $Sp(4, F_v) \times Z_v \subseteq GSp(4, F_v)$ – from the $Sp(4, F_v)$ representations obtained from the $(O(V), Sp(2n))$ correspondence [104], p. 470. This often allows us to decide whether σ_v^{ext} has a lift to $GSp(4, F_v)$ [104], Proposition 4.11. Inversely, the restriction from $G = GO(V)$ to $O(V)$ is multiplicity free. For this see [41], Lemma 7.2, and the remarks after it on p. 94 for the split and the anisotropic quaternary spaces V.

The Definite Theta Lift. Using the last remarks, the statements, Theorem 9.1, and the results obtained in [41], Sect. 10, carry over from $Sp(4)$ to $GSp(4)$. Hence, the local theta correspondence for $GSp(4, F_v)$ and $GO(V)$ is always nontrivial for anisotropic V. In other words, to every irreducible representation $\check{\sigma}_v$ of $\check{M}_v$ the theta lift $\Theta(\check{\sigma}_v)$ is nonzero for the theta correspondence with respect to some extension of $\check{\sigma}_v$ from $\check{M}_v = GSO(V)$ to $GO(V)$. Let $\mathcal{E}\Theta(\check{\sigma}_v)$ denote the set of isomorphism classes of irreducible admissible representations π_v of $G_v = GSp(4, F_v)$, for which $\check{\sigma}_v \times \pi_v$ is a quotient of the big Weil representation of the pair $GO(V) \times G_v$.

Lemma 5.1 (Density Lemma). *(i) $\mathcal{E}\Theta(\check{\sigma}_v) = \{\pi_-(\sigma_v)\}$, and for $\check{\sigma}_v \not\cong \check{\sigma}_v^*$ or $\dim(\check{\sigma}_v) > 1$ this is a cuspidal representation.*

(ii) Furthermore, given a finite set S of places w of a number field F and given unitary representations $\check{\sigma}_w$ of the discrete series of M_w and irreducible representations π_w of G_w in $\mathcal{E}\Theta(\check{\sigma}_w)$ for all $w \in S$, suppose the central characters ω_v of $\check{\sigma}_v$ are the components of a grössen character ω of F. Then there exists a corresponding automorphic irreducible cuspidal representation of $M(\mathbb{A}_F)$ and a nontrivial irreducible cuspidal automorphic theta lift π of σ, which realizes the given local representations π_w for all places in S.

Proof. See Corollary 4.17 for (i). For (ii) see page 171, noticing the following   □

Comment on Global Embeddings. One can enlarge S by a place w, and choose some $\check{\sigma}_w \not\cong \check{\sigma}_w^*$ with central character $\tilde{\omega}_w$ to obtain the situation described in Chap. 4, page 171. From the existence of very strong pseudocoefficients for $Gl(2)$, one can find a global cuspidal σ with central character ω realizing σ_v for $v \in S \cup \{w\}$.

Concerning the Central Character. Let l be the order of the group of roots of unity in F. If we choose the place w outside S of residue characteristic larger than l, we can always extend given unitary central characters $\omega_v, v \in S$ to a grössen character ω of F^*, which is unramified outside $\{w\} \cup S$ and induces the given local character up to an unramified character twist $\omega_v = \tilde{\omega}_v|.|_v^{it_v}$. For this, at one place $v \in S$ we can prescribe $\omega_v = \tilde{\omega}_v$. [We leave this as an exercise for the reader with the following hints: Use [102], p. 342ff. Notice that extending condition (A) of [102] can be solved by choosing a suitable character $\tilde{\omega}_w : \mathfrak{o}_w^* \to \mathbb{C}^*$, which is trivial on $1 + \wp_w$. Then the remaining extension of condition (B) of [102] for the parameters t_v can be solved. This defines an extension ω on the subgroup $(\prod_{v \in \{w\} \cup S} F_v^* \prod_{v \text{ else}} \mathfrak{o}_v^*)/(F_S^*)$ of $(\mathbb{A}_F^1/F^*) \times \mathbb{R}^*$. Since $\mathbb{C}^*$ is divisible, and hence injective, there is no obstruction to extending ω from this subgroup to $\mathbb{A}_F^1/F^*$. Notice there are only finitely many such extensions to a character unramified outside w and S and tame at w, since the S-ideal class group is finite].

Indefinite Theta Lift. If σ_v is an irreducible admissible representation of the group M_v and V is the split quaternary quadratic space, one can define a corresponding set $\mathcal{E}\Theta_+(\sigma_v)$. It consists of the classes of irreducible admissible representations π_v of G_v, for which $\sigma_v \times \pi_v$ is a quotient of the big Weil representation of the pair $GO(V) \times G_v$. From [96], proof of Theorem 3.1 on p. 366, there is at most one generic class of representations in $\mathcal{E}\Theta_+(\sigma_v)$. Otherwise one could easily show that the space of functionals of [96], p. 366, would have dimension 2 or more, which would give a contradiction. A representative of this generic class will be denoted $\theta_+(\sigma_v)$ if it exists. A generic $\theta_+(\sigma_v)$ exists in $\mathcal{E}\Theta_+(\sigma_v)$ if σ_v is in the discrete series of M_v (see Lemma 4.25). Hence, $\mathcal{E}\Theta_+(\sigma_v)$ is nonempty in this case. This argument also gives $\theta_+(\sigma_v) \cong \pi_+(\sigma_v)$ for discrete series representations σ_v of M_v (as in Theorem 4.4).

Definite vs. Indefinite Theta Lift. For σ_v in the discrete series of M_v, there exists a unique $\check{\sigma}_v$ for which for $\sigma_v = JL(\check{\sigma}_v)$. Then we have defined $\theta_+(\sigma_v)$ (indefinite theta lift) and we define $\theta_-(\sigma_v)$ to be the anisotropic theta lift $\theta_-(\check{\sigma}_v)$ attached to $\check{\sigma}_v$.

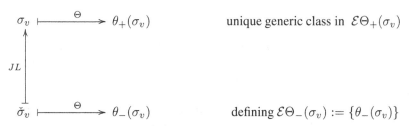

Proposition 5.1. *For irreducible, admissible representations σ_v in the discrete series of M_v the set $\mathcal{E}\Theta_-(\sigma_v)$ has cardinality 1. Furthermore, there is a unique generic representation in $\mathcal{E}\Theta_+(\sigma_v)$, and the endoscopic lift is given by these Weil representations*

$$r(\sigma_v) = \theta_+(\sigma_v) - \theta_-(\sigma_v).$$

Proof. The proof for arbitrary F_v remains the same as that for Corollary 4.16. However, recall that the proof of Corollary 4.16 depended on Lemma 5.6, which is proved later in this chapter. $\square$

Remark 5.1. $\mathcal{E}\Theta_+(\sigma_v)$ presumably has cardinality 1, although we have not been able to show this. Notice since we want to include places of residue characteristic 2, and since we consider the theta lift for the groups of similitudes, the result obtained by Waldspurger for the Howe duality cannot be directly applied in this situation. To overcome this – for the global applications later – we introduce the subset

$$\mathcal{E}\Theta_\pm^{glr}(\sigma_v) \subseteq \mathcal{E}\Theta_\pm(\sigma_v)$$

of "globally relevant" representations. It consists of all classes of representations π_v in $\mathcal{E}\Theta_\pm(\sigma_v)$ for which there exists a global irreducible cuspidal automorphic representation $\sigma \not\cong \sigma^*$ (or $\breve{\sigma} \not\cong \breve{\sigma}^*$), for which π_v is the local component of a weak lift π attached to σ. The notion of weak lift will be explained in Sect. 5.2. The density lemma, stated above, implies

$$\mathcal{E}\Theta_-^{glr}(\sigma_v) = \mathcal{E}\Theta_-(\sigma_v),$$

in the analogous sense. Later, in Corollary 5.1, with proof in Sect. 5.3, and more generally for $F \neq \mathbb{Q}$ in Sect. 5.4, it is shown that for discrete series representations σ_v

$$\mathcal{E}\Theta_+^{glr}(\sigma_v) = \{\theta_+(\sigma_v)\}.$$

So $\mathcal{E}\Theta_+^{glr}(\sigma_v)$ has cardinality 1.

5.2 The Global Situation

General Assumptions. Let F be a number field, and $\mathbb{A}_F$ its ring of adeles. Since we apply the findings of [56] (in the argument preceding Lemma 5.3) we assume F to be totally real. This restriction will not be needed otherwise (and therefore is most likely unnecessary). Let $G = GSp(4)$ and let $M = Gl(2) \times Gl(2)/\mathbb{G}_m^*$ be its proper elliptic endoscopic group. Let $\pi = \otimes_v \pi_v$ be an irreducible, unitary cuspidal automorphic representation of the group $G(\mathbb{A}_F)$. Recall an irreducible automorphic cuspidal representation π is a cuspidal representation associated with a parabolic subgroup (CAP representation) if it is weakly equivalent to an automorphic representation associated with an Eisenstein series. Two irreducible automorphic representations are called weakly equivalent if their local components are isomorphic at almost all places. In the following we assume π is not a CAP representation. Then, by the Langlands theory of Eisenstein series, π only contributes to the cuspidal part of the discrete spectrum

$$m_{cusp}(\pi) = m_{disc}(\pi).$$

Definition 5.1. An irreducible, unitary cuspidal automorphic representation π of the group $G(\mathbb{A}_F)$ is called a weak endoscopic lift (from $M(\mathbb{A}_F)$) if there exist two automorphic representations

$$\sigma_1 , \sigma_2$$

of $Gl(2, \mathbb{A}_F)$ such that σ_i are either induced from a pair of global grössen characters or are irreducible cuspidal such that the representations σ_i have the same central character $\omega_{\sigma_1} = \omega_{\sigma_2}$, and such that the spinor L-series of π satisfies

$$L_v(\pi_v, s) = L_v(\sigma_{1,v}, s) \cdot L_v(\sigma_{2,v}, s)$$

for almost all places v of F. We also call π a weak lift of the irreducible automorphic representation $\sigma = (\sigma_1, \sigma_2)$ of $M(\mathbb{A}_F)$ and $Gl(2, \mathbb{A}_F) \times Gl(2, \mathbb{A}_F)$ in this situation.

Suppose π is a weak endoscopic cuspidal lift as in the definition above, which is not CAP. Let σ be the corresponding automorphic representation of $Gl(2, \mathbb{A}_F) \times Gl(2, \mathbb{A}_F)$. Then $\sigma = (\sigma_1, \sigma_2)$ has to be a cuspidal representation. Otherwise either σ_1 or σ_2, say, σ_1, has L-factors of the form $L_v(\sigma_1, s) = L_v(\lambda, s)L_v(\lambda^{-1}\omega_{\sigma_2}, s)$ for an idele class character λ for $v \notin S$. Put $\tau = \sigma_2 \otimes \lambda^{-1}$. Then $L^S(\pi, s) = L^S(\lambda, s)L(\lambda\omega_\tau, s)L^S(\tau \otimes \lambda, s)$. If $\tau^S \cong \nu^S \times \mu^S$ (locally induced from the unramified characters ν_v, μ_v), then this is the L-series with the local parameters $\lambda_v, \lambda_v\nu_v, \lambda_v\mu_v, \lambda_v\nu_v\mu_v$ or from [97], p. 85, the L-series of the induced representation $\nu^S \times \mu^S \lhd \lambda^S$ of $GSp(4, \mathbb{A}^S)$ in the short notation used in Chap. 3, page 121, and induced from a Borel subgroup. Since $\nu^S \times \mu^S \lhd \lambda^S = \tau^S \lhd \lambda^S$ (induction in steps), we see that $L^S(\pi, s)$ would be the partial L-series of the automorphic form $\tau \boxtimes \lambda$ of the Levi component of the Siegel parabolic subgroup. Hence, the weak lift would be associated with an automorphic Eisenstein representation, which is weakly equivalent to π. Hence, π is CAP of Saito–Kurokawa type. This implies

$$\pi \text{ not CAP} \Rightarrow \sigma \text{ cuspidal.}$$

Notice that σ is not uniquely defined by π. If π is a weak lift of $\sigma = (\sigma_1, \sigma_2)$, then it is also a weak lift of $\sigma^* = (\sigma_2, \sigma_1)$. These are the only possibilities.

Proposition 5.2. *Suppose π is an irreducible cuspidal automorphic representation of the group $G(\mathbb{A}_F)$, which is not a CAP representation. If π is a weak endoscopic lift of the representations σ and $\tilde{\sigma}$ of $M(\mathbb{A}_F)$, then $\tilde{\sigma}$ is isomorphic to either σ or σ^* and σ is cuspidal.*

Proof. Write $\sigma = (\sigma_1, \sigma_2)$ and $\tilde{\sigma} = (\sigma_3, \sigma_4)$. Then $\sigma, \tilde{\sigma}$; hence, $\sigma_1, \sigma_2, \sigma_3, \sigma_4$ are cuspidal representations, as explained above. Then for a cuspidal representation ρ of $Gl(2, \mathbb{A})$ there exists a finite set S of places of F outside of which we have an equality of partial L-series

$$L^S(\sigma_1 \times \rho, s) \cdot L^S(\sigma_2 \times \rho, s) = L^S(\sigma_3 \times \rho, s) \cdot L^S(\sigma_4 \times \rho, s).$$

Since the σ_i are cuspidal, there exist complex numbers s_i for which $|.|^{s_i} \otimes \sigma_i$ becomes unitary. By a suitable (re)indexing, we may suppose $Re(s_1) = max_i(Re(s_i))$. For the contragredient ρ of σ_1 the function $L^S(\sigma_1 \times \rho, s)$ has a simple pole at $s = 1$, and $L^S(\sigma_2 \times \rho, s)$ does not vanish at $s = 1$ by statements (2.2) and (2.3) in [8], p. 200. Furthermore, if σ_1 is neither isomorphic to σ_3 nor isomorphic to σ_4, the right side is holomorphic at $s = 1$. This gives a contradiction. So, possibly by switching σ_3 and σ_4, we may assume $\sigma_1 \cong \sigma_3$. But then $L^S(\sigma_2, s) = L^S(\sigma_4, s)$, and the strong multiplicity 1 theorem for $Gl(2)$ implies $\sigma_2 \cong \sigma_4$. $\square$

Definition 5.2. Let π be an irreducible (unitary) cuspidal automorphic representation of the group $G(\mathbb{A}_F)$, which is not a CAP representation. Let π be a weak endoscopic lift. Suppose σ is a corresponding cuspidal automorphic representation of $M(\mathbb{A}_F)$. Then the set of equivalence classes of irreducible, automorphic representations π' of $G(\mathbb{A}_F)$, which are weakly equivalent to π, is called the *global L-packet* of π.

Since this global L-packet consists of all the weak endoscopic lifts attached to the given cuspidal representation σ, it will also be called the global L-packet attached to σ. Since π was assumed not to be CAP, all representations in the global L-packet are cuspidal.

Theorem 5.2 (Main Theorem). *Suppose π is an irreducible, cuspidal automorphic representation of $GSp(4, \mathbb{A}_F)$, and suppose π is not CAP. Suppose the global L-packet attached to the cuspidal irreducible automorphic representation σ of $Gl(2, \mathbb{A}_F)^2 / \mathbb{A}_F^*$ is nonempty and contains π as a weak endoscopic lift. Then:*

1. *σ_1 and σ_2 are not isomorphic as representations of $Gl(2, \mathbb{A}_F)$ for $\sigma = (\sigma_1, \sigma_2)$.*
2. *The restriction of π to $Sp(4, \mathbb{A}_F)$ is obtained as a Weil representation from the orthogonal group attached to a quaternary quadratic form of a square discriminant.*
3. *All local representations π_v of $\pi = \otimes' \pi_v$ belong to the local L-packets attached to σ_v. Hence,*
$$\pi_v \in \{\pi_+(\sigma_v), \pi_-(\sigma_v)\}$$
   if σ_v belongs to the discrete series, and $\pi_v = \pi_+(\sigma_v)$ otherwise.
4. *The multiplicity of any irreducible representation $\pi' = \otimes \pi'_v$ weakly equivalent to π is*
$$m(\pi') = \frac{1}{2}(1 + (-1)^{e(\pi')}),$$
   where $e(\pi')$ denotes the (finite) number of representations π'_v which do not have a local Whittaker model.
5. *Let $d(\sigma)$ be the number of local components σ_v of σ in the discrete series. The global L-packet attached to σ contains a single representation if $d(\sigma) < 2$ and contains $2^{d(\sigma)-1}$ representations, each occurring with multiplicity 1 otherwise.*

In Addition. *For any cuspidal irreducible automorphic representation $\sigma = (\sigma_1, \sigma_2)$ with $\sigma_1 \not\cong \sigma_2$ the global L-packet attached to σ is nonempty.*

Remark 5.2. For CAP representations a result analogous to statement (5) of the last theorem is Theorem 2.6 in [69].

Proof. The proof of the main theorem has several steps, and it is carried out in the remaining sections of this chapter. We deal separately with the case where π_∞ belongs to the discrete series and F is the field of rational numbers. This situation is easier to begin with since we can directly refer to the multiplicity formula of Corollary 4.2 in Chap. 4. For the general case we have to use the Arthur trace formula to prove an analogous multiplicity formula, and some of the arguments therefore become more complicated. See Sects. 5.4 and 5.5 for this. Although the arguments are primarily global, they simultaneously provide us with the necessary local information, e.g., on local Whittaker models. The relevant local information will be derived in the subsequent Lemmas 5.4–5.6. By some kind of bootstrap, we then improve this to complete the proof for general representations π_∞ and for general (totally real) number fields F in Sects. 5.4 and 5.5. □

We start with the proof of statements (1) and (2) of the main theorem.

Lemma 5.2 (Global Endoscopic = Theta Lift). *Suppose π is an irreducible cuspidal automorphic representation of $GSp(4, \mathbb{A}_F)$, which is a weak lift in the global L-packet attached to an irreducible automorphic representation $\sigma = (\sigma_1, \sigma_2)$ of $M(\mathbb{A}_F)$.*
   (A) Suppose π is not CAP. Then:

1. *σ is cuspidal and*
$$\sigma_1 \not\cong \sigma_2 \quad for \ \sigma = (\sigma_1, \sigma_2).$$

2. *Furthermore, all irreducible constituents of the restriction of π to $Sp(4, \mathbb{A}_F)$ are contained in the image of theta lifts of the form $\Theta : \mathcal{A}_{cusp}(H(\mathbb{A}_F)) \rightarrow \mathcal{A}(Sp(4, \mathbb{A}_F))$, where H are orthogonal groups of similitudes over F, which are inner forms of $Gl(2, F)^2/F^*$. Hence, $H(F) = D^* \times D^*/\{(x, x^{-1})|x \in F^*\}$, where D^* is either the multiplicative group of a quaternion algebra over F or $Gl(2, F)$.*

3. *For each local component π_v of π there exists a quadratic character μ_v such that $\pi_v \otimes \mu_v$ is contained in*
$$\mathcal{E}\Theta^{glr}_+(\sigma_v) \cup \mathcal{E}\Theta^{glr}_-(\sigma_v).$$

   (B) Conversely, suppose σ is cuspidal and $\sigma_1 \not\cong \sigma_2$. Then the weak lift π of σ is not CAP.

Concerning the notation. $\mathcal{A}$ will denote packets of irreducible automorphic representations.

Proof. Concerning (A) it is enough to prove the global statements (1) and (2). The local statement (3) then follows by Frobenius reciprocity [104], Proposition 4.12(c), using compact induction. So let us prove (A)(1) and (2). Let $\omega = \omega_\pi$ denote the central character of the irreducible representation π of $GSp(4, \mathbb{A}_F)$. The standard L-function of π corresponds to the four-dimensional standard representation st of the L-group ^LG of G. The alternating square of the standard representation is a six-dimensional representation $\bigwedge^2(st)$ of the L-group ^LG. The corresponding L-series of π outside a finite set S of ramified places is a product $\zeta^S(\pi, \bigwedge^2(st), s) = L^S(\omega, s) \cdot \zeta^S(\pi, \omega, s)$ of the Dirichlet L-series $L(\omega, s)$ for the character ω and a twist of the degree 5 standard L-series $\zeta(\pi, \omega, s)$ of the restriction of π to $Sp(4, \mathbb{A}_F)$. Here, by abuse of notation, we again let π denote any of the irreducible constituents of its restriction to $Sp(4, \mathbb{A}_F)$. Then in fact, the degree 5 L-series $\zeta(\pi, \omega, s)$ is the standard L-series attached to the representation $\pi \times \omega$ of $Sp(4, \mathbb{A}_F) \times Gl(1, \mathbb{A}_F)$, at least if we consider partial L-series at almost all unramified places. Since π is a weak lift, we get from this

The Partial L-series. If we omit a suitable chosen finite set of places S depending on σ and π, then

$$\zeta^S(\pi, \bigwedge^2 (st), s) = L^S(\omega_{\sigma_1}, s) \cdot L^S(\omega_{\sigma_2}, s) \cdot L^S(\sigma_1 \otimes \sigma_2, s).$$

Hence, from $\omega_{\sigma_1} = \omega_{\sigma_2}$

$$\begin{aligned} L^S(\omega_\pi \omega_{\sigma_2}^{-1}, s) \zeta^S(\pi, \omega_\pi \omega_{\sigma_2}^{-1}, s) &= \zeta^S(s)^2 \cdot L^S(\sigma_1 \otimes \sigma_2 \otimes \omega_{\sigma_2}^{-1}, s) \\ &= \zeta^S(s)^2 \cdot L^S(\sigma_1 \otimes \check{\sigma_2}, s) \end{aligned}$$

for the contragredient representation $\check{\sigma_2}$ of σ_2. For cuspidal σ_i the L-series $L^S(\sigma_1 \otimes \check{\sigma_2}, s)$ does not vanish at $s = 1$. By a character twist the σ_i become unitary. A twist of the cuspidal representations σ_i becomes unitary if and only if their central character ω_{σ_i} becomes unitary. Since this is a common character, one can take the same character twist in both cases. So without restriction of generality, we can assume that σ_1 and σ_2 are both unitary. Notice the representations have the same central character. If $\omega_\Pi \neq \omega_{\sigma_i}$, this implies that $\zeta^S(\pi, \omega_\pi \omega_{\sigma_2}^{-1}, s)$ has a pole of order 2 at $s = 2$. This is impossible, as will be explained below. Therefore,

$$\omega_\pi = \omega_{\sigma_2}$$

holds. Then the partial degree 5 standard L-function $\zeta^S(\pi, s) = \zeta(\pi, 1, s)$ of π is given by the formula

$$\zeta^S(\pi, s) = \zeta^S(s) \cdot L^S(\sigma_1 \otimes \sigma_2 \otimes \omega_{\sigma_2}^{-1}, s) = \zeta^S(s) \cdot L^S(\sigma_1 \otimes \check{\sigma_2}, s).$$

Then $L^S(\sigma_1 \otimes \check{\sigma_2}, 1) \neq 0$ holds at $s = 1$ for cuspidal unitary representations σ_i. Hence,

$$ord_{s=1} \zeta^S(\pi, s) \geq 1.$$

For $\sigma_1 \cong \sigma_2$ the pole order would be 2 or more. Hence, the first assertion of Lemma 5.2 comes from the next result obtained by Soudry.

Review of a Result Obtained by Soudry.

$\zeta^S(\pi, \chi, s)$ has at most a simple pole at $s = 1$ for unitary cuspidal π.

In fact, this is proved in [97] for a special character $\chi = \chi_T$. Indeed, this is the most difficult case. For other characters one can use the same approach. Then again as in [97], the order of the pole at $s = 1$ can be estimated by the pole order of an Eisenstein series. Up to a factor $L^S(\chi^2, 2s)L^S(\chi, s+1)$, which is irrelevant at $s = 1$, the L-series $\zeta^S(\pi, \chi_T\chi, s)$ is (except for an unimportant function $A(s)$ with $A(1) \neq 0, \infty$) an integral $\int \varphi(g)\theta^\phi_{T,\psi}(g, 1)E(f_{s,\chi}; g)dg$ by (2.2) and (2.5) in [97]. The poles of $E(f_{s,\chi}; g)$, therefore, give an upper bound for the poles of $\zeta^S(\pi, \chi\chi_T, s)$. From [97], Theorem 2.4, the poles of these Eisenstein series are contained in the poles of $L(\chi^2, 2s)L(\chi, s+1)$, $L(\chi^2, 2s)L(\chi, s)$, $L(\chi^2, 2s-1)L(\chi, s)$, $L(\chi^2, 2s-1)L(\chi, s-1)$ counted with multiplicity. There is no pole at $s = 1$ for $\chi^2 \neq 1$ and at most a simple pole for $\chi \neq 1$. The difficult case is $\chi = 1$. In this case there is no pole according to [97], Theorems 3.1 and 4.4(b). Therefore, the claim follows.

Now Apply Kudla–Rallis–Soudry Theorem 7.1 [56]. This is now the point where we have to restrict ourselves to the case of totally real number fields. This assumption regarding the number field is made in [56]. Under this assumption the existence of a pole for the partial degree 5 L-series $\zeta^S(\pi, s)$ at $s = 1$ implies that π – or more precisely any irreducible constituent of π as a representation of $Sp(4, \mathbb{A}_F)$ – is a constituent of a suitable theta lift

$$\Theta(\mathcal{A}(O(V', \mathbb{A}_F))$$

attached to a quadratic space V' of discriminant 1 over F:

$$\Delta(V') = \Delta(V_T) \cdot \Delta(V_{T'}) = 1.$$

This formula for the discriminants follows from $\chi_T\chi_{T'} = 1 = (\ .\ , \Delta(V_T)) \cdot (\ .\ , \Delta(V_{T'}))$ using Theorem 7.1(ii) in [56] if $\chi_T \neq 1$ or Theorem 7.1(i) if $\chi_T = 1$.

Remark 5.3. The assumption F is totally real is most likely unnecessary since it is needed only for the case of split discriminant. In fact, looking at the integral representation in [97], formula (2.2), it would suffice to show that the residues of the nonholomorphic Siegel Eisenstein series $E(f_s; g)$ at $s = 1$ ([97], 2(b)) are binary theta series. A partial result in this direction – namely, the local case of this statement – is Lemma 3.3 in [97].

Structure of V'. The quaternary quadratic F-spaces V' of discriminant $\Delta(V') = 1$ are classified by their local Hasse invariants, or in this case alternatively by the structure of their orthogonal groups of similitudes

$$GSO(V')(F) \cong D^* \times D^*/F^*.$$

Here D is the corresponding quaternion algebra or $Gl(2, F)$, having invariants determined by the signs of the local Hasse invariants of V'. This proves part (A) of Lemma 5.2, and it also proves statements (1) and (2) of the main theorem.

For the converse part (B) of Lemma 5.2 we may assume σ has unitary central character. Notice $L^S(\pi, s) = L^S(\sigma_1, s)L^S(\sigma_2, s)$ and $\zeta^S(\pi \otimes \chi, s) = L^S(\chi, s)L^S(\sigma_1 \otimes \sigma_2^\vee \otimes \chi, s)$ outside a suitable finite set S. The known CAP criteria for π in terms of poles of these L-series (Soudry and Piatetski-Shapiro) exclude the possibility that π is CAP for cuspidal σ, since the known analytic behavior of L-series of cuspidal forms for $Gl(2)$ or $Gl(2) \times Gl(2)$ excludes the existence of poles at $s = 3/2$ for the degree 4 L-series and poles at $s = 2$ for degree 5 L-series (for unitary central characters and unitary χ). This proves part (B), and completes the proof of Lemma 5.2. $\square$

Lemma 5.3. *We have*

$$\bigcup_{\sigma_v} \mathcal{E}\Theta_+(\sigma_v) \cap \bigcup_{\sigma_v} \mathcal{E}\Theta_-(\sigma_v) = \emptyset,$$

where the union is over all generic irreducible admissible representations σ_v of M_v on the left, and over all generic irreducible admissible discrete series representations σ_v of M_v on the right. In other words, the two theta lifts attached to M_v (split case) and $\check{M}_v$ (anisotropic case) have no irreducible representation of $GSp(4, F_v)$ in common. The same statement also holds globally or after restriction to the subgroup $Sp(4, F_v)$.

Proof. It is enough to consider the restrictions to $Sp(4, F_v)$. Here we refer to Howe and Piatetski-Shapiro [41], Theorem 9.4, where it is shown that there is at most one representation in common. In the proof of Lemma 4.26 we saw that the common representation corresponds to a one-dimensional representation σ_v. However, this is not a generic representation, and therefore it is excluded in the statement above. This proves Lemma 5.3. $\square$

Lemma 5.4 (Principle of Exchange). *Suppose the cuspidal irreducible automorphic representation $\pi = \otimes \pi_v$ of $GSp(4, \mathbb{A}_F)$ is a weak endoscopic lift, but is not CAP. Let σ denote the corresponding cuspidal representation of $M(\mathbb{A}_F)$:*

1. *Fix some place v_0. Then π_{v_0} is in $\mathcal{E}\Theta_\varepsilon^{glr}(\sigma_{v_0} \otimes \mu_{v_0})$ for some quadratic character μ_v and some $\varepsilon \in \{\pm\}$. Suppose $\pi'_{v_0} \not\cong \pi_{v_0}$ is an irreducible representation of $GSp(4, F_{v_0})$ and consider*

$$\pi' = \pi'_{v_0} \otimes \bigotimes_{v \neq v_0} \pi_v.$$

Make the assumption: *$\pi'_{v_0} \notin \mathcal{E}\Theta_\varepsilon^{glr}(\sigma_{v_0} \otimes \mu'_{v_0})$ for all quadratic characters μ'_v. Under this assumption π' has global multiplicity*

$$m_{disc}(\pi') = 0.$$

2. *If $d(\sigma) < 2$ ($d(\sigma)$ is defined in Theorem 5.2(5)), then $\pi_v = \theta_+(\sigma_v \otimes \mu_v)$ for all v and certain local quadratic characters μ_v. In particular, π_v is generic.*

Proof. Suppose that to the contrary $m(\pi') \geq 1$ holds. Then π' is in the global L-packet of π, and hence it is a weak endoscopic lift attached to the same irreducible representation σ of $M(\mathbb{A})$ as π. According to Lemma 5.2 all constituents of π'

and π, after restriction to the group $Sp(4, \mathbb{A}_F)$, are in the image of some theta lifts Θ. These theta correspondences lift automorphic forms from orthogonal groups arising from global simple algebras of rank 4 to automorphic forms on $GSp(4)$. In our case let D and D' be the corresponding algebras, respectively. From the known local properties of the theta lift it follows that

$$D_v \cong D'_v \qquad (\forall v \neq v_0),$$

since by Lemma 5.3 we would otherwise get a contradiction to the fact that

$$\pi_v \cong \pi'_v \qquad (\forall v \neq v_0).$$

But then $D_{v_0} \cong D'_{v_0}$, as a consequence of the theorem of Hasse–Brauer–Noether. $D_{v_0} \cong D'_{v_0}$ implies that the constituents of both π_v and π'_v (after restriction to $Sp(4, F_v)$) either both belong or both do not belong to the local theta lift $\mathcal{E}\Theta_\varepsilon^{glr}(\sigma_v)$. Since this contradicts the assumptions, the proof of the first part of Lemma 5.4 is complete.

The second part is shown similarly. Now D splits globally. Then [41], Theorem 5.7b or 8.1b, implies that the restriction of π_v to $Sp(4, F_v)$ is generic. By compact induction also $\pi_v \otimes \mu_v$, and hence π_v is generic. This implies $\pi_v \cong \theta_+(\sigma_v \otimes \mu_v) = \theta_+(\sigma_v) \otimes \mu_v$ for some quadratic character. This proves part (2) of the lemma. $\square$

The principle of exchange can be applied for an Archimedean place in the situation of Lemma 5.5, since in this case the assumption made in Lemma 5.4(1) is satisfied. In fact, assume $F_{v_0} = \mathbb{R}$ and let π_{v_0} and π'_{v_0} be representations in the local Archimedean L-packet $\{\pi_\infty^W, \pi_\infty^H\} = \{\pi_+(\sigma_\infty), \pi_-(\sigma_\infty)\}$ attached to a discrete series representation σ_∞ of M_∞. By Lemma 5.5 $\pi^W \in \mathcal{E}\Theta_+(\sigma_\infty)$ and $\pi^H \in \mathcal{E}\Theta_-(\sigma_\infty \otimes \mu_\infty)$; hence, $\pi^H \notin \mathcal{E}\Theta_+(\sigma_\infty \otimes \mu_\infty)$ for all μ_∞ by Lemma 5.3. Hence, the principle of exchange can be applied.

Lemma 5.5. *Let $F_v = \mathbb{R}$ and $\pi'_v = \pi_v^W$ and $\pi_v = \pi_v^H$, then the constituents of π'_v restricted to $Sp(4, \mathbb{R})$ have Whittaker models, whereas the constituents of π_v restricted to $Sp(4, \mathbb{R})$ do not have a Whittaker model. Furthermore, $\pi^W \otimes \mu_v \cong \pi^W$, $\pi^H \otimes \mu_v \cong \pi^H$ for all quadratic characters μ_∞, and also $\pi^W \in \mathcal{E}\Theta_+(\sigma_\infty \otimes \mu_\infty)$ and $\pi^H \in \mathcal{E}\Theta_-(\sigma_\infty \otimes \mu_\infty)$ for all quadratic characters μ_∞.*

Proof. Well known. $\square$

5.3 The Multiplicity Formula

The main results obtained in this section are the following: the local–global principle (Lemma 5.6) and the key formula. Under the assumption $F = \mathbb{Q}$ and the assumption that in the global L-packet of the representation π a representation exists for which the Archimedean component π_∞ belongs to the discrete series, Lemma 5.6 and the key formula are proven in this section. This of course suffices for the applications in

Chap. 2. The proof of Lemma 5.6 and key formula for general totally real number fields and arbitrary cusp forms π is contained in Sects. 5.4 and 5.5. In this section we also show that the local–global principle (Lemma 5.6) and the key formula imply for arbitrary F and π certain corollaries (Corollaries 5.1–5.5), from which the main Theorem 5.2 follows.

Let π be a cuspidal irreducible weak endoscopic lift attached to some irreducible cuspidal representation σ of $M(\mathbb{A}_F)$. Write $\pi = \pi_\infty \otimes \pi_f$, where π_∞ denotes the tensor product over all Archimedean components. Assume π is not CAP. Then σ is cuspidal and $\sigma \not\cong \sigma^*$ (Lemma 5.2(1)). For irreducible automorphic representations of the group $Gl(2, \mathbb{A}_F)$ cuspidal implies generic. So the same is true for the group $M(\mathbb{A}_F)$.

So Lemma 5.6 proves assertion (3) of the main theorem.

Lemma 5.6 (Local–Global Principle). *Let π be a weak endoscopic lift contained in the global L-packet attached to some generic, irreducible cuspidal automorphic representation σ of $M(\mathbb{A})$. Suppose π is cuspidal but not CAP. Then for all places v of F the local components π_v of π belong to the local L-packets $\{\pi_\pm(\sigma_v)\}$ and $\{\pi_+(\sigma_v)\}$ of σ_v.*

Proof of Lemma 5.6 for $F = \mathbb{Q}$ and π_∞ in the discrete series (for the general case see Sects. 5.4 and 5.5). By assumption π_∞ is contained in some local Archimedean L-packet $\{\pi^W, \pi^H\}$ attached to a discrete series representation of M_∞. From the topological trace formula we have the

Weak Multiplicity Formula (Corollary 4.2). For $F = \mathbb{Q}$ and π_f as above

$$m(\pi_\infty^W \otimes \pi_f) - m(\pi_\infty^H \otimes \pi_f) = n(\pi_f),$$

where $n(\pi_f)$ is zero if there exists a non-Archimedean place v for which π_v is not contained in the local L-packet of σ_v, and where $n(\pi_f)$ is equal to $(-1)^{e(\pi_f)}$ otherwise. Here $e(\pi_f)$ denotes the number of non-Archimedean places v for which $\pi_v \cong \pi_-(\sigma_v)$ holds. Since $e(\pi_f) < d(\sigma)$, this number is finite. *Some comments*: The multiplicity $m(\pi)$ is the multiplicity of π in the discrete spectrum of $G(\mathbb{A}_\mathbb{Q})$ or equivalently the multiplicity in the cuspidal spectrum since π is not CAP. We remark that the multiplicity formula above is a reformulation of Corollary 4.2. In fact we used Corollary 4.4 to identify the multiplicities $m_1(\pi_f), m_2(\pi_f)$ used in Corollary 4.2 with the multiplicities $m_1(\pi_f) = m(\pi_\infty^H \otimes \pi_f)$ and $m_2(\pi_f) = m(\pi_\infty^W \otimes \pi_f)$. We also used the property that the normalizing sign ε from Corollary 4.2 is $\varepsilon = -1$ (Lemma 8.4). Also recall that in Corollary 4.2 the value $n(\pi_f) = \prod_{v \neq \infty} n(\pi_v)$ is defined in terms of the multiplicities $n(\pi_v)$ of π_v in the local character expansion $r(\sigma(\sigma_v)) = \sum n(\pi_v) \cdot \pi_v$ for the local endoscopic lift $r : R_\mathbb{Z}(M_v) \to R_\mathbb{Z}(G_v)$. The expression for $n(\pi_f)$ in the form stated above follows from the structure of the local L-packets of the local endoscopic lift $r(\sigma_v)$ of a generic representation σ_v of M_v. This was described in Sect. 4.11 and also in Sect. 5.1 of this chapter in the non-Archimedean case and in Lemma 5.5 in the Archimedean case. The L-packets consist of two representations $\pi_+(\sigma_v), \pi_-(\sigma_v)$ if σ_v belongs to the discrete series, and of one representation $\pi_+(\sigma_v)$ if σ_v does not belong to the discrete series. Among these only the representations $\pi_-(\sigma_v)$ do not have Whittaker models. Hence, $n(\pi_v) = 0$ unless π_v is in the local L-packet of σ_v. Furthermore, $n(\pi_v)$ is $+1$ or -1 otherwise, and the sign depends on the existence of a Whittaker model.

If π_v is not in the local L-packet of σ_v, then v is non-Archimedean and the weak multiplicity formula implies $m(\pi_\infty^W \otimes \pi_f) = m(\pi_\infty^H \otimes \pi_f)$. But then $m(\pi) > 0$ and $m(\pi') > 0$ for $\pi = \pi_\infty^W \otimes \pi_f$ and $\pi' = \pi_\infty^H \otimes \pi_f$. But this contradicts the principle of exchange. By Lemma 5.5 the assumptions of Lemma 5.4 are satisfied for the place $v_0 = \infty$. Hence, Lemma 5.6 follows under the assumptions $F = \mathbb{Q}$ and π_∞ is in the discrete series.

Now let us return to the general case. In fact, suppose that Lemma 5.6 holds assuming the results obtained in Sects. 5.4 and 5.5. From Lemma 5.6 we now deduce the next three corollaries. First recall from Sect. 5.1 the global $(GSp(4), GO(V))$-theta correspondence: For discrete series representations σ_v we constructed a global irreducible automorphic cuspidal representation σ of $M(\mathbb{A}_F)$ and a nonvanishing global theta lift π of σ with local components π_v such that $\pi_v \in \mathcal{E}\Theta_-(\sigma_v)$ which do not have local Whittaker models. Hence, $\pi_v \not\cong \pi_+(\sigma_v)$ since $\pi_+(\sigma_v)$ has Whittaker models. This implies $\pi_v = \pi_-(\sigma_v)$ (Lemma 5.6). In particular, $\pi_-(\sigma_v)$ is in $\mathcal{E}\Theta_-(\sigma_v)$ and does not have local Whittaker models. Since we also know that $\pi_+(\sigma_v) = \theta_+(\sigma_v)$ is generic, we obtain from Lemma 5.6 $\quad\square$

Corollary 5.1 (Local Endoscopic Lift = Theta Lift). *Under the assumptions of Lemma 5.6 the representations $\pi_\pm(\sigma_v)$ of the local L-packet attached to a discrete series representation σ_v of M_v are the two Weil representations $\theta_\pm(\sigma_v)$:*

(a) $\Theta_+^{glr}(\sigma_v) = \{\pi_+(\sigma_v)\} = \{\theta_+(\sigma_v)\}$ *for split D.*
(b) $\Theta_-(\sigma_v) = \{\pi_-(\sigma_v)\} = \{\theta_-(\sigma_v)\}$ *for nonsplit D.*

In particular, $\pi_+(\sigma_v)$ has a local Whittaker model, whereas $\pi_-(\sigma_v)$ does not have local Whittaker models.

Notice Lemma 5.6 and its Corollary 5.1 imply Proposition 5.1.

Corollary 5.2 (Whittaker Models). *Suppose the assumptions of Lemma 5.6 are satisfied. Let v be a place where σ_v belongs to the discrete series. Then $\pi_+(\sigma_v) \in \Theta_+(\sigma_v)$ has local Whittaker models, whereas $\pi_-(\sigma_v) \otimes \mu_v \notin \Theta_+(\sigma)$ for all (quadratic) characters μ_v, and these representations are cuspidal and do not have local Whittaker models.*

Proof. Follows from Corollary 5.2 and Lemmas 4.17 and 5.3. $\quad\square$

Corollary 5.3 (Refined Principle of Exchange). *Suppose the assumptions of Lemma 5.6 are satisfied. For the weak automorphic lift $\pi = \otimes_v \pi_v$ of σ put $\pi' = \pi'_v \otimes_{v \neq v_0} \pi_v$ for an irreducible admissible representation π'_v of G_v. Then for $\pi'_v \not\cong \pi_v$*

$$m(\pi') = 0.$$

Proof. Suppose $m(\pi') > 0$ and $\pi'_v \not\cong \pi_v$. Then both π and π' are weak automorphic endoscopic lifts of σ. Hence, $\pi_v, \pi'_v \in \{\pi_+(\sigma_v), \pi_-(\sigma_v)\}$ by Lemma 5.6. Since they are not isomorphic, we can apply the principle of exchange (Lemma 5.4), because Corollaries 5.1 and 5.2 imply that all the assumptions of Lemma 5.4 are now satisfied. Hence, $m(\pi') = 0$ (Lemma 5.4), which completes the proof of Corollary 5.3. □

Proof of the Remaining Assertions (4) and (5) of the Main Theorem. Suppose π is an automorphic representation of $G(\mathbb{A}_F)$ as in the main theorem. π is irreducible, cuspidal automorphic but not CAP, and π is a weak lift attached to some irreducible automorphic representation σ of $M(\mathbb{A}_F)$. Then, as already shown, σ is cuspidal and its local components σ_v are generic. □

Conversely for an irreducible cuspidal automorphic representation $\sigma = (\sigma_1, \sigma_2)$ of $M(\mathbb{A}_F)$ consider the global L-packet of all weak endoscopic lifts attached to it. Since we are mainly interested in cuspidal automorphic representations, assume

$$\sigma_1 \not\cong \sigma_2$$

because otherwise the global L-packet attached to σ does not contain a cuspidal representation (Lemma 5.2). See also Corollary 5.5. In fact, recall from Theorem 4.3

The Base Point of the L-packet. *From a result obtained by Soudry this global L-packet is nonempty and contains the representation $\pi_+(\sigma) = \prod_v \pi_+(\sigma_v)$ as a cuspidal automorphic representation with multiplicity $m(\pi_+(\sigma)) = 1$.*

In fact the situation in Theorem 4.3 is slightly more restrictive. So for the convenience of the reader we sketch the argument.

Sketch of Proof. $\pi_+(\sigma)$ is the theta lift $\mathcal{E}\Theta_+(\sigma)$ (for the split case). By the assumptions regarding σ this theta lift does not vanish, and it is cuspidal since $\sigma_1 \not\cong \sigma_2$. Hence, $m(\pi_+(\sigma)) > 0$. Then from a result obtained by Soudry

$$m(\pi_+(\sigma)) = 1,$$

since every cuspidal representation which is isomorphic to $\pi_+(\sigma)$ has a global Whittaker model. This follows from another result obtained by Soudry, namely, since by construction $\pi_+(\sigma)$ is locally generic at all places v, it is enough to show the nonvanishing of $\zeta^S(\pi_+(\sigma), s)$ at $Re(s) = 1$ (if ω is normalized to be unitary). But this nonvanishing follows immediately from the computations made in the proof of Lemma 5.2, since under our assumptions the L-series $L^S(\sigma_1 \times \hat{\sigma}_2, s)$ does not vanish on the line $Re(s) = 1$. This completes the proof of the claim. □

Corollary 5.1 now implies that for every representation π in the global L-packet there exists a finite set of places S of F where π_v does not have a Whittaker model and such that

$$\pi = \bigotimes_{v \in S} \pi_-(\sigma_v) \otimes \bigotimes_{v \notin S} \pi_+(\sigma_v).$$

Let $e(\pi) = \#S$ be cardinality of this uniquely defined set

$$S = S(\pi).$$

This is a subset of the finite set of places, where σ_v belongs to the discrete series.

Corollary 5.4. *Suppose the assumptions of Lemma 5.6 are satisfied. Also suppose the key formula (stated below) holds. Then for any π in the global L-packet of a cuspidal irreducible representation σ with $\sigma \not\cong \sigma^*$ the multiplicity $m_{cusp}(\pi) = m_{disc}(\pi)$ is*

$$m(\pi) = \frac{1}{2}(1 + (-1)^{e(\pi)}), \qquad e(\pi) = \#S(\pi).$$

In Other Words: $m(\pi) = 1$ *for even* $e(\pi)$, *and* $m(\pi) = 0$ *otherwise.*

Proof. As in the proof of Lemma 5.4(2) this is true for $e(\pi) < 2$. So we may assume $e(\pi) \geq 2$. Pick two places $v_1 \neq v_2$ which contribute to $e(\pi)$. $\square$

If some Archimedean π_v belongs to the discrete series we assume v_1 to be Archimedean. More generally, let S be the set of places v where π_v is in the discrete series. Then we may assume v_1 to be some fixed place of S. Now consider the representation π' of $G(\mathbb{A}_F)$, which is obtained from π by replacing the two local representations π_{v_1} and π_{v_2} within their local L-packets. Since π_v is in the discrete series for $v = v_1, v_2$, the local L-packets both have cardinality 2 at v_1 and at v_2. With this notation we formulate the

Key Formula. *Let π be a weak endoscopic lift of a cuspidal automorphic representation σ with $\sigma^* \not\cong \sigma$. Suppose π is not a CAP representation and suppose $e(\pi) \geq 2$. Let π' be defined as above by replacements at two places v_1, v_2, where π belongs to the discrete series. Then $m(\pi) + m(\pi') > 0$ implies that $e(\pi)$ is even, and implies*

$$m(\pi) + m(\pi') = 2.$$

Continuation of the Proof of Corollary 5.4. It is clear that with this key formula we can prove Corollary 5.4 by induction on $e(\pi)$, by reduction to the known cases $e(\pi) = 0, 1$, since every set S' of even cardinality of the set of places where σ_v belongs to the discrete series can be obtained by a finite number of exchanges at two places of this set. It is of no harm to assume, in addition, that at one of the places, say, v_1 (for $F = \mathbb{Q}$, e.g., the Archimedean place, if π_∞ belongs to the discrete series), σ_{v_1} remains unchanged. This implies $m(\pi') = 1$ or $m(\pi') = 0$ by induction, depending on whether $e(\pi')$ is even or odd. This proves Corollary 5.4. $\square$

Proof of the Key Formula. We now prove it for $F = \mathbb{Q}$ and π_∞ in the discrete series. The general case is done in Sect. 5.4. Put $\pi = \pi_{\varepsilon_2}(\sigma_{v_2}) \otimes \pi_{\varepsilon_1}(\sigma_{v_1}) \otimes \pi^{v_1, v_2}$. Then by the weak multiplicity formula, applied twice,

$$\left(m\big(\pi_{\varepsilon_2}(\sigma_{v_2}) \otimes \pi_{\varepsilon_1}(\sigma_{v_1}) \otimes \pi^{v_1,v_2}\big) - m\big(\pi_{\varepsilon_2}(\sigma_{v_2}) \otimes \pi_{-\varepsilon_1}(\sigma_{v_1}) \otimes \pi^{v_1,v_2}\big)\right)$$

$$-\left(m\big(\pi_{-\varepsilon_2}(\sigma_{v_2}) \otimes \pi_{\varepsilon_1}(\sigma_{v_1}) \otimes \pi^{v_1,v_2}\big) - m\big(\pi_{-\varepsilon_2}(\sigma_{v_2}) \otimes \pi_{-\varepsilon_1}(\sigma_{v_1}) \otimes \pi^{v_1,v_2}\big)\right)$$

$$= (-1)^{e(\pi)} - (-1)^{e(\pi)\pm 1} = 2 \cdot (-1)^{e(\pi)} = 2 \cdot (-1)^{e(\pi)}.$$

By the principle of exchange, two of these multiplicities always vanish. If $m(\pi_{v_1 v_2}) > 0$ or $m(\pi) > 0$, then $m(\pi_{-\varepsilon_2}(\sigma_{v_2}) \otimes \pi_{\varepsilon_1}(\sigma_{v_1}) \otimes \pi^{v_1,v_2}) = m(\pi_{\varepsilon_2}(\sigma_{v_2}) \otimes \pi_{-\varepsilon_1}(\sigma_{v_1}) \otimes \pi^{v_1,v_2}) = 0$ vanishes (Corollary 5.3). Hence,

$$m\big(\pi_{\varepsilon_2}(\sigma_{v_2}) \otimes \pi_{\varepsilon_1}(\sigma_{v_1}) \otimes \pi^{v_1,v_2}\big) + m\big(\pi_{-\varepsilon_2}(\sigma_{v_2}) \otimes \pi_{-\varepsilon_1}(\sigma_{v_1}) \otimes \pi^{v_1,v_2}\big) = 2 \cdot (-1)^{e(\pi)}.$$

The left side is nonnegative by assumption, so $e(\pi)$ is even and $m(\pi) + m(\pi_{\infty,v}) = 2$. This proves the key formula, and hence Corollary 5.4 and the main theorem (in the special case). $\square$

Corollary 5.5. *Suppose that σ is a (generic) irreducible cuspidal representation of $M(\mathbb{A}_F) = Gl(2, \mathbb{A}_F) \times Gl(2, \mathbb{A}_F)/A_F^*$. Suppose $\sigma = JL(\breve{\sigma})$ is the Jacquet–Langlands lift of an irreducible representation $\breve{\sigma}$ of some inner form $D^*(\mathbb{A}) \times D^*(\mathbb{A})/A^*$ of $M(\mathbb{A})$, where D is a quaternion algebra. Then the corresponding theta lift $\Theta_D(\breve{\sigma})$ does not vanish.*

Remark 5.4. The corresponding statement for $D^* = Gl(2)$ is easier and follows from the existence of Whittaker models. See [41] or Chap. 4. See also the discussion preceding Corollary 5.4.

Proof of Corollary 5.5. Suppose $\sigma_1 \cong \sigma_2$. Consider the theta lift $\Theta_D(\sigma)$ and its zero Fourier coefficient with respect to the maximal parabolic Q, which is not the Siegel parabolic. In the classical theory of Siegel modular forms this corresponds to considering the Siegel ϕ-operator. As a representation of the $Gl(2, \mathbb{A}_F)$-factor of the Levi component, this Fourier coefficient essentially defines the Jacquet–Langlands lift attached to the automorphic representation $\breve{\sigma}_1 \cong \breve{\sigma}_2$ of $D^*(\mathbb{A}_F)$ (for details we refer to [41] and Chap. 4, Corollary 4.13). Since the Jacquet–Langlands lift is always nontrivial, this implies that the theta lift $\Theta_D(\sigma)$ is not trivial. In particular, it is not cuspidal under the assumption $\sigma_1 \cong \sigma_2$.

To prove the first assertion we can now assume $\sigma_1 \not\cong \sigma_2$. We will then show that the theta lift $\Theta_D(\sigma)$ does not vanish. Let S be the set of nonsplit places of the quaternion algebra D. As the cardinality of S is even, the multiplicity of the representation $\pi = \prod_{v \in S} \pi_-(\sigma_v) \otimes \prod_{v \notin S} \pi_+(\sigma_v)$ is 1 by Corollary 5.4. Hence, in particular, it is not zero, and π is therefore a cuspidal automorphic representation. As π is a weak endoscopic lift by definition, Proposition 5.1 implies that π is a theta lift. The set S of places where π_v does not have a Whittaker model is the set of nonsplit places of the corresponding algebra by Corollary 5.1. Hence, this algebra is D. But then by Proposition 5.2 π is a theta lift of type $\Theta_D(\sigma)$ or $\Theta_D(\sigma^*)$. But $\Theta_D(\sigma) = \Theta_D(\sigma^*)$ since the theta lift was originally defined by passing from $M = GSO(4)$ to $GO(4)$ (we refer to the remarks at the end of Sect. 5.1). Hence the claim follows. This proves Corollary 5.5. $\square$

5.4 Local and Global Trace Identities

We say a locally constant function f_v on G_v with compact support on G_v satisfies the *condition (RS)*, if its support is contained in the locus of regular points. A semisimple element of G_v is called elliptic if it is an element of an elliptic Cartan subgroup. The set of regular elliptic points is open in G_v. For locally constant functions f_v on G_v with compact support we say *condition (ES)* holds if the support is contained in the regular elliptic locus. Condition (ES) implies that all the orbital integrals $O_{\gamma_v}^{G_v}(f_v)$ of f_v vanish for G_v-regular nonelliptic elements $\gamma_v \in G_v = GSp(4, F_v)$

$$O_{\gamma_v}^{G_v}(f_v) = 0 \quad \text{for } \gamma_v \text{ regular and not elliptic.}$$

Hence, $f_v \in A(G_v)$ in the sense of [44]. And $f_v \in A(G_v)$ implies that f_v is *cuspidal* in the sense of [6], p. 538. By definition, this means $(f_v)_{M_v} = 0$ for all Levi subgroups $M_v \neq G_v$ of G_v, or equivalently by the adjunction formula for induced representations

$$tr\ \pi_v(f_v) = 0, \quad \text{for all } \pi_v = Ind_{P_v}^{G_v}(\sigma_v).$$

Here π_v runs over all induced representations of tempered representations σ_v of M_v for proper parabolic subgroups $P_v = M_v N_v$. See [5], p. 328.

We say *condition* $(*)_v$ holds for f_v if the stable orbital $SO_{\gamma_v}^{G_v}(f_v)$ of f_v vanishes for all regular semisimple $\gamma_v \in G_v$

$$(*)_v \qquad SO_{\gamma_v}^{G_v}(f_v) = 0 \qquad \text{(for all } \gamma_v \text{ regular semisimple).}$$

Since unstable tori of the group $G_v = GSp(4, F_v)$ are elliptic, the orbital integrals and stable orbital coincide for regular points outside the elliptic locus. Hence, condition $(*)_v$ implies $O_{\gamma_v}^{G_v}(f_v) = 0$ for all regular nonelliptic γ_v.

Let $\tilde{G}_v \subseteq G_v$ denote the subgroup of elements whose value under the similitude character is a square in $(F_v)^*$. Then $tr\ (\pi_v \otimes \mu_v)(f_v) = tr\ \pi_v(f_v)$ for all quadratic characters μ_v, if $supp(f_v) \subseteq \tilde{G}_v$. Note $\tilde{G}_v = Z_v \cdot Sp(4, F_v)$, and $\tilde{G}_v$ is open in G_v. Every element in G_v, which is stably conjugate to an element of $\tilde{G}_v$, is contained in $\tilde{G}_v$.

Remark 5.5. We later use the following auxiliary result, which is related to Proposition 5.3. For $F_v = \mathbb{R}$ there exists some σ_v in the discrete series and some K-finite infinitely differentiable function f_v with compact support satisfying $(*)_v$ and $tr\ \pi_-(\sigma_v)(f_v) > 0$. This is constructed by smooth truncation of a function $f_{\pi_+(\sigma_v)} - f_{\pi_+(\sigma_v)}$, where $f_{\pi_\pm(\sigma_v)}$ are pseudocoefficients. We leave this as an exercise. See also [91], (4.7.1).

Lemma 5.7 (Stability). *For σ_v in the discrete series of M_v, the character T of $\pi_+(\sigma_v) \oplus \pi_-(\sigma_v)$ is a stable distribution. In other words, for any locally constant function f_v on G_v, condition $(*)_v$ implies $T(f_v) = 0$.*

Proof of Lemma 5.7. In the Archimedean case this follows from [90], Lemma 5.2. In the non-Archimedean case one can argue as in [76], Proposition 12.5.3 and Corollary 12.5, to show that the stability of a distribution T is equivalent to the statement $< T, r(\rho_v) >_e = 0$ for all discrete series representations ρ_v of M_v. Since for $T = tr\, \pi_+(\sigma_v) + tr\, \pi_-(\sigma_v)$ the latter means $< \pi_+(\sigma_v) + \pi_-(\sigma_v), \pi_+(\rho_v) - \pi_-(\rho_v) >_e = 0$, this follows from the scalar product formulas proved in Chap. 4, Lemma 4.22. There are two cases that have to be distinguished. Either $\sigma_v \not\cong \sigma_v^*$. Then we can apply the orthogonality relations, since both $\pi_\pm(\sigma_v)$ belong to the discrete series, and since furthermore $\pi_\varepsilon(\sigma_v) \cong \pi_{\varepsilon'}(\rho_v)$ implies $\sigma_v \cong \rho_v$ and $\varepsilon = \varepsilon'$. Or $\sigma_v \cong \sigma_v^*$. Then $< \pi_\varepsilon(\sigma_v), \pi_{\varepsilon'}(\rho_v) >_e = 1$ for $\sigma_v \cong \rho_v$ and is zero otherwise as shown in Chap. 4. So this gives the proof.   □

Remark 5.6. The same argument proves stability for the traces $T = tr\, \pi_v$ of discrete series representations π_v of G_v for which π_v is not isomorphic to one of the representations $\pi_\pm(\sigma_v)$, σ_v in the discrete series.

Lemma 5.7 implies the existence of auxiliary functions $f_v \in A(G_v)$ as follows

Proposition 5.3 (Instability). *For σ_v in the discrete series of M_v, there exists a locally constant function f_v with condition (ES) and support in $\hat{G}_v$ such that $(*)_v$ holds and such that $tr\, \pi_-(\sigma_v) > 0$.*

Proof of Proposition 5.3. We can assume v is non-Archimedean. Recall $f_v \in A(G_v)$, and this implies $T(f_v) = 0$ for $T = tr\, \pi_+(\sigma_v) + tr\, \pi_-(\sigma_v)$ by Lemma 5.7. Therefore, $tr\, \pi_-(\sigma_v)(f_v) > 0$ and $tr\, r(\sigma_v)(f_v) < 0$ are equivalent statements. Hence, it is enough to show $tr\, \sigma_v(f_v^{M_v}) < 0$ for a matching function $f_v^{M_v}$. The character of σ_v does not vanish in any neighborhood of the identity element for at least one elliptic torus T_v in M_v. Such a T_v defines two conjugacy classes of tori in G_v, which are stable conjugate, together with admissible isomorphisms between T_v and these tori. Since regular points are smooth points of the conjugation map, one can easily construct matching functions f_v and $f_v^{M_v}$ with support in these tori sufficiently near to the identity such that $(*)_v$ holds for f_v, but such that $tr\, \sigma_v(f_v^{M_v}) < 0$. This is done by the implicit function theorem. Simply consider bump functions f_v with small support near $\gamma_v \in T_v$ such that $SO_{\gamma_v}^{G_v}(f_v) = 0$ and such that $O_{\gamma_v}^\kappa(f_v) = SO_{\gamma_v}^{M_v}(f_v^{M_v}) \neq 0$ for a corresponding bump function with sufficiently small support near γ_v. Then the support of f_v is in $\tilde{G}_v$, and f_v satisfies (ES). Furthermore, $tr\, r(f_v) < 0$ and $f_v \in A(G_v)$. This proves the claim.   □

We now prove the following results.

Proposition 5.4. *For irreducible σ_v in the discrete series $\mathcal{E}\Theta_-(\sigma_v) = \{\pi_-(\sigma_v)\}$.*

Proposition 5.5. *Suppose σ_v is irreducible and generic. Then $\pi_v \in \mathcal{E}\Theta_+^{glr}(\sigma_v)$ implies $\pi_v \cong \theta_+(\sigma_v)$.*

In some cases these statements are known already, for instance, in the Archimedean case. For p-adic fields (completions of $\mathbb{Q}$), Proposition 5.3 follows from the density lemma (Chap. 4, Sect. 5.1) and Lemma 5.6 in Sect. 5.3. Still another case: If σ_v is cuspidal non-Archimedean such that $\sigma_v \cong \sigma_v^*$, then $\mathcal{E}\Theta_+(\sigma_v) = \{\pi_+(\sigma_v)\}$ (see Chap. 4 and [106], p. 64, lines 10-15, and p. 55), which is stronger than Proposition 5.5. The distribution T of Lemma 5.7 is stable in this case for a trivial reason. Namely $\pi_+(\sigma_v)$ and $\pi_-(\sigma_v)$ are the two constituents of an induced representation called $1 \times \sigma_v$ (Lemma 4.19). Since the orbital integral of any $f_v \in A(G_v)$ has elliptic support, this implies $T(f_v) = 0$ since the trace of the induced representation $1 \times \sigma_v$ vanishes on the regular elliptic locus. In particular, we obtain

Auxiliary functions. For non-Archimedean p-adic local fields F_v and given an irreducible cuspidal representation σ_v with $\sigma_v \cong \sigma_v^*$, there exist functions $f_v \in A(G_v)$ with support in the regular elliptic point of $\tilde{G}_v$ such that $tr\ \pi_v(f_v) > 0$ holds for all $\pi_v \in \mathcal{E}\Theta_-(\sigma_v \otimes \mu_v)$, μ_v quadratic.

Proof of Proposition 5.4. The Archimedean case is well known, so assume v is non-Archimedean.

Suppose for some σ_v in the discrete series of M_v the assertion of Proposition 5.4 is false. Then there is a $\pi_v \in \mathcal{E}\Theta_-(\sigma_v)$ not isomorphic to $\pi_-(\sigma_v)$. Choose a global field F for the given local field F_v such that the central character of σ_v is induced from a grössen character ω of F. For reasons to become clear soon, we write v'' for the place v from now on.

We choose additional auxiliary non-Archimedean places v, v', for which $F/\mathbb{Q}$ is split. Choose $\sigma_v, \sigma_{v'}$ at these auxiliary places in the discrete series such that $\sigma_v \cong \sigma_v^*$ and $\sigma_{v'} \cong \sigma_{v'}^*$ with central characters $\omega_v\ \omega_{v'}$, respectively. Then we fix auxiliary functions $f_v \in A(G_v)$ and $f_{v'} \in A(G_{v'})$ with condition (ES) and support in $\tilde{G}_v$ and $\tilde{G}_{v'}$, respectively, as constructed above (Proposition 5.3).

By the density lemma formulated in Sect. 5.1 and the appropriate choice of central characters, we find a global cuspidal automorphic representation σ of $M(\mathbb{A}_F)$ such that $\sigma \not\cong \sigma^*$ realizes the given discrete series representations σ_v, $\sigma_{v'}$, and $\sigma_{v''}$ at the places v, v', and v''. Consider the set C of classes of irreducible cuspidal automorphic representation π' in the global L-packet of σ, which specialize to the given representations $\pi_-(\sigma_v), \pi_-(\sigma_{v'})$, and $\pi_{v''}$ at the places v, v', and v''. Such a π' is never CAP, and by the density lemma we can assume C is nonempty.

We now apply Lemma 5.8 from Sect. 5.5 for the auxiliary functions $f_v, f_{v'}$ chosen above. Hence, conditions (ii), (iv), (v), and (vi) of Lemma 5.8 hold for $w = v$. For our choice of π (i) and (iii) are also satisfied. Furthermore, $f_{v''}$ may be arbitrary. Then all assumptions for Lemma 5.8 are satisfied. We get – since $f_{v''}$ is now arbitrary –

$$\sum_{\pi',(\pi')^{v,v'}=\pi^{v,v'}} m(\pi') \cdot tr\ \pi'_v(f_v) \cdot tr\ \pi'_{v'}(f_{v'}) = 0.$$

The right side vanishes, since $\pi_{v''} \not\cong \pi_\pm(\sigma_{v''})$ at the local place v'', where we originally started from. Furthermore, as we know, the summation is over theta lifts.

Hence, every π'_v is in $\mathcal{E}\Theta_\pm(\sigma_v)$ up to a quadratic character twist (Lemma 5.2, part 1), and similarly for $\pi'_{v'}$.

Since the places v, v' are split in F, they are p-adic. Hence $\mathcal{E}\Theta_-(\sigma_v) = \{\pi_-(\sigma_v)\}$ from Sect. 5.3, and similarly $\mathcal{E}\Theta_-(\sigma_{v'}) = \{\pi_-(\sigma_{v'})\}$. From our choice of $f_v, f_{v'}$ (Proposition 5.4), therefore,

$$tr\ \pi'_v(f_v) \cdot tr\ \pi'_{v'}(f_{v'}) > 0$$

holds for all relevant π' in the sum, unless either $\pi_v \in \mathcal{E}\Theta_+(\sigma_v)$ or $\pi_{v'} \in \mathcal{E}\Theta_+(\sigma_{v'})$ holds up to quadratic character twists. But then the principle of exchange at v or v', respectively (Lemma 5.4), implies that π_v and $\pi_{v'}$ must be both in the plus space. From the assumption $\sigma_v \cong \sigma_v^*$ and $\sigma_{v'} \cong \sigma_{v'}^*$, we know from [106] that $\mathcal{E}\Theta_+(\sigma_v) = \{\theta_+(\sigma_v)\}$ and $\mathcal{E}\Theta_+(\sigma_{v'}) = \{\theta_+(\sigma_{v'})\}$. Hence, $tr\ \pi'_v(f_v) = tr\ \pi'_v \otimes \mu'_v(f_v) = tr\ \pi_+(\sigma_v)(f_v) = -tr\ \pi_-(\sigma_v)(f_v)$, and similarly at the place v'. Since the signs of v and v' cancel, again

$$tr\ \pi'_v(f_v) \cdot tr\ \pi'_{v'}(f_{v'}) > 0.$$

In other words, this term is positive for all relevant π' appearing in the trace formula stated above. Furthermore, $m(\pi') \geq 0$ and $m(\pi) \geq 1$. Since the total sum of the terms $tr\ \pi'_v(f_v) \cdot tr\ \pi'_{v'}(f_{v'})$ is zero, this gives a contradiction and completes the proof of Proposition 5.4. $\square$

Proof of Proposition 5.5. Fix a global cuspidal representation $\sigma \not\cong \sigma^*$ of $M(\mathbb{A}_F)$ for some number field F. An irreducible cuspidal automorphic representation π in the global L-packet of σ will be called *strange* if for some place it is strange locally, i.e., if $\pi_v \in \mathcal{E}\Theta_+^{glr}(\sigma_v)$ for $\pi_v \not\cong \pi_+(\sigma_v)$. Notice such a v is never Archimedean, since in this case the theta lift is understood well enough. We therefore have to show that strange representations π cannot arise at non-Archimedean places.

For any π' in the global L-packet attached to σ let $S = S(\pi')$ be the set of places v where $\pi'_v \in \mathcal{E}\Theta_-(\sigma_v)$ holds up to some quadratic character twist. If strange π exist in this L-packet, we choose π to be minimal with respect to the cardinality of $S(\pi)$ among all the strange π in this L-packet. For this π there exists a non-Archimedean place v'' where $\pi_{v''}$ is strange locally.

We claim the cardinality of $S(\pi)$ is 2 or more. Otherwise the underlying quaternion algebra $D = D(\pi)$ would be split and $S = \emptyset$ (second assertion of Lemma 5.4). In this situation the corresponding theta lift is locally and also globally generic according to results obtained by Howe and Piatetski-Shapiro [41] and also by Soudry. But this implies $\pi_v \cong \theta_+(\sigma_v)$ for all v, which contradicts the strangeness assumption. For this recall from [96] and Sect. 5.1 that there is at most one class of irreducible generic representation in $\Theta_+(\sigma_v)$. It follows that $\#S(\pi) \geq 2$.

Fix two different places v, v' in $S = S(\pi)$. Choose $f_v, f_{v'}$ to be cuspidal with $supp(f_v) \subseteq \tilde{G}_v$ and $supp(f_{v'}) \subseteq \tilde{G}_{v'}$ such that $f_v, f_{v'}$ satisfy $(*)_v$ ad $(*)_{v'}$, respectively. Furthermore, suppose $tr\ \pi_v(f_v) > 0$ and $tr\ \pi_{v'}(f_{v'}) > 0$ (Proposition 5.3).

Furthermore, choose an auxiliary non-Archimedean place $w \neq v, v', v''$ of residue characteristic different from 2, where σ_w and $\pi_w = \pi_+(\sigma_w)$ are unramified.

In particular, $w \notin S(\pi)$. Choose f_w with regular support contained in $\tilde{G}_w$ such that $tr\ \pi_+(\sigma_w)(f_w) > 0$. For example, take a bump function with support in the maximal split torus concentrated at a regular point near the origin, where the character of the unramified representation $\pi_+(\sigma_w)$ is nontrivial. Then, in particular, condition (RS) holds.

With these data fixed we apply Lemma 5.8 from Sect. 5.5. This gives

$$\sum_{\pi',(\pi')^{v,v',w}=\pi^{v,v',w}} m(\pi') \cdot tr\ \pi'_v(f_v) \cdot tr\ \pi'_{v'}(f_{v'}) \cdot tr\ \pi'_w(f_w) = 0.$$

The right side vanishes since $\pi_{v''} \not\cong \pi_\pm(\sigma_{v''})$ at the place v'' where $\pi_{v''}$ is locally strange.

The minimality of π and Propositions 5.3 and 5.4 imply $\pi'_v \in \mathcal{E}\Theta_-(\sigma_v)$ (up to quadratic character twists) for all π' which contribute to the trace formula above, i.e., those π' which are isomorphic to π outside v, v'. The same assertion is true at the place v'. Hence, $tr\ \pi'_v(f_v) = tr\ \pi_-(f_v) > 0$ and also for the place v', by our choice of test functions. It remains for us to consider the auxiliary place w. Here $\pi'_w \otimes \mu_w \in \mathcal{E}\Theta^+(\sigma_w)$ holds up to some quadratic character μ_w. The unramified Sp–O Howe correspondence matches unramified representations with unramified representations. Hence, the restriction of π'_w to $Sp(4, F_w)$ contains an unramified representation since σ_w was unramified by assumption. Hence, up to a quadratic character twist both π_w and π'_w are constituents of $Ind^{\tilde{G}_w}_{Z_w \cdot Sp(4,F_w)}(\tilde{\pi}'_w)$ for the same irreducible unramified representation $\tilde{\pi}'_w$ of $\tilde{G}_w$. This implies $\pi'_w = \chi_w \otimes \pi_w$ for some quadratic character χ_w [104], p. 480, line 5. Therefore, $tr\ \pi'_w(f_w) = tr\ \pi_w(\sigma_w) > 0$ holds independently of π', because the support f_w is contained in $\tilde{G}_w$. This gives a contradiction, since in the trace formula above all these nonnegative terms should sum to zero with at least one of them being positive (for π itself). Hence, there are no strange representations π in the global L-packet of σ. Proposition 5.5 is proven. $\square$

To complete the proof of Theorem 5.2 and also to obtain all the results stated in Sects. 5.2 and 5.3 in full generality, it remains to prove the key formula and the local–global principle (Lemma 5.6) stated in Sect. 5.3. The essential part of the local–global principle is stated in Propositions 5.4 and 5.5. What remains to obtain Lemma 5.6 will be shown below together with the key formula (as stated in Sect. 5.3). So once more we exploit the trace formula.

Proof. Let $\sigma \not\cong \sigma^*$ be a cuspidal irreducible automorphic representation of $M(\mathbb{A}_F)$. Then any π in the global L-packet attached to σ is not CAP. Fix π and suppose $e(\pi) \geq 2$. Choose two places $\{v, v'\}$ in $S(\pi)$ and consider Lemma 5.8 in Sect. 5.5 in the special case where the assumptions $(*)_v$ and $(*)_{v'}$ hold simultaneously. In particular, the functions $f_v, f_{v'}$ are cuspidal. We choose f_v and $f_{v'}$ so that $a_+ \neq 0$ and $a'_+ \neq 0$, respectively, and condition (ES) holds in addition with supports in $\tilde{G}_v$ and $\tilde{G}_{v'}$, respectively. This is possible by Proposition 5.3. Lemma 5.7 implies $a_- = -a_+ \neq 0$ and $a'_- = a'_+ \neq 0$, where we used the abbreviations $a_\pm := tr\ \pi'_{\pm\varepsilon}(\sigma_v)(f_v)$ and $a'_\pm := tr\ \pi'_{\pm\varepsilon'}(\sigma_{v'})(f_{v'})$. To satisfy assumption (vi) in Sect. 5.5, if both v and v' are Archimedean, we also choose some additional non-Archimedean place w and some function f_w as in the proof of Proposition 5.5.

The semilocal trace identity of Lemma 5.8 in this case is simplified considerably. This means that in the sum $\sum' a_{disc}(\pi'_v \pi'_{v'} \pi'_w \pi^{vv'w}) \cdot tr \; \pi'_v(f_v) \cdot tr \; \pi'_{v'}(f_{v'}) \cdot tr\pi'_w(f_w)$ only those representations π' contribute whose local components are in the local L-packet of σ at least up to a quadratic character twist. In particular, π_w is unramified up to a quadratic character twist. In fact all this follows from the last two propositions, and the arguments used for their proof. By the support conditions of f_v, $f_{v'}$, and f_w, quadratic character twists do not have an effect on the trace. Therefore, the contribution $tr \; \pi'_w(f_w) = tr \; \pi_+(f_w) \neq 0$ is independent of π' and can be canceled from both sides of the semilocal identity. Similarly, the nonvanishing terms $a_+ = -a_- \neq 0$ and $a'_+ = -a'_- \neq 0$ are independent of π'. Canceling these terms therefore gives $M_{++} + M_{+-} + M_{-+} + M_{--} = 2(-1)^{e(\pi)}$.

Here $M_{\varepsilon,\varepsilon'}$ is the sum over all multiplicities $m(\pi')$ over all π' such that $\pi' \otimes \mu_v \mu_{v'} \mu_w \cong \pi_\varepsilon(\sigma_v)\pi_{\varepsilon'}(\sigma_v)\pi^{vv'}$ for some local quadratic characters μ_v, $\mu_{v'}$, and μ_w. Since the right side $2(-1)^{e(\pi)}$ equals the left side, which is zero or more, we get $2|e(\pi)$. Furthermore, by the principle of exchange, the condition $M_{++} > 0$ (note $m_{++} > 0$ by assumption) implies $M_{-+} = M_{+-} = 0$. Hence, $M_{++} + M_{--} = 2$ and $e(\pi)$ is even. This almost completes the proof of the key formula. In fact we know $M(\pi') = 1$ from Lemma 5.4 part 2, if the cardinality of $S(\pi')$ is $0, 1$. So by induction on this cardinality, the argument surrounding the statement of the key formula in Sect. 5.3 proves $M_{++} = M_{--} = 1$. It only remains for us to show $m(\pi') = 0$ unless π_v and $\pi_{v'}$ are in the local L-packet of σ_v and $\sigma_{v'}$, respectively. For $\pi_v \in \mu_v \otimes \Theta_+(\sigma_v)$ this follows with the same argument used in the proof of Proposition 5.5. In particular, we always get $\pi'_w \cong \pi_+(\sigma_w)$ at the auxiliary place w. (This could also be seen by moving this place around, i.e., changing to some other place where σ_w is unramified and where $\pi' = \pi_w(\sigma_w)$ holds.) So by the symmetry of v, v' it only remains to show $m(\pi') = 0$ whenever $\pi'_v \cong \mu_v \otimes \pi_-(\sigma_v)$ holds for some quadratic character μ_v but $\pi'_v \not\cong \pi_-(\sigma_v)$. Assume this were not true. To obtain a contradiction, again apply the trace formula with $f_{v'}, f_w$ as above, but now with f_v satisfying only condition (ES) but being arbitrary otherwise. Still the assumptions of Lemma 5.8 in Sect. 5.5 are satisfied. From this we obtain a contradiction, provided the characters χ_1 and χ_2 of $\mu_v \otimes \pi_-(\sigma_v)$ and $\pi_-(\sigma_v)$ are linear-independent on the regular elliptic locus of G_v (the support of f_v). Notice the first character, χ_1, appears on the left side of the trace identity, whereas the other character, χ_2, appears on the endoscopic left side of the assertion in Lemma 5.8. So it remains for us to show that linear independence holds on the elliptic locus. For $\sigma_v \not\cong \sigma_v^*$ both representations are cuspidal and the claimed linear independence follows from the orthogonality relations for cuspidal characters with respect to the elliptic scalar product. If $\sigma_v \cong \sigma_v^*$, both representations do not belong to the discrete series. But in this case the distribution T of Lemma 5.7 vanishes on the regular elliptic locus. Hence, $-2\chi_1 = r(\mu_v \otimes \sigma_v)$ and $-2\chi_2 = r(\sigma_v)$ holds on the regular elliptic locus, so the linear independence of χ_1, χ_2 on the elliptic locus follows from the corresponding linear independence of $\mu_v \otimes \sigma_v$ and σ_v. Since these representations are in the discrete series, and since they are not isomorphic by our assumptions, they are linear-independent on the regular elliptic locus of M_v. This contradiction implies that any π' in the global L-packet with $m(\pi') > 0$ has its local components π'_v in the local L-packet attached to σ_v for all places v of F. In other words, the local–global principle holds. This, together with the multiplicity result obtained above, implies the key formula. □

Remark 5.7. If we relax the assumptions on $\sigma = (\sigma_1, \sigma_2)$ in definition 5.1, we can similarly consider the case where σ_1 is one-dimensional. If σ_2 is in the discrete series, this is the situation considered in [3]. With the results on L-series shown in [69] (instead of the results obtained by Soudry used in the non-CAP case), the coefficients a_{disc} can be worked out directly, and the arguments from above should extend to give a characterization of the CAP representation in terms of the local Arthur packets. This indeed would give a refined description of the Saito–Kurokawa lift in term of local Arthur packets.

5.5 Appendix on Arthur's Trace Formula

In this appendix we apply the Arthur trace formula. We deduce from it certain local character identities, whose coefficients contain global information. We therefore refer to them as semilocal character relations. This semilocal character identities are useful for several reasons. First, they are the starting point for the proof of the local character identities of the endoscopic lift r in Sect. 5.1. Next, we used them in Sect. 5.4 to show local statements like the stability lemma. Finally, they provide global information, once the local concepts are well understood. For this reason, and to be flexible enough for all these applications, we formulate a number of technical conditions. Under these conditions we prove the semilocal character identities in Lemma 5.8.

Make the following assumptions (i)–(vi):

(i) Let σ be an irreducible automorphic representation of $M(\mathbb{A}_F)$ with $\sigma_v, \sigma_{v'}$ in the discrete series for at least two different places v, v'.

(ii) Suppose that σ is *cuspidal* and $\sigma \not\cong \sigma^*$. Then the global L-packet of σ contains cuspidal representations. These are not CAP representations, since otherwise there would exist poles for partial degree 4 or 5 L-functions of π, which contradict the cuspidality of σ. See Sect. 5.2.

Choose signs $\varepsilon, \varepsilon' \in \{\pm\}$. Consider the representations $\pi = \pi_\varepsilon(\sigma_v) \otimes \pi_{\varepsilon'}(\sigma_{v'}) \otimes \pi^{v,v'}$ and $\pi' = \pi_{-\varepsilon}(\sigma_v) \otimes \pi_{-\varepsilon'}(\sigma_{v'}) \otimes \pi^{v,v'}$ in the global L-packet of the weak lift of σ. We write $\pi = \pi_{++}$, $\pi' = \pi_{--}$ in the following. Let $m = m_{++} = m(\pi)$ and $m' = m_{--} = m(\pi')$ denote their multiplicity in the discrete spectrum of $G(\mathbb{A}_F)$. Assume:

(iii) $m_{++} + m_{--} > 0$.

From (iii) it follows that either π or π' is an automorphic representation. By switching $\varepsilon, \varepsilon'$ into their negatives, we can assume without loss of generality $m(\pi) > 0$.

Then the restriction of π to $Sp(4, \mathbb{A}_F)$ is a theta lift and is cuspidal (Lemma 5.2). It is not CAP by assumption (i). So π and π' only contribute to the cuspidal spectrum. Every automorphic representation in the discrete spectrum isomorphic to π or π' belongs to the cuspidal spectrum. This follows from Langlands results on spectral decomposition and the fact that π is not CAP. Therefore, put $m = m_{++} = m_{cusp}(\pi)$ and $m' = m_{--} = m_{cusp}(\pi')$.

Now the essential conditions:

(iv) Let $f_v, f_{v'}$ be locally compact functions with compact support on G_v and $G_{v'}$, respectively, which are in the Hecke algebra in the sense of [6]. Assume that both functions have vanishing orbital integrals at regular nonelliptic semisimple elements. In particular, they are cuspidal.

(v) Assume that either condition $(*)_v$ or condition $(*)_{v'}$ holds.

Furthermore, suppose:

(vi) Condition (RS) holds for the test function f_w at least at one auxiliary non-Archimedean place w, and the case $w \in \{v, v'\}$ is not excluded.

Lemma 5.8. *Suppose assumptions (i)–(vi) hold. Then*

$$\sum_{\pi'}{}' a_{disc}(\pi'_v \pi'_{v'} \pi'_w \pi^{vv'}) \cdot tr\ \pi'_v(f_v) \cdot tr\ \pi'_{v'}(f_{v'}) \cdot tr\ \pi'(f_w)$$

is either zero if for a place $v'' \neq v, v', w$ the class of $\pi_{v''}$ satisfies $\pi_{v''} \notin \{\pi_{\pm}(\sigma_{v''})\}$, or equal to

$$\frac{1}{2}(-1)^{e(\pi)}(a_+ - a_-)(a'_+ - a'_-) \cdot \tilde{T}(f_w)$$

otherwise. Here $\tilde{T} = tr\ \pi_+(\sigma_w) - tr\ \pi_-(\sigma_w)$ if σ_w is in the discrete series and $\tilde{T} = tr\ \pi_+(\sigma_w)$ otherwise.

Concerning the Notation. In these formulas the summation $\sum'$ is over all classes of global representations π' in the global L-packet of σ for which $(\pi')^{v,v',w} \cong \pi^{v,v',w}$ holds. Here $a_{\pm}$ and $a'_{\pm}$ are abbreviations for $a_{\pm} = tr\ \pi'_{\pm\varepsilon}(\sigma_v)(f_v)$ and $a'_{\pm} = tr\ \pi'_{\pm\varepsilon'}(\sigma_{v'})(f_{v'})$, respectively. The coefficients $a_{\varepsilon\varepsilon'}$ are zero unless the corresponding multiplicities $m_{\varepsilon\varepsilon'}$ of the weak lifts $\pi_{\varepsilon}(\sigma_v)\pi_{\varepsilon'}(\sigma_{v'})\pi^{v,v'}$ in the discrete spectrum are nonzero. In fact they are equal to the *multiplicities* of these representations. This follows from [6] (see the references given below) and the Langlands theory of spectral decomposition, since by assumption (ii) all these representations are not CAP, as explained above. In fact $a_{++} := a_{disc}^G(\pi) = m_{cusp}(\pi)$ for any π in the weak lift of σ, which is in the discrete spectrum and is similar for all π' in the global L-packet of σ.

Proof. For simplicity we suppose for the proof $m(\pi) > 0$, as above.

We apply the results obtained by Arthur [6] concerning the trace formula for $G(\mathbb{A}_F)$. Choose any test functions f_λ for the places $\lambda \neq v, v', w$ and put $f =$

$\prod_\lambda f_\lambda$. By assumption f_v and f'_v are cuspidal. So the orbital integrals of these two functions vanish outside the regular elliptic locus and these functions are cuspidal in the sense of [6], p. 538. Since f is cuspidal at the two different places v, v', the Arthur trace formula is simplified. The spectral side is a sum over the traces of the discrete spectrum suitably ordered using the Archimedean infinitesimal characters ([6], Theorem 7.1 and Corollary 7.2) using the notation from [6]

$$\sum_{\pi'} a^G_{disc}(\pi') \cdot I_G(f).$$

Notice $I_G(\pi', f) = tr\, \pi'(f)$ in the notation in [5], p. 325. The coefficients $a^G_{disc}(\pi')$ are complex numbers, and by grouping together the linear combinations of weighted characters defined in [6], formula (4.3), we have

$$\sum_{M_0 \subseteq L_0 \subseteq G_0} |W_0^{L_0}||W_0^G|^{-1} \sum_s |det(s-1)|_{a_{L_0}^G}^{-1} \cdot tr\left(M_{Q_0|sQ_0}(0) \circ \rho_{Q_0,t}(s, 0, f^1)\right).$$

Here $\rho_{Q_0,t}$ is an induced representation, induced from the part $L^2_{disc,t}(L_0(F)A_{L_0,\infty} \setminus L_0(\mathbb{A}_F))$ of the discrete spectrum of the Levi subgroup L_0. If π' is cuspidal but not CAP, only $L_0 = G_0$ contributes to the coefficient $a_{disc}(\pi)$ and the sum becomes the trace of f on $L^2_{disc,t}(L_0(F)A_{L_0,\infty} \setminus L_0(\mathbb{A}_F))$. Hence, $a_{disc}(\pi) = m_{cusp}(\pi)$ holds in this case. In general, of course, $a_{disc}(\pi) = 0$ unless $m_{disc}(\pi') \neq 0$.

Concerning the geometric side of this trace formula, we obtain from [6], Corollary 7.2,

$$\sum_{\gamma \in (G(F))_{G,S}} a^G(S, \gamma) I_G(\gamma, f).$$

Here $I_G(\gamma, f)$ is the global orbital integral $O^G_\gamma(f) = \prod_v O^{G_v}_{\gamma_v}(f_v)$ of f (see [6], p. 325). Moreover, by our assumption (vi) regarding f_w, the orbital integral at w vanishes unless γ_w is regular semisimple. This implies γ is regular semisimple. Moreover, by our assumption (iv) regarding $f_v, f_{v'}$, the geometric side only involves regular elliptic terms (as in [6], Corollary 7.4). But in this case one can express $a^G(S, \gamma)$ explicitly. As in [6], Corollary 7.4, one obtains the simpler expression for the geometric side

$$\sum_{\gamma \in (G(F)_{ell})} vol(G(F,\gamma)A_{G,\infty} \setminus G(\mathbb{A}_F, \gamma)) \int_{G(\mathbb{A}_F,\gamma) \setminus G(\mathbb{A}_F)} f(x^{-1}\gamma x) dx.$$

Here $G(F,\gamma) = Zent(\gamma, G^0)(F)$. In our case $G = G^0$. Furthermore, in our case G_{der} is simply connected.

Stabilization of the elliptic terms of the geometric side as in [53], using the fundamental lemma proved in Chaps. 6–9, gives for the geometric side of the Arthur trace formula a rearrangement in terms of stable orbital integrals for G and stable orbital integrals for the endoscopic group M

$$\sum_{\pi'} a_{disc}(\pi') \cdot tr \; \pi'(f) \; = \; ST^{G,**}(f) + \frac{1}{4} ST^{M,**}(f^M).$$

The $**$-condition on central terms in [53] can be ignored by the regular support condition (vi) at the places w. Moreover, the global stable orbital $ST^{G,**}(f) = ST^G(f) = 0$ vanishes by the local assumption $(*)_v$ or $(*)_{v'}$ of assumption (v)

$$ST^{G,**}(f) \; + \; \frac{1}{4} ST^{M,**}(f^M) \; = \; \frac{1}{4} ST^{M,**}(f^M).$$

Since M is quasisplit, the terms omitted in $ST^{M,**}(f^M)$ are again the central terms of the stable (semisimple) trace $ST^M(f^M)$ or preferably of a suitable stable trace on the z-extension $\tilde{M} = Gl(2) \times Gl(2)$. One of the functions $f_v, f_{v'}, f_w$ satisfies condition (RS). Without restriction of generality, suppose it is $f_{v'}$. Then there is a matching function $f_{v'}^M$ with regular support by the implicit function theorem using the smoothness of the regular locus of an elliptic torus. Furthermore, we can assume that for the two places v, v' the corresponding functions $f_v^M, f_{v'}^M$ have vanishing orbital integrals for regular nonelliptic elements. Then

$$ST^{M,**}(f^M) \; = \; ST^M(f^M).$$

The z-extension $\tilde{M}$ of M does not have nontrivial endoscopy. The strong cuspidal condition (iv) is inherited by f^M, as well as condition (vi). So again the geometric side of Arthur's trace formula for $f^{\tilde{M}}$ is simple, and in particular only involves elliptic regular terms. Hence, stabilization gives

$$\frac{1}{4} ST^{M,**}(f^M) \; = \; \frac{1}{4} T^{\tilde{M}}(f^{\tilde{M}}).$$

If we compare the geometric terms $T^{\tilde{M}}(f^{\tilde{M}})$ with the spectral side, the simple form of the Arthur trace formula now applied for $\tilde{M}$ yields the character expansion

$$\frac{1}{4} T^{\tilde{M}}(f^{\tilde{M}}) \; = \; \frac{1}{4} \sum_{\sigma} a_{disc}(\otimes_v' \sigma_v) \cdot \prod_v tr \; r(\sigma_v)(f_v)$$

for the geometric side. The right side is a sum with σ running over the discrete spectrum of $\tilde{M}(\mathbb{A}_F)$, suitably ordered. To obtain this formula we used the local character identities $tr \; \sigma_v(f_v^{\tilde{M}_v}) = r(\sigma_v)(f_v)$. By the multiplicity 1 theorem for $Gl(2)$ and $\tilde{M}$ and the spectral theory for $Gl(2)$, we get $a_{disc}(\sigma) = 1$ for all cuspidal representations $\sigma = \otimes_v' \sigma_v$. See [29, 42].

Notice the expansion for $\frac{1}{4} T^{\tilde{M}}(f^{\tilde{M}})$ above is a character expansion in terms of representations of $G(\mathbb{A}_F)$. It involves only representations π which are weak lifts coming from M. All local components π_v that appear are in the local L-packet of an underlying global representation σ in the discrete spectrum of $M(\mathbb{A}_F)$. Comparing this character expansion with the one obtained from the trace formula for $G(\mathbb{A}_F)$ gives the

Formula (CharIdent).

$$\sum_{\pi'} a_{disc}(\pi') \cdot tr\ \pi'(f) \ = \ \frac{1}{4} \cdot \sum_{\sigma} a_{disc}(\sigma) \cdot \prod_{v} tr\ r(\sigma_v)(f_v).$$

A fixed representation $\pi^{v,v',w}$ of $G(\mathbb{A}_F^{v,v',w})$ belongs to the global L-packet of a fixed pair σ, σ^* of representations of $M(\mathbb{A}_F)$. This follows from Proposition 5.2 and the strong multiplicity 1 theorem for $\tilde{M}$. So the last character identity should imply – so to say from the linear independence of characters of $G(\mathbb{A}_F^{v,v',w})$ and by separating the component $\pi^{v,v',w}$ – the following *semilocal identity*: The term

$$\sum_{}^{'} a_{disc}(\pi'_v \pi'_{v'} \pi'_w \pi^{vv'w}) \cdot tr\ \pi'_v(f_v) \cdot tr\ \pi'_{v'}(f_{v'}) \cdot tr\ \pi'_w(f_w)$$

is equal to

$$= \frac{1}{2}(-1)^{e(\pi)} a_{disc}(\sigma) \cdot (a_+ - a_-) \cdot (a'_+ - a'_-) \cdot \tilde{T}(f_w)$$

if all local components of $\pi^{v,v',w}$ are in the local L-packets, and it is zero otherwise.

Since σ and σ^* were supposed to be not isomorphic, and since both σ and σ^* contribute (Proposition 5.2), we got the factor $\frac{1}{2}$ instead of the factor $\frac{1}{4}$ from the sum over the σ on the right side. In the sum the representations vary over all π' with $\pi^{v,v',w}$ fixed up to isomorphism.

Since the Arthur trace formula is not known to converge absolutely, an easy argument which implies the linear independence of characters in the sense above is not known at present. However, assumptions (vi) or (vi) put us into a situation where the above semilocal identity can nevertheless be extracted from the global trace formula.

To extract the semilocal identity stated above from the global Arthur trace formula in this case one uses multipliers at the Archimedean places.

Multipliers. Let $f_\infty \in C_c^\infty(G_\infty, K_\infty)$ be a K_∞-finite test function at the Archimedean places. A W-invariant distribution α with compact support on the Lie algebra of the standard Cartan subgroup h^1 is called a multiplier. In our cases h^1 is the Lie algebra of a maximal split torus; for the general case see [6], Sect. 6. Typical examples are elements in the center of the universal enveloping algebra or W-invariant smooth functions with compact support. Multipliers α act on $C_c^\infty(G_\infty, K_\infty)$ in a natural way as shown by Arthur. Let $f_\infty \mapsto (f_\infty)_\alpha$ denote this action. For an irreducible admissible unitary representation π_∞ of G_∞ let ν_π denote its infinitesimal character viewed as a W-orbit in h^1. Let t_π denote the length of its imaginary part with respect to a suitable norm on h^1 [6]. Then $\pi_\infty((f_\infty)_\alpha)) = \hat{\alpha}(\nu_\pi)\pi_\infty(f_\infty)$ for the Fourier transform $\hat{\alpha}$ of the distribution α. Indeed this formula uniquely characterizes the action.

Notice $(f_\infty)_\alpha$ is cuspidal if f_∞ is cuspidal. This is true since cuspidality is characterized by the vanishing of the traces $tr\ \pi_\infty(f_\infty)$ for all representations π_∞ which are properly induced from irreducible tempered representations. Since $tr\ \pi_\infty((f_\infty)_\alpha)) = \hat{\alpha}(\nu_\pi)tr\ \pi_\infty(f_\infty)$, this property is preserved.

Furthermore, the condition $(*)_\infty$ is preserved by the action of multipliers. See [97] and [6], Definition 1.2.1. Without restriction of generality $F_v = \mathbb{R}$. Then $(*)_\infty$ is equivalent to $T(f) = 0$ for all characters $T = tr\ \pi_\infty$ of discrete series representations of G_v not isomorphic to $\pi_\pm(\sigma_v)$, and for all $T = tr\ \pi_+(\sigma_v) + tr\ \pi_-(\sigma_v)$ attached to discrete series representations σ_v of M_v and all characters T of representations properly induced from tempered representations. This follows from [90], Lemma 5.3. Furthermore, for an irreducible unitary representation σ_∞ the infinitesimal character ν_{σ_∞} determines the infinitesimal character $\nu_\pm$ of the representations $\pi_\pm(\sigma_\infty)$. Indeed

$$\nu_+ = \nu_- = \xi(\nu_{\sigma_\infty})$$

for a suitable "linear" map ξ, up to some shift. This follows from the description of the endoscopic lift in terms of the theta correspondence or from [91], Lemma 4.2.1. The precise nature of the map ξ is of no importance here. For $\nu_v = \xi(\nu_{\sigma_v})$ this implies for the Archimedean places v

$$T((f_v)_\alpha) = \hat{\alpha}(\xi(\nu_v)) \cdot T(f_v) = \widehat{\tilde{\alpha}}(\nu_v) \cdot T(f_v)$$

for $\tilde{\alpha} = \alpha \circ \xi$.

Multipliers act on the global K-finite test function $f = \prod_v f_v$ in $C_c^\infty(G_\infty \times G((\mathbb{A}_F)_f))$ via their action on the Archimedean component f_∞. Hence, the non-Archimedean condition (vi) is preserved by the action of multipliers for trivial reasons.

The infinitesimal character $\nu = \nu_{\pi_\infty}$ of a unitary irreducible admissible representation π_∞ defines a W-orbit. Given ν, a smooth multiplier α is constructed on p. 182ff of [8] such that for $\alpha_m = \alpha * \ldots * \alpha$ (m-fold convolution) the following holds:

$$lim_{m\to\infty} \sum_{\pi'} a_{disc}(\pi') \cdot tr\ \pi'(f_{\alpha_m}) = \sum_{\pi',\ \nu_{\pi'}=\nu} a_{disc}(\pi') \cdot tr\ \pi'(f).$$

Recall the left side is the spectral side of the Arthur trace formula in a simple form (for our purposes f is supposed to satisfy the assumption of the trace hypothesis; under the second assumption of condition (vi) this assumption is stable under the action of multipliers). This sum is not necessarily absolutely convergent, so summation is with respect to a suitable ordering using the parameter $t_{\pi'}$. The sum on the right is absolutely convergent owing to the admissibility statement [6], Lemma 4.1. Hence, linear independence of the characters involved holds in the sense of [42], Lemma 16.1.1. We recall that for the above limit formula it suffices to know that $\hat{\alpha}(\nu) = 1$ and $|\hat{\alpha}(\nu_{\pi'})| < 1$ unless the W-orbits of $\nu_{\pi'}$ and ν coincide.

A similar separation of the infinitesimal character can be obtained for the spectral side of the simple trace formula for the endoscopic group. In fact, one can do better.

For an irreducible unitary representation σ_∞ the infinitesimal character ν_{σ_∞} determines the infinitesimal character $\nu_\pm$ of the representations $\pi_\pm(\sigma_\infty)$, as explained above: $\nu_+ = \nu_- = \xi(\nu_{\sigma_\infty})$ for a suitable "linear" map ξ. For $\nu_v = \xi(\nu_{\sigma_v})$ this implies for the Archimedean places v

$$tr\, r(\sigma_v((f_v)_\alpha)) = \hat\alpha(\xi(\nu_v)) \cdot tr\, r(\sigma_v)(f_v) = \hat\alpha(\xi(\nu_v)) \cdot tr\, \sigma_v((f_v)^{M_v}) = tr\, \sigma_v((f_v)_{\tilde\alpha})$$

for $\tilde\alpha = \alpha \circ \xi$. If we identify the domains of the functions α and $\tilde\alpha$, as we may do, we can symbolically write

$$\left((f_v)_\alpha\right)^{M_v} = \left(f_v^{M_v}\right)_\alpha.$$

In other words, the action of multipliers commutes with the endoscopic matching condition. Since the Weyl group of M_v is a subgroup of the Weyl group of G_v, the smooth multiplier α for G_v can therefore be considered as a smooth ($*$-invariant) multiplier for M_v. This being said, we can consider the formulas (CharIdent) from above for the various test functions f_{α_m}. From the spectral limit formulas both for G and for M we obtain for $m \to \infty$

$$\sum_{\pi',\nu_{\pi'}=\xi(\nu)} a_{disc}(\pi') \cdot tr\, \pi'(f) = \frac{1}{4} \cdot \sum_{\sigma,\,\nu_\sigma=\nu} a_{disc}(\sigma) \cdot \prod_v tr\, \sigma_v((f_v)^{M_v})$$

$$= \frac{1}{4} \cdot \sum_{\sigma,\,\nu_\sigma=\nu} a_{disc}(\sigma) \cdot \prod_v tr\, r(\sigma_v)(f_v).$$

From this formula [6], Lemma 4.1, and [42], Lemma 16.1.1, the semilocal identity follows. This completes the proof of Lemma 5.8. □

Proof of formula (0). See page 180. We now explain how to obtain formula (0) in the non-p-adic case as a complement to the proof of Lemma 5.8. For this choose v, v' as in Lemma 5.8 to be Archimedean for a suitable chosen auxiliary number field F, and a suitably chosen auxiliary global irreducible cuspidal automorphic representation $\sigma \ncong \sigma^*$ of $M(\mathbb{A}_F)$, and we choose w to be some auxiliary "harmless" place, where the global representation σ is unramified and for which the norm of w is sufficiently large. Furthermore, we choose F and σ so that for some additional auxiliary non-Archimedean place w', where $F/\mathbb{Q}$ splits with residue characteristic different from 2, $\sigma_{w'}^* \ncong \sigma_{w'}$ holds and is cuspidal. Then $\pi_\pm(\sigma_{w'})$ are cuspidal and up to character twists the only representations in $\mathcal{E}\Theta_\pm(\sigma_{w''})$ (see Proposition 5.1 for the notation). Indeed, the statement on the theta lift uses Waldspurger's proof of the Howe duality for the dual pair $Sp(4) \times O(4)$, whereas the cuspidality statement uses what we already considered the case of local fields, which are completions of $\mathbb{Q}$, in Theorem 4.5. For the Archimedean places we choose f_v to be some fixed auxiliary cuspidal function in the Hecke algebra satisfying $(*)_v$ such that $a_- = -a_+ = tr\, \pi_-(\sigma_v)(f_v) > 0$ (notation as in Sect. 5.4). At the Archimedean place v' we choose two functions $f_{v'}$ such that $tr\, \pi_+(\sigma_{v'})(f_{v'}) = 0$ and $tr\, \pi_-(\sigma_{v'})(f_{v'}) = 1$ or vice versa. At the place w' we choose a function $f_{w'}$

in the Hecke algebra whose traces separate the finitely many cuspidal representations of G_v which appear in the packets $\mathcal{E}\Theta_\pm(\sigma_{w'})$ of the theta lift twisted by quadratic characters. For this we can assume $f_{w'}$ to be a linear combination of matrix coefficients of these finitely many cuspidal representations. Finally f_w is chosen with regular support in a maximal split torus of G_w. With these choices the global trace formula is considerably simplified, as observed by Deligne–Kazhdan, by Arthur and by Henniart. In particular, on the spectral side only the cuspidal automorphic spectrum contributes, owing to the cuspidal matrix coefficients $f_{w'}$ at the non-Archimedean place w'. This makes the use of multipliers, which were necessary in the more complicated situation of Lemma 5.8, superfluous. Secondly, the $f_{w'}$ chosen completely suffice for the detection of all local constituents of global endoscopic lifts at the place w'. This is due to Lemma 5.2. The next observation is that "moving" the unramified auxiliary place w to become "sufficiently large" allows us to get rid of the influence of f_w in the trace formula. This follows from well-known finiteness results (i.e., apply the same trick as in the proof of Lemma 5.6 and of the key formula at the end of Sect. 5.4). Furthermore, similarly as for the place w', Lemmas 5.2–5.5 in Sect. 5.2 control the Archimedean constituents of the global endoscopic lift. So the $f_{v'}$ chosen again completely detect all local constituents of global endoscopic lifts at the place v' (Lemmas 5.2 and 5.5 in Sect. 5.2). With these choices made, we can now follow the arguments in the proof of Lemma 5.8 mutatis mutandis to obtain

$$\sum_{\pi'} m(\pi'_v \otimes (\pi')^v) \cdot tr\, \pi'(f_v) \;=\; \frac{1}{2}(a_+ - a_-) \prod_{w \neq v} n(\sigma_w, \pi'_w),$$

where $\pi' = \pi'_v \otimes (\pi')^v$ runs over all global cuspidal representations, which are weak endoscopic lifts (in the sense of Sect. 5.2) of our fixed auxiliary global automorphic representation σ with fixed $(\pi')^v$ outside v. Since $\pi'_v \in \{\pi_\pm(\sigma_\infty)\}$ by Lemma 5.5, we then obtain $tr\, \pi'(f_v) = a_\pm = \pm a_+$. Canceling $a_+ = -a_- \neq 0$ from the formula leaves us with a formula which is the precise analogue of formula (0) with the unique Archimedean place ∞ of $F = \mathbb{Q}$ now replaced by the fixed chosen Archimedean place v of the number field F. To make a comparison with the situation above we change the notation, and let v denote ∞ from now on. Then for any local non-Archimedean local field F_v of characteristic zero and any irreducible admissible representation σ_v of M_v in the discrete series we may have chosen F and σ such that they extend F_v and σ_v (maybe up to a local twist by a character). With this additional choice made, the analogue of formula (0) has now been established. $\square$

Chapter 6
A Special Case of the Fundamental Lemma I

6.1 Introduction

In the following we prove the endoscopic fundamental lemma essentially used in Chap. 3 for the group of symplectic similitudes

$$G(F) = GSp(4, F),$$

i.e., the group of all 4×4-matrices g satisfying

$$g^t \begin{pmatrix} 0 & -E \\ E & 0 \end{pmatrix} g = \lambda(g) \begin{pmatrix} 0 & -E \\ E & 0 \end{pmatrix}, \qquad E = \begin{pmatrix} 1 & 0 \\ 0 & 1 \end{pmatrix}$$

with coefficients in a non-Archimedean local field F of residue characteristic different from 2. Let K denote the maximal compact subgroup of $G(F)$ defined by the matrices g with coefficients in the ring of integers $o_F \subseteq F$.

In [36] the existence of matching functions (f, f^M) for endoscopic orbital integrals is shown for the group $GSp(4, F)$ and its unique endoscopic group M. However, the proof in [36] only gives the existence of matching functions. It does not explicitly describe the correspondence between the functions that have matching orbital integrals. However, it is this additional information which is relevant for many applications. In particular, it is important to know this correspondence for all functions f in the K-spherical Hecke algebra of all K-bi-invariant functions with compact support on $G(F)$. The fundamental lemma asserts that there exists a specific ring homomorphism b between the spherical Hecke algebras of the groups $G(F)$ and $M(F)$ for which the pairs $(f, f^M) = (f, b(f))$ define matching functions (if the transfer factors and measures are suitably normalized). Using the trace formula, one can reduce this assertion to the case of one particular function in the spherical Hecke algebra, the unit element $f = 1_K$. In this special case, of course, $f^M = b(1_K) = 1_{K_M}$. The reduction of the fundamental lemma to this special case can be found in [35] and also in Chap. 9. So it is enough to prove that $f = 1_K$ and $f^M = 1_{K_M}$ are functions with matching orbital integrals.

R. Weissauer, *Endoscopy for GSp(4) and the Cohomology of Siegel Modular Threefolds*, Lecture Notes in Mathematics 1968, DOI: 10.1007/978-3-540-89306-6_6, © Springer-Verlag Berlin Heidelberg 2009

As already mentioned, there exists only one proper elliptic endoscopic group M. This endoscopic group is isomorphic to the quotient of the group $Gl(2)^2$ divided by $\mathbb{G}_m$. See page 54 for details. The endoscopic matching condition is roughly the following. Consider a maximal F-torus $T = T_M$ of M and an admissible embedding of T into G. Then one has to show a relation between the stable orbital integral $SO_t^M(1_{K_M})$ of 1_{K_M} for all sufficiently regular elements $t \in T_M(F)$, and a certain linear combination of orbital integrals $O_{\eta'}^G(1_K)$

$$SO_t^M(1_{K_M}) = \sum_{\eta'} \Delta(\eta', t) O_{\eta'}^G(1_K),$$

where the sum is over all elements η' stably related to t by a norm mapping. For the rather complicated details of these notions and, in particular, for the definition of the transfer factors $\Delta(\eta', t)$ in the general case we refer to [60]. In the present situation all this will become quite simple and explicit. In fact, the transfer factor satisfies

$$\Delta(\eta', t) = \kappa(\eta', \eta) \cdot \Delta(\eta, t)$$

for some fixed η, where $\kappa(\eta', \eta) = 1$ if η' is conjugate to η under $G(F)$ and $\kappa(\eta', \eta) = -1$ if η' is not conjugate to η under $G(F)$.

In Sect. 4.5 we used the notation

$$\Delta(\eta, t) = \Delta_{IV}(\eta, t) \cdot \tau(\eta, t),$$

where $\Delta_{IV}(\eta, t)$ is a simple volume factor defined in [60].

There are four types of maximal F-tori T_M in the group M. In the list on page 94 these are the tori of types (4)–(7). Among these, the tori are contained in an F-rational parabolic subgroup of M in cases (6) and (7) where the matching condition is very easy to verify by parabolic descent. We leave these two cases as an exercise. What remains are the two essential cases (4) and (5). Here the torus T_M is uniquely characterized in M, up to conjugation by $M(F)$, by the property $T_M(F) \cong (L_1^* \times L_2^*)/F^*$, where L_1 and L_2 are quadratic field extensions of F. For $L_1 \cong L_2$ this gives elliptic case I, and for $L_1 \not\cong L_2$ this gives elliptic case II. Although both cases are similar, several things have to be distinguished. Therefore, to avoid confusion, we treat them separately: elliptic case I for $L = L_1 = L_2$ in this chapter; elliptic case II in the next chapter.

It will finally turn out for suitable choices to be given in detail in the following, in particular for a suitable choice of η, that there exist $D_0 \in F^*$ and $A_0 \in F^*$ such that $L = F(\sqrt{D_0})$ and $E^+ = F(\sqrt{A_0})$ are algebras over F for which we can view η as defining an element $x \in (E^+ \otimes_F L)^*$ such that

$$\tau(\eta, t) = \begin{cases} 1 & \text{nonelliptic } T_M, \\ \chi_{L/F}\big((x - x^\sigma)(x' - x'^\sigma)\big) & \text{elliptic cases I}, \\ \chi_{L/F}\big(A_0(x - x^\sigma)(x^\tau - x^{\tau\sigma})\big) & \text{elliptic cases II}. \end{cases}$$

Here $\chi_{L/F}$ denotes the quadratic character attached to the extension L/F by class field theory. In the elliptic cases L/F is a quadratic field extension of F with automorphism σ. E^+/F is a field extension in elliptic case II with automorphism τ, and splits as an F-algebra $E^+ \cong F^2$ in elliptic case I. In fact, these formulas are compatible in a strong sense. In particular, the local factors satisfy a product formula for global data (global property). This will be shown in Sect. 7.15.

Conventions. For the local field F let o_F be the ring of integers and q be the number of elements in the residue field of F. Let L be a quadratic extension field of F. Let π_F and π_L denote prime elements of F and L, respectively. We assume $|\pi_F| = q^{-1}$ and $ord(\pi_F) = 1$. If L/F is ramified, we may assume π_L to be chosen such that $\pi_L^2 = -\pi_F \in F^*$ since the residue characteristic is different from 2.

6.2 The Torus T

In this section F is a local field of residue characteristic not equal to 2. Consider the subgroup $H(F) \subseteq GSp(4, F)$ of all matrices

$$(s, s') = \begin{pmatrix} \alpha & 0 & \beta & 0 \\ 0 & \alpha' & 0 & \beta' \\ \gamma & 0 & \delta & 0 \\ 0 & \gamma' & 0 & \delta' \end{pmatrix}, \qquad s = \begin{pmatrix} \alpha & \beta \\ \gamma & \delta \end{pmatrix}, \qquad s' = \begin{pmatrix} \alpha' & \beta' \\ \gamma' & \delta' \end{pmatrix}$$

for $s, s' \in Gl(2, F)$. $H(F)$ is isomorphic to the subgroup $(Gl(2, F)^2)^0$ of $Gl(2, F)^2$ of all elements (s, s') for which $det(s) = det(s')$:

$$H(F) \cong (Gl(2, F)^2)^0.$$

The exponent 0 indicates the determinant condition, or later a similar norm condition.

The Torus T. For a fixed quadratic field extension L/F consider the algebraic F-torus $T \subseteq Res_{L/F}(\mathbb{G}_m) \times Res_{L/F}(\mathbb{G}_m)$ determined by

$$T(F) = (L^* \times L^*)^0 = \{(x, x') \in L^* \times L^* | Norm(x/x') = 1\}.$$

Let $o_T \subseteq T(F)$ denote the subgroup of $t = (x, x') \in T(F)$ such that $x, x' \in o_L^*$ are units in the ring of integers o_L of L.

Standard Embeddings. Fix $D = D_0$ such that $L = F(\sqrt{D})$. For $D' \in F$ such that $L = F(\sqrt{D}) = F(\sqrt{D'})$, we have $\sqrt{D}/\sqrt{D'} = \alpha$ for some $\alpha = \alpha(\sqrt{D}, \sqrt{D'}) \in F^*$. For simplicity now assume D, D' to be *integral*. Such a choice of $(\sqrt{D}, \sqrt{D'})$ defines an F-rational embedding of T into H, called a *standard embedding*, as follows. For $t = (x, x')$ in $T(F)$ write

$$x = a + b\sqrt{D} \in L, \qquad x' = a' + b'\sqrt{D'} \in L.$$

The embedding will be defined by mapping t to the pair $\eta = (s, s') \in H(F) \subseteq G(F)$, where

$$s = \phi_{\sqrt{D}}(x) = \begin{pmatrix} a & Db \\ b & a \end{pmatrix}, \qquad s' = \phi_{\sqrt{D'}}(x') = \begin{pmatrix} a' & D'b' \\ b' & a' \end{pmatrix}.$$

Here we use the notation (s, s') for elements of $H(F)$ introduced above. Notice that $\phi_{-\sqrt{D}}(x) = \phi_{\sqrt{D}}(x^\sigma)$ definitely depends on the choice of the square root. σ denotes the generator of the Galois group of the field extension L/F. The image defines a maximal F-torus T_G of H and G.

Since any two F-embeddings $Res_{L/F}(\mathbb{G}_m) \hookrightarrow Gl(2)$ are conjugate by an element in $Gl(2, F)$, the first assertion of the following lemma is an immediate consequence.

Lemma 6.1. *Any F-rational embedding of T into H or G is conjugate under $H(F)$ or under $G(F)$, respectively, to a standard embedding with fixed $\sqrt{D}$. A second such standard embedding*

$$\phi_{\sqrt{D}}(x) = \begin{pmatrix} a & Db \\ b & a \end{pmatrix}, \qquad \phi_{\sqrt{D''}}(x') = \begin{pmatrix} a' & D'\theta b' \\ \theta^{-1}b' & a' \end{pmatrix}$$

is conjugate to the given one under $H(F)$ or under $G(F)$, respectively, as an embedding of T into H or into G, respectively, if and only if the quotient $\theta = \alpha(\sqrt{D}, \sqrt{D''})/\alpha(\sqrt{D}, \sqrt{D'})$ is in $Norm(L^*)$.

Proof. Since embeddings of maximal F-tori T into a reductive group G, up to conjugation by $G(F)$, are parameterized by $ker(H^1(F, T) \to H^1(F, G))$, the second assertion is an immediate consequence of $H^1(F, T) = F^*/Norm(L^*)$ (Hilbert 90) and $H^1(F, H) = H^1(F, G) = 1$. Since $F^*/Norm(L^*) \cong \mathbb{Z}/2\mathbb{Z}$, Lemma 6.1 shows that up to conjugation by $G(F)$ there exist exactly two different embeddings of T into G. It is an easy exercise to show that θ gives the relevant parameter in $F^*/Norm(L^*)$. $\square$

That the standard embeddings defined above are admissible embeddings in the sense of [60] is shown in Chap. 8.

Example 6.1. Let $\chi_{L/F}$ denote the quadratic character of F^* such that $ker(\chi_{L/F}) = Norm(L^*)$. Suppose $\chi_{L/F}(-1) = -1$. Then $\alpha(\sqrt{D}, \sqrt{D'})/\alpha(\sqrt{D}, -\sqrt{D'}) = -1$ is not in $Norm(L^*)$. Hence, the standard embeddings defined by $(\sqrt{D}, \sqrt{D'})$ and $(\sqrt{D}, -\sqrt{D'})$, respectively, are not $G(F)$-conjugate as embeddings, although these two different standard embeddings define the same image torus in G. In fact this change of embedding amounts to a change $(x, x') \mapsto (x, x'^\sigma)$ for the automorphism σ of L/F.

6.2.1 Orbital Integrals

Assume F is a local field with residue characteristic different from 2. For $\eta \in G(F)$ let G_η be the centralizer of η in G. For the orbital integral

$$O_\eta^G(1_K) = \int_{G_\eta(F)\backslash G(F)} 1_K(g^{-1}\eta g)dg/dg_\eta$$

of the *unit element* 1_K of the Hecke algebra (equivalent to a characteristic function of the group K of unimodular symplectic similitudes) for the group $G = GSp(4)$ define the κ-orbital integral by

$$\boxed{O_\eta^\kappa(1_K) = O_\eta^G(1_K) - O_{\eta'}^G(1_K)},$$

where

$$\eta' = (\phi_{\sqrt{D_0}}(x), \phi_{\theta \cdot \sqrt{D_0'}}(x')), \qquad \eta = (\phi_{\sqrt{D_0}}(x), \phi_{\sqrt{D_0'}}(x')),$$

and where $\theta \in F^*$ is such that $F^* = Norm(L^*) \cup \theta \cdot Norm(L^*)$. We normalize measures such that $vol_G(K) = 1$ and we assume $(\sqrt{D_0}, \sqrt{D_0'})$ is fixed. We also assume *regularity* in the form $x \neq x^\sigma$ and $x' \neq (x')^\sigma$ and $x' \neq x, x^\sigma$, where σ denotes the nontrivial automorphism of L/F. Assuming this, we get $G_\eta = T_G \cong T$. We normalize the measure on T_G such that $vol_T(o_T) = 1$ for o_T as defined earlier.

Further Assumptions. We choose D_0, D_0' to be normalized of minimal order 0 or 1. Notice this does not mean $D' = D_0'\theta^2$ has the same property. In fact, in the unramified case we have to choose θ such that $ord(\theta) = 1$.

6.3 The Endoscopic Group M

For the convenience of the reader we now describe the endoscopic group M and its L-embedding. In fact, we will also introduce some explicit isomorphism between tori, which will be essential for the computations made in the following.

Tori in M. The torus $T = T_M$ (elliptic case I) is the pullback of the maps $Norm : Res_{L/F}(\mathbb{G}_m) \times Res_{L/F}(\mathbb{G}_m) \to \mathbb{G}_m \times \mathbb{G}_m$ and the diagonal embedding $\mathbb{G}_m \to \mathbb{G}_m \times \mathbb{G}_m$

$$
\begin{array}{ccc}
T & \longrightarrow & Res_{L/F}(\mathbb{G}_m)^2 \\
\downarrow & & \downarrow {\scriptstyle Norm} \\
\mathbb{G}_m & \xrightarrow{\ diag\ } & \mathbb{G}_m^2
\end{array}
$$

The dual group $\hat{T}$ is the pushout of the diagonal map $diag : \mathbb{C}^* \times \mathbb{C}^* \to (\mathbb{C}^*)^2 \times (\mathbb{C}^*)^2$ and the multiplication map $m : \mathbb{C}^* \times \mathbb{C}^* \to \mathbb{C}^*$

$$
\begin{array}{ccc}
\hat{T} & \longleftarrow & (\mathbb{C}^*)^2 \times (\mathbb{C}^*)^2 \\
\uparrow & & \uparrow {\scriptstyle diag} \\
\mathbb{C}^* & \longleftarrow{\scriptstyle m} & (\mathbb{C}^*)^2
\end{array}
\quad .
$$

Thus, $\hat{T} = (\mathbb{C}^*)^2 \times (\mathbb{C}^*)^2 / \{(t, t, t^{-1}, t^{-1})\}$. The generator σ of $Gal(L/F)$ acts by permuting the first and second and third and fourth components, respectively. The map

$$
(z_1, z_2, z_3, z_4)/(t, t, t^{-1}, t^{-1}) \mapsto (z_1 z_4, z_2/z_1, z_3/z_4)
$$

induces an isomorphism $\hat{T} \cong (\mathbb{C}^*)^3$. By transport of structure σ now acts on $(\mathbb{C}^*)^3$ by

$$
\sigma(x, \nu, \mu) = (x\nu\mu, \nu^{-1}, \mu^{-1}).
$$

This defines the L-group $^L T$ as the semidirect product of $\hat{T}$ and the Weil group W_F with the action of W_F factorizing over the quotient map $W_F \to Gal(L/F)$.

L-embeddings.   We define an embedding of L-groups

$$
\psi : {}^L T \to {}^L GSp(4, F) = GSp(4, \mathbb{C}) \times W_F
$$

by $\psi(x, \nu, \mu) = diag(x, x\nu, x\mu\nu, x\mu)$ for $(x, \nu, \mu) \in \hat{T}$, and

$$
\psi(1 \times \sigma) = i \cdot \begin{pmatrix} 0 & E \\ -E & 0 \end{pmatrix} \times \sigma.
$$

Notice the action of W_F factorizes over $\Gamma = Gal(L/F)$. A better, but more technical choice of ψ is given in Chap. 8. We identify $\hat{T}$ with its image under ψ. Then the center $Z(\hat{G})$ of $\hat{G}$ is the group of all $(x, \nu, \mu) = (x, 1, 1)$ in $\hat{T}$ and the fixed group $\hat{T}^\Gamma$ is

$$
\hat{T}^\Gamma = Z(\hat{G}) \cup Z(\hat{G})\kappa
$$

for $\kappa = (1, -1, -1)$. The centralizer of

$$
s = \psi(\kappa) = diag(1, -1, 1, -1)
$$

defines the group

$$
\hat{M} = \hat{M}^0 = \left\{ \begin{pmatrix} * & 0 & * & 0 \\ 0 & * & 0 & * \\ * & 0 & * & 0 \\ 0 & * & 0 & * \end{pmatrix} \subseteq GSp(4, \mathbb{C}) \right\}.
$$

We consider $^LM = \hat{M} \times W_F$ as a subgroup of $^LGSp(4, F)$. The inclusion map $^LM \hookrightarrow {}^LG$ will be denoted ξ. The morphism ψ factorizes through the embedding ξ

$$\psi : {}^LT \to {}^LM \hookrightarrow {}^LGSp(4, F).$$

^LM is the L-group of the reductive group $M = Gl(2)^2/\mathbb{G}_m$, where

$$M(F) = Gl(2, F) \times Gl(2, F)/\{(t, t, t^{-1}, t^{-1}) \mid t \in F^*\}.$$

In fact $\xi : {}^LM \hookrightarrow {}^LG$ defines up to equivalence the unique elliptic endoscopic datum $(M, {}^LM, \xi, s)$ of G which is not equivalent to $(G, {}^LG, id, 1)$ in the sense of Langlands and Shelstad.

Admissible Embeddings. The choice of some regular representation of $L^* \hookrightarrow Gl(2, F)$ defines an embedding $T_M \hookrightarrow M$ and

$$T_M(F) = (L^* \times L^*)/F^* \hookrightarrow M(F) = (Gl(2, F)^2)/F^*.$$

Define isomorphisms

$$\rho : (L^* \times L^*)/F^* \cong L^* \times (L^*/F^*) \cong L^* \times N^1(L) \cong (L^* \times L^*)^0 \cong T(F)$$
$$\cong T_G(F) \subseteq GSp(4, F),$$

by the maps

$$(t_1, t_2)/(t^{-1}, t) \mapsto (t_1 t_2, t_2 \bmod t) \mapsto (t_1 t_2, t_2^\sigma/t_2) \mapsto (t_1 t_2, t_1 t_2 (t_2^\sigma/t_2))$$
$$\mapsto (\phi_{\sqrt{D}}(t_1 t_2), \phi_{\sqrt{D'}}(t_1 t_2^\sigma)).$$

The composite map defines an admissible embedding (see Chap. 8) of the torus T_M into $GSp(4, F)$

$$T_M(F) = (L^* \times L^*)/F^* \to T(F) \cong T_G(F) \subseteq GSp(4, F),$$

$$t = (t_1, t_2) \bmod (t^{-1}, t) \mapsto \eta = (\phi_{\sqrt{D}}(x), \phi_{\sqrt{D'}}(x')),$$

where $x = t_1 t_2$ and $x' = t_1(t_2)^\sigma$. We often identify T with its image T_G.

Notice, the induced map

$$\tilde{\rho} : L^* \times L^* \to (L^* \times L^*)/F^* \to T(F)$$

does not necessarily map $o_L^* \times o_L^*$ onto o_T, as we will see in the next section.

We now describe our choice of the admissible embedding $T_M \hookrightarrow T_G$. We considered the following chain of embeddings:

$$M = Gl(2, F)^2/F^* \hookleftarrow T_M(F) = (L^*)^2/F^* \xrightarrow{\iota} E^0 \subseteq E = L \oplus L,$$

where $E = L \oplus L$, and where $E^0 \subseteq E$ denotes the elements, whose components in the two summands of E have equal norm. Furthermore,

$$E^0 \stackrel{\phi_{\sqrt{D_0}}}{\to} Gl(2, F \oplus F)^0 \stackrel{\phi}{\to} T_G(F) \subseteq H(F) \subseteq GSp(4, F).$$

The maps from left to right are as follows. The first embedding is induced from $(L^*)^2 \hookrightarrow Gl(2, F)^2$ via the regular representation, which identifies $Gl(2, F)$ and $Gl(L/F)$. This map is unique up to conjugation. The map ι with image in E^0 is defined by

$$\iota : (t_1, t_2)/ \sim \quad \mapsto \quad (x, x') = (t_1 t_2, t_1 t_2^\sigma).$$

The map $\phi_{\sqrt{D_0}}$ is

$$\phi_{\sqrt{D_0}} : \quad (x, x') \mapsto (s, s') = \left(\begin{pmatrix} a & D_0 b \\ b & a \end{pmatrix}, \begin{pmatrix} a' & D_0 b' \\ b' & a' \end{pmatrix} \right),$$

with coefficients $a, b, a', b' \in F$ determined by $t_1 t_2 = a + b\sqrt{D_0}$ and $t_1 t_2^\sigma = a' + b'\sqrt{D_0}$ in $L = F(\sqrt{D_0})$. Finally, we have the isomorphism ϕ onto $H(F)$, which maps (s, s') to the symplectic similitude matrix

$$\eta = \begin{pmatrix} a & 0 & D_0 b & 0 \\ 0 & a' & 0 & D_0 b' \\ b & 0 & a & 0 \\ 0 & b' & 0 & a \end{pmatrix}.$$

The torus T_G in $GSp(4, F)$ is the image of the torus T_M in M under $\phi \circ \phi_{\sqrt{D_0}} \circ \iota$

$$T_G = \phi \circ \phi_{\sqrt{D_0}} \circ \iota(T_M).$$

Keep in mind $E = L \oplus L \ni (x, x')$. On E the field automorphism σ acts on each summand. The fixed algebra is $E^+ = F \oplus F$, which contains F diagonally. Flipping both summands defines another automorphism $\tau : E \to E$, whose fixed algebra is the quadratic extension field $L = F(\sqrt{D_0})$ of F, diagonally embedded into E. The fixed subalgebra L' under $\sigma \circ \tau$ is isomorphic to the field L.

The Factor Δ_{IV}. For comparison with [60] we record the following formulas for $D_G(\eta) = \prod_\alpha |\alpha(\eta) - 1|^{1/2}$ (product over all roots):

$$D_G(\eta) = |1 - x'/x||1 - x/x^\sigma||1 - x'/(x')^\sigma||1 - x/(x')^\sigma|$$

since η is conjugate to the diagonal matrix $diag(x, x', x^\sigma, (x')^\sigma)$ over $\overline{F}$. Also notice $|x| = |x'|$. Similarly,

$$D_M(t_1, t_2) = |1 - t_1/t_1^\sigma||1 - t_2/t_2^\sigma| = |1 - x/(x')^\sigma||1 - x/x'|,$$

since (t_1, t_2) is conjugate to $diag(t_1, t_1^\sigma), diag(t_2, t_2^\sigma)$. This shows $\Delta_{IV} = D_G(\eta)/D_M(t_1, t_2) = |(x - x^\sigma)(x' - (x')^\sigma)|/|xx'|$. Since $(x - x^\sigma)(x' - (x')^\sigma)$ is invariant under σ, it is contained in F. Furthermore, $|xx'| = |x^2 \frac{x'}{x}| = |x|^2 \in q^{\mathbb{Z}} \subseteq \mathbb{Q}^*$. Hence,

Lemma 6.2. $D_G(\eta)/D_M(t_1, t_2)$ has $\mathbb{Q}$-rational values.

6.4 Orbital Integrals for M

Since M_{der} is not simply connected, it is convenient to use the z-extension

$$p : M_1 = Gl(2)^2 \to M$$

and to consider orbital integrals for M_1 instead of stable orbital integrals for M. See [49], Lemma 3.1. The preimage T_{M_1} of T_M in $M_1 = Gl(2)^2$ satisfies $T_{M_1} = (M_1)_{(t_1, t_2)}$ by our regularity assumption. Furthermore, $T_{M_1}(F) \cong (L^*)^2$. Put $K_{M_1} = Gl(2, o_F)^2$ and normalize measures such that $vol_{M_1}(K_{M_1}) = vol_{K_{M_1}}(K_{M_1}) = 1$. We normalize measures on T_{M_1} by $vol_{T_{M_1}}(o_L^* \times o_L^*) = 1$.

Lemma 6.3. *We have exact sequences*

$$0 \to F^* \longrightarrow L^* \times L^* \overset{\tilde{\rho}}{\longrightarrow} T(F) \to 0$$

and

$$0 \to o_F^* \longrightarrow o_L^* \times o_L^* \overset{\tilde{\rho}}{\longrightarrow} o_T \to Q \to 0,$$

where $Q \cong \mathbb{Z}/(e_{L/F}\mathbb{Z})$ for the ramification index $e_{L/F}$.

Proof. Postponed to the beginning of Sect. 6.6. ☐

Remark 6.1. The image K_M of K_{M_1} in $M(F)$ is not a maximal compact subgroup of $M(F)$ in general. If L/F is ramified, choose π_L such that $\pi_L^2 \in F^*$. Then the image of $(t_1, t_2) = (\pi_L, \pi_L^{-1})$ in $T_M(F)$ is not contained in K_M, but generates together with K_M a compact group of $M(F)$. Furthermore, $\tilde{\rho}(\pi_L, \pi_L^{-1}) = (-\pi_F, \pi_F)$ and $\tilde{\rho}(o_L^* \times o_L^*)$ generate o_T.

From [49], Lemma 3.1, the stable orbital integral $SO^M_{(t_1, t_2)}(1_K)$ is

$$SO^M_{(t_1, t_2)}(1_{K_M}) = c \cdot \int_{T_M(F) \backslash M(F)} 1_{K_M}(g^{-1}(t_1, t_2)g)dg/dt.$$

Measures are normalized such that $vol_M(K_M) = 1$ and on T_M such that $vol_{T_M}(o_T) = 1$, by transport of structure. In that case, $c = 1$. For a change of measure on T_M such that $vol_{T_M}(\tilde{\rho}(o_L^* \times o_L^*)) = 1$, we have to make an adjustment by $c = \#Q = e_{L/F}$.

We can express this in the form

$$SO^M_{(t_1, t_2)}(1_{K_M}) = e_{L/F} \cdot \int_{T_{M_1}(F) \backslash M_1(F)} 1_{K_{M_1} \cdot F^*}(g^{-1}(t_1, t_2)g)dg/dl,$$

with measures on M_1 chosen as above. On M_1 we should work with the unit element of the Hecke algebra of K_{M_1}-bi-invariant functions f on M_1 with the property $f(zg) = f(g)$ for all $z \in Kern(p)(F) \cong F^*$. Observe $1_{K_{M_1} F^*}(g^{-1}(t_1, t_2)g) \neq 0$ holds if and only if there exists a $z \in Kern(p)$ such that $(t_1, t_2) = z(t_1', t_2')$ and

$1_{K_{M_1}}(g^{-1}(t'_1, t'_2)g) \neq 0$ holds. This is clear because $Kern(p)/(Kern(p) \cap K_{M_1})$ is injected into $X_*(M_1^{ab})$. We call such (t'_1, t'_2) *adjusted*. The last observation has the consequence that for adjusted pairs (t_1, t_2) we have

$$O_{(t_1, t_2)}^{M_1}(1_{K_{M_1} F^*}) = O_{(t_1, t_2)}^{M_1}(1_{K_{M_1}}) = O_{t_1}^{Gl(2,F)}(1_K) \cdot O_{t_2}^{Gl(2,F)}(1_K).$$

Since the element $\eta = \eta(t_1, t_2)$ and the orbital integral $O_{(t_1, t_2)}^{M_1}(1_{K_{M_1} F^*})$ do not change under the replacement $(t_1, t_2) \mapsto z(t_1, t_2)$, we can restrict ourselves to adjusted pairs (t_1, t_2) in the proof of Theorem 6.1.

In particular, to compute $O_{(t_1, t_2)}^{M_1}(1_{K_{M_1} \cdot F^*})$, we may now use the well-known formulas.

Lemma 6.4.

$$\int_{L^* \backslash Gl(2,F)} 1_K(g^{-1}t_i g) dg/dl = 1_{o_L^*}(t_i) \cdot \left(-\frac{1}{q-1} + \frac{q|A_0|^{1/2}}{q-1}|t_i - t_i^\sigma|^{-1}\right)$$

in the case of ramified extensions L/F and

$$\int_{L^* \backslash Gl(2,F)} 1_K(g^{-1}t_i g) dg/dl = 1_{o_L^*}(t_i) \cdot \left(-\frac{2}{q-1} + \frac{|A'_0|^{1/2}(q+1)}{q-1}|t_i - t_i^\sigma|^{-1}\right)$$

in the case of unramified extensions L/F. Here $K = Gl(2, o_F)$. Measures on $Gl(2, F)$ and L^* are normalized such that $vol_{L^*}(o_L^*) = 1$ and $vol_{Gl(2,F)}(K) = 1$.

Theorem 6.1. *Let $x = t_1 t_2$ and $x' = t_1 t_2^\sigma$ be regular in the sense that $x \neq x^\sigma$ and $x' \neq x, x^\sigma, (x')^\sigma$. Put $\eta = (\phi_{\sqrt{D_0}}(x), \phi_{\sqrt{D_0}}(x')) \in T_G(F)$ with "image" $(t_1, t_2) \in T_M(F)$ under ρ^{-1}. Assume the residue characteristic of F is different from 2. Then*

$$\Delta(\eta, (t_1, t_2)) \cdot O_\eta^\kappa(1_K) = SO_{(t_1, t_2)}^M(1_K),$$

for the transfer factor

$$\Delta(\eta, (t_1, t_2)) = |xx'|^{-1} \cdot (\chi_{L/F} \cdot |)((x - x^\sigma)(x' - (x')^\sigma)),$$

where $\chi_{L/F}$ is the quadratic character of F^* attached to the field extension L/F.

Proof. Sections 6.5–6.13. □

Since $\chi_{L/F}$ is a quadratic character, Lemma 6.2 implies

Corollary 6.1. *The transfer factor $\Delta(\eta, (t_1, t_2))$ has $\mathbb{Q}$-rational values, and satisfies*

$$\Delta(\lambda \eta, (\lambda t_1, t_2)) = \Delta(\eta, (t_1, t_2))$$

for all $\lambda \in F^*$.

Lemma 6.5. *For the involution* $(t_1, t_2)^* = (t_2, t_1)$ *we have* $\Delta(\eta, t^*) = \chi_{L/F}(-1)\Delta(\eta, t)$.

Proof. The change from t to t^* replaces (x, x') by (x, x'^{σ}).
   Of course (x, x'^{σ}) is $G(F)$-conjugate to (x, x') if and only if $\chi_{L/F}(-1)=1$. □

6.5 Reduction to H

An essential ingredient of the whole computation is Schröder's representatives [81] for the double cosets $H(F) \backslash G(F)/K$. A generalization of this double coset decomposition to symplectic groups of arbitrary genus $g \geq 2$ is contained in Chap. 12, Theorem 12.1. Together with the measures considered in Sect. 6.16, this double coset decomposition leads to an expression as an infinite (actually then finite) sum

$$O_\eta^G(1_K) = \sum_{g_i} \frac{vol_G(g_i K g_i^{-1})}{vol_H(H \cap g_i K g_i^{-1})} O_\eta^H(1_{K_i}) = \sum_{i=0}^{\infty} \frac{vol_G(K)}{vol_H(K_i)} O_\eta^H(1_{K_i}),$$

where $K_i = K_H = K \cap H(F)$ for $i = 0$, and where $K_i = H(F) \cap g_i K g_i^{-1}$ for $i \geq 1$ is

$$K_i = \left\{ (h_1, h_2) \in K_H \subseteq Gl(2, o_F)^2 \,\middle|\, h_1 = h_2^\sigma \bmod \pi_F^i, det(h_1) = det(h_2) \right\},$$

i.e., one of the groups denoted H_D on page 349ff. Observe the abbreviation

$$h^\sigma = \begin{pmatrix} 1 & 0 \\ 0 & -1 \end{pmatrix} h \begin{pmatrix} 1 & 0 \\ 0 & -1 \end{pmatrix}.$$

This notion is compatible with the notation for the Galois action of $\sigma \in Gal(L/F)$ in the sense that $h \mapsto h^\sigma$ induces the automorphism σ on $\phi_{A'}(L^*)$. Finally,

$$O_\eta^H(1_{K_i}) = \int_{T_G(F)\backslash H(F)} 1_{K_i}(h^{-1}\eta h)dh/dt.$$

Integration is over all elements in $h \in H(F)$

$$h = (h_1, h_2), \qquad det(h_1) = det(h_2), \qquad h_i \in Gl(2, F).$$

We use measures on G and H such that the volumes of the compact subgroups $vol_G(K) = 1$ and $vol(K_H) = 1$ are 1. This means

$$vol_G(K)/vol_H(K_i) = [K_0 : K_i].$$

Remember, measures on $G_\eta \cong T$ were normalized by $vol(o_T) = 1$.

6.6 Preliminary Considerations for (x, x')

Fix some $(x, x') \in T(F)$. The coordinates a, b and a', b' of $x = a + b\sqrt{D_0}$ and $x' = a' + b'\sqrt{D_0'}$ depend on the choice of the parameters $\sqrt{D_0}$ and $\sqrt{D_0'}$. From now on we fix the notation such that these coordinates will be reserved for some fixed choice of normalized $\sqrt{D_0}$ and $\sqrt{D_0'}$.

Proof of Lemma 6.3. For x, x' in L^* with $Norm(x) = Norm(x')$, elements $t_1, t_2 \in L^*$ with $x = t_1 t_2$ and $x' = t_1 t_2^\sigma$ can be found by solving the equation $t_2/t_2^\sigma = x/x'$. The element x/x' is in the norm-1-group $N_L^1 \subseteq L^*$. Therefore, this equation can be solved for $t_2 \in L^*$, unique up to some multiple in F^*. Then $t_1 = x/t_2$ defines the second parameter. The pair (t_1, t_2) is uniquely determined up to some multiple $z = (t, t^{-1})$ for $t \in F^*$. This allows us to assume that the pair (t_1, t_2) can be chosen to be "adjusted" in the sense of Sect. 6.4.

Integral Solutions. For the proof of Theorem 6.1 the only relevant elements (x, x') are those where x and x' are units in o_L^*. This is the obvious eigenvalue constraint for an element to be conjugate to a unimodular matrix. So it is natural to ask whether one can express $x, x' \in o_L^*$ as $x = t_1 t_2, x' = t_1 t_2^\sigma$ in terms of units $t_1, t_2 \in o_L^*$. This is not always possible. Of course, it would be enough to find $t_2 \in o_L^*$ such that $t_2/t_2^\sigma = x/x'$. The image of o_L^* under $t \mapsto t/t^\sigma$ is $N_L^1 \cap (1 + \pi_L o_L)$ in the ramified case, and N_L^1 in the unramified case, since the residue characteristic is different from 2 by our assumptions. Every element in N_L^1 is in o_L^* and congruent to $\pm 1 \bmod \pi_L$ in the ramified case. If

$$x/x' = -1 \bmod(\pi_L)$$

we do not get a solution with $t_2 \in o_L^*$. For $x/x' = 1 \bmod(\pi_L)$ we do, which implies $Q = o_T/\tilde{\rho}((o_L^*)^2) \cong \mathbb{Z}/(e_{L/F}\mathbb{Z})$ and completes the proof of Lemma 6.3.   □

Conclusion. Suppose that x and x' are *units* in L^* with $x = x' \bmod \pi_L$ in the case where L/F is ramified. Then the equations

$$x = t_1 t_2, \qquad x' = t_1 t_2^\sigma$$

are solvable with *units* t_1, t_2 in L^* (and this gives an adjusted solution).

Some Notation. For $x, x' \in o_F^*$ with equal norm define the following integers

$$\chi = ord(a - a'), \qquad f = ord(b), \qquad f' = ord(b'), \qquad N = min(f, f').$$

With this notation we get $b = \pi_F^f B$ and $b' = \pi_F^{f'} B'$ for units B, B' in o_F^* and

$$q^{f+f'} = |\frac{x - x^\sigma}{\sqrt{D_0}}|^{-1}|\frac{x' - (x')^\sigma}{\sqrt{D_0'}}|^{-1}.$$

Furthermore,

$$\chi = ord(Tr(x) - Tr(x')) = ord(t_1 t_2 + t_1^\sigma t_2^\sigma - t_1 t_2^\sigma - t_1^\sigma t_2) = ord((t_1 - t_1^\sigma)(t_2 - t_2^\sigma)).$$

Now assume $x = x' \bmod (\pi_L)$ in the case where L/F is ramified. Then $|t_1| = |t_2| = 1$ by Lemma 6.3.

From now on put $\sqrt{D_0'} = \sqrt{D_0}$. Under these assumptions

$$|t_2 - (t_2)^\sigma| = |x - x'| = |(a - a')^2 - (b - b')^2 D_0|^{1/2},$$

and

$$|t_1 - (t_1)^\sigma| = |x - (x')^\sigma| = |(a - a')^2 - (b + b')^2 D_0|^{1/2}.$$

Our assumption regarding x, x' earlier is an automatic consequence of another

Assumption. Assume $x, x' \in o_F^*$ and $Norm(x/x') = 1$ such that $\chi > 0$.

This assumption implies for ramified L/F that $x' = x \bmod (\pi_L)$, since then $D_0 = \pi_L$ and $x - x' = a - a' \bmod (\pi_L)$. Since $x'/x = \pm 1 \bmod (\pi_L)$, the condition $x' = x \bmod (\pi_L)$ is equivalent to $\chi > 0$ (the residue characteristic is not 2). It will turn out in the final computation in Sect. 6.12 that only those x, x' will play a role for which $\chi > 0$ holds.

Lemma 6.6. *Suppose* $x, x' \in o_L^*$ *with* $Norm(x/x') = 1$ *and* $\chi > 0$. *Also suppose* $\sqrt{D_0} = \sqrt{D_0'}$. *Then:*

1. $\{|D_0|^{1/2}|t_1 - t_1^\sigma|^{-1}, |D_0|^{1/2}|t_2 - t_2^\sigma|^{-1}\} = \{q^N, q^{\chi - N - ord(D_0)}\}$.
2. $\chi = ord(D_0(b^2 - (b')^2))$ *unless* $f = f' = 0$ *and* $ord(D_0) = 0$ (L/F *is unramified*).
3. *For* $(f, f', ord(D_0)) = (0, 0, 0)$ *we have* $\chi \leq ord(D_0(b^2 - (b')^2))$.
4. $\chi \geq ord(D_0) + 2N$ *with equality in the case* $f \neq f'$ *for* $(f, f', ord(D_0)) \neq (0, 0, 0)$.

Proof. Recall $ord(D_0) = 0$ if L/F is unramified and $ord(D_0) = 1$ if L/F is ramified. From the identity above we get

$$0 < 2 \cdot \chi = ord[(a - a')^2 - (b - b')^2 D_0)] + ord[(a - a')^2 - (b + b')^2 D_0].$$

If for both signs $ord(b \pm b') < ord(a - a')$, this implies (2)

$$2\chi = ord(D_0(b + b')^2) + ord(D_0(b - b')^2) = 2ord(D_0(b^2 - (b')^2)).$$

Since the residue characteristic is different from 2, we have

$$N = min(ord(b), ord(b')) = min(ord(b - b'), ord(b + b')).$$

Together with (2) this implies (1). Now assume $ord(b \pm b') \geq ord(a - a')$ for one choice of the sign. Since $\chi = ord(a - a') > 0$, this gives $ord[(a - a')^2 -$

$D_0(b \pm b')^2] \geq 2ord(a - a') = 2\chi$. Hence, the equality above implies for the other choice of sign $ord[(a - a')^2 - D_0(b \mp b')^2] = 0$. Since $\chi = ord(a - a') > 0$, $ord(D_0(b \mp b')^2) = 0$. In particular, $ord(b \mp b') = 0$ and $ord(D_0) = 0$. Then $N = min(ord(b), ord(b')) = min(ord(b - b'), ord(b - b')) = 0$ and $ord(b) = ord(b')=0$. This implies claims (1) and (2). Claim (3) is an immediate consequence of the argument above. Claim (4) is an immediate consequence of $\chi = ord(D_0(b^2 - (b')^2)$ (assertion 2). $\square$

6.7 The Residue Rings $R = o_F/\pi_F^i o_F$

Assume $i \geq 1$. Let $R = o_F/\pi_F^i o_F$ denote the residue ring and let D, D' be elements in R. Suppose matrices s, s' in $Gl(2, R)$ are given by

$$ s = \begin{pmatrix} a & bD \\ b & a \end{pmatrix}, \qquad s' = \begin{pmatrix} a' & b'D' \\ b' & a' \end{pmatrix}. $$

Let

$$ N_D \subseteq R^* $$

be the group of invertible elements of the form $a^2 - b^2 D$ for a, b in R. Then

$$ (R^*)^2 \subseteq R_D \subseteq R^*. $$

Notice that elements congruent to 1 mod (π_F) are contained in $(R^*)^2$ by the assumption regarding the residue characteristic.

Lemma 6.7. *The centralizer $Gl(2, R)_s$ of s in $Gl(2, R)$ is either $Gl(2, R)$ if $b = 0$ in R, or is the group of all matrices*

$$ y \in \left\{ \begin{pmatrix} u & vD + Ann(b) \\ v & u + Ann(b) \end{pmatrix} \;\middle|\; u^2 - v^2 D \in R^* \right\}. $$

Furthermore, such $y \in Gl(2, R)_s$ have unique decompositions into a product of two invertible matrices of the form

$$ y = \begin{pmatrix} u & vD \\ v & u \end{pmatrix} \begin{pmatrix} 1 & g \\ 0 & 1 + g' \end{pmatrix}, \qquad g, g' \in Ann(b). $$

Remark 6.2. This does not define a semidirect product decomposition of $Gl(2, R)_s$.

Proof. One direction is clear from

$$ \begin{pmatrix} a & bD \\ b & a \end{pmatrix} \begin{pmatrix} u & vD \\ v & u \end{pmatrix} = \begin{pmatrix} u & vD \\ v & u \end{pmatrix} \begin{pmatrix} a & bD \\ b & a \end{pmatrix} $$

and

$$\begin{pmatrix} a & bD \\ b & a \end{pmatrix} \begin{pmatrix} 1 & g \\ 0 & 1+g' \end{pmatrix} = \begin{pmatrix} a & ag+bD \\ b & a+ag' \end{pmatrix} = \begin{pmatrix} 1 & g \\ 0 & 1+g' \end{pmatrix} \begin{pmatrix} a & bD \\ b & a \end{pmatrix}.$$

The other direction follows in the course of the proof of Lemma 6.8 using $y \in Gl(2, R)$. $\square$

Lemma 6.8. *s and s' are conjugate*

$$y^{-1}sy = s',$$

by an element y in $Sl(2, R)$ if and only if s and s' are conjugate in $Gl(2, R)$ by a matrix y of the form

$$y = \begin{pmatrix} 1 & 0 \\ 0 & \epsilon \end{pmatrix}, \qquad \epsilon \in N_D \subseteq R^*,$$

in other words if and only if $a = a'$, $b' = b/\epsilon$, $b'D' = bD\epsilon$ holds for some $\epsilon \in N_D$.

Proof. One direction is clear. That the second assertion implies the first follows by conjugation with some element

$$y = g^{-1} \begin{pmatrix} 1 & 0 \\ 0 & \epsilon \end{pmatrix}, \qquad \epsilon \in R^*$$

in $Sl(2, R)$, where $g \in Gl(2, R)_s$ is suitably chosen such that $det(g) = \epsilon$.

For the reverse direction we may assume $b'|b$ by symmetry. If $b' = 0$, then $b = 0$ is trivial, so we may also assume $b' \neq 0$ in R. Then $y^{-1}sy = s'$ implies $Tr(s) = 2a = 2a' = Tr(s')$. Since $2 \in R^*$, this implies $a = a'$. For $y = \begin{pmatrix} u & v \\ z & w \end{pmatrix}$, $sy = ys'$ means $vb' = zDb$, $ub'D' = wbD$, $wb' = ub$, and $zb'D' = vb$. Therefore, $v = z(Db/b') + f$ for some $f \in Ann(b')$ and $w = u(b/b') + f'$ for some $f' \in Ann(b')$. This implies

$$y = \begin{pmatrix} u & z(Db/b') + f \\ z & u(b/b') + f' \end{pmatrix}.$$

Now $b' \neq 0$ implies $Ann(b') \subseteq (\pi_F)$. Hence, the assumption $det(y) = 1$ implies $(u^2 - Dz^2)(b/b') \in (1 + Ann(b')) \subseteq 1 + (\pi_F) \subseteq (R^*)^2$. This proves

$$\epsilon = b/b' \in N_D \subseteq R^*$$

and $u^2 - Dz^2 \in R^*$. Hence,

$$y = \begin{pmatrix} u & zD + f \\ z & u + f' \end{pmatrix} \begin{pmatrix} 1 & 0 \\ 0 & \epsilon \end{pmatrix} \qquad \blacklozenge$$

for some other $f, f' \in Ann(b')$. Since the left matrix is contained in the centralizer of s, we are done. This proves the last lemma.

In fact we now also get the product decomposition of the elements

$$\begin{pmatrix} u & zD + f \\ z & u + f' \end{pmatrix} = \begin{pmatrix} u & zD \\ z & u \end{pmatrix} \begin{pmatrix} 1 & g \\ 0 & 1 + g' \end{pmatrix}$$

for some $g, g' \in Ann(b')$ (used for the proof of Lemma 6.7) by solving the equation

$$\begin{pmatrix} u & zD \\ z & u \end{pmatrix} \begin{pmatrix} g \\ g' \end{pmatrix} = \begin{pmatrix} f \\ f' \end{pmatrix}.$$

This equation has a unique solution since $u^2 - Dz^2 \in R^*$. Hence Lemmas 6.7 and 6.8 have been proven. $\square$

Now consider a field extension $L = F(\sqrt{D})$ for some integer $D \in o_F$. By abuse of notation let D also denote its residue class in $R = o_F/(\pi_F^i)$. In this case either $D \in (\pi_F)$; then $N_D = (R^*)^2$. Or $D \in R^*$; then L/F is unramified and $N_D = R^*$ (once again using the property that the residue characteristic is not equal to 2). Therefore, either $[R^* : N_D] = 2$ or $[R^* : N_D] = 1$.

For such D we deduce from Lemma 6.7

Corollary 6.2. *In the situation of Lemma 6.7 we have for $R = o_F/(\pi_F^i)$*

$$\#Sl(2, R)_s = \#Sl(2, R), \qquad ord(b) = \nu \geq i$$

if $b = 0$ in R, or

$$\#(Sl(2, R)_s) = 2q^{2\nu + i}, \qquad ord(b) = \nu < i$$

for ramified L/F and

$$\#(Sl(2, R)_s) = \begin{cases} 2q^{2\nu + i} & ord(D) \neq 0 \\ (q + 1)q^{2\nu + i - 1} & ord(D) = 0 \end{cases}, \qquad ord(b) = \nu < i$$

for unramified L/F.

Proof. For $ord(b) < i$, by Lemma 6.7, the cardinality of $Sl(2, R)_s$ is

$$\#(Ann(b)) \cdot \#\left(\left\{ x \in Image(\phi_{\sqrt{D}}(o_L^*)) \mid det(x) \in 1 + Ann(b) \right\}\right).$$

$Ann(b) = (\pi_F^{i-\nu}) \mod (\pi_F^i)$ has cardinality q^ν and $[R^* : (1 + Ann(b))] = (q - 1)q^{i - \nu - 1}$. Furthermore, the image of $det(\phi_{\sqrt{D}}(o_L^*))$ in R^* is N_D. Hence,

$$\#Sl(2, R)_s = q^\nu \frac{[R^* : N_D]}{(q - 1)q^{i - 1 - \nu}} \#Image\left(\phi_{\sqrt{D}} : o_L^* \to Gl(2, R)\right).$$

Since $\phi_{\sqrt{D}}(o_L^*)$ has cardinality $(q - 1)q^{2i - 1}$ for $\pi_F | D$ and cardinality $(q^2 - 1)q^{2i - 2}$ for $D \in R^*$ (also see Sects. 6.14, 6.15), the claim follows. $\square$

6.8 On $T \setminus H$-Integration

For this section recall that we assume $D_0 = D_0'$ and θ to be chosen normalized of order 0 or 1. Then $ord(D_0) = 0$ if and only if L/F is unramified.

In Sect. 6.9 we describe representatives r of a disjoint double coset decomposition

$$H(F) = \coprod_{r \in R} T(F) \cdot r \cdot K_0,$$

where for certain $j, j' \in \mathbb{Z}$ and $\epsilon_0 \in o_F^*$ the representatives r are of the form

$$r \in \left(\phi_{\sqrt{D_0}}(L^*), \phi_{\theta\sqrt{D_0}}(L^*) \right) \cdot \left(\begin{pmatrix} 1 & 0 \\ 0 & \pi_F^j \end{pmatrix}, \begin{pmatrix} 1 & 0 \\ 0 & \pi_F^{j'-ord(\theta)}\epsilon_0 \end{pmatrix} \right).$$

Hence, $\int_{H(F)} f(h)dh = \sum_r [o_T : (T(F) \cap rK_0r^{-1})] \int_{T(F)} dt \int_{K_0} f(trk)dk$. Thus,

$$\int_{T(F) \setminus H(F)} f(h)dh/dt = \sum_r [o_T/(T(F) \cap rK_0r^{-1})] \int_{K_0} f(rk)dk.$$

For regular elements

$$\eta = \left(\phi_{\sqrt{D_0}}(x), \phi_{\theta\sqrt{D_0}}(x') \right)$$

with centralizer T this allows us to calculate the orbital integrals for $H(F)$ by integrations over K_0. Using the results given in Sect. 6.5, one obtains

Lemma 6.9.

$$O_\eta^G(1_K) = \sum_{i \geq 0} \sum_{r \in R} [o_T/(T(F) \cap rK_0r^{-1})][K_0 : K_i] \int_{K_0} 1_{K_i}(k^{-1}r^{-1}\eta rk)dk.$$

Observe $a' + b'\sqrt{D_0'} = a' + (b'/\theta)\sqrt{\theta^2 D_0}$. Thus, for $\Theta = \theta/\pi_F^{ord(\theta)} \in o_F^*$,

$$\eta_r := r^{-1}\eta r = \left(\phi_{\pi_F^j \sqrt{D_0}}(x), \phi_{\Theta\pi_F^{j'}\epsilon_0\sqrt{D_0}}(x') \right)$$

since

$$\begin{pmatrix} 1 & 0 \\ 0 & \pi_F^{ord(\theta)-j'}\epsilon_0^{-1} \end{pmatrix} \begin{pmatrix} a' & b'\theta D_0 \\ b'\theta^{-1} & a' \end{pmatrix} \begin{pmatrix} 1 & 0 \\ 0 & \pi_F^{j'-ord(\theta)}\epsilon_0 \end{pmatrix}$$

$$= \begin{pmatrix} a' & \frac{b'}{\Theta\epsilon_0\pi_F^{j'}}(\Theta\epsilon_0\pi_F^{j'})^2 D_0 \\ \frac{b'}{\Theta\epsilon_0\pi_F^{j'}} & a' \end{pmatrix}.$$

Therefore, $T(F) \cap rK_0r^{-1} \subseteq o_T = \left\{ (x, x') \in o_L^* \times o_L^* \mid Norm(x/x') = 1 \right\}$ is the group

$$\left\{ (x, x') = (a + b\sqrt{D_0}, a' + b'\sqrt{D_0}) \,\middle|\, Norm(x) \right.$$
$$\left. = Norm(x') \in o_F^*; a, a', \pi_F^{-j} b, \pi_F^{-j'} b' \in o_F \right\},$$

which does not depend on the choice of (normalized) D_0 or θ.

Notation. For normalized D_0 we define the orders

$$o_L(j) = o_F + o_F \pi_F^j \sqrt{D_0} \subseteq o_L$$

for $j > 0$. With this notation we obtain

Lemma 6.10 (constraints for fixed r). $(x, x') \in T(F) \cap r K_0 r^{-1}$ means $x \in o_L(j)$ and $x' \in o_L(j')$.

This follows from the equivalent condition $\eta_r \in K_0$ and the formula for β_r from above.

Hence, the *index* $[o_T : (T(F) \cap r K_0 r^{-1})]$ is given by the formula

Lemma 6.11.

$$[o_T : (T(F) \cap r K_0 r^{-1})]$$
$$= [o_L^* : o_L(j)^*][o_L^* : o_L(j')^*][Norm(o_L^*)/Norm(o_L(min(j, j'))^*)]^{-1}.$$

The constraint above must be satisfied for the nonvanishing of the orbital integral

$$[K_0 : K_i] \int_{K_0} 1_{K_i}(k^{-1} \eta_r k) dk.$$

In fact $k^{-1} \eta_r k \in K_i$ implies $\eta_r \in K_0$, and hence $(x, x') \in T(F) \cap r K_0 r^{-1}$.

Suppose this r-constraint is satisfied. Then to compute the integral

$$[K_0 : K_i] \int_{K_0} 1_{K_i}(k^{-1} \eta_r k) dk, \qquad \eta_r = (s_r, s_r')$$

means counting all cosets $k K_i \subseteq K_0$ for which $k^{-1} \eta_r k \in K_i$. This count gives 1 for $i = 0$. For $i \geq 1$ it gives the number of all elements $y \in Sl(2, o_F/(\pi_F^i))$ with

$$y^{-1} s_r y \equiv (s_r')^\sigma \mod (\pi_F^i), \qquad y \in Sl(2, o_F/(\pi_F^i)).$$

By Lemma 6.8 this is either zero, or is equal to the order of the centralizer $Sl(2, o_F/(\pi_F^i))_{s_r}$.

6.9 Double Cosets in H

For normalized D_0 we have [42]

$$Gl(2, F) = \coprod_{j \geq 0} \phi_{\sqrt{D_0}}(L^*) \begin{pmatrix} 1 & 0 \\ 0 & \pi_F^j \end{pmatrix} Gl(2, o_F).$$

For $(h, h') \in H(F)$ this gives decompositions

$$h = \phi_{\sqrt{D_0}}(l) \begin{pmatrix} 1 & 0 \\ 0 & \pi_F^j \end{pmatrix} k, \qquad k \in Gl(2, o_F),\ l \in L^*,$$

$$h' = \phi_{\theta\sqrt{D_0}}(l') \begin{pmatrix} 1 & 0 \\ 0 & \pi_F^{j'-ord(\theta)} \end{pmatrix} k', \qquad k' \in Gl(2, o_F),\ l' \in L^*,$$

for *uniquely* defined integers $j \geq 0$ and $j' \geq 0$. Since $det(h) = det(h')$, we have

$$det(k)/det(k') \in Norm(L^*) \cdot \pi_F^{j'-j-ord(\theta)}.$$

In the case where L/F is *ramified*, we choose once for all $\pi_F = -D_0$ such that $\pi_F = Norm(\pi_L)$. Then $det(k)/det(k') \in Norm(L^*) \cap o_F^* = (o_F^*)^2$. Changing l by some unit in o_F^* allows us to assume $det(k) = det(k')$, i.e., $(k, k') \in K_0 \subseteq H$. Hence, we can write $H(F)$ as a disjoint union of double cosets

$$\coprod_{r \in R} T(F) \cdot r \cdot K_0, \qquad R = \mathbb{N}_0 \times \mathbb{N}_0$$

for the representatives

$$r = r_{j,j'} = \left(\phi_{\sqrt{D_0}}(\pi_L^{-j}), \phi_{\sqrt{D_0}}(\pi_L^{-j'}) \right) \left(\begin{pmatrix} 1 & 0 \\ 0 & \pi_F^j \end{pmatrix}, \begin{pmatrix} 1 & 0 \\ 0 & \pi_F^{j'} \end{pmatrix} \right).$$

Now consider the case where L/F is *unramified*. Now $\pi_F^{j'-j-ord(\theta)} \in Norm(L^*)o_F^*$, and hence

$$j' - j - ord(\theta) \equiv 0 \bmod 2.$$

Secondly, to achieve $det(k) = det(k')$ we are only allowed to change k or k' by elements in $\phi_{\pi_F^j \sqrt{D_0}}(L^*) \cap Gl(2, o_F)$ and $\phi_{\theta\pi_F^{j'} \sqrt{D_0}}(L^*) \cap Gl(2, o_F)$, respectively. So $det(k)/det(k')$ can be changed within the group $Norm\left(o_L^*(j)o_L^*(j') \right)$. This group is $(o_F^*)^2$ unless $min(j, j') = 0$. Hence, $H(F)$ is a disjoint union of double cosets

$$\coprod_{r \in R} T(F) \cdot r \cdot K_0,$$

where $R \subseteq \left((o_F^* / (o_F^*)^2) \times \mathbb{N} \times \mathbb{N} \right) \cup \left(1 \times \mathbb{N}_0 \times 0 \right) \cup \left(1 \times 0 \times \mathbb{N} \right)$ is the subset

of all those elements $(\epsilon_0, j, j') \in \left(o_F^* / (o_F^*)^2 \right) \times \mathbb{N}^2$ with the additional property

$$j - j' - ord(\theta) \equiv 0 \bmod 2.$$

The representatives r are

$$r = r_{\epsilon_0, j, j'}$$

$$= \left(\phi_{\sqrt{D_0}}(\pi_L^{-[j/2]}), \phi_{\theta\sqrt{D_0}}(\pi_L^{-[j'-ord(\theta)/2]}\epsilon_0') \right) \left(\begin{pmatrix} 1 & 0 \\ 0 & \pi_F^j \end{pmatrix}, \begin{pmatrix} 1 & 0 \\ 0 & \pi_F^{j'-ord(\theta)}\epsilon_0 \end{pmatrix} \right),$$

where $\epsilon_0' \in o_L^*$ is chosen such that $Norm(\epsilon_0') = \epsilon_0^{-1}$ for $\epsilon_0 \in o_F^* \setminus (o_F^*)^2$.

6.10 Summation Conditions

Put $\theta = \Theta \cdot \pi_F^{ord(\theta)}$. If L/K is unramified we put $\theta = \pi_F$. Lemma 6.9 gave a formula for the orbital integral $O_\eta^G(1_K)$ as a sum with certain indices $(i, r) \in \mathbb{N}_0 \times R$. From Lemma 6.8 we get conditions for whether a summation index $(i, r) \in \mathbb{N}_0 \times R$ gives a nonzero contribution to the orbital integral. Put $b/\pi_F^j = B\pi_F^\nu$, $b'/\pi_F^{j'} = B'\pi_F^{\nu'}$ for units $B, B' \in o_F^*$. Then, in fact, we have to apply Lemma 6.8 to the elements

$$s_r = \begin{pmatrix} a & \frac{b}{\pi_F^j}(\pi_F^{j'})^2 D_0 \\ \frac{b}{\pi_F^j} & a \end{pmatrix} = \begin{pmatrix} a & B\pi_F^{\nu+2j}D_0 \\ B\pi_F^\nu & a \end{pmatrix}$$

and with an additional sign from the definition of the groups K_i involving a twist by σ (see the end of Sect. 6.8)

$$(s_r')^\sigma = \begin{pmatrix} a' & -\frac{b'}{\Theta\epsilon_0\pi_F^{j'}}(\Theta\epsilon_0\pi_F^{j'})^2 D_0 \\ -\frac{b'}{\Theta\epsilon_0\pi_F^{j'}} & a' \end{pmatrix} = \begin{pmatrix} a' & -\frac{B'}{\Theta\epsilon_0}(\Theta\epsilon_0)^2\pi_F^{\nu'+2j'}D_0 \\ -\frac{B'}{\Theta\epsilon_0}\pi_F^{\nu'} & a' \end{pmatrix},$$

which gives the summation conditions $a \equiv a'$,

$$b/\pi_F^j \equiv -\epsilon\frac{b'}{\Theta\epsilon_0\pi_F^{j'}}, \quad \text{and} \quad -\frac{b'}{\Theta\epsilon_0\pi_F^{j'}}(\Theta\epsilon_0\pi_F^{j'})^2 D_0 \equiv \epsilon\frac{b}{\pi_F^j}(\pi_F^j)^2 D_0$$

in $R = o_F/(\pi_F^i)$ for some $\epsilon \in R_{\pi_F^{2j}D_0}$. Notice $R_{\pi_F^{2j}D_0} = (o_F^*)^2$ unless $j = 0$ and $ord(D_0) = 0$, in which case this is o_F^*.

Notice that we use the following notation $N = min(f, f')$, $\nu = f - j$, $\nu' = f' - j'$, $\chi = ord(a - a')$. With this notation the first two summation conditions of Lemma 6.8 become the next conditions (1) and (2), whereas the third condition becomes – using the other two conditions – condition (3).

The Summation Conditions.

1. $0 \leq i \leq \chi$.
2. $\nu' \geq i \Longleftrightarrow \nu \geq i$. Furthermore, if $\nu < i$, then $\nu = \nu'$ must hold, and – unless L/K is unramified and $\nu = f - \Theta\epsilon B'/B \in (R^*)^2$ must also hold.
3. $2f + ord(D_0) - \nu \geq i \Longleftrightarrow 2f' + ord(D_0) - \nu' \geq i$. If $2f + ord(D_0) - \nu < i$, then $f = f'$ and $\nu = \nu'$ and $i \leq \chi - \nu$ must hold.

The third condition needs some explanation. For this observe that the conditions

$$(b/\pi_F^j)\pi_F^{2j} D_0 \equiv -(b'/\pi_F^{j'})\pi_F^{2j'} D_0\Theta\epsilon_0\epsilon^{-1} \ mod \ \pi_F^i$$

(third congruence) and

$$(b/\pi_F^j) \equiv -(b'/\pi_F^{j'}\Theta\epsilon_0)\epsilon \ mod \ \pi_F^i$$

for some $\epsilon \in N_{\pi_F^{2j} D_0}$ (second congruence) can be combined and thus simplified.

Since $(b/\pi_F^j)\pi_F^{2j} D_0 = B\pi_F^{\nu+(2f-2\nu)} D_0 = BD_0\pi_F^{2f-\nu}$ holds, and similarly to the other case, the inequality $2f + ord(D_0) - \nu < i$ implies $2f' + ord(D_0) - \nu' < i$ and vice versa by the third congruence. Suppose this holds. Then the third congruence even implies the equality $2f + ord(D_0) - \nu = 2f' + ord(D_0) - \nu'$. Since $\nu \leq f$ and therefore $\nu \leq 2f - \nu + ord(D_0)$, the assumed condition $2f + ord(D_0) - \nu < i$ implies $\nu < i$; hence, $\nu = \nu'$ by condition (2). Combined with $2f + ord(D_0) - \nu = 2f' + ord(D_0) - \nu'$ this gives equivalently $f = f'$. Hence, for $2f + ord(D_0) - \nu < i$ the third condition becomes: $f = f'$ and $\nu = \nu'$ plus the condition

$$\epsilon \equiv -\Theta\epsilon_0 B'/B \ mod \ \pi_F^{i-2f-ord(D_0)+\nu},$$

where

$$\epsilon^{-1} \equiv -(\Theta\epsilon_0)^{-1} B'/B \ mod \ \pi_F^{i-\nu}.$$

Since $i - 2f - ord(D_0) + \nu \leq i - \nu$, the extra condition is $(B'/B)^2 = 1 \ mod \ \pi_F^{i-2f-ord(D_0)+\nu}$, or in other words

$$i - 2f - ord(D_0) + \nu = i - 2N - ord(D_0) + \nu \leq ord\left(B^2 - (B')^2\right).$$

Since $\nu \leq f$, the inequality $2f + ord(D_0) - \nu < i$ forces $i > 0$; hence, $\chi = ord(a - a') > 0$ by condition (1). Now from $\chi > 0$ we get $\chi = ord(D_0(b^2 - (b')^2))$ by Lemma 6.6, except for $f = f' = ord(D_0) = 0$ and therefore $\nu = 0$. Since we already know $f = f'$, Lemma 6.6 implies

$$\chi = 2N + ord(D_0(B^2 - (B')^2)).$$

Therefore, the extra condition can be reformulated into

$$2f + ord(D_0) - \nu < i \quad \text{implies} \quad f = f' \, \nu = \nu' \text{ and } i \le \chi - \nu.$$

In the exceptional case $f = f' = ord(D_0) = \nu = 0$, Lemma 6.6 only gives the inequality $\chi \le 2N + ord(D_0(B^2 - (B')^2))$. Since $i \le \chi$ by condition (1), this implies that the extra condition $i \le 2N + ord(D_0(B^2 - (B')^2)) - \nu$ automatically holds. This provides evidence that conditions (1)–(3) are the correct summation conditions.

6.11 Résumé

We now express the results of the section on $T(F) \setminus H(F)$ integration in terms more suitable for summation. From the summation conditions (1)–(3) in Sect. 6.10, the domain of possible values i can be divided into three disjoint ranges:

(A) **Summation conditions for** i (apart from $0 \le i \le \chi$):
   Range of small i: $0 \le i \le \nu$ and $0 \le i \le \nu'$.
   Middle range i: $\nu < i \le 2f + ord(D_0) - \nu$ and $\nu' < i \le 2f' + ord(D_0) - \nu'$.
   In the middle range $\nu = \nu'$ must be imposed by summation condition (2) in the last section.
   Large range i: $2f + ord(D_0) - \nu < i \le \chi - \nu$ and $2f' + ord(D_0) - \nu' < i \le \chi - \nu'$.
   In the large range $\nu = \nu'$ and $f = f'$ must be imposed by summation condition (3). However, since $0 \le 2f + ord(D_0) < \chi$, the extra condition $f = f'$ can be dropped, since it automatically holds (Lemma 6.6(4)).

(B) **Summation conditions for** ν, ν', **and** θ:
   In the domain of possible values ν, ν' we have, in addition to restrictions mentioned above, the conditions

$$0 \le \nu \le f, \qquad 0 \le \nu' \le f',$$

$$\nu - \nu' = f - f' + ord(\theta) \bmod 2 \qquad \text{(only in the unramified case)}.$$

Furthermore, in the middle and in the large range, with respect to the variable i, we have contributions to the orbital integral only for

$$-\Theta\epsilon_0 B'/B \in (o_F^*)^2,$$

unless both conditions $\nu = f$ and L/F is unramified hold.

(C) **Contributions from the** $index(\nu, \nu') = [o_T : (T(F) \cap rK_0r^{-1})]$: See Lemma 6.11 and Sects. 6.14 and 6.15.
   Since $j = f - \nu, j' = f' - \nu'$, the relevant contributions are

$$1 \qquad (\nu, \nu') = (f, f'),$$

$$\frac{q+1}{q} q^{f+f'-\nu-\nu'} \qquad \nu = f, \nu' \ne f' \text{ or } \nu' = f', \nu \ne f,$$

$$\frac{1}{2}\frac{(q+1)^2}{q^2}q^{f+f'-\nu-\nu'}\qquad \nu < f, \nu' < f',$$

in the *unramified* case and

$$q^{f+f'-\nu-\nu'}$$

in the *ramified* case.

(D) **Contributions from the** $centralizer(\nu, i)$: See Corollary 6.2, Sect. 6.8, and Sects. 6.14 and 6.15.

$$(q^2 - 1)q^{3i-2} \text{ for } i \geq 1, \text{ otherwise } 1 \quad (i \leq \nu \text{ small range}),$$

$$2q^{2\nu+i} \quad (i > \nu \text{ middle and large range}),$$

unless

$$(q+1) \cdot q^{2\nu+i-1} \quad (\nu = f, \nu' = f', L/F \text{ unramified}).$$

In principle we could now compute the orbital integrals. We leave this computation as an exercise to the reader. We only give the formula for the κ-orbital integrals. Since this is a difference of orbital integrals, some of the summation terms cancel, and hence the formula becomes simpler.

6.12 The Summation (L/F-Ramified)

Assume $\eta = (s, s')$ satisfies $x, x' \in o_L^*$, since otherwise the κ-orbital integral is zero. For the κ-orbital integral the contributions from the small range of the i-summation cancel in the *ramified* case, since their contribution is stable (i.e., does not depend on θ). Only the middle and the large range contributions remain. The whole summation is over the empty set unless $\chi > 0$, i.e., unless $x = x' \bmod \pi_L$ holds in addition to $x, x' \in o_L^*$. For this notice that $\chi = 0$ implies $i = 0$, which is in the stable range. So we are in the situation where we can apply Lemma 6.6.

The quadratic character $\chi_{L/F}$ of F^* attached to L/F is trivial on $(o_F^*)^2$ with $\chi_{L/F}(\Theta) = -1$. Notice $ord(D_0) = 0$ and $\epsilon_0 = 1$ and $\theta = \Theta \in o_F^*$.

Instability enters via $-\Theta B'/B \in (o_F^*)^2$ imposed by summation condition (B). Since B'/B is fixed, only one of the two orbital integrals contributes to the κ-orbital integral with the corresponding sign $\chi_{L/F}(-B/B')$. Summing up the terms $q^{f+f'-\nu-\nu'} \cdot 2q^{2\nu+i} = 2q^{f+f'}q^i$ over all indices $i, r_{j,j'} \in R$ for $j = f - \nu, j' = f' - \nu'$ from the middle and large range (therefore $\nu = \nu'$) gives

$$O_\eta^\kappa(1_K) = 2q^{f+f'}\chi_{L/F}(-B/B') \cdot 1_{o_L^* \times (1+\pi_L o_L)}(x, x/x') \sum_{0 \leq \nu \leq N} \sum_{\nu < i \leq \chi - \nu} q^i.$$

The double sum on the right gives

$$(q-1)^{-1} \sum_{0 \le \nu \le N} (q^{\chi+1-\nu} - q^{\nu+1})$$

$$= (q-1)^{-1}[q^{\chi+1}(q^{-N-1}-1)/(q^{-1}-1) - q(q^{N+1}-1)/(q-1)]$$

$$= \frac{q}{(q-1)^2}(1-q^{N+1})(1-q^{\chi-N+1}) = q \int_{Gl(2,F)^2/(L^*)^2} 1_K(g^{-1}(t_1,t_2)g)dg/dl$$

by Lemma 6.6(1) and Lemma 6.4. Recall $b = \pi_F^f B$ for $b = (x - x^\sigma)/2\sqrt{D_0}$ and $\pi_F \in Norm(L^*)$, and similarly for b'. Hence, $O_\eta^\kappa(1_K)$ is equal to

$$2\Big|\frac{x-x^\sigma}{2\sqrt{D_0}}\Big|^{-1}\Big|\frac{x'-(x')^\sigma}{2\sqrt{D_0}}\Big|^{-1}\chi_{L/F}\Big(\frac{x^\sigma-x}{2\sqrt{D_0}}\Big)\chi_{L/F}\Big(\frac{x'-(x')^\sigma}{2\sqrt{D_0}}\Big) \cdot q \times$$

$$\int_{Gl(2,F)^2/(L^*)^2} 1_K(g^{-1}(t_1,t_2)g)dg/dl.$$

This implies Theorem 6.1 for ramified L/F. The factor 2 disappears since in the ramified case $[o_T : \rho((o_L^*)^2)] = 2$. Furthermore, $|D_0|q = 1$ and $\chi_{L/F}(D_0) = \chi_{L/F}(-1)$ since $-D_0 = Norm_{L/F}(\sqrt{D_0})$ and $|xx'| = 1$ holds on the support. Therefore, the factor in front becomes $(|.|\chi_{L/F})\big((x-x^\sigma)(x'-x'^\sigma)\big)$.

6.13 The Summation (L/F-Unramified)

To compute the κ-orbital integral we concentrate on the case $\chi > 0$. The case $\chi = 0$ is very simple because only the small range summation contributes with the value $(-q)^{f+f'}$. Let us therefore turn to the essential case $\chi > 0$. From the résumé we get the following contributions:

1. **Small range:** Its contribution will be

$$SRC = \sum_{0 \le \nu \le f} \sum_{0 \le \nu' \le f'} \sum_{0 \le i \le min(\nu,\nu')} index(\nu,\nu')\#centralizer(i)$$

$$= \sum_{0 \le i \le N} \#centralizer(i) \left(\left(\sum_{i \le \nu < f} \sum_{i \le \nu' < f'} \frac{1}{2} \cdot 2 \cdot \frac{(q+1)^2}{q^2}(-q)^{f+f'-\nu-\nu'} \right) \right.$$

$$+ \left(\sum_{i \le \nu' < f'} \frac{q+1}{q}(-q)^{f'-\nu'} \right)$$

$$\left. + \left(\sum_{i \le \nu < f} \frac{q+1}{q}(-q)^{f-\nu} \right) + 1 \right).$$

The representatives ϵ_0 appearing in R give the additional factor 2 in the first sum. So

$$SRC = \sum_{0 \leq i \leq N} \#centralizer(i) \cdot S(i, f, f'),$$

where $\#centralizer(i) = 1$ and $(q^2 - 1)q^{3i-2}$ for $i > 0$ and $S(i, f, f') = (S(i, f) + 1)(S(i, f') + 1)$ with

$$S(i, f) = \frac{q+1}{q} \sum_{i \leq \nu < f} (-q)^{f-\nu} = (-q)^{f-i} - 1.$$

This gives

$$SRC = (-q)^{f+f'} \left(1 + \frac{q^2 - 1}{q^2} \sum_{0 < i \leq N} q^i \right) = (-q)^{f+f'} \left(1 + \frac{q+1}{q}(q^N - 1) \right).$$

2. **Middle range contribution:** Now summation condition (B) chooses one of the representatives ε_0. From the summation condition $\nu = \nu'$ we get

$$MRC = \sum_{0 \leq \nu \leq N} \sum_{\nu < i \leq 2N - \nu} (-1)^{f+f'-2\nu} \cdot index(\nu) \cdot \#centralizer(\nu, i).$$

The summation condition forces $\nu < N$; hence,

$$MRC = \sum_{0 \leq \nu < N} \sum_{\nu < i \leq 2N - \nu} \left(\frac{1}{2} \frac{(q+1)^2}{q^2} (-q)^{f+f'-2\nu} \right) \cdot \left(2q^{2\nu+i} \right)$$

$$= (-q)^{f+f'} \frac{(q+1)^2}{q^2} \sum_{0 \leq \nu < N} \sum_{\nu < i \leq 2N - \nu} q^i$$

$$= (-q)^{f+f'} \frac{(q+1)^2}{(q-1)^2 q} \left(q^{2N+1} - q^{N+1} - q^N + 1 \right).$$

3. **Large range contribution:** We have $f = f' = N$ and $\nu = \nu'$. This gives

$$LRC = \sum_{0 \leq \nu \leq N} \sum_{2N - \nu < i \leq \chi - \nu} (-1)^{2N+2\nu} \cdot index(i) \cdot \#centralizer(\nu, i)$$

$$= \sum_{0 \leq \nu < N} \sum_{2N - \nu < i \leq \chi - \nu} \left(\frac{1}{2} \frac{(q+1)^2}{q^2} q^{2N-2\nu} \right) \cdot \left(2q^{2\nu+i} \right) + \sum_{N < i \leq \chi - N} 1 \cdot \frac{q+1}{q} q^{2N+i}$$

$$= \frac{(q+1)(-q)^{2N}}{q} \left(\frac{q+1}{q} \left(\sum_{0 \leq \nu < N} \frac{q^{\chi-\nu+1} - q^{2N-\nu+1}}{q-1} \right) + \left(\frac{q^{\chi-N+1} - q^{N+1}}{q-1} \right) \right)$$

$$
= (-q)^{f+f'} \frac{q+1}{(q-1)^2 q} \Big((q-1)(q^{\chi-N+1} - q^{N+1}) + (q+1)(q^{\chi+1}
$$
$$
+ q^{N+1} - q^{\chi-N+1} - q^{2N+1}) \Big).
$$

The sum of the contributions from (1)–(3) is

$$
SRC + MRC + LRC = \frac{(-q)^{f+f'}}{(q-1)^2} \Big((q+1)q^N - 2 \Big) \Big((q+1)q^{\chi-N} - 2 \Big).
$$

Since $\chi > 0$, by Lemma 6.6(1) and Lemma 6.4 the product over the two $Gl(2, F)$-orbital integrals is equal to

$$
\frac{1}{(q-1)^2} \Big((q+1)q^N - 2 \Big) \Big((q+1)q^{\chi-N} - 2 \Big).
$$

The factor $(-q)^{f+f'}$ in front is $(|.|\chi_{L/F})^{-1}(\frac{x-x^\sigma}{\sqrt{D_0}} \frac{x-x^\sigma}{\sqrt{D_0}})$. Since $D_0 \in o_F^*$ and therefore $(|.|\chi_{L/F})(D_0) = 1$, this proves Theorem 6.1 in the unramified case.

6.14 Appendix on Orders (L/F-Ramified)

Let the situation be as above. In particular, the residue characteristic is different from 2. Then the orders $o_L(n) = o_F + o_F \pi_F^n \sqrt{D_0} = o_F + \pi_F^n o_L \subseteq o_L$ (for integers $n > 0$ by assumption) have the following properties:

1. $Norm(o_L(n)^*) = (o_F^*)^2$.
   This is clear from $(o_F^*)^2 \subseteq Norm(o_L(n)^*) \subseteq Norm(o_L^*) \subseteq (o_F^*)^2$.
2. $[o_L^*(n) : (1 + \pi_F^n o_L)] = [o_F^* : (o_F^* \cap (1 + \pi_F^n o_L)] = [o_F^* : (1 + \pi_F^n o_F)] = (q-1)q^{n-1}$.
3. $[o_L^* : o_L^*(n)] = [o_L^* : (1 + \pi_F^n o_L)]/[o_L^*(n) : (1 + \pi_F^n o_L)] = (q-1)q^{2n-1}/(q-1)q^{n-1} = q^n$.
4. $[o_T : (T(F) \cap rK_0 r^{-1})] = [o_L^* : o_L^*(j)][o_L^* : o_L^*(j')] = q^{j+j'}$, since $[Norm(o_L^*) : Norm(o_L(min(j, j')))] = 1$.
   Finally, for $R = o_F/\pi_F^i o_F$ and $i > 0$:
5. $\#(Sl(2, R)) = (q-1)(q+1)q^{3i-2} = (q^2-1)q^{3i-2}$.
6. $\#Image \Big(\phi_{\sqrt{D}} : o_L^* \to Gl(2, R) \Big) = q^i \cdot (q-1)q^{i-1} = (q-1)q^{2i-1}$ for $i > 0$.

6.15 Appendix on Orders (L/F-Unramified)

Let the situation be as above with residue characteristic different from 2. Then the orders $o_L(n) \subseteq o_L$ have the following properties:

1. $Norm(o_L(n)^*) = (o_F^*)^2$ if $n > 0$ and $Norm(o_L^*) = o_F^*$ and $[o_F^* : (o_F^*)^2] = 2$.
   This is clear from $(o_F^*)^2 \subseteq Norm(o_L(D)^*) \subseteq Norm(o_L^*) = (o_F^*)$.

2. $[o_L^*(n) : (1 + \pi_F^n o_L)] = [o_F^* : (o_F^* \cap (1 + \pi_F^n o_L)] = [o_F^* : (1 + \pi_F^n o_F)]$
   $= (q - 1)q^{n-1}$ for $n > 0$.

3. $[o_L^* : o_L^*(n)] = [o_L^* : (1 + \pi_F^n o_L)]/[o_L^*(n) : (1 + \pi_F^n o_L)] = (q^2 - 1)q^{2(n-1)}/(q - 1)q^{n-1} = (q + 1)q^{n-1}$ for $n > 0$.

From (1) and (3) we obtain:

4. $[o_T : (T(F) \cap rK_0 r^{-1})] = [o_L^* : o_L(j)^*][o_L^* : o_L^*(j')][Norm(o_F^*) : Norm(o_L^*(min(j, j')))]^{-1}$ is

$$1 \quad \text{for} \quad j = j' = 0,$$

$$(q + 1)q^{j+j'-1} \quad \text{if exactly one of the } j, j' \text{ is zero},$$

$$\frac{1}{2}(q + 1)^2 q^{j+j'-2} \quad \text{for} \quad j, j' > 0.$$

Finally for $R = o_F/\pi_F^i o_F$ and $i > 0$:

5. $\#(Sl(2, R)) = (q - 1)(q + 1)q^{3i-2} = (q^2 - 1)q^{3i-2}$.

6. $\#Image\left(\phi_{\sqrt{D}} : o_L^* \to Gl(2, R)\right) = \#\left\{(u, v) \in R \mid Norm(u + v\sqrt{D}) \in R^*\right\}$. For $i > 0$ this number is

$$(q - 1)q^{2i-1}, \qquad \pi_F | D,$$

$$(q^2 - 1)q^{2i-2}, \qquad D \in o_F^*.$$

6.16 Appendix on Measures

Let G be a locally compact unimodular totally disconnected group, H a closed subgroup, and K compact open in G. Let

$$G = \coprod H \cdot x \cdot K$$

be a disjoint double coset decomposition with countably many representatives x. Let dg, dh, dk be Haar measures on G, H, K such that $vol_G(K) = 1$ and $vol_K(K) = 1$. Then according to [109], p. 478

$$\int_{HxK} f(g)dg = const' \int_{H/H \cap (xKx^{-1})} dh/dh_x \int_K f(hxk)dk$$

$$= const \int_H (\int_K f(hxk)dk)dh.$$

The constants depend on the choice of measures dh_x on $H \cap (xKx^{-1})$ and dh on H. Evaluating with the function $f(g) = 1_{xK}(g)$ gives

$$1 = const \int_H 1_{H \cap xKx^{-1}}(h)dh = const \cdot vol_H(H \cap xKx^{-1}).$$

If dh is *normalized* such that $vol_H(K_H) = 1$, then

$$const = vol_H(H \cap xKx^{-1})^{-1} = [K_H : (H \cap xKx^{-1})].$$

It follows that

$$\int_{T \backslash H} 1_K(g^{-1}\eta g)\frac{dg}{dt} = \sum_x [K_H : (H \cap xKx^{-1})] \cdot \int_{T \backslash G} 1_{H \cap xKx^{-1}}(h^{-1}\eta h)\frac{dh}{dt}$$

since $x^{-1}h^{-1}\eta hx \in K \iff h^{-1}\eta h \in H \cap xKx^{-1}$, and since

$$\frac{vol_G(K)}{vol_H(H \cap xKx^{-1})} = [K_H : (H \cap xKx^{-1})].$$

Chapter 7
A Special Case of the Fundamental Lemma II

7.1 The Torus T

In this section F is a non-Archimedean local field of residue characteristic not equal
to 2. Let E/F be a field extension of degree 4 with an involution σ which has fixed
field E^+ such that $[E : E^+] = 2$. Let q be the number of elements in the residue
field of E^+. Prime elements will be denoted π_{E^+}. We write $E^+ = F(\sqrt{A})$ for
some element $A \in F^*$ assuming $A = A_0$ to be *normalized*, i.e., chosen integral
with minimal possible order. Then $o_{E^+} = o_F + o_F\sqrt{A}$.

The Group H. Consider the subgroup $H(F) \subseteq GSp(4, F)$ of all matrices

$$
\eta = \begin{pmatrix}
\alpha_1 & \alpha_2 A^{-1} & \beta_1 & \beta_2 \\
\alpha_2 & \alpha_1 & \beta_2 & \beta_1 A \\
\gamma_1 & \gamma_2 A^{-1} & \delta_1 & \delta_2 \\
\gamma_2 A^{-1} & \gamma_1 A^{-1} & \delta_2 A^{-1} & \delta_1
\end{pmatrix}.
$$

This defines an algebraic F-group H such that

$$\phi : Gl(2, E^+)^0 = \{g \in Gl(2, E^+) \mid det(g) \in F^*\} \cong H(F).$$

The inverse isomorphism ϕ^{-1} maps the above matrix $\eta = \phi(s)$ to the matrix

$$s = \begin{pmatrix} a & b \\ c & d \end{pmatrix} \in Gl(2, E^+)^0,$$

where $a = \alpha_1 + \alpha_2\sqrt{A^{-1}}, b = \beta_1 + \beta_2\sqrt{A^{-1}}, c = \gamma_1 + \gamma_2\sqrt{A^{-1}}, d = \delta_1 + \delta_2\sqrt{A^{-1}}$.

Maximal Tori. Consider the F-torus $T \subseteq Res_{E/F}(\mathbb{G}_m)$ such that

$$T(F) = \{x \in E^* \mid Norm_{E/E^+}(x) = xx^\sigma \in F^*\}.$$

R. Weissauer, *Endoscopy for GSp(4) and the Cohomology of Siegel Modular Threefolds*,   239
Lecture Notes in Mathematics 1968, DOI: 10.1007/978-3-540-89306-6_7,
© Springer-Verlag Berlin Heidelberg 2009

T can be embedded into H and G over the field F. Its image in $G = GSp(4)$ is a maximal torus denoted T_G. Any F-embedding of T into G is conjugate under $G(F)$ to one of the following

Standard Embeddings. Choose $D \in E^+$ such that $E = E^+(\sqrt{D})$. For simplicity assume D to be *integral*. Embed T by the D-regular embedding, mapping elements

$$x = a + b\sqrt{D}, \qquad a, b \in E^+$$

of $T(F)$ to

$$s = \phi_{\sqrt{D}}(x) = \begin{pmatrix} a & Db \\ b & a \end{pmatrix} \in Gl(2, E^+)^0.$$

Via the isomorphism above, we can view T as a maximal F-subtorus of H and G.

Lemma 7.1. *Two such standard embeddings are conjugate as embeddings into G or H if and only if the quotient $\theta = \sqrt{D'}/\sqrt{D}$ is in $F^* \cdot Norm_{E/E^+}(E^*)$.*

Proof. There is an exact sequence $T_1 \hookrightarrow T \twoheadrightarrow \mathbb{G}_m$, where T_1 is the kernel of the surjection $Norm_{E/E^+} : Res_{E/F}(\mathbb{G}_m) \to Res_{E^+/F}(\mathbb{G}_m)$. So $H^1(F, T_1) = (E^+)^*/Norm_{E/E^+}(E^*)$. By Hilbert 90 then $H^1(F, T) = H^1(F, T_1)/F^* = (E^+)^*/(F^* \cdot Norm_{E/E^+}(E^*))$. Similarly to Lemma 6.1 $H^1(F, H) = H^1(F, G) = 1$; hence, the embeddings of T into H and G up to conjugation are parameterized by $H^1(F, T)$, any embedding is conjugate to a standard embedding, and two standard embeddings are conjugate in G under $G(F)$ if and only if they are conjugate in H under $H(F)$. Now it is easy to see that two standard embeddings $T \hookrightarrow H$ are $H(F)$-conjugate if and only if θ is in $F^* \cdot Norm_{E/E^*}(E^*)$. $\square$

Orbital Integrals. Our aim is to calculate the $GSp(4, F)$ orbital integral

$$O_\eta^G(1_K) = \int_{G_\eta \backslash G} 1_K(g^{-1} \eta g) dg/dg_\eta$$

for the *unit element* 1_K of the Hecke algebra (equivalent to characteristic functions of the group $K = K_G$ of the unimodular symplectic similitudes). Measures are normalized by $vol_G(K) = 1$ and $vol_T(o_T) = 1$, where o_T is the subgroup of all $x \in T(F)$ with $x \in o_E^*$. By transport of structure this defines a unique Haar measure on $T \cong T_G \cong G_\eta$. Here we assume *regularity* in the sense that

$$E = F(x).$$

Then we calculate the κ-orbital integrals: The subgroup $F^* \cdot Norm_{E/E^+}(E^*)$ has at most index 2 in $(E^+)^*$. We are interested only in the unstable case, where the two groups are not equal. Assume this is the case

$$(E^+)^* = F^* \cdot Norm_{E/E^+}(E^*) \quad \uplus \quad \theta \cdot F^* \cdot Norm_{E/E^+}(E^*).$$

Then

$$O_\eta^\kappa(1_K) = O_\eta^G(1_K) - O_{\eta'}^G(1_K)$$

defines the κ-orbital integrals $O_\eta^\kappa(1_K)$, where

$$\eta = \phi \circ \phi_{\sqrt{D_0}}(x), \qquad \eta' = \phi \circ \phi_{\theta\sqrt{D_0}}(x), \qquad (x \in T),$$

and where $\theta \in (E^+)^*$ is chosen as above.

The Unstable Case. For local fields E^* considered above $(E^+)^* = F^* \cdot Norm_{E/E^+}(E^*)$ unless E/F is a noncyclic Galois extension. Since we are only interested in κ-orbital integrals, we will restrict ourselves from now on to that particular case.

By assumption, the residue characteristic of F is different from 2; hence, the non-cyclic Galois extension E/F is the composite field of the three nonisomorphic quadratic extension fields of F in $\overline{F}$. Exactly one of these three extension fields L, L', E^+ is unramified over F, and hence exactly one of the two extensions E/E^+, E^+/F is ramified or unramified, respectively.

The Galois group $Gal(E/F)$ has four elements $1, \sigma, \tau,$ and $\sigma\tau = \tau\sigma$. Consider fixed fields. Let us denote

$$L = E^{\langle\tau\rangle} = F(\sqrt{D_0}),$$

$$L' = E^{\langle\tau\sigma\rangle} = F(\sqrt{A_0 D_0}),$$

and $E^+ = E^{\langle\sigma\rangle}$ as the three fixed fields of E. If one of the extensions L/F or L'/F is unramified, we can assume that L/F is the *unramified* extension. Notice the apparent asymmetry of this normalization!

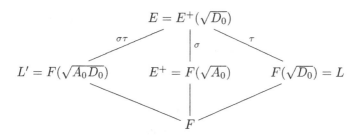

We remark:

- E/E^+ is ramified if and only if E^+/F is unramified if and only if L'/F and L/F are ramified.
- Furthermore, if E/E^+ is unramified, then L/F is unramified (by our choice) and E^+/F and L'/F are ramified.
- In particular, L'/F is always ramified.

Notation. Choose $D = D_0 \in F^*$ to be normalized of minimal order 0 or 1. This does not imply that $D = D_0\theta^2$ is also normalized or in F^*, since $ord(\theta) = 1$ if E/E^+ is unramified. We fix prime elements $\pi_{E^+} = -D_0$, and $\pi_{E^+} = \sqrt{A_0}$ according to whether E/E^+ is ramified or not. The valuation on F is chosen such that $|\pi_F| = q_0^{-1}$, where q_0 is the cardinality of the residue field of F. Let q denote the cardinality of the residue class field of E^+. We extend the valuation of F to $\overline{F}$.

7.2 The Endoscopic Group M

The torus T is the pullback of the maps $Norm : E^* \to (E^+)^*$ and the embedding $F^* \to (E^+)^*$:

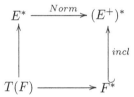

Therefore, the dual group $\hat{T}$ is the pushout

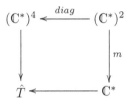

of the diagonal map $(\mathbb{C}^*)^2 \to (\mathbb{C}^*)^4$ and the multiplication map $m : (\mathbb{C}^*)^2 \to \mathbb{C}^*$. This gives $\hat{T} \cong (\mathbb{C}^*)^4/\{(t, t, t^{-1}, t^{-1})\}$, where the Galois group $G(E/F)$ acts by the generator σ of $Gal(E/E^+)$ permuting, say, the first and second and the third and fourth components, respectively. Let $\tau \neq \sigma$ be the second involution in $G(E/F)$ with fixed field $L = F(\sqrt{D_0})$, where we assumed $D_0 \in F^*$ to be normalized. Then τ acts by permuting, say, the first and third and the second and fourth components. The map $(z_1, z_2, z_3, z_4)/(t, t, t^{-1}, t^{-1}) \mapsto (z_1 z_4, z_2 z_4, z_3/z_4)$ induces an isomorphism $\hat{T} \cong (\mathbb{C}^*)^3$. The action of σ is transported to

$$\sigma(t_1, t_2, t_3) = (t_2 t_3, t_1 t_3, t_3^{-1}).$$

The action of τ is transported to

$$\tau(t_1, t_2, t_3) = (t_2 t_3, t_2, t_1/t_2).$$

This defines an embedding of L-groups

$$\psi : {}^L T \to {}^L GSp(4, F) = GSp(4, \mathbb{C}) \times W_F$$

by

$$\psi(t_1, t_2, t_3) = diag(t_1, t_2, t_2 t_3, t_1 t_3)$$

and

$$\psi(\sigma) = i \begin{pmatrix} 0 & E \\ -E & 0 \end{pmatrix} \times \sigma \quad , \quad \psi(\tau) = \begin{pmatrix} 0 & 0 & i & 0 \\ 0 & 1 & 0 & 0 \\ -i & 0 & 0 & 0 \\ 0 & 0 & 0 & 1 \end{pmatrix}.$$

Identify $\hat{T}$ with its image under ψ. Then the center $Z(\hat{G})$ of the dual group $\hat{G}$ is the group of all $(t_1, t_2, t_3) = (x, x, 1)$ and $\hat{T}^\Gamma = Z(\hat{G}) \cup Z(\hat{G})\kappa$, where $\kappa = (1, -1, -1) \in \hat{T}$. The centralizer of $s = \psi(\kappa) = diag(1, -1, 1, -1)$ is the group

$$\hat{M} = \hat{M}^0 = \left\{ \begin{pmatrix} * & 0 & * & 0 \\ 0 & * & 0 & * \\ * & 0 & * & 0 \\ 0 & * & 0 & * \end{pmatrix} \subseteq GSp(4, \mathbb{C}) \right\}.$$

Consider $^L M = \hat{M} \times \Gamma$ as a subgroup of $^L GSp(4, F)$ in the trivial way. The embedding $^L M \hookrightarrow {}^L G$ is denoted ξ. This defines an elliptic endoscopic datum $(M, {}^L M, s, \xi)$. The group M is $Gl(2, F) \times Gl(2, F)/\{(t, t, t^{-1}, t^{-1}) \mid t \in F^*\}$.

Since $\psi(\sigma), \psi(\tau)$ are contained in $^L M$, the morphism ψ factors in the evident way:

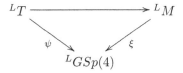

Consider tori $T_M \subseteq M$, uniquely defined up to conjugation by $M(F)$ by $T_M(F) = (L^* \times (L')^*)/F^*$ in $M(F) = (Gl(2, F)^2)/F^*$. The isomorphism

$$(L^* \times (L')^*)/F^* \cong T(F)$$

defined by the map $(t_1, t_2)/(t^{-1}, t) \mapsto t_1 t_2 \mapsto \phi \circ \phi_{\sqrt{D}}(t_1 t_2)$ gives an admissible embedding of the torus, which on the level of F-rational points is given by

$$\rho : T_M(F) \cong (L^* \times (L')^*)/F^* \to T(F) \to T_G(F) \subseteq GSp(4, F).$$

Let us describe the admissible embedding $T_M \hookrightarrow T_G$. We have the following chain of embeddings:

$$M(F) = Gl(2, F)^2/F^* \hookleftarrow T_M(F) \cong T(F) = (L^*) \times (L')^*/F^* \xrightarrow{\iota} E^0 \subseteq E,$$

$$E^0 \xrightarrow{\phi_{\sqrt{D_0}}} Gl(2, E^+)^0 \xrightarrow{\phi} H(F) \subseteq G(F) = GSp(4, F).$$

The exponent ^{0} indicates that the relative norm from E to E^+ or the determinant has values in F. The maps involved are, from left to right, as follows. Firstly, the embedding defined by a regular representation identifying $Gl(2, F)$ and $Gl(L/F)$, which is unique up to conjugation. Then the map ι defined by

$$\iota : (t_1, t_2)/\sim \quad \mapsto \quad x = t_1 t_2$$

which has its image in E^0. Thirdly, the map

$$\phi_{\sqrt{D_0}} : x \mapsto s = \begin{pmatrix} a & D_0 b \\ b & a \end{pmatrix},$$

where the coefficients $a, b \in E^+$ are determined by writing $x = t_1 t_2 = a + b\sqrt{D_0}$, where $L = F(\sqrt{D_0})$ and all elements x, t_1, t_2 are viewed as elements of the field $E = LL'$ of degree 4 over F. Finally, the isomorphism ϕ of $Gl(2, E^+)^0$ onto its image in $GSp(4, F)$. As a result we obtain the element $\eta \in GSp(4, F)$ attached to (t_1, t_2).

The torus T_G in $GSp(4)$ is defined by

$$T_G = \phi \circ \phi_{\sqrt{D_0}} \circ \iota(T_M).$$

Δ_{IV}. For comparison with [60] we record for $x = t_1 t_2 \in T(F)$

$$D_G(x) = |1 - x/x^\tau||1 - x/x^\sigma||1 - x^\tau/x^{\tau\sigma}||1 - x^\tau/x^\sigma|,$$

$$D_M(t_1, t_2) = |1 - t_1/t_1^\sigma||1 - t_2/t_2^\sigma| = |1 - x/x^{\sigma\tau}||1 - x/x^\tau|.$$

Hence, $\Delta_{IV} = D_G(x)/D_M(t_1, t_2) = |1 - x/x^\sigma||1 - x^\tau/x^\sigma| = |(1 - x/x^\sigma)(1 - x/x^\sigma)^\tau|$. Since $(1 - x/x^\sigma)(1 - x/x^\sigma)^\tau| \in E^{\langle\tau\rangle} = L$ and since L/F is unramified, we conclude

Lemma 7.2. $D_G(x)/D_M(t_1, t_2)$ *has values in* $\mathbb{Q}$.

7.3 Orbital Integrals on M

Orbital integrals for M are stable:

$$SO^M_{(t_1, t_2)}(1_K) = \int_{T_M(F)\backslash M(F)} 1_{K_M}(g^{-1}(t_1, t_2)g)dg/dt.$$

Here 1_{K_M} is the unit element of the Hecke algebra of the group K_M defined in Chap. 6. The measure on $M(F)$ is chosen as in Chap. 6, on T_M by transport of structure from the unique Haar measure on T, such that $vol_T(o_T) = 1$.

We will show by explicit computation

Theorem 7.1. *Assume the residue characteristic of F is different from 2. Suppose $t_1 \in L^*, t_2 \in (L')^*$. Assume that $x = t_1 t_2 \in T(F)$ is regular in the sense that $E = F(x)$. Then $\eta = \phi \circ \phi_{\sqrt{D_0}}(x) \in T_G(F)$ is regular with image $(t_1, t_2) \in T_M(F)$ under ρ^{-1}. Then*

$$\Delta(\eta, (t_1, t_2)) O_\eta^\kappa (1_{K_G}) = SO_{(t_1, t_2)}^M (1_{K_M}),$$

for the transfer factor

$$\Delta(\eta, (t_1, t_2)) = \chi_{L/F}(A_0) |N_{E/F}(x)|^{-1/2} (\chi_{L/F} \cdot |)((x - x^\sigma)(x^\tau - x^{\sigma\tau})).$$

$\chi_{L/F}$ *is the quadratic character of F^* attached to the field extension L/F.*

Since $\chi_{L/F}$ is quadratic, we obtain from Lemma 7.2

Corollary 7.1. *The transfer factor $\Delta(\eta, (t_1, t_2))$ has values in $\mathbb{Q}$ and satisfies*

$$\Delta(\lambda\eta, (\lambda t_1, \lambda t_2)) = \Delta(\eta, (t_1, t_2))$$

for all $\lambda \in F^*$.

Lemma 7.3. *The transfer factor $\Delta(\eta, t)$ only depends on $x = t_1 t_2$, and hence is invariant under the involution $t = (t_1, t_2) \mapsto t^* = (t_2, t_1)$. Furthermore, the admissible embedding map $T_M(F) \ni t \mapsto \eta \in T_G(F)$ defines an admissible embedding $T_M^*(F) \ni t^* \mapsto t \mapsto \eta \in T_G(F)$.*

Remark 7.1. Notice T_M^* is not $M(F)$-conjugate to T_M in M since $L_1 \not\cong L_2$.

Remark 7.2. Instead of working with stable orbital integrals on $M(F)$, it is better to work with orbital integrals on $M_1(F) = Gl(2, F)^2$ via the z-extension $p : M_1 \to M$. As in Chap. 6 we get

$$SO_{(t_1, t_2)}^M (1_{K_M}) = [o_T : \rho(o_L^* \times o_{L'}^*)] \int_{(L^* \times (L')^*) \backslash Gl(2, F)^2} 1_{K_{M_1}} (g^{-1}(t_1, t_2)g) dg/dl,$$

for a suitable adjusted choice of (t_1, t_2) in the cosets of $M_1(F)/Kern(p)$. Measures are chosen such that $vol_{L^* \times (L')^*}(o_L^* \times o_{L'}^*) = 1$ and $vol_{M_1}(K_{M_1}) = 1$, where $K_{M_1} = Gl(2, o_F)^2$.

We use this, always tacitly assuming that (t_1, t_2) is chosen to be adjusted, to rewrite the stable orbital integral on M as a product of two orbital integrals on $Gl(2, F)$, namely, of

$$\int_{L^* \backslash Gl(2, F)} 1_K (g^{-1} t_1 g) dg/dl = 1_{o_L^*}(t_1) \left(-\frac{f(L : F)}{q_0 - 1} + \frac{|D_0|^{1/2}(q_0 - 1 + f(L : F))}{q_0 - 1} |t_1 - t_1^\sigma|^{-1} \right)$$

for L/F and similarly for $\int_{(L')^* \backslash Gl(2, F)} 1_K (g^{-1} t_2 g) dg/dl$

$$1_{o_{L'}^*}(t_2) \left(-\frac{f(L' : F)}{q_0 - 1} + \frac{|D_0 A_0|^{1/2}(q_0 - 1 + f(L' : F))}{q_0 - 1} |t_2 - t_2^\sigma|^{-1} \right).$$

Here $f(L/F)$ and $f(L'/F)$ denote the degrees of the residue field extension. Furthermore, $K = Gl(2, o_F)$, and measures are chosen analogously as for M_1.

7.4 Reduction to H

We use the following double coset decomposition due to Schröder ([81], Satz A.19.3(2)):

$$G = \coprod_{i=0}^{\infty} H(F) \cdot z(i) \cdot GSp(4, o_F),$$

where (see Theorem 12.2)

$$z(i) = \begin{pmatrix} 1 & 0 & -\mu^{-1} & 0 \\ 0 & 1 & 0 & 0 \\ 0 & 0 & 1 & 0 \\ 0 & 0 & 0 & 1 \end{pmatrix}$$

with the integers $\mu = \mu(i) = A \cdot \pi_F^i$ for $i = 0, 1, 2, \ldots$.

Obviously

$$k = z(i)^{-1} \cdot \eta \cdot z(i) = \begin{pmatrix} \alpha_1 + \gamma_1 \frac{\alpha_2+\gamma_2}{A} & (\beta_1 - \alpha_1 + \delta_1 - \gamma_1)\mu^{-1} & (\beta_2 + \delta_2)\mu^{-1} \\ \alpha_2 & \alpha_1 & (\beta_2 - \alpha_2)\mu^{-1} & \beta_1 A\mu^{-1} \\ \mu\gamma_1 & \mu\frac{\gamma_2}{A} & \mu(\delta_1 - \gamma_1)\mu^{-1} & \mu\delta_2\mu^{-1} \\ \mu\frac{\gamma_2}{A} & \mu\frac{\gamma_1}{A} & \mu\frac{\delta_2-\gamma_2}{A}\mu^{-1} & \mu\delta_1\mu^{-1} \end{pmatrix}$$

for $a = \alpha_1 + \alpha_2\sqrt{A^{-1}}, b = \beta_1 + \beta_2\sqrt{A^{-1}}, c = \gamma_1 + \gamma_2\sqrt{A^{-1}}, d = \delta_1 + \delta_2\sqrt{A^{-1}}$, and

$$\eta = \phi\left(\begin{pmatrix} a & b\mu^{-1} \\ \mu c & \mu d\mu^{-1} \end{pmatrix}\right), \qquad s = \begin{pmatrix} a & b \\ c & d \end{pmatrix}$$

using a slight reparameterization compared with the notion on page 239 in Sect. 7.1. Notice η and $\phi(s)$ are $H(F)$-conjugate in $H(F)$, i.e., $\eta = Ad(\phi(\begin{smallmatrix} 1 & 0 \\ 0 & \mu \end{smallmatrix}))(\phi(s))$.

Lemma 7.4. *For the isomorphism $\check{\phi}$ defined by the commutative diagram*

$$
\begin{array}{ccc}
\check{s} \in Gl(2, E^+)^0 & \xrightarrow[\simeq]{Ad((\begin{smallmatrix} 1 & 0 \\ -1 & 1 \end{smallmatrix})(\begin{smallmatrix} 1 & 0 \\ 0 & \sqrt{A} \end{smallmatrix}))} & Gl(2, E^+)^0 \ni s \\
\check{\phi} \downarrow \simeq & & \simeq \downarrow Ad((\begin{smallmatrix} 1 & 0 \\ 0 & \mu \end{smallmatrix})) \\
\eta \in H(F) & \xleftarrow[\phi]{\simeq} & Gl(2, E^+)^0
\end{array}
$$

defined by ϕ we have

$$\check{s} \in Gl\big(2, o_{E^+}(i)\big)^0 \iff \eta = \check{\phi}(\check{s}) \in z(i) \cdot GSp\big(4, o_F\big) \cdot z(i)^{-1}.$$

Here $o_{E^+}(i)$ denotes the *order* $o_F + \pi_F^i o_{E^+}$ in o_{E^+} for $i > 0$. The index ^{0} indicates that determinants should be in o_F^*.

Remark 7.3. Since $det(h) = \sqrt{A} \notin F^*$, conjugation by $h = \big(\begin{smallmatrix} 1 & 0 \\ -1 & 1 \end{smallmatrix}\big)\big(\begin{smallmatrix} 1 & 0 \\ 0 & \sqrt{A} \end{smallmatrix}\big) \in Gl(2, E^+)$ does not preserve the F-rational conjugacy classes of $Gl(2, E^+)^0$. It only preserves the stable conjugacy classes in $Gl(2, E^+)^0$. The defect, of whether the conjugacy class is preserved or not, is measured by $Norm_{E^+/E}(\sqrt{A})$ in $F^*/Norm_{E^+/F}((E^+)^*)$. $Norm_{L/F}(L^*)$ and $Norm_{E^+/F}((E^+)^*)$ generate F^* with intersection $Norm_{E/F}(E^*)$. Therefore, the norm map $Norm_{E^+/F}$ induces an isomorphism

$$Norm_{E^+/F} : (E^+)^*/Norm_{E/E^+}(E^*) \cong F^*/Norm_{E^+/F}((E^+)^*);$$

hence, the defect is measured by $\chi_{L/F}(\sqrt{A_0}^{-1}\sqrt{A_0}^{-\tau}) = \chi_{L/F}(-A_0)$.

Proof of Lemma 7.4. Put $\mu = \mu(i)$. In the notation above, if $z(i)^{-1}\eta_\mu z(i)$ is in $GSp(4, o_F)$, then $\alpha_1, \alpha_2, \delta_1, \delta_2, \alpha_1 + \gamma_1 \in o_F$; hence, $\gamma_1 \in o_F$. Furthermore, $\alpha_2 = -\gamma_2 = -\delta_2 \bmod (A)$; hence, $\gamma_2 \in o_F$. Finally,

$$\beta_1 \in (\mu/A),$$

$$-\delta_2 = \beta_2 = \alpha_2 \bmod (\mu),$$

$$\beta_1 - \alpha_1 + \delta_1 - \gamma_1 = 0 \bmod (\mu).$$

Hence, $\beta_1, \beta_2 \in o_F$. Notice μ and $\mu/A \in o_F$; hence, all $\alpha_i, \beta_i, \gamma_i, \delta_i \in o_F$. Now put

$$\tilde{s} = \begin{pmatrix} \tilde{\alpha} & \tilde{\beta} \\ \tilde{\gamma} & \tilde{\delta} \end{pmatrix} = \begin{pmatrix} 1 & 0 \\ 1 & 1 \end{pmatrix}\begin{pmatrix} a & b \\ c & d \end{pmatrix}\begin{pmatrix} 1 & 0 \\ -1 & 1 \end{pmatrix} = \begin{pmatrix} a - b & b \\ a + c - b - d & b + d \end{pmatrix}.$$

Put $\tilde{\alpha} = \tilde{\alpha}_1 + \tilde{\alpha}_2\sqrt{A^{-1}}, \ldots, \tilde{\delta} = \tilde{\delta}_1 + \tilde{\delta}_2\sqrt{A^{-1}}$. Then the conditions above become $\tilde{\alpha}_2, \tilde{\delta}_2 \in (\mu), \tilde{\gamma}_1 \in (\mu)$ and $\tilde{\beta}_1 \in (\mu/A)$ and $\tilde{\gamma}_2 \in (A)$. This uses $\tilde{\gamma}_2 = \tilde{\alpha}_2 + \gamma_2 - \delta_2 \in (A)$ since $\tilde{\alpha}_2 \in (\mu) \subseteq (A)$ and $\alpha_2 + \delta_2 = \tilde{\alpha}_2 + \tilde{\delta}_2 \in (\mu) \subseteq (A)$. If we replace $\tilde{s}$ by

$$\check{s} = \begin{pmatrix} 1 & 0 \\ 0 & \sqrt{A} \end{pmatrix}^{-1}\tilde{s}\begin{pmatrix} 1 & 0 \\ 0 & \sqrt{A} \end{pmatrix} = \begin{pmatrix} \tilde{\alpha}_1 + \frac{\tilde{\alpha}_2}{A}\sqrt{A} & \tilde{\beta}_2 + \tilde{\beta}_1\sqrt{A} \\ \tilde{\gamma}_2 + \frac{\tilde{\gamma}_1}{A}\sqrt{A} & \tilde{\delta}_1 + \frac{\tilde{\delta}_2}{A}\sqrt{A} \end{pmatrix},$$

the condition $z(i)^{-1}\eta_\mu z(i) \in GSp(4, o_F)$ becomes equivalent to

$$\check{s} \in Gl(2, o_{E^+}(i)), \qquad det(\check{s}) \in o_F^*.$$

In other words, for $\mu = A \cdot \pi_F^i$, the isomorphism

$$\check{\phi} : Gl(2, E^+)^0 \cong H(F) \subseteq GSp(4, F)$$

identifies $(\check{\phi})^{-1}\Big(H(F) \cap z(i)GSp(4, o_F)z(i)^{-1}\Big)$ with the following subgroup of $Gl(2, o_{E^+})^0$

$$K_i = Gl(2, o_{E^+}(i))^0. \quad \square$$

For a matrix $\check{s} \in K_i$ let $k = z(i)^{-1} \cdot \eta \cdot z(i)$ be its image in $GSp(4, o_F)$. Then we have

Lemma 7.5. *For a matrix* $\check{s} \in K_i$ *let* k *be its image in* $GSp(4, o_F)$. *Then the congruence* $k \equiv id \bmod (\pi_F)$ *implies the congruence* $\check{s} \equiv id \bmod (\pi_{E^+})$ *in* $Gl(2, o_{E^+}(i)/\pi_{E^+})^0$.

Proof. The unramified case is obvious. Hence, it suffices to consider the case where E^+/F is ramified. Then the reduction of $\check{s} \bmod \pi_{E^+}$ is $\begin{pmatrix} \tilde{\alpha}_1 & \tilde{\beta}_2 \\ \tilde{\gamma}_2/A & \tilde{\delta}_1 \end{pmatrix} \bmod (\pi_F)$. If the reduction of $k \bmod \pi_F$ is 1, this implies $\alpha_1 \equiv \delta_1 \equiv 1$, $\delta_2 \equiv \alpha_2 \equiv 0$. Therefore, $\beta_2 \equiv 0$ since $\beta_2 + \delta_2 \in (\pi_F\mu)$, and $\beta_1 \equiv 0$ since $\beta_1 A/\mu \equiv 0 \bmod (\pi_F)$. It also implies $(\alpha_2 - \beta_2)/A \equiv 0$ and $(\gamma_2 - \delta_2)/A \equiv 0 \bmod (\pi_F)$. These conditions imply $\check{s} \equiv id \bmod (\pi_{E^+})$ since $\tilde{\alpha}_1 = \alpha_1 - \beta_1$, $\tilde{\beta}_2 = \beta_2$, $\tilde{\gamma}_2/A = (\alpha_2 - \beta_2)/A + (\gamma_2 - \delta_2)/A$, and $\tilde{\delta}_1 = \beta_1 + \delta_1$. $\quad \square$

Orbital Integrals. We use measures on $G(F)$, $H(F)$, and $G_\eta(F) \cong T(F)$ such that the volumes of the following compact subgroups $vol_G(K) = 1$, $vol(K_H) = 1$, and $vol_T(o_T) = 1$ are 1. This implies

$$vol_G(K)/vol_H(K_i) = [K_0 : K_i].$$

From the double coset representatives for $H(F) \backslash G(F)/K$ and Sect. 6.16 we get as an immediate consequence of Lemma 7.4

$$O_{\check{\phi}(\check{s})}^G(1_K) = \sum_{i=0}^{\infty} \frac{vol_G(K)}{vol_H(K_i)} \int_{\check{\phi}^{-1}(T(F)) \backslash Gl(2, E^+)^0} 1_{K_i}(h^{-1}\check{s}h)dh/dt,$$

where

$$K_i = Gl(2, o_{E^+}(i))^0 \subseteq Gl(2, o_{E^+})^0 = K_0 = K_H$$

is mapped under $\check{\phi}$ isomorphically onto $\check{\phi}(K_i) = H(F) \cap z(i)K_Gz(i)^{-1}$ (Lemma 7.4). Recall $o_{E^+}(i)$ is the *order* $o_F + \pi_F^i o_{E^+}$ in o_{E^+} for $i > 0$, and the index ^{0} means that determinants should be in o_F^*.

An analogous formula holds for the κ-orbital integral. In Sect. 7.1 we defined elements $\eta = \phi(s) \in H(F)$. Above we also defined $\check{\phi}(s) \in H(F)$. Notice that $\phi(s)$ and $\check{\phi}(s)$ are stably conjugate in $H(F)$ (and hence are stably conjugate in $G(F)$), but are not necessarily conjugate. Hence, for the computation of the κ-orbital integrals $O_\eta^\kappa(1_K)$, one has to take care of a possible sign change. The sign depends on the residue class of $\sqrt{A}, A = A_0$ in $(E^+)^*/F^*Norm_{E/E^+}(E^*)$. This latter group is equal to $(E^+)^*/Norm_{E/E^+}(E^*)$ by local class field theory and the assumption $E = LL'$. Hence, this sign is $\chi_{E/E^+}(\sqrt{A_0}^{-1})$. Since $Norm(L^*)$

and $Norm(E^*)$ generate F^* with the intersection $Norm(E^*)$, we can apply the norm map $Norm_{E^+/F}$ to see that this sign again equals $\chi_{L/F}(\sqrt{A_0}^{-1}\sqrt{A_0}^{-\tau}) = \chi_{L/F}(-A_0)$. By abuse of notation we will therefore write s instead of $\check{s}$ to simplify the notation. This gives an extra sign $\chi_{L/F}(-A_0)$ for the κ-orbital, which will have to be considered at the end of the computation, since $O_{\phi(s)}^{\kappa}(1_K) = O_{\phi(\check{s})}^{\kappa}(1_K) = \chi_{L/F}(-A_0) \cdot O_{\phi(s)}^{\kappa}(1_K)$.

7.5 The Elements of $T(F)$

By definition $T(F) \subseteq E^*$ such that for $x \in T(F)$ we have $xx^{\sigma} \in F^*$. Recall $L = F(\sqrt{D_0})$; hence, $\tau(\sqrt{D_0}) = \sqrt{D_0}$. Choose coordinates $a, b \in E^+$ of x

$$x = a + b\sqrt{D_0},$$

giving coordinates $a' = a^{\tau}, b' = b^{\tau}$ of

$$x^{\tau} = a' + b'\sqrt{D_0}.$$

Of course the coordinates depend on the choice of $D = D_0$. Recall $D_0 \in F^*$ was chosen normalized such that $o_L = o_F + o_F\sqrt{D_0}$.

Assumption. Suppose $x \in T(F)$ is a *unit* in E^*. Furthermore, assume $x = x^{\tau} \mod \pi_E$ (in later notation $\chi > 0$) if E/E^+ is ramified.

Claim 7.1. Then

$$x = t_1 t_2$$

is solvable with units $t_1 \in o_L^*$ and $t_2 \in o_{L'}^*$.

Proof. x can be written in the form $x = T_1 T_2$ for certain $T_1 \in L^*$ and $T_2 \in (L')^*$ as we saw in Sect. 7.2. Then $x/x^{\sigma\tau} = T_1/T_1^{\sigma}$ and $x/x^{\tau} = T_2/T_2^{\sigma}$.

$x \in T(F) \cap o_E^*$ implies $x/x^{\sigma\tau} \in N_L^1 \cap o_L^*$. Namely, $x \in T(F)$ implies $xx^{\sigma} = x^{\tau}x^{\sigma\tau} \in F^*$; hence, $x/x^{\sigma\tau} \in E^{\langle\tau\rangle} = L$. Furthermore, $Norm_{L/F}(x/x^{\sigma\tau}) = xx^{\sigma}/x^{\sigma\tau}x^{\tau} = 1$. If $t_1/t_1^{\sigma} = x/x^{\sigma\tau}$ has a solution $t_1 \in o_L^*$, then one can replace T_1 by t_1, since then $T_1/t_1 \in F^*$. Also replace T_2 by $t_2 = T_2 T_1/t_1$ within $(L')^*$ such that $x = t_1 t_2$. Since $t_2 = x/t_1 \in o_E^* \cap (L')^* = o_{L'}^*$, our claim is true.

Similarly $x/x^{\tau} \in L' = E^{\langle\sigma\tau\rangle}$ and $N_{L'/F}(x/x^{\tau}) = 1$. Therefore, $x/x^{\tau} \in N_{L'}^1$. If $t_2/t_2^{\sigma} = x/x^{\tau}$ has a solution with $t_2 \in o_{L'}^*$, again our claim is true.

By assumption, the residue characteristic is odd. Therefore, the image of o_L^* under $t \mapsto t/t^{\sigma}$ is $N_L^1 \cap (1 + \pi_L o_L)$ if L/F ramifies and is equal to the norm-1-group N_L^1 of L^* otherwise. Similarly for L'. Therefore, if one of the two extensions L/F or L'/F is unramified, our claim holds by the arguments above. For the remaining case, both L/F and L'/F are ramified. Then E^+/F is unramified, and hence E/E^+

is ramified. Then our assumption implies $x/x^\tau \in 1 + \pi_{L'}o_{L'}$. Hence, we can solve $t_2/t_2^\sigma = x/x^\tau$ with $t_2 \in o_{L'}^*$. Therefore, in all possible cases $x = t_1 t_2 \in o_L^* o_{L'}^*$. $\square$

Notation. Recall that we defined an isomorphism $T(F) \cong T_G(F)$. Define o_T to be the image of $o_E^* \cap T(F)$ under this isomorphism. Then the arguments above in fact show $[o_T : \rho(o_L^* \times o_{L'}^*)] \le e_{E/E^+}$. Since for $e_{E/E^+} = 2$ the element $x = \pi_{L'}/\pi_L$ is in $T(F)$, but not in $o_L^* o_{L'}^*$, where $\pi_L = \sqrt{D_0}$ and $\pi_{L'} = \sqrt{A_0 D_0}$, we obtain

Lemma 7.6. $[o_T : \rho(o_L^* \times o_{L'}^*)] = e_{E/E^+}$.

Notation. Assume the valuation group of E^+ is normalized to be $\mathbb{Z} \cup \infty$, and assume $|\pi_F|^{-1}$ to be the cardinality of the residue field of o_F for a prime element π_F of F. Then $|x| = q^{-ord(x)/2}$ for $x \in E^+$. Put

$$\chi = ord(a - a'), \qquad N = ord(b) = ord(b').$$

Therefore, $b = \pi_{E^+}^N B, b' = \pi_{E^+}^N B'$ for units B, B' in $o_{E^+}^*$

$$q^N = |\frac{x - x^\sigma}{\sqrt{D_0}}|^{-2}.$$

By the regularity assumption $E = F(x)$ the coordinates of x and x^τ satisfy $a \ne a'$. Hence, $\chi < \infty$. Furthermore, $x = t_1 t_2$ with $t_1 = t_1^\tau \in L$ and $t_2 = t_2^{\tau\sigma} \in L'$. Thus,

$$\chi = ord(Tr_{E/E^+}(x) - Tr_{E/E^+}(x^\tau)) = ord(t_1 t_2 + t_1^\sigma t_2^\sigma - t_1 t_2^\sigma - t_1^\sigma t_2) = ord((t_1 - t_1^\sigma)(t_2 - t_2^\sigma)).$$

Hence, for $|t_1| = 1$

$$|t_2 - t_2^\sigma| = |t_2 - t_2^\sigma||t_1| = |x - x^\tau| = |(a - a')^2 - (b - b')^2 D_0|^{1/2},$$

and for $|t_2| = 1$

$$|t_1 - t_1^\sigma| = |t_1 - t_1^\sigma||t_2| = |x - x^{\sigma\tau}| = |(a - a')^2 - (b + b')^2 D_0|^{1/2}.$$

Notice $x^{\sigma\tau} = a' - b'\sqrt{D_0}$ since a', b' are fixed by σ and $\sigma\sqrt{D_0} = -\sqrt{D_0}$.

Lemma 7.7. *Suppose $x \in o_E^*$ and $\chi > 0$. Then by the claim above $x = t_1 t_2$ for $t_1 \in o_L^*$ and $t_2 \in o_L^*$, such that:*

1. $\{|D_0|^{1/2}|t_1 - t_1^\sigma|^{-1}, |D_0|^{1/2}|t_2 - t_2^\sigma|^{-1}\} = \{q^{N/2}, q^{(\chi - N - ord(D_0))/2}\}$.
2. $\chi = ord(D_0(b^2 - (b')^2))$ *except for $ord(b) = ord(b') = ord(D_0) = 0$.*
3. $\chi \ge ord(D_0) + 2N$.

Parity Rules. Suppose E/E^+ is unramified. Then $ord(D_0) = 0$. Since $a - a' \in E^+ \setminus F$ and E^+/F ramifies, χ is odd. In particular, $\chi > 0$. Furthermore, $ord((t_1 - t_1^\sigma)/2\sqrt{D_0})$ is even since L/F is unramified. Then $ord((t_2 - t_2^\sigma)/2\sqrt{D_0})$ is odd since χ is odd (using Lemma 7.7(1)).

Proof of Lemma 7.7. $\chi > 0$ implies $x = t_1 t_2$ for some $t_1 \in o_L^*$ and $t_2 \in o_{L'}^*$, by the claim above. Therefore,

$$2\chi = ord[(a - a')^2 - (b - b')^2 D_0)] + ord[(a - a')^2 - (b + b')^2 D_0].$$

Suppose $ord(b \pm b')^2 D_0 \geq ord(a - a')^2 := 2\chi > 0$ for some choice of sign. Then $ord(b \mp b')^2 D_0 = 0$ for the other choice. Then $ord(b \pm b')^2 D_0 = ord(a - a')^2 = 2\chi > 0$, and hence $N = ord(b) = ord(b') = ord(D_0) = 0$. Hence, claims (1) and (2) follow immediately, and claim (3) is trivial. Otherwise for both signs $ord((b \pm b')^2 D_0) < ord(a - a')^2$. Hence,

$$2\chi = ord(-D_0(b + b')^2) + ord(D_0(b - b')^2) = 2 \cdot ord(D_0(b^2 - (b')^2)).$$

This proves claim (2). Since $N = min(ord(b - b'), ord(b + b'))$, claim (2) then implies claim (3). Furthermore, $|t_i - t_i^\sigma| = |D_0(b \pm b')^2|^{1/2}$; hence, the product of $|D_0|^{1/2}|t_i - t_i^\sigma|^{-1}$ for $i = 1, 2$ is $|(b^2 - b'^2)|^{-1} = |\pi_{E+}|^{\chi - ord(D_0)} = q^{(\chi - ord(D_0))/2}$ by claim (2). This implies claim (1). $\square$

7.6 The Residue Rings $S = o_{E+}/\pi_F^i$

Assume $i \geq 1$. Let S denote some residue ring o_{E+}/π_F^i and R the corresponding ring $R = o_F/\pi_F^i$. Obviously

$$S = R[X]/(X^2 - A) = R + R\sqrt{A}$$

for $A = A_0$ and $\sqrt{A} = X \bmod (X^2 - A)$. The image of the order $o_{E+}(i)$ in S is R. Let $Gl(2, S)^0$ denote the group of matrices with determinant in R^*. Then the image of K_i in $Gl(2, S)^0$ is $Gl(2, R)$.

For $s \in T(F)$ the expression $[K_0 : K_i] \int_{K_0} 1_{K_i}(k^{-1}sk)dk$ is zero unless $s \in T(F) \cap K_0$. Suppose this is the case. Then

$$[K_0 : K_i] \int_{K_0} 1_{K_i}(k^{-1}sk)dk = \#\mathcal{S}$$

is the cardinality $\#\mathcal{S}$ of the set $\mathcal{S}$ of left cosets $kK_i \subseteq K_0$ for which $k^{-1}sk \in K_i$, or equivalently the number of left cosets $y \cdot Gl(2, R) \subseteq Gl(2, S)^0$ for which

$$y^{-1}sy \in Gl(2, R).$$

Assumption. Now suppose s is of the form

$$s = \begin{pmatrix} \bar{a} & \bar{b}D \\ \bar{b} & \bar{a} \end{pmatrix},$$

for $\bar{a}, \bar{b} \in S$ and some D such that $D_0 | D$.

Remark 7.4. We later apply this when $\bar{a}$ is the residue class of a and $\bar{b}$ is the residue class of a certain element $(\Theta \epsilon_0)^{-1} B \cdot \pi_{E+}^{\nu}$.

Lemma 7.8. *Let* $e = e_{E/E+}$ *be the ramification index. Then the following assertions are equivalent:*

1. *There exists* $y \in Gl(2, S)^0$ *such that* $y^{-1}sy \in Gl(2, R)$.
2. *There exists a unit* $\epsilon \in (S^*)^e$ *such that*

$$\begin{pmatrix} 1 & 0 \\ 0 & \frac{\epsilon}{\epsilon^{\tau}} \end{pmatrix} s \begin{pmatrix} 1 & 0 \\ 0 & \frac{\epsilon}{\epsilon^{\tau}} \end{pmatrix}^{-1} = s^{\tau}.$$

3. *There exists a unit* $\epsilon \in (S^*)^e$ *such that*

$$\begin{pmatrix} 1 & 0 \\ 0 & \epsilon \end{pmatrix} s \begin{pmatrix} 1 & 0 \\ 0 & \epsilon \end{pmatrix}^{-1} \in Gl(2, R).$$

In other words, $\bar{a} \in R, \epsilon \bar{b} \in R, \epsilon^{-1} \bar{b} D \in R$.

Proof. For $s = diag(\bar{a}, \bar{a})$ the lemma is obvious; hence, we can assume $\bar{b} \neq 0$. Secondly, the equivalence (2)$\Longleftrightarrow$(3) is trivial. To show that assertion (3) implies assertion (1) consider $y \in Gl(2, S)$ such that $ysy^{-1} \in Gl(2, R)$. If

$$det(y) = r \cdot det(z), \quad r \in R^*, \quad z \in Gl(2, S)_s,$$

then by replacing y with

$$z^{-1}y \begin{pmatrix} 1 & 0 \\ 0 & r \end{pmatrix}^{-1}$$

assertion (1) follows. Hence, it remains for us to show that $\epsilon \in R^* \cdot det(Gl(2, S)_s)$. We claim $(S^*)^e \subseteq R^* \cdot det(Gl(2, S)_s)$: In fact, for $e = 1$ the extension E/E^+ is unramified and E^+/F is ramified. Hence, $R^* \cdot (S^*)^2 = S^* \subseteq R^* \cdot det(Gl(2, S)_s)$. For $e = 2$ we have $(S^*)^e \subseteq det(Gl(2, S)_s)$, since $diag(\lambda, \lambda) \in Gl(2, S)_s$. Hence, assertion (3) implies assertion (1).

The remaining implication (1)$\Longrightarrow$(2):
$$y^{-1}sy \in Gl(2, R) \iff y^{-1}sy = y^{-\tau}s^{\tau}y^{\tau} \iff xsx^{-1} = s^{\tau} \text{ for}$$

$$x = y^{\tau}y^{-1}, \quad x^{\tau} = x^{-1},$$

where x uniquely determines the left coset $yGl(2, R)$ of y.
$det(x) = 1$ and $x^{\tau} = x^{-1}$ implies

$$x = \begin{pmatrix} x_1 & x_2 \\ x_3 & x_4 \end{pmatrix}, \quad \begin{pmatrix} x_1^{\tau} & x_2^{\tau} \\ x_3^{\tau} & x_4^{\tau} \end{pmatrix} = \begin{pmatrix} x_4 & -x_2 \\ -x_3 & x_1 \end{pmatrix}.$$

Hence,

$$x = \begin{pmatrix} x_1 & r_2\sqrt{A} \\ r_3\sqrt{A} & x_1^\tau \end{pmatrix}, \qquad r_2, r_3 \in R.$$

Observe that assertion (1) implies the trace condition $\bar{a} = \bar{a}^\tau$. This and $xs = s^\tau x$ gives the conditions $x_1^\tau \bar{b} = \bar{b}^\tau x_1$, $x_1 \bar{b} D = x_1^\tau \bar{b}^\tau D^\tau$, and $\sqrt{A} r_2 \bar{b} = \sqrt{A} r_3 \bar{b}^\tau D^\tau$. Therefore (notice $ord(\bar{b}) = ord(\bar{b}^\tau)$):

- $\frac{\bar{b}}{\bar{b}^\tau} x_1^\tau = x_1 + Ann_S(\bar{b})$.
- $\frac{\bar{b}^\tau}{\bar{b}} x_1^\tau D^\tau = x_1 D + Ann_S(\bar{b})$.
- $\frac{\bar{b}}{\bar{b}^\tau} r_2\sqrt{A} = r_3\sqrt{A} D^\tau + Ann_S(\bar{b})$.

Since $det(x) = 1$, we have $1 = x_1 x_1^\tau - r_2 r_3 \sqrt{A}^2$. Hence,

$$\frac{\bar{b}}{\bar{b}^\tau} = (\frac{\bar{b}}{\bar{b}^\tau} x_1^\tau) x_1 - (r_2\sqrt{A}\frac{\bar{b}}{\bar{b}^\tau}) r_3\sqrt{A} = x_1^2 + Ann_S(\bar{b}) - r_3^2 A D^\tau + Ann_S(\bar{b}).$$

$Ann_S(\bar{b}) \subseteq \pi_{E^+} S$ since $\bar{b} \neq 0$. Since E/F is always ramified, $\pi_F | A_0 D_0 | A D^\tau$. Hence,

$$\bar{b}/\bar{b}^\tau = x_1^2 \ mod \ \pi_{E^+} S.$$

Therefore, $x_1 \in S^*$. By $x_1^\tau \bar{b} = \bar{b}^\tau x_1$, therefore, $\bar{b}/\bar{b}^\tau = x_1/x_1^\tau + Ann_S(\bar{b})$. So $x_1 \equiv x_1^\tau \ mod \ Ann_S(\bar{b})$. Hence,

$$x_1 x_1^\tau = 1 \ mod \ \pi_{E^+}.$$

Furthermore, since $r_2\sqrt{A} = r_3\sqrt{A}\frac{\bar{b}^\tau}{\bar{b}} D^\tau$ and since $D^\tau x^\tau \equiv (\bar{b}/\bar{b}^\tau) x_1 D \equiv (x_1^2/x_1^\tau) D \ mod \ Ann_S(\bar{b})$,

$$x \in \begin{pmatrix} x_1 & r_3\sqrt{A} D^\tau \frac{x_1^\tau}{x_1} + Ann_S(\bar{b}) \\ r_3\sqrt{A} & x_1^\tau \end{pmatrix}$$

and, therefore,

$$x \in \begin{pmatrix} 1 & 0 \\ 0 & \frac{x_1^\tau}{x_1} \end{pmatrix} \begin{pmatrix} x_1 & \frac{x_1}{x_1^\tau} r_3\sqrt{A} D + Ann_S(\bar{b}) \\ \frac{x_1}{x_1^\tau} r_3\sqrt{A} & x_1 \end{pmatrix} \subseteq \begin{pmatrix} 1 & 0 \\ 0 & \frac{x_1^\tau}{x_1} \end{pmatrix} \cdot Gl(2,S)_s.$$

This proves

$$xsx^{-1} = \begin{pmatrix} 1 & 0 \\ 0 & \frac{x_1^\tau}{x_1} \end{pmatrix} s \begin{pmatrix} 1 & 0 \\ 0 & \frac{x_1^\tau}{x_1} \end{pmatrix}^{-1} = s^\tau$$

for $x_1 \in S^*$. Hence,

$$\begin{pmatrix} 1 & 0 \\ 0 & x_1 \end{pmatrix}^{-1} s \begin{pmatrix} 1 & 0 \\ 0 & x_1 \end{pmatrix} \in Gl(2,R).$$

To prove assertion (3) it only remains to show $x_1 \in (S^*)^e$ if $e = e_{E/E^+} = 2$. Then E/E^+ is ramified and E^+/F is unramified. It is enough to show that $x_1 \bmod \pi_{E^+}$ is a square in the residue field $\kappa(E^+)$. Since $x_1 x_1^\tau = 1 \bmod \pi_{E^+}$ the residue of x_1 has norm 1 for the quadratic residue field extension $\kappa(E^+)/\kappa(F)$. Since $\kappa(E^+)$ is cyclic, and since the norm map $\kappa(E^+)^* \to \kappa(F)^*$ is surjective, we see that the kernel of the norm is contained in $(\kappa(E^+)^*)^2$. Here we used the fact that 2 divides the cardinality of $\kappa(F)^*$, since the residue characteristic is different from 2. This implies $x_1 \in (S^*)^2$ for $e = 2$ and proves the lemma. $\square$

The Sets $\mathcal{S}$ and $\mathcal{C}$. In the situation of Lemma 7.8 consider the sets $\mathcal{S} = \mathcal{S}(s)$

$$\mathcal{S} = \{y \in Gl(2, S)^0/Gl(2, R) \mid y^{-1}sy \in Gl(2, R)\}$$

and $\mathcal{C} = \mathcal{C}(s)$

$$\mathcal{C} = \{x \in Sl(2, S) \mid x^\tau = x^{-1}, \quad xsx^{-1} = s^\tau\}$$

for

$$s = \begin{pmatrix} \bar{a} & \bar{b}D \\ \bar{b} & \bar{a} \end{pmatrix}.$$

Lemma 7.9. Suppose the set $\mathcal{S}$ is nonempty. Then $\#\mathcal{S}$ is the cardinality of $Sl(2, S)/Sl(2, R)$, if $\bar{b} = 0$ in S, or otherwise

$$\boxed{\#\mathcal{S} = e_{E/E^+} \cdot \#R \cdot \#Ann_S(\bar{b})}.$$

Proof. Since the case $\bar{b} = 0$ is obvious, we will assume $\bar{b} \neq 0$.

The map $y \mapsto x = y^\tau y^{-1}$ defines an injection

$$i : \mathcal{S} \hookrightarrow \mathcal{C}.$$

This map is not surjective in general. However, if $e_{E/E^+} = 2$, then $R \to S$ is etale and $Sl(2)$ is smooth and connected. Therefore, the map i is surjective, and hence is bijective. If $e_{E/E^+} = 1$, then E^+/F is ramified. Therefore,

$$y^\tau y^{-1} = id \bmod \pi_{E^+}.$$

$x^\tau = x^{-1}$ implies $x^2 = id \bmod \pi_{E^+}$; hence, only

$$x = \pm \bmod \pi_{E^+}.$$

Indeed for $x \in \mathcal{C}$ also $-x \in \mathcal{C}$. If $x = id \bmod \pi_{E^+}$ for $x \in \mathcal{C}$, then one shows easily $x = y^\tau y^{-1}$ for $y \in Gl(2, S)^0$ by recursion modulo powers of π_{E^+}, and $x \in \mathcal{C}$ gives $y \in \mathcal{S}$. Therefore, $\mathcal{C} = i(\mathcal{S}) \cup -i(\mathcal{S})$. We get

$$\boxed{\#\mathcal{S} = \tfrac{1}{2}e(E^+/E) \cdot \#\mathcal{C}}.$$

So in our counting problem at the beginning of this section we are reduced to counting the elements in the set $\mathcal{C}$, i.e., all $x \in Sl(2, S)$ with $x = x^* = (x^\tau)^{-1}$ and $xsx^{-1} = s^\tau$.

If $\mathcal{C}$ is Nonempty. Put $\bar{b} = B\pi_{E^+}^\nu$ for a unit $B \in S^*$ and

$$inv(\bar{b}, D) = (\bar{b}/\bar{b}^\tau)D \in S.$$

In fact, this is only well-defined modulo $Ann_S(\bar{b})$, but by abuse of notation we choose some fixed representatives in S. Suppose $\mathcal{C} \neq \emptyset$. This implies

$$inv(\bar{b}, D)^\tau = inv(\bar{b}, D) \mod Ann_S(\bar{b}).$$

Namely, $(\bar{b}^\tau/\bar{b})D^\tau = (x_1/x_1^\tau)D \mod Ann_S(\bar{b})$ and $(x_1/x_1^\tau)D = (\bar{b}/\bar{b}^\tau)D \mod Ann_S(\bar{b})$ as shown in the course of the proof of Lemma 7.8. So we can choose the representative

$$inv(\bar{b}, D) \in R.$$

Cardinality of $\mathcal{C}$. Again from the proof of Lemma 7.8

$$\mathcal{C} = \left\{ x = \begin{pmatrix} x_1 & r_2\sqrt{A} \\ r_3\sqrt{A} & x_1^\tau \end{pmatrix} \,\middle|\, det(x) = 1 \,;\, x_1^\tau = x_1(\bar{b}^\tau/\bar{b}) \mod Ann_S(\bar{b}) \,; \right.$$

$$\left. r_2\sqrt{A} = inv(\bar{b}, D)r_3\sqrt{A} + a \text{ for } a \in Ann_S(\bar{b}) \cap \sqrt{A}R \right\}.$$

We get from $\bar{b} = B\pi_{E^+}^\nu$ modulo $Ann_S(\bar{b})$

$$\bar{b}^\tau/\bar{b} \equiv (B^\tau/B)(-1)^{e\nu}.$$

This gives, modulo $Ann_S(\bar{b})$, the relation $(x_1/B)^\tau \equiv (-1)^{e\nu}(x_1/B)$. We claim $(-1)^{e\nu} = 1$. This is obvious for $e = 2$. For $e = 1$ the extension E^+/F is ramified. If ν were odd, then $(x_1/B) \in S^* \cap \sqrt{A}R = \emptyset$ would give a contradiction. Hence,

$$(x_1/B)^\tau = (x_1/B) \quad \text{or} \quad x_1 \in B \cdot R^*$$

modulo elements in $Ann_S(\bar{b})$. This gives for $\mathcal{C}$

$$\left\{ x = \begin{pmatrix} r_1B(1+\delta) & inv(\bar{b}, D)r_3\sqrt{A} + a \\ r_3\sqrt{A} & r_1B^\tau(1+\delta)^\tau \end{pmatrix} \,\middle|\, det(x) = 1; a, \delta \in Ann_S(\bar{b}); a^\tau \right.$$

$$\left. = -a \right\} / \cong,$$

where we consider solutions $x = (r_1, r_3, \delta, a)$ and $x' = (r_1', r_3', \delta', a')$ to be equivalent $x \cong x'$ if $r_3 = r_3'$, $a = a'$ and $r_1(1+\delta) = r_1'(1+\delta')$. So we have a description of $\mathcal{C}$ in terms of equivalence classes of elements $(r_1, r_3, \delta, a) \in R^2 \times Ann_S(\bar{b})^2$, which satisfy the equations

$$a^\tau = -a$$

and $r_1^2 BB^\tau (1+\delta)(1+\delta)^\tau - r_3^2 A \cdot inv(\bar{b}, D) - ar_3 \sqrt{A} = 1$.

Now choose $\delta \in Ann_S(\bar{b}), \alpha \in Ann_S(\bar{b}\sqrt{A}) \cap R$ and $r_3 \in R$ arbitrarily. Put $a = \sqrt{A}\alpha$. Then there are exactly two solutions $r_1 \in R^*$ of the equation

$$r_1^2 BB^\tau (1+\delta)(1+\delta)^\tau - r_3^2 A \cdot inv(\bar{b}, D) - \alpha r_3 = 1.$$

This holds since BB^τ is a square in R^* or otherwise $\mathcal{C} = \emptyset$. Furthermore, $A \cdot inv(\bar{b}, D)$ and αA are in the maximal ideal of S. Bringing them to the right side, we obtain a unit congruent to 1, which is a square by our assumption on the residue characteristic. So extracting the root gives us exactly two solutions for $r_1 \in R^*$. Hence,

$$\#\mathcal{C} = 2\#R\#Ann_S(\bar{b})\#(Ann_S(\bar{b}) \cap R\sqrt{A})\#(Ann_S(\bar{b}) \cap R)^{-1}.$$

For this notice that we count equivalence classes of solutions. The equivalence relation identifies solutions which differ by an element in $1 + (Ann_S(\bar{b}) \cap R)$. If $\bar{b} \neq 0$ and $\mathcal{C} \neq \emptyset$, the last two terms cancel; hence,

$$\#\mathcal{C} = 2\#R\#Ann_S(\bar{b}).$$

This cancelation is obvious if E/E^+ is ramified, or for unramified E/E^+ if $ord(\bar{b}) = \nu$ is even. However, this latter condition is forced by $(-1)^{e\nu} = 1$, which follows from $\mathcal{C} \neq \emptyset$. This proves the lemma since $\#\mathcal{S} = \frac{1}{2}e(E^+/E) \cdot \#\mathcal{C}$. □

7.7 Integration over $T(F) \setminus H(F)$

In this section assume $D_0 \in E^+$ and $\theta \in E^+$ to be normalized of order 0 or 1. Consider regular elements

$$\eta = \phi_{\theta^2 D_0}(x) \in H(F)$$

with centralizer $\phi_{\theta^2 D_0}(T(F))$. Later we write $H(F)$ as a disjoint union of double cosets

$$\coprod_{r \in R} \phi_{\theta^2 D_0}(T(F)) \cdot r \cdot K_0,$$

where the representatives $r \in Gl(2, E^+)^0$ are of the form

$$r \in \phi_{\theta^2 D_0}(T(F)) \cdot \begin{pmatrix} 1 & 0 \\ 0 & \pi_{E^+}^{j-ord(\theta)}\epsilon_0 \end{pmatrix}$$

for certain $j \in \mathbb{N}$ and $\epsilon_0 \in o_{E^+}^*$. This implies

$$\int_{H(F)} f(h)dh = \sum_r [o_T : (T(F) \cap rK_0r^{-1})] \int_{T(F)} dt \int_{K_0} f(trk)dk;$$

hence,

$$\int_{T(F)\setminus H(F)} f(h)dh/dt = \sum_r [o_T/(T(F) \cap rK_0r^{-1})] \int_{K_0} f(rk)dk.$$

It allows us to calculate the orbital integrals for H by integrations over K_0 and gives

$$O_\eta^G(1_K) = \sum_{i \geq 0} \sum_{r \in R} [o_T/(T(F) \cap rK_0r^{-1})][K_0 : K_i] \int_{K_0} 1_{K_i}(k^{-1}r^{-1}\eta rk)dk.$$

Observe $x = a + b\sqrt{A_0} = a + (b/\theta)\sqrt{\theta^2 D_0}$; hence, for $\Theta = \theta/\pi_{E+}^{ord(\theta)}$

$$\eta_r := r^{-1}\eta r = \phi_{\Theta^2 D_0 \pi_{E+}^{2j} \epsilon_0^2}(x)$$

since $\eta_r = \phi(s_r)$, where $s_r \in H(F) = Gl(2, E^+)^0$ is

$$\begin{pmatrix} 1 & 0 \\ 0 & \pi_{E+}^{ord(\theta)-j}\epsilon_0^{-1} \end{pmatrix} \begin{pmatrix} a & b\theta D_0 \\ b\theta^{-1} & a \end{pmatrix} \begin{pmatrix} 1 & 0 \\ 0 & \pi_{E+}^{j-ord(\theta)}\epsilon_0 \end{pmatrix}$$

$$= \begin{pmatrix} a & \dfrac{b}{\Theta\epsilon_0\pi_{E+}^j}(\Theta\epsilon_0\pi_{E+}^j)^2 D_0 \\ \dfrac{b}{\Theta\epsilon_0\pi_{E+}^j} & a \end{pmatrix}.$$

Therefore, $T(F) \cap rK_0r^{-1} \subseteq o_T = \{x \in o_E^* \mid Norm_{E/E+}(x) \in o_F^*\}$ is the group

$$\{x = a + b\sqrt{D_0} \mid Norm_{E/E+}(x) \in o_F^*; a, \pi_{E+}^{-j}b \in o_{E+}\}.$$

It does not depend on the choice of (normalized) D_0 or θ.

Constraints (for fixed r and corresponding j). The condition $x \in T(F) \cap rK_0r^{-1}$ is equivalent to $x \in o_E(j) \cap T(F)$.

Furthermore, the *index* is

$$[o_T/(T(F) \cap rK_0r^{-1})] = [o_E^* : o_E(j)^*][(Norm_{E/E+}(o_E^*) \cap o_F^*) : (Norm_{E/E+}(o_E(j)^*) \cap o_F^*)]^{-1}.$$

Certain constraints have to be satisfied for the orbital integral

$$[K_0 : K_i] \int_{K_0} 1_{K_i}(k^{-1}\eta_r k)dk$$

not to vanish. Namely, $k^{-1}\eta_r k \in K_i$ implies $\eta_r \in K_0$; hence, $x \in T(F) \cap rK_0r^{-1}$.

Suppose now these "r-constraints" are satisfied. Then, computing

$$[K_0 : K_i] \int_{K_0} 1_{K_i}(k^{-1}\eta_r k)dk$$

amounts to counting all cosets $kK_i \subseteq K_0$ for which $k^{-1}\eta_r k \in K_i$. This is the number of all cosets $y \in Gl(2, o_{E^+}/\pi_F^i o_{E^+})^0/Gl(2, o_{E^+}(i))^0$ in $Gl(2, o_{E^+}/\pi_F^i o_{E^+})$ such that

$$y^{-1}\eta_r y \in Gl(2, o_{E^+}(i))^0 \bmod \pi_F^i$$

holds (Lemma 7.4). This number is 1 for $i = 0$ and is computed in Lemmas 7.8 and 7.9 for $i \geq 1$. To apply these lemmas it is enough to show $K_i/K(\pi_F^i) \cong Gl(2, R)$, $K_0/K(\pi_F^i) \cong Gl(2, S)^0$, and $Gl(2, o_{E^+})/K(\pi_F^i) \cong Gl(2, S)$ for the normal subgroup $K(\pi_F^i)$ of all matrices in $Gl(2, o_{E^+})$, which are congruent 1 modulo π_R^i. Use the property that the residue characteristic is not 2; therefore, $(1 + \pi_F^i o_{E^+})^2 = 1 + \pi_F^i o_{E^+}$.

7.8 Double Cosets in $H(F)$

For normalized D_0 we use as in Sect. 6.9

$$Gl(2, E^+) = \coprod_{j \geq 0} \phi_{D_0}(E^*) \cdot \begin{pmatrix} 1 & 0 \\ 0 & \pi_{E^+}^j \end{pmatrix} \cdot Gl(2, o_{E^+}).$$

This immediately gives a decomposition (with *unique* integers $j \geq 0$) for the elements $h \in H(F)$

$$h = \phi_{\theta^2 D_0}(l) \cdot \begin{pmatrix} 1 & 0 \\ 0 & \pi_{E^+}^{j-ord(\theta)} \end{pmatrix} \cdot k \quad \text{for} \quad k \in Gl(2, o_{E^+}), \, l \in E^*.$$

To "adjust" this decomposition to the group $H(F) = Gl(2, E^+)^0 \subseteq Gl(2, E^+)$, we are (only) allowed to change k to $k_0 k$ for some $k_0 \in Gl(2, o_{E^+})$ with the property

$$k_0 \in \begin{pmatrix} 1 & 0 \\ 0 & \pi_{E^+}^{j-ord(\theta)}\epsilon_0 \end{pmatrix}^{-1} \cdot \phi_{\theta^2 D_0}(E^*) \cdot \begin{pmatrix} 1 & 0 \\ 0 & \pi_{E^+}^{j-ord(\theta)}\epsilon_0 \end{pmatrix},$$

which is equivalent to the condition

$$k_0 \in o_E^*(j).$$

This allows us to choose k_0 in K_0 if $Norm(o_E^*(j)) \cdot o_F^* = o_{E^+}^*$. This is the case only if $j = 0$ and E/E^+ is unramified. In all other cases the group $Norm(o_E^*(j)) \cdot o_F^* = (o_{E^+}^*)^2$ has index 2 in $o_{E^+}^*$.

$h \in Gl(2, E^+)$ with the condition $det(h) \in F^*$ implies

$$(*) \qquad det(k)\pi_{E^+}^{j-ord(\theta)} \equiv 1 \quad \text{in} \quad (E^+)^*/(F^* \cdot Norm_{E/E^+}(E^*)) \cong H^1(F, T).$$

Obviously the condition $h \in H(F)$ imposes conditions on $det(k)$. Let us distinguish two cases.

First Case. If E/E^+ is *ramified*, we have chosen the prime elements in the form $\pi_{E^+} = Norm_{E/E^+}(\pi_E) = -D_0$ to be in $Norm_{E/E^+}(E^*)$. Then (*) implies $det(k) \in Norm(L^*) \cap o_{E^+}^* = (o_{E^+}^*)^2$. Changing k by some k_0, which has determinant in $(o_{E^+}^*)^2$, allows us to assume $det(k) \in o_F^*$; hence, $k \in K_0 \subseteq H(F)$.

As explained above, $H(F)$ is a disjoint union of double cosets

$$r = \coprod_{r \in R} \phi_{\theta^2 D_0}(T) \cdot r \cdot K_0, \qquad R = \mathbb{N}_0$$

with representatives $r = r_j$ indexed by $j \in \mathbb{N}_0$

$$r_j = \phi_{\theta^2 D_0}(\pi_E^{-j+ord(\theta)}) \cdot \begin{pmatrix} 1 & 0 \\ 0 & \pi_{E^+}^{j-ord(\theta)} \end{pmatrix} \quad (\pi_{E^+} \in F^* \cap Norm_{E/E^+}(E^*)).$$

Second Case. Now let E/E^+ be *unramified*. Then $det(k)\pi_{E^+}^{j-ord(\theta)} \in F^* \cdot Norm_{E/E^+}(E^*)$ implies

$$j - ord(\theta) = 0 \mod 2.$$

Secondly, to achieve $det(k) \in o_F^*$ in the decomposition written above, we are only allowed to change $det(k)$ by elements from the group $Norm_{E/E^+}(o_E^*(j))$. This group is $(o_F^*)^2$ unless $j = 0$. Let ϵ_0 be in $o_{E^+}^* \setminus (o_{E^+}^*)^2$. Then $H(F)$ is a disjoint union of double cosets

$$\coprod_{r \in R} \phi_{\theta^2 D_0}(T(F)) \cdot r \cdot K_0,$$

where the set of representatives $R \subseteq \big((o_F^*/(o_F^*)^2) \times \mathbb{N} \cup (1 \times 0)\big)$ consists of all $(\epsilon_0, j) \in (o_F^*/(o_{F*})^2) \times \mathbb{N}_0$ with the property $j = ord(\theta) \mod 2$. The representatives $r = r_{\epsilon_0, j}$ are

$$r_{\epsilon_0, j} = \phi_{\theta^2 D_0}(\pi_E^{(-j+ord(\theta))/2} \epsilon_0') \cdot \begin{pmatrix} 1 & 0 \\ 0 & \pi_{E^+}^{j-ord(\theta)} \epsilon_0 \end{pmatrix} \cdot$$

Here $\epsilon_0' \in o_E^*$ is chosen such that $Norm_{E/E^+}(\epsilon_0') = \epsilon_0^{-1}$ holds for $\epsilon_0 \in o_{E^+}^*/(o_{E^+}^*)^2$. Furthermore, in the unramified case we choose the prime element to be $\pi_{E^+} = \sqrt{A_0}$.

7.9 Summation Conditions for ν

Our final summation will be indexed by certain integers ν (related to representatives in $R \subseteq H(F)$) and the integers i (coming from $H(F) \setminus G(F)/K$ double cosets). There are several summation conditions controlled among others by the parameters N and $\chi = ord(a - a')$, which were attached to $x \in T(F)$.

Put $\theta = \Theta \pi_{E+}^{ord(\theta)}$. If E/E^+ is unramified, put $\theta = \pi_{E+}$. Let us introduce the new parameter $0 \leq \nu = N - j \leq N$. It is introduced since we have to consider the matrix

$$
s_r = \begin{pmatrix} a & \dfrac{b}{\Theta\epsilon_0\pi_{E+}^j}(\Theta\epsilon_0\pi_{E+}^j)^2 D_0 \\ \dfrac{b}{\Theta\epsilon_0\pi_{E+}^j} & a \end{pmatrix}
$$

as a matrix with coefficients in the residue ring $S = o_{E+}/(\pi_F^i)$. With this new variable we get $b/\pi_{E+}^j = B\pi_{E+}^\nu, b'/\pi_{E+}^j = B'\pi_{E+}^\nu$ with units $B, B' \in o_{E+}^*$.

Warning (Concerning i). From now on we change the notation. We replace i by $i/e_{E+/F}$. This has the advantage that π_F^i becomes π_{E+}^i in all formulas. On the other hand, we get from now on the extra condition that i is divisible by 2 if $e_{E/E^+} = 1$.

From Lemmas 7.8 and 7.9, applied to the matrix s_r above, we obtain conditions on the summation indices $(i, r) \in \mathbb{N}_0 \times R$ with respect to the representative $r = (\epsilon_0, j) = (\epsilon_0, N - \nu)$, which have to be satisfied for these data to give a nonzero contribution to the twisted orbital integral. These conditions are:

1. $0 \leq i \leq \chi$, or equivalently $a \equiv a' \bmod \pi_{E+}^i$. This is the condition $\bar{a} \in R$ of Lemma 7.8(3).
2. If $\nu < i$ (or equivalently $b/\pi_{E+}^j \neq 0$ in $S = o_{E+}/\pi_F^i$), then

$$
(\pi_{E+}^\nu B\Theta^{-1}\epsilon_0^{-1})/(\pi_{E+}^\nu B\Theta^{-1}\epsilon_0^{-1})^\tau = \epsilon/\epsilon^\tau \bmod \pi_{E+}^{i-\nu}
$$

   must hold for some $\epsilon \in (o_{E+}^*)^{e_{E/E^+}}$. This again comes from Lemma 7.8(3) applied for s_r, namely, the condition $\epsilon\bar{b} \in R$.
3. If $2N + ord(D_0) - \nu < i$, then $i \leq \chi - \nu$ has to hold. (Notice that for $2N + ord(D_0) - \nu$ the inequality $i \leq \chi - \nu$ automatically holds by Lemma 7.7(3).)

Concerning the third condition, the conditions $\epsilon\bar{b} \in R$ and $\epsilon^{-1}\bar{b}D$ of Lemma 7.8(3) (for some $\epsilon \in (o_{E+}^*)^{e_{E/E^+}}$) are in explicit form

$$
(*) \qquad (\pi_{E+}^\nu B\Theta^{-1}\epsilon_0^{-1})/(\pi_{E+}^\nu B\Theta^{-1}\epsilon_0^{-1})^\tau = \epsilon/\epsilon^\tau \bmod \pi_{E+}^{i-\nu}
$$

and

$$
(**) \qquad (\pi_{E+}^{2N-\nu} B\Theta\epsilon_0 D_0)/(\pi_{E+}^{2N-\nu} B\Theta\epsilon_0 D_0)^\tau = \epsilon^\tau/\epsilon \bmod \pi_{E+}^{i-2N-ord(D_0)+\nu}.
$$

The left and right sides are in o_F. Both congruences can be combined and in this way simplified to give condition (3) in the presence of conditions (1) and (2).

To show this first observe $\pi_{E+}^2 \in F$, since either $\pi_{E+} = Norm_{E/E^+}\sqrt{D_0} = -D_0$ or $\pi_{E+} = \sqrt{A_0}$. Let us now multiply (*) and (**). From the way these conditions arise in Lemma 7.8, it is clear that (*) is a congruence condition of order greater than or equal to the order of the congruence condition (**). Therefore, the multiplied form (***) is still equivalent to (**).

Since $(\pi_{E+}^2)^\tau = (\pi_{E+}^2)$, this gives

$$(***) \qquad (B^2 D_0)/(B^2 D_0)^\tau = 1 \; mod \; \pi_{E+}^{i-2N-ord(D_0)+\nu}.$$

In other words (since multiplication with $B^2 D_0$ shifts the congruence by $ord(D_0)$),

$$(****) \qquad i - 2N + \nu \leq ord\Big(D_0 B^2 - (D_0 B^2)^\tau\Big).$$

Condition (***) is meaningless unless $i - 2N - ord(D_0) + \nu > 0$. This explains the formulation of condition (3) above. If $i > 2N + ord(D_0) - \nu$ holds, this implies $i > 0$, since $N - \nu = j \geq 0$ and $ord(D_0) \geq 0$. Now $i > 0$ implies $\chi = ord(a - a') \geq i > 0$ by condition (1) above. For $\chi > 0$, Lemma 7.7 can be applied. Lemma 7.7 implies $\chi = ord(D_0(b^2 - (b')^2))$ except for $ord(b) = ord(D_0) = 0$ and therefore $\nu = N = 0$. In this exceptional case $i \leq \chi - \nu = \chi$ by condition (1). In the other cases we get from $b = B\pi_{E+}^N$, $b' = B'\pi_{E+}^N$ and $\chi = ord(D_0(b^2 - (b')^2))$

$$\chi = 2N + ord\Big(D_0 B^2 - (D_0 B^2)^\tau\Big).$$

Therefore, condition (****) can be stated as

$$2N + ord(D_0) - \nu < i \quad \text{implies} \quad i \leq \chi - \nu.$$

This is condition (3) stated above.

7.10 Résumé

We now express the results of the section on $T(F) \backslash H(F)$ integration in terms more suitable for summation. We have to evaluate a sum indexed by $0 \leq \nu \leq N$ and $0 \leq i$ subject to further conditions. We have gathered everything in the following list. In fact, the domain of possible values i is divided into two parts subject to the reparameterization condition that i must be divisible by $e_{E+/F}$:

(A) **Conditions imposed on i (apart from $0 \leq i \leq \chi$, which is only relevant for $\chi = 0$):**
   Range of small i: $0 \leq i \leq \nu$.
   Large range i: $\nu < i \leq \chi - \nu$.
(B) **Conditions imposed on ν and θ:** In the domain of possible values ν we have, in addition to possible restrictions mentioned above, the conditions:

   (i) $0 \leq \nu \leq N$.
   (ii) $\nu = N + ord(\theta) \; mod \; 2$ \qquad if $e_{E/E+} = 1$.

Furthermore, in the large range (with respect to the variable i) we have contributions only if

(iii) $\pi_{E^+}^{\nu} \cdot B\Theta^{-1}\epsilon_0^{-1} \in (o_{E^+}^*)^{e_{E/E^+}} \cdot o_F \bmod \pi_{E^+}^{i}$

is satisfied (see condition (2) above). This forces $\nu = 0 \bmod 2$ in the large range for $e_{E/E^+} = 1$, since up to a unit $\pi_{E^+}^{\nu} \in o_F$.

(C) **Contributions from the $Index(\nu) = [o_T : (T(F)) \cap rK_0r^{-1}]$**: The relevant contributions are

$$1 \qquad (\nu = N),$$

$$\frac{(q+1)}{2q} q^{N-\nu} \qquad (\nu < N),$$

in the case where E/E^+ is *unramified* and

$$q^{N-\nu}$$

in the case where E/E^+ is *ramified*. This is shown in the Sects. 7.13 and 7.14.

(D) **Contributions from Counting the Number $sol(\nu, i) = \#S$ of Solutions**:
In the last section we made a reparameterization of the parameter i which implies that i has to be divisible by 2 if $e_{E/E^+} = 1$. Therefore, we set $sol(\nu, i) = 0$ unless this condition is satisfied.

If the condition is satisfied, put $S = o_{E^+}/\pi_F^{i/e_{E^+/F}}$ and $R = o_F/\pi_F^{i/e_{E^+/F}}$. Then $sol(\nu, i)$ is given by the number of solutions computed in Lemma 7.9, namely,

$$1 \quad (i = 0)$$

and for $i \geq 1$

$$\#Sl(2, S)/\#Sl(2, R) \qquad \text{for } i \leq \nu \text{ (small range)}$$

$$e_{E/E^+} \cdot \#R \cdot \#Ann_S(\pi_{E^+}^{\nu}) \qquad \text{for } i > \nu \text{ (large range)}.$$

In principle we now could compute the orbital integrals. For simplicity we only compute the κ-orbital integrals. This is done in Sects. 7.11 and 7.12.

7.11 The Summation (E/E^+-Ramified)

For η corresponding to $x = t_1t_2$, both $t_1 \in L^*$ and $t_2 \in (L')^*$ can be assumed to be units, since otherwise the κ-orbital integral would be zero. This follows from the r-constraints and the following fact: For the κ-orbital integral we do not get any contribution from the small range of the i-summation in the *ramified* case since this contribution is stable (i.e., does not depend on θ). This is true because $\chi = 0$ implies $i = 0$, which is in the stable range. Hence, only the large-range contributions remain. Notice condition B(iii) in the large range depends on θ. But in the large range $\chi > 2\nu \geq 0$; hence, $\chi > 0$. The κ-orbital integral is therefore zero unless $\chi > 0$ holds.

Since E/E^+ is ramified, D_0 is not a unit. Hence, $x - x' \equiv a - a' \mod \pi_E$. Since $\chi = ord(a - a') > 0$, this implies $x \equiv x' \mod \pi_E$ in addition to $x \in o_E^*$. Therefore, Lemma 7.6 and its proof imply $x = t_1 t_2, t_1 \in o_L^*, t_2 \in o_{L'}^*$.

Let χ_0 be the quadratic character of $(E^+)^*$ attached to E/E^+. χ_0 is trivial on $(o_{E+}^*)^2$ and satisfies $\chi_0(\theta) = -1$. Observe $index(\nu) = q^{N-\nu}$ and $e(E/E^+) = 2$, $\#R = q_0^i$ and $\#Ann_S(\pi_{E+}^\nu) = q^\nu$. Hence, the summands are $q^{N-\nu} \cdot (2q_0^i q^\nu) = 2q^N q_0^i$ for $q_0 = q^{1/2}$ (cardinality of the residue field of F). We therefore obtain

$$O_{\check{\phi}(s)}^\kappa(1_K) = \chi_0(B) \cdot 1_{o_L^* \times o_{L'}^*}(t_1, t_2) 2q^N \cdot \sum_{0 \le \nu \le N} \sum_{\nu < i \le \chi - \nu} q_0^i,$$

since $\epsilon_0 = 1$. Condition B(iii), i.e., $B/\Theta \in (o_{E+}^*)^2$, gives the factor $\chi_0(B)$. The double sum is

$$(q_0 - 1)^{-1} \sum_{0 \le \nu \le N} (q_0^{\chi+1-\nu} - q_0^{\nu+1})$$

$$= (q_0 - 1)^{-1}[q_0^{\chi+1}(q_0^{-N-1} - 1)/(q_0^{-1} - 1) - q_0(q_0^{N+1} - 1)/(q_0 - 1)]$$

$$= \frac{q_0}{(q_0 - 1)^2}(1 - q_0^{N+1})(1 - q_0^{\chi-N}) = q_0 \int_{(L^* \times L'^*) \backslash Gl(2,F)^2} 1_K(g^{-1}(t_1, t_2)g)dg/dl.$$

Since E^*/F is unramified, both L/F and L'/F are ramified. Hence, the last equation follows from Sect. 7.3 and Lemma 7.7(1), since A_0 is a unit.

Hence, $O_\eta^{\check{\phi}(s)}(1_K)$ is equal to

$$2\left|\frac{x - x^\sigma}{2\sqrt{D_0}}\right|^{-2} \chi_0\left(\frac{x^\sigma - x}{2\sqrt{D_0}}\right) \cdot q_0 \int_{(L^* \times L'^*) \backslash Gl(2,F)^2} 1_K(g^{-1}(t_1, t_2)g)dg/dl.$$

For this notice $q^N = |(x - x^\sigma)/2\sqrt{D_0}|^{-2}$; hence, $B\pi_{E+}^N = (x - x^\sigma)/2\sqrt{D_0}$, by the formulas preceding Lemma 7.7(1). By assumptions regarding the choice of prime elements, $\pi_{E+} = -D_0$ is in $Norm_{E/E^+}(E^*)$ since E/E^+ is ramified. Therefore, $\chi_0(\frac{x^\sigma - x}{2\sqrt{D_0}}) = \chi_0(B)$.

For the final comparison notice that the factor 2 in the formula disappears by our normalization of measures, since in the ramified case $[o_T : \rho((o_L^*)^2)] = 2$ (Lemma 7.6). Furthermore, q_0 cancels $|2\sqrt{D_0}|^2$. Also $\chi_0(\frac{x - x^\sigma}{2\sqrt{D_0}}) = \chi_{L/F}$ $((x - x^\sigma)(x^\tau - x^{\sigma\tau}))\chi_{L/F}(2^{-2}D_0^{-1})$ by taking norms, and $\chi_{L/F}(2^{-2}D_0^{-1}) = \chi_{L/F}(-1)$ since $-D_0 = Norm_{L/F}(\sqrt{D_0})$. Finally remember $O_\eta^\kappa(1_K) = O_{\check{\phi}(s)}^\kappa(1_K) = \chi_{L/F}(-A_0)O_{\check{\phi}(s)}^\kappa(1_K)$. Hence, Theorem 7.1 follows if E/E^+ is ramified.

7.12 The Summation (E/E^+-Unramified)

First of all there is the support condition that t_1, t_2 have to be units. To compute the κ-orbital integral we concentrate on the case $\chi > 0$. The case $\chi = 0$ is excluded by the parity rule since now χ is always odd. From the résumé we get the following contributions:

1. **Small-range Contributions.** The contribution will be

$$SRC = \sum_{0 \leq \nu \leq N} \sum_{0 \leq i \leq \nu} (-1)^{N-\nu} \#(\epsilon_0 - reprs.) \cdot index(\nu) \cdot sol(i)$$

$$= \sum_{0 \leq i \leq N} sol(i) \left(\sum_{i \leq \nu < N} (-1)^{N-\nu} \cdot 2 \cdot \frac{(q+1)}{2q} q^{N-\nu} + \sum_{\nu=N} 1 \right).$$

The representatives ϵ_0 appearing in R give the factors 2 in the first inner sum. The second inner sum has only one term. The computation of the inner sum gives

$$SRC = \sum_{0 \leq i \leq N} sol(i)(-q)^{N-i} .$$

The residue fields of E^+ and F have the same cardinality q. Therefore, for even i the number $sol(i) = \#Sl(2, S)/\#Sl(2, R)$ becomes $q^{dim(Sl(2))i/2} = q^{3i/2}$. Therefore, $sol(i) = q^{3i/2}$ or 0 depending on the parity of i (see résumé part (D)). Thus,

$$SRC = (-q)^N \sum_{0 \leq j \leq N/2} q^j = \frac{(-q)^N}{(q-1)^2}[(q-1)q^{1+[N/2]} - (q-1)].$$

Here $[x]$ is the largest integer $n \leq x$.

2. **Large-range Contribution.** We get

$$LRC = \sum_{\nu < i \leq \chi - \nu} \sum_{0 \leq \nu \leq N} (-1)^{N-\nu} \cdot \#(\epsilon_0 - reprs.) \cdot index(\nu) \cdot sol(\nu, i),$$

where the summation is over all even integers ν. See résumé B(iii) in the case $e(E/E^+) = 1$ for the large range. The number $sol(\nu, i) = 0$ unless i is even. Furthermore, for even i we have $e(E/E^+) = 1$, $\#R = q^{i/2}$, and $\#Ann_S(\pi_{E^+}^\nu) = q^\nu$. Therefore, $sol(\nu, i) = q^{i/2+\nu}$ for even i. Hence,

$$LRC = \sum_{\nu < i \leq \chi - \nu, i \text{ even}} \left(\sum_{0 \leq \nu < N, \nu \text{ even}} (-1)^N \cdot 2 \cdot \frac{q+1}{2q} q^{N-\nu} \cdot q^{i/2+\nu} \right)$$

$$+ \left(\sum_{N < i \leq \chi - N, i, N \text{ even}} (-1)^N \cdot 1 \cdot q^{i/2+N} \cdot 1 \right).$$

Summation over i and then ν gives

$$\frac{(-q)^N}{(q-1)^2} \left((q+1)q^{\frac{\chi+1}{2}} - (q+1)q^{\frac{\chi-N}{2}} - (q+1)q^{\frac{N+1}{2}} + (q+1) \right)$$

in the case where N is *odd* and

$$\frac{(-q)^N}{(q-1)^2}\left((q+1)q^{\frac{\chi+1}{2}} - 2q^{\frac{\chi-N+1}{2}} - (q^2+1)q^{\frac{N}{2}} + (q+1)\right)$$

in the case where N is *even*.

3. **Total Contribution.** Adding together the contributions from (1) and (2) gives

$$SRC + LRC = \frac{(-q)^N}{(q-1)^2}\left(-2 + (q+1)q^{\frac{ev}{2}}\right)\left(-1 + qq^{-1/2}q^{\frac{odd}{2}}\right),$$

where ev is the even number among $N, \chi - N$ and odd is the odd one (by the parity rules). But $|D_0| = 1$, $|A_0| = q^{-1/2}$, $q = q_0$, $(-1)^N = (-1)^{ord(\pi_{E^+}^N B)} = \chi_0(\frac{(x-x^\sigma)}{2\sqrt{D_0}}) = \chi_{L/F}((x-x^\sigma)(x^{\sigma\tau} - x^\tau))$. In the present case the first of the two $Gl(2, F)$-orbital integrals corresponds to the unramified extension L/F, and the second to the ramified extension L'/F. Finally, take into account the factor $\chi_{L/F}(-A_0)$, which comes from our change from $\phi(s)$ to $\check{\phi}(s)$. If we multiply together the two $Gl(2)$-orbital integrals and consider the transfer factor, the result coincides with the right side of the formula above for $SRC + LRC$. Hence Theorem 7.1 is now also proved for the unramified case.

7.13 Appendix I on Orders (E/E^+-Ramified)

Let the situation be as above. In particular, the residue characteristic is different from 2. Then the orders $o_E(j) = o_{E^+} + \pi_{E^+}^j o_E \subseteq o_E$ (where $j > 0$ by assumption) have the following properties:

1. $Norm(o_E(n)^*) = (o_{E^+}^*)^2$.

This is clear from $(o_{E^+}^*)^2 \subseteq Norm(o_E(n)^*) \subseteq Norm(o_E^*) \subseteq (o_{E^+}^*)^2$.

2. $[o_E^*(n) : (1 + \pi_{E^+}^n o_E)] = [o_{E^+}^* : (o_{E^+}^* \cap (1 + \pi_{E^+}^n o_E)] = [o_{E^+}^* : (1 + \pi_{E^+}^n o_{E^+})] = (q-1)q^{n-1}$.

3. $[o_E^* : o_E^*(n)] = [o_E^* : (1+\pi_{E^+}^n o_E)]/[o_E^*(n) : (1+\pi_{E^+}^n o_E)] = (q-1)q^{2n-1}/(q-1)q^{n-1} = q^n$.

4. The *index* is $[o_T : (T(F) \cap rK_0r^{-1})] = [o_L^* : o_L^*(j)] = q^j$, since $(o_E^*)^{(\sigma+1)(\tau-1)} = (o_E(j)^*)^{(\sigma+1)(\tau-1)}$ follows from (1).

Finally, for $S = o_{E^+}/\pi_{E^+}^i o_{E^+}$ and $i > 0$:

5. $\#(Sl(2, S)) = (q-1)(q+1)q^{3i-2}$.

7.14 Appendix II on Orders (E/E^+-Unramified)

Let the situation be as above. Let $o_E(j) = o_{E^+} + \pi_{E^+}^j o_E \subseteq o_E$ for $j > 0$:

1. $Norm(o_E(j)^*) = (o_{E^+}^*)^2$ if $j > 0$ and $Norm(o_E^*) = o_{E^+}^*$.

This is clear from $(o_{E+}^*)^2 \subseteq Norm(o_E^*(j)) \subseteq Norm(o_E^*) = (o_{E+}^*)$ and $[o_{E+}^* : (o_{E+}^*)^2] = 2$ and $o_E^*(j) \subseteq o_{E+}^*(1 + \pi_E o_E)$:

2. For $n > 0$ we have $[o_E^*(n) : (1 + \pi_{E+}^n o_E)] = [o_{E+}^* : (o_{E+}^* \cap (1 + \pi_{E+}^n o_E)] = [o_{E+}^* : (1 + \pi_{E+}^n o_{E+})] = (q - 1)q^{n-1}$.

3. For $n > 0$ we have $[o_E^* : o_E^*(n)] = [o_E^* : (1 + \pi_{E+}^n o_E)]/[o_E^*(n) : (1 + \pi_{E+}^n o_E)] = (q^2 - 1)q^{2(n-1)}/(q - 1)q^{n-1} = (q + 1)q^{n-1}$.

Using (1) and (3) we get for the *index*:

4. $[o_T : (T(F) \cap rK_0 r^{-1})]$ is 1 or $\frac{1}{2}(q + 1)q^{j-1}$. This follows because $Norm_{E/E+}(o_E^*) \cap o_F^* = Norm_{E/E+}(o_E(j)^*) \cap o_F^*$ for $j = 0$ and has index 2 for $j > 0$. Namely, in this case $(o_{E+}^*)^2 \cap o_F^* = (o_F^*)^2$ has index 2 in $o_{E+}^* \cap o_F^* = o_F^*$.

This value is

$$1 \qquad (j = 0),$$

$$\frac{1}{2}(q + 1)q^{j-1} \qquad (j > 0).$$

7.15 The Global Property

We now show that the transfer factor behaves well under degenerations. For this consider separable quadratic algebras L/F and L'/F over F with automorphisms τ' and τ, respectively, whose fixed algebra is F. τ' and τ induce commuting automorphisms of the F-algebra

$$E = L \otimes_F L'$$

with fixed algebra F. The algebra isomorphism $\rho : L \otimes_F L' \cong E$ induced by $\rho(l \otimes l') = l \cdot l'$ defines an isomorphism

$$\rho : (L^* \times (L')^*)/F^* \cong E^*$$

such that $\rho(t_1, t_2) = t_1 t_2$. Put $\sigma = \tau' \otimes \tau$ and let E^+ be the fixed algebra of σ in E. E^+ is a separable quadratic algebra over F. For all $x \in E$ we have

$$(x - x^\sigma)(x^\tau - x^{\sigma\tau}) = (x - x^{\tau\tau'})(x^\tau - x^{\tau'}) \in F,$$

since it is invariant under τ and τ'. Notice the apparent asymmetry for τ and τ' in this expression.

Notation. Since $L \cong F[X]/(X^2 - D_0)$ and $E^+ \cong F[X](X^2 - A_0)$ for some $A_0, D_0 \in F^*$, we simply write $L = F(\sqrt{D_0})$ and $E^+ = F(\sqrt{A_0})$. The element $\pm\sqrt{A_0}\sqrt{D_0}$ generates the separable algebra $L' = F(\sqrt{A_0 D_0})$ in E. If $L \not\cong L'$ are field extension, this is precisely the situation from Sect. 7.1. If $E^+ \cong F^2$ or equivalently $A_0 \in (F^*)^2$, then $L \cong L'$ and $E^* \cong (L^*)^2/F^*$. As we will see below, this leads us to elliptic case I if L/F is a field extension. If $L \cong F^2$ or $L' \cong F^2$, then $E^* \cong F^* \times (E^+)^*$ for $E^+ \cong L'$ and $E^+ \cong L$, respectively. This leads to cases (6) and (7) in the list of tori on page 94, as we will see in a moment.

The Degeneration $A_0 \in (F^*)^2$. Without restriction of generality, $A_0 = 1$. Then, under the isomorphism $E^+ \cong F^2$, an element $x = \alpha_1 + \alpha_2\sqrt{A_0} \in E^+$ corresponds to $(\alpha, \alpha') = (\alpha_1 + \alpha_2, \alpha_1 - \alpha_2)$, possibly up to a permutation by $\tau(\alpha, \alpha') = (\alpha', \alpha)$. Under the specialization $A = A_0 = 1$, the group $H(F) \subseteq G(F)$ defined in Sect. 7.1 becomes the group $H^{II}(F)$ of all matrices

$$\begin{pmatrix} \alpha_1 & \alpha_2 & \beta_1 & \beta_2 \\ \alpha_2 & \alpha_1 & \beta_2 & \beta_1 \\ \gamma_1 & \gamma_2 & \delta_1 & \delta_2 \\ \gamma_2 & \gamma_1 & \delta_2 & \delta_1 \end{pmatrix}.$$

Conjugation by $g = diag\left(\left(\begin{smallmatrix} 1 & 1 \\ 1 & -1 \end{smallmatrix}\right), \left(\begin{smallmatrix} 1 & 1 \\ 1 & -1 \end{smallmatrix}\right)\right) \in G(F)$ gives the group $gH^{II}(F)g^{-1} = H^I(F)$ of all matrices

$$\begin{pmatrix} \alpha & 0 & \beta & 0 \\ 0 & \alpha' & 0 & \beta' \\ \gamma & 0 & \delta & 0 \\ 0 & \gamma' & 0 & \delta' \end{pmatrix}$$

considered in Sect. 6.2 for elliptic case I. In fact the elements η defined as in Theorem 6.1 or as in Theorem 7.1 become similarly related by conjugation

$$g \begin{pmatrix} \alpha_1 & \alpha_2 & \beta_1 D_0 & \beta_2 D_0 \\ \alpha_2 & \alpha_1 & \beta_2 D_0 & \beta_1 D_0 \\ \beta_1 & \beta_2 & \alpha_1 & \alpha_2 \\ \beta_2 & \beta_1 & \alpha_2 & \alpha_1 \end{pmatrix} g^{-1} = \begin{pmatrix} \alpha & 0 & \beta D_0 & 0 \\ 0 & \alpha' & 0 & \beta' D_0 \\ \gamma & 0 & \delta & 0 \\ 0 & \gamma' & 0 & \delta' \end{pmatrix}$$

under the degeneration $A_0 \mapsto 1$. Now look at the transfer factor $\Delta(\eta, t)$ obtained in Theorem 7.1 in elliptic case II. Under the degeneration $A_0 \mapsto 1$ it becomes the transfer factor $\Delta(\eta, t)$ obtained in Theorem 6.1 for elliptic case I. The same holds for the volume factor Δ_{IV}.

The Degeneration $L \cong F^2$. In this case $\chi_{L/F}$ becomes trivial. Hence, the transfer factor from Theorem 7.1 specializes to the volume factor Δ_{IV}, which is the correct transfer factor for the nonelliptic maximal F-tori coming from M. Since these tori are stable, it is obvious that the $G(F)$-conjugacy class of η specializes in a unique way.

The Degeneration $L' \cong F^2$. This case is implicitly contained in the last case. Nevertheless – in view of the asymmetry in which L and L' appear in the transfer factor – it is instructive to see that for $x = (u, v) \in L \oplus L \cong E$ the expression $(x - x^{\tau\tau'})(x^\tau - x^{\tau'}) = (u - v^\sigma, v - u^\sigma)(v - u^\sigma, u - v^\sigma)$ is (ζ, ζ) for $\zeta = -(u-v^\sigma)(u-v^\sigma)^\sigma \in F$, where σ is the nontrivial F-automorphism of L/F. Hence,

$\tau(\eta, t) = \chi_{L/F}\big(A_0(x - x^\sigma)(x^\tau - x^{\sigma\tau})\big)$ becomes $\chi_{L/F}(-A_0)$. Now $A_0 D_0 \in (F^*)^2$, since L' splits. Therefore, $\chi_{L/F}(-A_0) = \chi_{L/F}(-D_0) = 1$, since $-D_0 \in F^*$ is a norm from $L = F(\sqrt{D_0})$. Again this gives the correct transfer factor Δ_{IV} for the nonelliptic maximal F-torus coming from M.

Global Case. Now let L/F and L'/F be defined as above for a number field F with the corresponding global elements A_0 and D_0. Every global semisimple conjugacy class η in $G(F)$ obtained globally by a standard embedding as described in Sect. 7.1 gives such a conjugacy class in $G(F_v)$ locally for each place v of F. This defines local transfer factors $\Delta_v(\eta, t)$ for each place

$$\Delta_v(\eta, t) = \chi_{L_v/F_v}\big(A_0(x - x^{\tau\tau'})(x^\tau - x^{\tau'})\big) \cdot \Delta_{IV,v}(\eta, t).$$

As explained above, for non-Archimedean fields the fundamental lemma holds for these transfer factors. In Chap. 8 we will see that for Archimedean fields (where only elliptic case I exists) this transfer factor coincides with the Langlands–Shelstad transfer factor. Since $A_0(x - x^{\tau\tau'})(x^\tau - x^{\tau'}) \in F^*$ is a global element and since $\chi_{L/F}$ is an idele class character, we obtain from the product formula $\prod_v \Delta_v(\eta, t) = \prod_v \Delta_{IV,v}(\eta, t) = 1$.

Lemma 7.10. *If the Archimedean local transfer factor is defined to be the one of Langlands–Shelstad, the non-Archimedean local transfer factors defined in Theorems 6.1 and 7.1 satisfy the product formula*

$$\prod_v \Delta_v(\eta, t) = 1$$

for all global sufficiently regular semisimple elements $\eta \in G(F)$.

Finally Some Remarks on the Embeddings ϕ. Although the image $H(F)$ of ϕ is not stable under transposition of matrices, this is nevertheless almost true. In fact

Lemma 7.11. $\phi(g)^t = Ad_{diag(1,A,1,A^{-1})} \circ \phi(g^t)$ *for all* $g \in Gl(2, E)^0$.

Proof. An easy calculation.   □

For $g = s$ the transposed matrix s^t is conjugate to s in $Gl(2, E)^0$ by $s^t = Ad_{diag(1,D_0)} \circ s$. Hence, by the last lemma every sufficiently regular semisimple matrix h in $H(F)$ is conjugate in $G(F)$ to its transposed matrix $h^t \in G(F)$. Obviously by the same argument, this is also true in elliptic case I. Hence, every sufficiently regular semisimple element g in $G(F)$ is conjugate under $G(F)$ to its transposed g^t. Since $g^{-1} = \lambda(g)^{-1} \cdot J^{-1} g^t J$ for the matrix J defining the symplectic form, it follows for the character of an irreducible representation $\chi_{\pi^*}(g) = \chi_\pi(g^{-1}) = \omega(\lambda(g))^{-1}\chi_\pi(g)$. Notice it is enough to know this for all sufficiently regular semisimple elements by the theory of Harish-Chandra. Hence,

Corollary 7.2. *For an irreducible admissible representation π of $GSp(4, F)$ over a local field let π^* denote its contragredient representation. Then $\pi^* \cong \pi \otimes (\omega_\pi \circ \lambda)^{-1}$, where ω_π denotes the central character and λ the similitude homomorphism.*

Chapter 8
The Langlands–Shelstad Transfer Factor

In this chapter we give an explicit formula for the Archimedean transfer factor of
Langlands and Shelstad [60]. Since it makes no essential difference, we do this for
type I elliptic tori over an arbitrary local field of characteristic zero (elliptic tori
of type II do not occur over Archimedean local fields). The Archimedean transfer
factor is important for the global property of the transfer factors (product formula)
necessary for the stabilization of the trace formula. The Archimedean transfer factor,
on the other hand, determines the endoscopic lift at the Archimedean place by the
results obtained by Shelstad [90]. For us this is crucial since it implicitly determines
the relevant sign ε that appears in Corollary 4.1 to be $\varepsilon = -1$ (via the theory of
Whittaker models in Sect. 4.10 and some global multiplicity arguments).

8.1 An Admissible Homomorphism

For the construction of transfer factors one needs to specify an admissible homomorphism $T_M \hookrightarrow F_G$ between the maximal tori T_G and T_M. In this section we
explicitly construct this homomorphism.

Let F be a local field of characteristic zero and $GSp(4, F)$ the group of symplectic similitudes in four variables. Let $M(F) = Gl(2, F)^2/F^*$ be the proper elliptic
endoscopic group. The subgroup F^* divided out is embedded as the set of all matrices $\{(t \cdot id, t^{-1} \cdot id)\}$. In the following we consider elliptic tori T_M of M of the
form $T_M(F) \cong (L_1^* \times L_2^*)/F^*$ for quadratic extension fields L_1, L_2 of F where
$L_1 = L_2$. Let L be a quadratic extension field of F with a nontrivial field automorphism σ fixing F. L defines a conjugacy class of tori $T_M \subseteq M$ by the regular
embedding. Fix T_M. For elements $t \in T_M(F) = (L^*)^2/F^*$ we write $t = (t_1, t_2)$
for $t_1, t_2 \in L^*$ by abuse of notation, in the sense that (t_1, t_2) and $(t_1 t, t_2 t^{-1})$ for
$t \in F^*$ define the same element in $T_M(F)$. Then

$$x = t_1^\sigma t_2^\sigma \in L^*, \qquad x' = t_1 t_2^\sigma \in L^*$$

R. Weissauer, *Endoscopy for GSp(4) and the Cohomology of Siegel Modular Threefolds*,
Lecture Notes in Mathematics 1968, DOI: 10.1007/978-3-540-89306-6_8,
© Springer-Verlag Berlin Heidelberg 2009

are well defined. Notice this notation differs from the one used in Chap. 6 since x is now replaced by x^σ.

The Split Tori T_M^s and their Dual $\hat{T}_M^s$. Over the algebraic closure $\overline{F}$ the torus T_M is conjugate to the diagonal torus T_M^s of all elements

$$\begin{pmatrix} t_1 & * \\ 0 & t_1^\sigma \end{pmatrix} \times \begin{pmatrix} t_2 & 0 \\ * & t_2^\sigma \end{pmatrix} \quad \mathrm{mod}\ (t \cdot id, t^{-1} \cdot id),$$

the stars in the diagonal being zero (the stars indicate our fixed choice of a reference Borel group containing the split torus T_M^s over $\overline{F}$). The corresponding positive roots are $R_M^+ = \{t_1/t_1^\sigma, t_2^\sigma/t_2\} = \{x/(x')^\sigma, x'/x\}$ and we get root data for M:

$$X_*(T_M^s) = \mathbb{Z}^2 \oplus \mathbb{Z}^2/(1, 1, -1, -1)\mathbb{Z},$$
$$\Delta^+ = \Delta_+ = \{(1, -1, 0, 0), (0, 0, -1, 1)\},$$
$$X^*(T_M^s) = \{(a, b, c, d) \in \mathbb{Z}^2 \oplus \mathbb{Z}^2 \mid a + b = c + d\}.$$

The Galois group acts by the generator σ of $Gal(L/F)$ and permutes the coordinates of each $\mathbb{Z}^2$-block. The dual torus $\hat{T}_M^s$ of T_M is the complex torus $X^*(T_M^s) \otimes \mathbb{C}^* = \{(z_1, z_2, z_3, z_4) \in (\mathbb{C}^*)^4 \mid z_1 z_2 = z_3 z_4\}$, on which σ acts by permuting z_1 and z_2 and z_3 and z_4, respectively. The induced positive coroots are $t \mapsto (t, t^{-1}, 1, 1)$ and $t \mapsto (1, 1, t^{-1}, t)$.

The Fixed Reference Tori $\hat{T}_M = \hat{T}_G$. We view the dual group $\hat{M}$ as the subgroup of $\hat{G} = GSp(4, \mathbb{C})$ of all elements

$$\begin{pmatrix} * & 0 & * & 0 \\ 0 & * & 0 & * \\ * & 0 & * & 0 \\ 0 & * & 0 & * \end{pmatrix}.$$

Specify the diagonal tori $\hat{T}_M = \hat{T}_G$ as reference tori, and the reference Borel subgroups $\hat{B}_M = \hat{B}_G \cap \hat{M}$ and

$$\hat{B}_G = \{\begin{pmatrix} * & 0 & * & * \\ * & * & * & * \\ 0 & 0 & * & * \\ 0 & 0 & 0 & * \end{pmatrix}\}.$$

We identify $\hat{T}_M^s$ and $\hat{T}_M$ by the isomorphism

$$\psi : (z_1, z_2, z_3, z_4) \in \hat{T}_M^s \mapsto diag(z_1, z_4, z_2, z_3) \in \hat{T}_M = \hat{T}_G.$$

This isomorphism preserves the positive roots and coroots. The positive coroots for $\hat{T}_M$ in $\hat{M}$ are defined by the cocharacters $t \mapsto (t, 1, t^{-1}, 1)$ and $t \mapsto (1, t, 1, t^{-1})$.

The Tori T_G^s and its Dual $\hat{T}_G^s$. Consider the split diagonal torus T_G^s in G over $\overline{F}$ and choose the Borel group B_G^s containing T_G^s as the group of the matrices of the form as above but now with entries in $\overline{F}$ instead of $\mathbb{C}$. The diagonal entries

$$g = diag(x_1, x_1', x_2, x_2') \in G(\overline{F})$$

satisfy $x_1 x_2 = x_1' x_2'$. Hence,

$$X^*(T_G^s) = \{(m_1, m_2, m_1', m_2') \in (\mathbb{Z}^2 \oplus \mathbb{Z}^2)/(1,1,-1,-1)\mathbb{Z}\},$$
$$X_*(T_G^s) = \{(m_1, m_2, m_1', m_2') \in \mathbb{Z}^2 \oplus \mathbb{Z}^2 \mid m_1 + m_2 = m_1' + m_2'\},$$
$$X^*(T_G^s) \supseteq \Phi^+ = \{(-1,0,1,0), (1,-1,0,0), (1,0,0,-1), (0,0,1,-1)\}.$$

Remark 8.1. Notice the reordering (!) such that (m_1, m_2, m_1', m_2') corresponds to the character, which maps $diag(x_1, x_1', x_2, x_2')$ to $x_1^{m_1} x_2^{m_2} x_1'^{m_1'} x_2'^{m_2'}$.

The first two positive roots listed in $\Phi^+ = \{\alpha_1, \alpha_2, \alpha_2 + \alpha_1, \alpha_2 + 2\alpha_1\}$ form a basis $\Delta = \{\alpha_1, \alpha_2\}$ of the root system such that $\alpha_1(g) = x_1'/x_1$ and $\alpha_2(g) = x_1/x_2$ (long root).

The Tori T_G and their Duals $\hat{T}_G$. Write $L = F(\sqrt{D_0})$ for some D_0 and consider the torus $T_G \subseteq G$ defined by the elements

$$\eta = \begin{pmatrix} a & 0 & bD_0 & 0 \\ 0 & a' & 0 & b'D_0 \\ b & 0 & a & 0 \\ 0 & b' & 0 & a' \end{pmatrix}$$

in $G(F)$. The element

$$h = \begin{pmatrix} -\sqrt{D_0} & 0 & -\sqrt{D_0} & 0 \\ 0 & -\sqrt{D_0} & 0 & -\sqrt{D_0} \\ -1 & 0 & 1 & 0 \\ 0 & -1 & 0 & 1 \end{pmatrix}$$

is in $G(\overline{F})$, but not in $G_{der}(\overline{F})$. For later use (see the definition of $\lambda_{T_{sc}}$) notice $h = h_{der} \cdot z$ for some z in the center $Z_G(\overline{F})$ and some $h_{der} \in G_{sc}(\overline{F})$. Then $h^{-1} T_G h = T_G^s$ and

$$h^{-1} \eta h = diag(x, x', x^\sigma, (x')^\sigma),$$

and similarly for h_{der}, where $x = a + b\sqrt{D_0}$ and $x' = a' + b'\sqrt{D_0}$.

The Galois group acts on $X^*(T_G)$ via the quotient group $Gal(L/F)$. By transport of structure from the action on $X^*(T_G)$, the diagonalization induces an action on $X^*(T_G^s)$ such that σ acts on $X^*(T_G^s)$ by $\sigma(m_1, m_2, m_1', m_2') = (m_2, m_1, m_2', m_1')$. This also induces an action on the dual

$$X_*(T_G^s) = \{(m_1, m_2, m_1', m_2') \mid m_1 + m_2 = m_1' + m_2'\},$$

$$(\Phi^\vee)^+ = \{(-1,1,1,-1),(1,-1,0,0),(1,-1,1,-1),(0,0,1,-1)\}.$$

The cobase $\Delta^\vee = \{\alpha_1^\vee, \alpha_2^\vee\}$ of the root system defined by T_G^s, B_G^s is defined by the first two elements of the above set $(\Phi^\vee)^+ = \{\alpha_1^\vee, \alpha_2^\vee, (\alpha_2 + \alpha_1)^\vee, (\alpha_2 + 2\alpha_1)^\vee\}$.

We identify $\hat{T}_G$ with $\hat{T}_G^s$ as groups. However, the Galois action differs, as explained above. The dual tori $\hat{T}_G^s$ and $\hat{T}_G$ are therefore in both cases

$$\{(w_1, w_2, w_1', w_2') \in (\mathbb{C}^*)^2 \times (\mathbb{C}^*)^2\}/\{(t,t,t^{-1},t^{-1})\}.$$

For $\hat{T}_G^s$ the Galois action is trivial: for $\hat{T}_G^s$ the Galois action of σ permutes the coordinates in the two $(\mathbb{C}^*)^2$-blocks.

The Isomorphisms between $\hat{T}_G$ and $\hat{T}_G^s$ and $\hat{T}_G$. To define an admissible isomorphism

$$T_G \cong T_M$$

in the sense of [60], (1.3), its dual has to be the composite of certain isomorphisms

$$\hat{T}_M \xrightarrow{\hat{\alpha}^{-1}} \hat{T}_M^s \xrightarrow{can} \hat{\mathbf{T}}_M \hookrightarrow \hat{\mathbf{T}}_G \xrightarrow{can} \hat{T}_G^s \xrightarrow{\hat{\beta}} \hat{T}_G,$$

where $\hat{\alpha}$ and $\hat{\beta}$ are induced by conjugation maps α and β, respectively, in $M(\overline{F})$ and $G(\overline{F})$, respectively, which preserve the chosen Borel subgroups B_M^s and B_G^s. β is defined by conjugation with h defined above. α is conjugation by $\left(\begin{smallmatrix} -\sqrt{D_0} & -\sqrt{D_0} \\ -1 & 1 \end{smallmatrix}\right) \times \left(\begin{smallmatrix} -\sqrt{D_0} & -\sqrt{D_0} \\ -1 & 1 \end{smallmatrix}\right)$ in $M(\overline{F})$.

For our purposes it suffices to identify $\hat{T}_M$ with $\hat{T}_M^s$ and $\hat{T}_G$ with $\hat{T}_G^s$, so $\hat{\alpha}$ and $\hat{\beta}$ become the identity map, except that we use the twisted Galois action on $\hat{T}_M^s$ and $\hat{T}_G^s$, which comes by transport of structure from $\hat{T}_G$ and $\hat{T}_M$, respectively. Besides $\hat{\alpha}$ and $\hat{\beta}$ the other isomorphisms are also required to preserve positivity of roots with respect to the chosen Borel groups. The canonical isomorphism $\psi : \hat{T}_M^s \cong \hat{\mathbf{T}}_M$ has been specified already:

$$(z_1, z_2, z_3, z_4) \in \hat{T}_M^s \mapsto diag(z_1, z_4, z_2, z_3) \in \hat{\mathbf{T}}_M = \hat{\mathbf{T}}_G.$$

The Canonical Isomorphism. The inverse $\hat{T}_G^s \to \hat{\mathbf{T}}_G$ of the canonical isomorphism $\hat{\mathbf{T}}_G \cong \hat{T}_G^s$ maps

$$(w_1, w_2, w_1', w_2') \bmod (t,t,t^{-1},t^{-1}) \in \hat{T}_G^s$$

to the diagonal matrix $diag(z_1, z_4, z_2, z_3)$ given by

$$diag(w_2 w_1', w_1 w_1', w_1 w_2', w_2 w_2') \in \hat{\mathbf{T}}_G.$$

In other words, $z_1 = w_2 w_1'$, $z_2 = w_1 w_2'$, $z_3 = w_2 w_2'$, and $z_4 = w_1 w_1'$ if we identify $\hat{\mathbf{T}}_G$ and $\hat{\mathbf{T}}_M$.

Proof. This canonical isomorphism is defined by the property that it maps the basis of positive roots defined by $(-1, 1, 1, -1), (1, -1, 0, 0) \in X_*(T_G^s) = X^*(\hat{T}_G^s)$ (this corresponds to the characters $(w_1, w_2, w_1', w_2') \mapsto w_2 w_1'/w_1 w_2'$ and w_1/w_2, respectively) to the basis of positive roots of $\hat{\mathbf{T}}_G$ defined by the characters $z_1/z_2, z_4/z_1$. Similarly, the basis of coroots $t \mapsto (t^{-1}, 1, t, 1)$ and $t \mapsto (t, t^{-1}, 1, 1)$, of $\hat{T}_G^s$ maps to the basis of coroots $t \mapsto diag(t, 1, t^{-1}, 1)$ and $t \mapsto diag(t^{-1}, t, t, t^{-1})$, respectively, of $\hat{\mathbf{T}}_G$. The composed homomorphism $\hat{T}_G^s \to \hat{\mathbf{T}}_G = \hat{\mathbf{T}}_M \to \hat{T}_M^s$ maps (w_1, w_2, w_1', w_2') to $diag(z_1, z_2, z_3, z_4) = diag(w_2 w_1', w_1 w_2', w_2 w_2', w_1 w_1')$. □

Hence, we have

Lemma 8.1. *The dual isomorphism* $\phi : T_M^s(\overline{F}) \to T_G^s(\overline{F})$ *is*

$$\phi\big((t_1, t_1^\sigma, t_2, t_2^\sigma) \mod (t, t, t^{-1}, t^{-1})\big) = diag\big(x, x', x^\sigma, (x')^\sigma\big),$$

and defines an admissible isomorphism of T_M *onto* T_G *where*

$$x = t_1^\sigma t_2^\sigma, \qquad x' = t_1 t_2^\sigma.$$

In the degenerate case $L \cong F^2$

$$\phi : (diag(t_1, t_1'), diag(t_2, t_2')) \mod F^* \mapsto diag(t_2' t_1', t_2' t_1, t_2 t_1, t_2 t_1')$$

is an admissible F-*isomorphism between the split diagonal tori* T_M^s *and* T_G^s.

Proof. Once we know that the dual isomorphism of ϕ gives the composed map $\hat{T}_G^s \to \hat{T}_M^s$ described above, from the description above it is clear that ϕ is an admissible homomorphism. ϕ induces $X^*(\phi) : X^*(T_G^s) \to X^*(T_M^s)$ defined by

$$
\begin{array}{lll}
(1,0,0,0) \longleftrightarrow x & \mapsto & t_1^\sigma t_2^\sigma \longleftrightarrow (0,1,0,1) \\
(0,1,0,0) \longleftrightarrow x^\sigma & \mapsto & t_1 t_2 \longleftrightarrow (1,0,1,0) \\
(0,0,1,0) \longleftrightarrow x' & \mapsto & t_1 t_2^\sigma \longleftrightarrow (1,0,0,1) \\
(0,0,0,1) \longleftrightarrow (x')^\sigma & \mapsto & t_1^\sigma t_2 \longleftrightarrow (0,1,1,0).
\end{array}
$$

Hence, $(a, b, c, d) \mod \mathbb{Z}(1, 1, -1, -1)$ maps to $(b + c, a + d, b + d, a + c)$ under $X^*(\phi)$, and hence the dual of ϕ is $\hat{\phi}(w_1, w_2, w_1', w_2') = (w_2 w_1', w_1 w_2', w_2 w_2', w_1 w_1') = (z_1, z_2, z_3, z_4) \in \hat{T}_M^s$ for (w_1, w_2, w_1', w_2') representing an element in $\hat{T}_G^s$. □

Résumé. We considered tori T_G, T_G^s, T_M, T_M^s and their duals and the reference tori $\hat{\mathbf{T}}_M = \hat{\mathbf{T}}_G$. The torus $T_G \subseteq GSp(4, F)$ was identified with the tori $T_M \subseteq M(F)$ by an admissible isomorphism ϕ. Over the algebraic closure we can diagonalize both tori to obtain T_G^s and T_M^s. In particular, we had

$$h^{-1} \begin{pmatrix} a & 0 & bD_0 & 0 \\ 0 & a' & 0 & b'D_0 \\ b & 0 & a & 0 \\ 0 & b' & 0 & a' \end{pmatrix} h = \begin{pmatrix} x & 0 & 0 & 0 \\ 0 & x' & 0 & 0 \\ 0 & 0 & x^\sigma & 0 \\ 0 & 0 & 0 & (x')^\sigma \end{pmatrix},$$

where $x = a + b\sqrt{D_0}$ and $x' = a' + b'\sqrt{D_0}$ and h was defined above by the choice of some square root $\sqrt{D_0}$. Then the admissible isomorphism "in diagonalized form" is

$$\begin{pmatrix} t_1 & * \\ 0 & t_1^\sigma \end{pmatrix} \times \begin{pmatrix} t_2 & 0 \\ * & t_2^\sigma \end{pmatrix} \quad mod \ (t \cdot id, t^{-1} id) \mapsto \begin{pmatrix} x & 0 & * & * \\ * & x' & * & * \\ 0 & 0 & x^\sigma & * \\ 0 & 0 & 0 & (x')^\sigma \end{pmatrix}$$

for $x = t_1^\sigma t_2^\sigma$, $x' = t_1 t_2^\sigma$, the stars being zero (only reminding us of the chosen position of the Borel subgroups). The admissible isomorphism between the tori T_M and T_G is defined by the composition with the diagonalization isomorphisms (i.e., $g \mapsto hgh^{-1}$ maps (t_1, t_2) to an element $\eta \in T_G = hT_G^s h^{-1}$). To be more precise, we also had to fix some embedding of T_M into M and some conjugation isomorphism, defined over $\overline{F}$, from T_M to the split torus T_M^s in M. This allows us to identify the element $(t_1, t_2) \in T_M(F)$ with the diagonal matrix in $T_M^s(\overline{F})$ considered above.

8.2 Statement of Result

With data and notation chosen as in Sect. 8.1, we now determine the transfer factor of Langlands and Shelstad. By definition it is a product of several terms, listed below. Let $\chi_{L/F}$ denote the quadratic character of the local field F^*, attached to the extension L/F by class field theory. Suppose $t = (t_1, t_2) \longleftrightarrow x \longleftrightarrow \eta$ correspond to each other by the admissible homomorphisms defined in Sect. 8.1. Then

Lemma 8.2. *The transfer factor of Langlands and Shelstad [60], (3.7), in the quasisplit situation described above is the product of the factors*

$$\Delta_I(\eta, t_1, t_2) = \chi_{L/F}(D_0) = \chi_{L/F}(-1),$$

$$\Delta_{II}(\eta, t_1, t_2) = \chi_0\left(\frac{x/x^\sigma - 1}{\sqrt{D_0}}\right)\chi_0\left(\frac{x'/(x')^\sigma - 1}{\sqrt{D_0}}\right),$$

where χ_0 is an arbitrary character on L^* inducing on the subgroup F^* the quadratic character $\chi_{L/F}$. For trivial reasons

$$\Delta_{III_1}(\eta, t_1, t_2) = 1$$

under our assumptions, which implies $\gamma = \gamma_G$ in the sense of [60], (3.4).

$$\Delta_{III_2}(\eta, t_1, t_2) = \chi_0(x^\sigma/x').$$

Finally, from Sect. 6.3

$$\Delta_{IV}(\eta, t_1, t_2) = |xx'|^{-1}|(x^\sigma - x)(x'^\sigma - x')|.$$

Remark 8.2. Hence, $\Delta(\eta, t_1, t_2) = (\chi_{L/F}|.|)((x - x^\sigma)(x' - x'^\sigma))|xx'|^{-1}$ since $\chi_0(x'x'^\sigma) = \chi_{L/F}(x'x'^\sigma) = 1$. Notice $x'x'^\sigma \in Norm_{L/F}(L^*) \subseteq F^*$.

In Chap. 6 the different assignment $x = t_1 t_2$ and $x' = t_1 t_2^\sigma$ was used instead of $x = t_1^\sigma t_2^\sigma$ and $x' = t_1 t_2^\sigma$ used presently. This amounts to replacing x by x^σ, which changes $(\chi_{L/F}|.|)((x - x^\sigma)(x' - x'^\sigma))|xx'|^{-1}$ by the sign $\chi_{L/F}(-1)$. But also η changes to some new element η'. Notice $\Delta(\eta', t_1, t_2) = \chi_{L/F}(-1)\Delta(\eta, t_1, t_2)$. So the Langlands–Shelstad transfer factor picks up the sign $\chi_{L/F}(-1)$, which corresponds to the change of η to η' within the stable conjugacy class (or the change of Δ_{III_1} caused by this). Hence, in the notation of Chap. 6, writing η now again instead of η', Lemma 8.2 implies

Corollary 8.1. *For $x = t_1 t_2, x' = t_1 t_2^\sigma$ and η defined as in Sect. 6.3, the Langlands–Shelstad transfer factor is*

$$\Delta(\eta, t_1, t_2) = (\chi_{L/F}|.|)((x - x^\sigma)(x' - (x')^\sigma))|xx'|^{-1}.$$

Proof of Lemma 8.2. The proof is a lengthy unraveling of the definitions and covers most of the remaining part of this chapter. The definition of the Langlands–Shelstad transfer factor (which is denoted Δ_0 in [60], p. 248, and is defined as a product of five factors in the quasisplit case) is sophisticated, and will not be reproduced here in detail. □

It requires many choices, among them the choice of a-data and χ-data. Following [60], we start with a-data and χ-data.

8.2.1 a-Data and χ-Data

These are numbers and characters attached to roots in the root system R_G. For the general definition of a-data and χ-data see [60], (2.2) and (2.5):

1. The positive roots R_G^+ of T_G are the characters which assign to $diag(x, x', x^\sigma, x'^\sigma)$ the values $x'/x, x/x^\sigma, x'/(x')^\sigma, x/(x')^\sigma$. Remember the Galois group acts via its quotient $Gal(L/F)$ such that $\sigma(\lambda) = -\lambda$ for all roots λ since $xx^\sigma = x'(x')^\sigma$.
   x/x^σ and $x'/(x')^\sigma$ are the positive roots, which do not occur in M. Hence ([60], (3.6)) the factor $\Delta_{IV}(\eta; (t_1, t_2))$ is

$$|(1 - x'/x)(1 - x/x^\sigma)(1 - x'/(x')^\sigma)(1 - x/(x')^\sigma)|/|(1 - t_1/t_1^\sigma)(1 - t_2^\sigma/t_2)|$$

$$= |(1 - x/x^\sigma)(1 - x'/(x')^\sigma)| = |(x - x^\sigma)(x' - (x')^\sigma)|/|xx'|.$$

2. Define a-data for the root system R_G

$$a_\lambda = \sqrt{D_0} \quad \text{for} \quad \lambda \in R_G^+, \qquad a_\lambda = -\sqrt{D_0} \quad \text{for} \quad \lambda \in R_G^-,$$

i.e., these numbers satisfy $a_\lambda \in \overline{F}^*$ and $\sigma(a_\lambda) = a_{\sigma(\lambda)}$ and $a_{-\lambda} = -a_\lambda$, as required in the definition [60], (2.2).

3. *Similarly we get the χ-data.* Fix a quasicharacter $\chi_0 : L^* \to \mathbb{C}^*$ such that the restriction of χ_0 from L^* to F^* is the quadratic character $\chi_{L/F}$ of F^* attached to L/F via class field theory. Then we get χ-data by defining

$$\chi_\lambda = \chi_0 \quad \text{for} \quad \lambda \in R_G^+, \qquad \chi_\lambda = \chi_0^{-1} \quad \text{for} \quad \lambda \in R_G^-.$$

The required conditions $\chi_{\sigma(\lambda)} = \chi_\lambda \circ \sigma^{-1}$ and $\chi_{-\lambda} = \chi_\lambda^{-1}$ of [60], (2.5), are satisfied since $\chi_0 \circ \sigma^{-1} = \chi_0^{-1}$. In the following χ_0 will be viewed as a character of the Weil group $W_+ = W_{L/F_+} = W_{L/L} \cong L^*$ as well.

4. *Notations from [60], (2.5).* $F_{+\lambda}$ denotes the fixed field of $\Gamma_{+\lambda} = \{\sigma \in \Gamma \mid \sigma(\lambda) = \lambda\}$ and $F_{\pm\lambda}$ is the fixed field of $\Gamma_{\pm\lambda} = \{\sigma \in \Gamma \mid \sigma(\lambda) = \pm\lambda\}$. Γ is the absolute Galois group of F. In our case $F_+ = L$ and $F_\pm = L$ for all λ, and the index is always 2.

5. The positive roots α of G given by the characters x/x^σ and $x'/(x')^\sigma$, which do not occur in M, define, following [60], (3.3), a product of corresponding terms $\chi_\alpha(\frac{\alpha(\eta)-1}{a_\alpha})$, which is the factor $\Delta_{II}(\eta; (t_1, t_2))$

$$\boxed{\Delta_{II}(\eta; (t_1, t_2)) = \chi_0\left(D_0^{-1}\left(x/x^\sigma - 1\right)\left(x'/(x')^\sigma - 1\right)\right)}.$$

8.2.2 The Element $\omega_T(\sigma)$

This section refers to [60], (2.1) and (2.3). Consider the split torus T_G^s and the Borel group B_G^s above. Then conjugation with the matrix h, defined on page 273, gave $h^{-1}(T_G^s, B_G^s)h = (T_G, B_G)$ with a Borel group B_G defined over $\overline{F}$. We get $(h^{-1}h^\sigma)^{-1}(T_G^s, B_G^s)(h^{-1}h^\sigma) = (T_G^s, B_G^s)$. The element $\omega_T(\sigma)$ is defined as $\omega_T(\sigma) = Int(h^{-1}h^\sigma)$ in [60] p. 231. It is in the normalizer of T_G^s. In our case

$$h^{-1}h^\sigma = h^{-\sigma}h = \begin{pmatrix} 0 & -E \\ -E & 0 \end{pmatrix}.$$

Hence, $\omega_T(\rho) \equiv 1$ for $\rho \in \Gamma_L$ and $\omega_T(\rho) \equiv w_{max}$ for $\rho \in \sigma\Gamma_L$ in the Weyl group for the longest element w_{max} in the Weyl group of T_G^s.

8.2.3 Lifting of Weyl Group Elements

This section refers to [60], (2.1) and (2.3), where a cocycle $t(\theta_1, \theta_2)$ is defined, that somehow measures the obstruction for lifting elements of the Weyl group $W = N(T)/T$ to the normalizer $N(T)$.

The simple roots of (T_G^s, B_G^s) are $\alpha_1 = 2e_1$ and $\alpha_2 = e_2 - e_1$. We get the embedded root subgroups

$$Sl(2)_{\alpha_1} \ni \begin{pmatrix} a & b \\ c & d \end{pmatrix} \longmapsto \begin{pmatrix} a & 0 & b & 0 \\ 0 & 1 & 0 & 0 \\ c & 0 & d & 0 \\ 0 & 0 & 0 & 1 \end{pmatrix}$$

and

$$Sl(2)_{\alpha_2} \ni \begin{pmatrix} a & b \\ c & d \end{pmatrix} \longmapsto \begin{pmatrix} d & -c & 0 & 0 \\ -b & a & 0 & 0 \\ 0 & 0 & a & b \\ 0 & 0 & c & d \end{pmatrix},$$

which map $\begin{pmatrix} 1 & * \\ 0 & 1 \end{pmatrix}$ to the Borel group B_G^s. The standard element of the Weyl group of $Sl(2)$ therefore maps to

$$n(\sigma_{\alpha_1}) = \begin{pmatrix} 0 & 0 & 1 & 0 \\ 0 & 1 & 0 & 0 \\ -1 & 0 & 0 & 0 \\ 0 & 0 & 0 & 1 \end{pmatrix}, \qquad n(\sigma_{\alpha_2}) = \begin{pmatrix} 0 & 1 & 0 & 0 \\ -1 & 0 & 0 & 0 \\ 0 & 0 & 0 & 1 \\ 0 & 0 & -1 & 0 \end{pmatrix}.$$

The element w_{max} in the Weyl group of $GSp(4)$ can be written as a product of simple reflections of length 4 $w_{max} = \sigma_{\alpha_2}\sigma_{\alpha_1}\sigma_{\alpha_2}\sigma_{\alpha_1}$. Hence, $n(w_{max}) := n(\sigma_{\alpha_2})n(\sigma_{\alpha_1})n(\sigma_{\alpha_2})n(\sigma_{\alpha_1})$ is

$$n(w_{max}) = \begin{pmatrix} 0 & 1 & 0 & 0 \\ 0 & 0 & -1 & 0 \\ 0 & 0 & 0 & 1 \\ 1 & 0 & 0 & 0 \end{pmatrix}^2 = \begin{pmatrix} 0 & -E \\ E & 0 \end{pmatrix}.$$

The Cocycle $t(\theta_1, \theta_2)$. For θ_1, θ_2 in the subgroup of the Weyl group Ω generated by 1 and w_{max}, this lifting to the normalizer defines a cocycle $t(\theta_1, \theta_2)$ ([60], Lemma 2.1A). Its value is 1 except for $\theta_1 = \theta_2 = w_{max}$, where $t(w_{max}, w_{max})$ is -1. The first property follows from the definition of t, the second from our computation above in view of the definition of $t(w_{max}, w_{max})n(w_{max}^2) = n(w_{max})n(w_{max})$ in [60], p. 228.

8.2.4 Coroot Modification

In this section we compute the class λ_T and $\lambda_{T_{sc}}$ defined in [60], (2.3):

1. The coroots of T_G^s are defined by the cocharacters which map $t \in F^*$ to the matrices

$$diag(t^{-1}, t, t, t^{-1}), \; diag(t, 1, t^{-1}, 1), \; diag(1, t, 1, t^{-1}), \; diag(t, t, t^{-1}, t^{-1}).$$

The first two are the coroots of $(e_2 - e_1)$ and $2e_1$, the last two those of $2e_2$ and $e_2 + e_1$. Langlands and Shelstad ([60], (2.3)) defined

$$x(\sigma_T) = \prod_{\alpha \in R_G^+,\, \sigma_T^{-1}\alpha \in R_G^-} (a_\alpha)^{\alpha^\vee}$$

for $\sigma_T \in \Gamma$ via the a-data a_α. $x(\sigma_T)$ only depends on σ_T modulo Γ_L. Notice $x(1) = 1$ and $x(\sigma) = diag(t, t^3, t^{-1}, t^{-3})_{t=\sqrt{D_0}}$ is the product of all four positive coroots evaluated at $a_\lambda = \sqrt{D_0}$.

2. The cocycle $\sigma_T \mapsto m(\sigma_T) = x(\sigma_T)n(\omega_T(\sigma_T))$ defined in [60], (2.3), has values in the $\overline{F}$-valued points of the normalizer of T_G^s. Modulo elements of $T_G^s(\overline{F})$ this coincides with $\sigma_T \mapsto h^{-\sigma}h$. Therefore, $\sigma_T \mapsto m(\sigma_T)h^{-\sigma}h \in T_G^s(\overline{F})$ defines a 1-cocycle with values in $T_G^s(\overline{F})$. The Galois action on this torus is understood to be transferred from the action of the Galois group on $T_G(\overline{F})$. Hence, σ acts on $T_G^s(\overline{F})$ by

$$\xi^\sigma := w_{max}(\sigma(\xi)).$$

$m(\sigma_T)$ is determined by its value on the generator σ of $Gal(L/F)$, which corresponds to $\sigma_T = w_{max} \times \sigma$. The computations above therefore give the value $\lambda := m(\sigma_T) = diag(t, t^3, -t^{-1}, -t^{-3})_{t=\sqrt{D_0}}$. Observe the cocycle condition $\lambda^\sigma \lambda = 1$. The class of λ in $H^1(\Gamma, T_G^s(\overline{F})_{trans})$ will be denoted λ_T.

Remark 8.3. With respect to the transferred action on $T_G^s(\overline{F})$ the map $Int(h)$ defines a Galois equivariant isomorphism between $T_G(\overline{F})$ and $T_G^s(\overline{F})$. This allows us to view λ_T as a cocycle with values $T_G(\overline{F})$ alternatively. Hence, by Lemma 6.1

$$H^1(\Gamma, T_G^s(\overline{F})_{trans}) = H^1(Gal(L/F), T_G(\overline{F})) \cong F^*/Norm_{L/F}(L^*) \cong \{\pm 1\}.$$

The class λ_T of the cocycle defined by λ is trivial if and only if there exists a $\rho \in T_G^s(\overline{F})$ or $T_G^s(L)$ such that $\lambda = w_{max}(\sigma(\rho))\rho^{-1}$.

Remark 8.4. The class $\lambda_{T_{sc}}$ is defined as above with h_{der} instead of h. Therefore, the image of $\lambda_{T_{sc}} \in H^1(\Gamma, T_{sc}(\overline{F}))$ in $H^1(\Gamma, T(\overline{F}))$ is λ_T since both cocycles differ by a coboundary $z^\sigma z^{-1} \in B^1(\Gamma, Z_G(\overline{F})) \subseteq B^1(\Gamma, T_G(\overline{F}))$.

3. For the root $\alpha_1 : T_G^s(L) \to L^*$ defined by $\rho = diag(\rho_1, \rho_2, \kappa\rho_1^{-1}, \kappa\rho_2^{-1}) \mapsto \rho_2/\rho_1$ we have $\alpha_1(\rho^\sigma \rho^{-1}) = Norm_{L/F}(\alpha_1(\rho))^{-1} \in F^*$ since

$$\rho^\sigma = w_{max}\sigma(\rho) = diag\big(\sigma(\kappa/\rho_1), \sigma(\kappa/\rho_2), \sigma(\rho_1, \sigma(\rho_2)\big).$$

The cocycle condition $\lambda^\sigma \lambda = 1$ for $\lambda = diag(\lambda_1, \lambda_2, \kappa\lambda_1^{-1}, \kappa\lambda_2^{-1})$ therefore reads $\kappa = \lambda_1/\sigma(\lambda_1) = \lambda_2/\sigma(\lambda_2)$. Hence, $\alpha_1(\lambda) = \lambda_2/\lambda_1 \in F^*$. Conversely, any $\lambda_2 \in F^*$ comes from a cocycle. For example, put $\lambda = (1, \lambda_2, 1, \lambda_2^{-1})$ for $\lambda_1 = 1$ and $\kappa = 1$. This defines an isomorphism

$$\alpha_1 : H^1(\Gamma, T_G^s(\overline{F})_{trans}) \cong F^*/Norm_{L/F}(F^*),$$

which maps the cohomology class of λ to $\alpha_1(\lambda) \in F^*/Norm_{L/F}(L^*)$. Composed with the isomorphism $\chi_{L/F} : F^*/Norm_{L/F}(L^*) \cong \{\pm 1\}$, the class of the cocycle λ obtained above can be identified with the sign $\lambda_T \in H^1(\Gamma, T_G(\overline{F})) \cong \{\pm 1\}$ defined by

$$\lambda_T = \chi_{L/F}(\alpha_1(\lambda)) = \chi_{L/F}(D_0).$$

Definition of Δ_I. According to [60], (3.2), the factor $\Delta_I(\eta; (t_1, t_2))$ is defined to be

$$\langle \lambda_{T_{sc}}, \mathbf{s_T} \rangle,$$

where $\mathbf{s_T} \in \pi_0((\hat{T}_{G,sc})^\Gamma)$ is the image of $s \in \hat{T}_G$. In our case $s \in (\hat{T}_M)^\Gamma$; see page 55. Hence, $\mathbf{s_T}$ is the image of the class $\mathbf{s} \in \pi_0(\hat{T}_G^\Gamma)$ under the natural map induced from the inclusion $i : T_{G,sc} \to T_G$.

4. Since $s \in Z(\hat{M})^\Gamma$ in our case, one can work with λ_T instead of $\lambda_{T_{sc}}$ to define the factor Δ_I since functoriality of the Tate–Nakayama pairing

$$\langle \cdot, \cdot \rangle : H^1(\Gamma, T(\overline{F})) \times \pi_0(\hat{T}^\Gamma) \to \mathbb{C}^*$$

with respect to $i : T_{G,sc} \to T_G$ implies $\langle \lambda_{T_{G,sc}}, \mathbf{s_T} \rangle = \langle i_*(\lambda_{T_{G,sc}}), \mathbf{s} \rangle$. Since $i_*(\lambda_{T_{G,sc}}) = \lambda_{T_G}$ according to Remark 8.4, the calculations above show

$$\boxed{\Delta_I(\eta; (t_1, t_2)) = \chi_{L/F}(D_0)}.$$

8.2.5 p-Gauges

As in [60], (2.5), choose a gauge p. For this let Γ denote the absolute Galois group of F and let $\{1, \epsilon\}$ be the group generated by the reflection ϵ at the origin, both acting on the root system R_G of T_G. The product Σ of these two groups acts on R_G with four disjoint orbits, each consisting of a pair $\alpha, -\alpha$, where α runs over the positive roots R_G^+. So we choose as a p-gauge the positive root gauge p on each orbit. See [60], (2.1) and (2.5). Then $n = 1$ and $\sigma_1 = id$ and $w_1 = id$ and $u_1(w) = w$ in the notation in [60], (2.5).

8.2.6 Weil Groups

Consider Weil groups as in [60], (2.5) or [101]. The Weil group W_F of F comes with a map $\pi : W_F \to \Gamma = \Gamma_F$ such that $W_L = \pi^{-1}(\Gamma_L)$, where Γ_L denotes the absolute Galois group of L. Then $W_F/W_L = \Gamma_F/\Gamma_L = \Gamma_{L/F}$. The relative Weil groups are defined by $W_{L/F} = W_F/\overline{[W_L, W_L]}$. One has exact sequences

$$0 \to L^* \to W_{L/L} \to \Gamma_{L/L} = 1,$$
$$0 \to L^* \to W_{L/F} \to \Gamma_{L/F} \to 0.$$

We choose representatives $v_0 = 1$ and v_1 in $W_{L/F}$ of 1 and σ in $\Gamma_{L/F}$. Then for $u \in W_{\pm} = W_{L/F}$ we have

$$v_0(u) = \begin{cases} u & u \in W_+ = W_{L/L} = L^* \\ uv_1^{-1} & u \notin W_+ = L^* . \end{cases}$$

8.2.7 The 1-Chain r_p for $\hat{G}$

To define the Δ_{III_2} factor one has to determine a certain 1-chain $r_p(w)$ attached to the gauge p. Consider $X = X_*(\hat{T}_G^s) = X^*(T_G^s)$. The roots $R = R_X$ are by definition the roots of T_G^s in G or the coroots of $\hat{T}_G^s$ in $\hat{G}$. Consider $X \otimes \mathbb{C}^*$. Langlands and Shelstad defined in [60], p. 236,

$$r_p(w)_{orbit} = \prod_{i=1}^n \chi_{\lambda_1}(v_0(w))^\lambda = \chi_0(v_0(w))^\lambda$$

for each orbit, i.e., positive coroot λ of $\hat{T}_G^s$. Here we have chosen the "positive" gauge on each orbit. The product over the contributions of all (four) orbits then defines

$$r_p(w) = \prod_{\lambda \in R^+} \chi_0(v_0(w))^\lambda \in X_*(\hat{T}_G^s, \hat{G}) \otimes \mathbb{C}^* = \hat{T}_G^s(\mathbb{C})$$

for $w \in W_{L/F}$. In our case this gives

$$r_p(w) = diag(t, t^3, t^{-1}, t^{-3})\big|_{t = \chi_0(v_0(w))} \in \hat{T}_G^s.$$

8.2.8 Definition of ξ_χ^G

As Langlands and Shelstad explain ([60], p. 238), $r_p(\omega)$ defines an L-extension $\xi = \xi_\chi^G$

$$\xi_\chi^G : {}^L T \to {}^L G$$

of the given embedding

$$\hat{T}_G \to \hat{T}_G^s \to \mathbf{T}_G \subseteq \hat{G},$$

which depends on the given χ-data, and which is defined by

$$\xi_\chi^G(1 \times w) = r_p(w) n^{\hat{G}}(\omega_T(w)) \times w, \qquad w \in W_F.$$

For this notation recall that L-groups are defined as semidirect products $\hat{G} \times W_F$ or $\hat{G} \times W_{L/F}$.

8.2.9 The Similar r_p for $\hat{M}$

This section refers to [60], (2.5), repeating for M what was done for G. Recall the definition of $\hat{T}^s_M$, $\hat{B}^s_M$ for $\hat{M}$, which is needed now. The inclusion $\xi : \hat{M} \to \hat{G}$ maps the (positive) coroots of $\hat{T}^s_M \subseteq \hat{M}$ to the (positive) coroots of $\hat{T}^s_G \subseteq \hat{G}$. Our choice of χ-data for G therefore immediately induces χ-data for M such that

$$\chi_\lambda = \begin{cases} \chi_0 & \text{for } \lambda \text{ being a positive coroot of } \hat{T}^s_M, \\ \chi_0^{-1} & \text{for } \lambda \text{ being a negative coroot of } \hat{T}^s_M. \end{cases}$$

The positive coroots of $\hat{T}^s_M$ are given by $\beta_1^\vee(t) = diag(t, 1, t^{-1}, 1)$ and by $\beta_2^\vee(t) = diag(1, t, 1, t^{-1})$. For the corresponding positive roots β_1 and β_2 of $\hat{T}^s_M$ in $\hat{M}$ obviously

$$n(\sigma_{\beta_1}) = \begin{pmatrix} 0 & 0 & 1 & 0 \\ 0 & 1 & 0 & 0 \\ -1 & 0 & 0 & 0 \\ 0 & 0 & 0 & 1 \end{pmatrix}, \qquad n(\sigma_{\beta_2}) = \begin{pmatrix} 1 & 0 & 0 & 0 \\ 0 & 0 & 0 & 1 \\ 0 & 0 & 1 & 0 \\ 0 & -1 & 0 & 0 \end{pmatrix}.$$

The product of these matrices defines

$$n^{\hat{M}}(w_{max}) = \begin{pmatrix} 0 & E \\ -E & 0 \end{pmatrix}$$

for the longest element w_{max} in the Weyl group of $\hat{T}^s_M$. Again the group Σ generated by the reflection at the origin and Γ acts via the T_M-twisted action on the roots of T^s_M and the coroots of $\hat{T}^s_M$, respectively, with two orbits $\beta_1, -\beta_1$ and $\beta_2, -\beta_2$. We can therefore choose the positive R^+_M gauge on each orbit. Then $n = 1$ and $\sigma_1 = id$ and $u_1(w) = w$ on each orbit in the notation in [60], (2.5). Consider $X = X_*(\hat{T}^s_M, \hat{B}^s_M, \hat{M})$ and R to be the set of coroots of $\hat{T}^s_M$, $\hat{B}^s_M$. Hence,

$$r_p(w) = diag(t, 1, t^{-1}, 1)diag(1, t, 1, t^{-1})_{t=\chi_0(v_0(w))} = diag(t, t, t^{-1}, t^{-1})_{t=\chi_0(v_0(w))}.$$

8.2.10 Definition of ξ_χ^M

Our choice of χ-data and the gauge for M defines an extension

$$\xi_\chi^M : {}^L T_M \to {}^L M$$

of the given embedding

$$\hat{T}_M \to \hat{T}^s_M \to \mathbf{T}_M \subseteq \hat{M}.$$

It factorizes over the quotient $W_{L/F}$ of the Weil group, and it is defined by

$$\xi_\chi^M(1 \times w) = diag(t, t, t^{-1}, t^{-1})\big|_{t=\chi_0(v_0(w))} \cdot n^{\hat{M}}(\omega_T(\sigma_w)) \times w.$$

8.2.11 The Corestriction Map and the Langlands Reciprocity Law

Let T be a torus over F, split by the extension field L. Then we have the following chain of identifications

$$Hom_{cont}(T(F), \mathbb{C}^*) \cong Hom_{cont}((L^* \otimes X_*(T))^{\Gamma_{L/F}}, \mathbb{C}^*) \cong Hom(L^*, X^*(T) \otimes \mathbb{C}^*)_{\Gamma_{L/F}}$$

$$\cong Hom(L^*, X_*(\hat{T}) \otimes \mathbb{C}^*)_{\Gamma_{L/F}} \cong Hom(L^*, \hat{T}(\mathbb{C}))_{\Gamma_{L/F}},$$

and the commutative diagram

$$
\begin{array}{ccc}
Hom_{cont}(T(F), \mathbb{C}^*) & \xrightarrow{\;\;\simeq\;\;} & Hom(L^*, \hat{T}(\mathbb{C}))_{\Gamma_{L/F}} \\
\Big\uparrow{\scriptstyle LR} & & \Big\uparrow \\
H^1_{cont}(W_{L/F}, \hat{T}(\mathbb{C})) & \xleftarrow{\;\;cores\;\;} & Hom(L^*, \hat{T}(\mathbb{C})).
\end{array}
$$

The left vertical isomorphism is the "Langlands" reciprocity map for tori. It is functorial in T. The right vertical map is the obvious surjection. The lower horizontal map is the corestriction map. Details can be found in Milne [63], p. 135ff and Theorem 8.6. The corestriction map is defined as follows: For $f \in Hom(L^*, \hat{T}(\mathbb{C}))$ define $cores(f) \in H^1(W_{L/F}, \hat{T}(\mathbb{C}))$ by the class of the cocycle

$$cores(f)(w) = \sum_g w_g f(w_g^{-1} w w_{g'}),$$

where w_g are representatives of $\Gamma_{L/F}$ in $W_{L/F}$ and $ww_{g'} = w_g \mod L^*$ determines g' and $w_{g'}$, respectively, for given w and g. For further details see [63], p. 129.

Now specialize to the quadratic extension field L over F. In this case we have already specified representatives $w_1 = v_0 = 1$ and $w_\sigma = v_1$ in the Weil group $W_{L/F}$. Put $\theta = v_1^2$. Notice $\theta \in F^*$. We then get

$$cores(f)(\lambda) = f(\lambda) v_1 f(v_1^{-1} \lambda v_1) = f(\lambda) v_1 f(\lambda^\sigma), \qquad \lambda \in L^* \subseteq W_{L/F},$$

$$cores(f)(\lambda v_1) = f(\lambda v_1 \cdot v_1) v_1 f(v_1^{-1} \lambda v_1) = f(\lambda \theta) v_1 f(\lambda^\sigma), \qquad \lambda \in L^* \subseteq W_{L/F}.$$

8.2.12 Langlands Reciprocity for the Norm-1-Torus $T = N_L^1$

Notation. Let N denote the torus $Res_{L/F}(\mathbb{G}_m)$ and N^1 the subtorus $N^1 \subseteq N$, which is the kernel of the norm $Norm_{L/F}$. The dual group of N^1 is denoted $\hat{T} = \mathbb{C}_-^*$, which is $\mathbb{C}^*$ as a Lie group. But the Galois or Weil group acts by $z^\sigma = z^{-1}$, $z \in \mathbb{C}_-^*$, where σ is the generator of $Gal(L/F)$.

A cocycle $a \in Z^1(W_F, \hat{N}^1)$ becomes a character once it is restricted to $W_L \subseteq W_F$. This character is trivial on $[W_L, W_L]$, and hence it factorizes over $W_{L/F} = W_F/[W_L, W_L]$:

$$0 \to L^* \to W_{L/F} \to \Gamma_{L/F} \to 0.$$

Choose $v_1 \in W_{L/F}$, but $v_1 \notin L^*$. Then $v_1(v_1^2)v_1^{-1} = (v_1)^2 \in F^*$. One can show $v_1^2 \notin Norm_{L/F}(L^*)$. An easy computation shows that the cocycle conditions for a are equivalent to:

1. The restriction of a to L^* is a continuous character with values in $\mathbb{C}_-^*$.
2. $a(\sigma(\lambda)) = a(\lambda^{-1})$ for $\lambda \in L^*$.
3. $a(v_1^2) = a(v_1)v_1(a(v_1)) = a(v_1)/a(v_1) = 1$.

Therefore, conditions (2) and (3) are equivalent to $a(\lambda) = 1, \lambda \in F^*$. If we modify a by a coboundary, this does not change the restriction of a to L^*, whereas the restriction to $L^* \cdot v_1$ can in this way be multiplied by an arbitrary constant in $\mathbb{C}^*$. Therefore, we formulate the last cocycle condition as:

4. $a(\lambda v_1) = const \cdot a(\lambda)$. Hence, the cohomology class $[a]$ of a is uniquely determined by the restriction of a to L^*, which is an element in $Hom_{cont}(L^*/F^*, \mathbb{C}^*)$.

To make the Langlands reciprocity map explicit, suppose we have a continuous character $\chi \in Hom_{cont}(N^1(F), \mathbb{C}^*)$. We can extend χ to a continuous character $\chi' : L \to \mathbb{C}^*$ via the embedding $N^1(F) \subseteq L^*$. The map

$$Hom(L^*, \hat{N}^1) \to Hom(L^*, \hat{N}^1)_\Gamma \cong Hom((L^* \otimes \mathbb{Z})^\Gamma, \mathbb{C}^*) \cong Hom(N^1(F), \mathbb{C}^*)$$

is the restriction map

$$(\chi' : L^* \to \mathbb{C}^*) \mapsto (\chi'|N^1(F) : N^1(F) \to \mathbb{C}^*).$$

This describes the right and then the upper way through the diagram

$$
\begin{array}{ccc}
Hom_{cont}(N^1(F), \mathbb{C}^*) & \overset{\simeq}{\rule{3cm}{0.4pt}} & Hom(L^*, \mathbb{C}_-^*)_{\Gamma_{L/F}} \\
\uparrow{\scriptstyle LR} & & \uparrow \\
H^1_{cont}(W_{L/F}, \mathbb{C}_-^*) & \overset{cores}{\longleftarrow} & Hom(L^*, \mathbb{C}_-^*)
\end{array}
$$

and in fact maps χ' to χ. The image of χ' under the lower horizontal corestriction map is a cohomology class, which is completely determined by its "restriction" to $L^* = W_L^{ab}$. This restriction is given by the character

$$cores(\chi')(\lambda) = \chi'(\lambda)v_1(\chi'(\lambda^\sigma)) = \chi'(\lambda/\lambda^\sigma) = \chi(\lambda/\lambda^\sigma), \qquad \lambda \in L^* = W_L^{ab},$$

since $\lambda/\lambda^\sigma \in N^1(F)$. Conversely, every $x \in N^1(F)$ can be written in the form $x = \lambda/\lambda^\sigma$, where $\lambda \in L^*$ is uniquely determined up to an element in F^*. Therefore, Langlands reciprocity (i.e., the inverse of the left vertical map) is explicitly given by

$$Hom_{cont}(N^1(F), \mathbb{C}^*) \ni \chi \mapsto a(\lambda) = \chi(\lambda/\lambda^\sigma),$$

in the sense that the image class $[a] \in H^1(W_F, \mathbb{C}^*_-)$ is determined by its restriction $a(\lambda) \in Hom_{cont}(W_L^{ab}, \mathbb{C}^*_-)$ specified above for $\lambda \in W_L^{ab} \cong L^*$.

Example 8.1. For $s \in (\hat{N}^1)^\Gamma = \{\pm 1\}$ and the homomorphism $ord : W_F \to W_F^{ab} \cong F^* \overset{ord}{\to} \mathbb{Z}$ define the 1-cocycle $t^s(w) = s^{ord(w)}$ in $Z^1(W_F, \hat{N}^1)$. The map

$$L^* \cong W_L^{ab} \to W_F^{ab} \cong F^*,$$

induced by the inclusion map $W_L \hookrightarrow W_F$ is the norm $Norm_{L/F} : L^* \to F^*$ ([101], diagram (1.2.2)). The restriction of the cocycle t^s to W_L^{ab} is therefore $t^s(\lambda) = (-1)^{ord(Norm_{L/F}(\lambda))}$. By Langlands reciprocity, t^s corresponds to a continuous character $\chi(x)$ of $N^1(F)$ given by:

- L/F is unramified: χ is trivial.
- L/F is ramified: $\chi(x) = (-1)^{ord(\lambda)}$, where $x = \lambda/\lambda^\sigma$. If the residue characteristic is odd, then $\chi(x) = x \bmod (\pi_L) \in \{\pm 1\}$ for $x \in N^1(F) \subseteq o_L^*$.

Example 8.2. Let now χ' be the character χ_0 of L^* used in the definition of χ-data. Its restriction to $N^1(F) \subseteq L^*$ will be denoted χ. From $\chi_0(x^\sigma) = \chi_0(x)^{-1}$ we obtain

$$cores(\chi)(\lambda) = \chi_0^2(\lambda), \qquad \lambda \in L^* \cong W_L^{ab}.$$

Therefore, $\chi_0 \in Hom_{cont}(N^1(F), \mathbb{C}^*)$ corresponds via Langlands reciprocity to the class of the cocycle $a(w) = \chi_0^2(v_0(w))$ of $W_{L/F}$ in $\hat{N}^1$.

8.2.13 The Definition of Δ_{III_2}

According to [60], (3.5), the factor Δ_{III_2} is defined by a pairing between η and a cocycle a, where a is obtained from a comparison of the L-homomorphisms $\xi \circ \chi_\chi^M$ and ξ_χ^G: Suppose T_M is identified with T_G as in Lemma 8.1. This corresponds to an identification of $\hat{T}_M$ and $\hat{T}_G$ and also $^L T_M$ and $^L T_G$. Using this identification, we get two different homomorphisms from $^L T_M = {}^L T_G$ to $^L G$, namely, $\xi \circ \xi_\chi^M$ and ξ_χ^G. They agree on $\hat{T}_M = \hat{T}_G$, and hence differ by a 1-cocycle $a(w)$ of $W_{L/F}$ with values in $\hat{\mathbf{T}}_G$ (twisted Galois action induced from $\hat{T}_G$!). In other words,

$$\xi \circ \xi_\chi^M(w) = a(w) \cdot \xi_\chi^G(w), \qquad w \in W_{L/F}.$$

Lemma 8.3. *Let χ_0 be the character defined by our specified choice of χ-data. Then*

$$a(w) = (-1)^{i(w)} \cdot diag\left(1, \chi_0^{-2}(v_0(w)), 1, \chi_0^2(v_0(w))\right) \in \hat{\mathbf{T}}_G.$$

By definition, the cocycle condition $a(w_1 w_2) = a(w_1) a(w_2)^{n^{\hat{G}}(\sigma(w_1))}$ holds.

Proof. Let $i(w)$ be 0 or 1 according to whether $w \in L^* \subseteq W_{L/F}$ or not. From the formulas for ξ_χ^M and ξ_χ^G obtained above, we now compute $a(w)$ as follows:

$$\left(\begin{smallmatrix} 0 & E \\ -E & 0 \end{smallmatrix} \right)^{i(w)} \cdot diag(t, t, t^{-1}, t^{-1}) \Big|_{t = \chi_0(v_0(w))}$$

$$= a(w) \cdot \left(\begin{smallmatrix} 0 & -E \\ E & 0 \end{smallmatrix} \right)^{i(w)} \cdot diag(t, t^3, t^{-1}, t^{-3}) \Big|_{t = \chi_0(v_0(w))}.$$

The matrix can be canceled, which implies the claim. $\square$

The Class a_T. Let $[a]$ be the cohomology class in $H^1(W_{L/F}, \hat{\mathbf{T}}_G)$ of the cocycle $a(w)$ obtained in Lemma 8.3. Since the Galois action is defined by transfer, $[a]$ naturally defines a cohomology class

$$a_T \in H^1(W_{L/F}, \hat{T}_G).$$

By Langlands reciprocity a_T corresponds to a continuous character $\langle a_T, . \rangle$ of $T_G(F)$, which evaluated at $\eta \in T_G(F)$ defines, following [60], (3.5), the factor

$$\boxed{\Delta_{III_2}(\eta, t_1, t_2) = \langle a_T, \eta \rangle}.$$

1. We have isomorphisms $\hat{T}_G^s \to \hat{\mathbf{T}}_G \to \hat{N} \times \hat{N}^1$ defined by the maps $\hat{T}_G^s \to \mathbf{T}_G$

$$diag(w_1, w_2, w_1', w_2') \bmod (t, t, t^{-1}, t^{-1}) \mapsto diag(z_1, z_4, z_2, z_3)$$
$$= diag(w_2 w_1', w_1 w_1', w_1 w_2', w_2 w_2')$$

and $\hat{\mathbf{T}}_G \to \hat{N} \times \hat{N}^1$

$$diag(z_1, z_4, z_2, z_3) \mapsto (z_1, z_2) \times z_3/z_2 = (w_2 w_1', w_1 w_2') \times w_2/w_1$$

with inverse $\hat{N} \times \hat{N}^1 \to \hat{\mathbf{T}}_G$

$$(u, v) \times w \mapsto (u, uw^{-1}, v, vw) \in \hat{\mathbf{T}}_G.$$

By $z_1 z_2 = z_3 z_4$ this is an isomorphism. Endow T_G^s with the twisted Galois action and $\hat{\mathbf{T}}_G$ with the induced twisted action, which is the twisted action transferred from $\hat{T}_M$ to $\hat{\mathbf{T}}_M = \hat{\mathbf{T}}_G$. With these conventions, the isomorphisms above are equivariant with respect to the Weil groups.

2. The dual isomorphism $N(F) \times N^1(F) \to T_G^s(F)$ is

$$(y, z) \in L^* \times N^1(F) \mapsto diag(x_1, x_1', x_2, x_2') = diag(y^\sigma z^{-1}, y, yz, y^\sigma) \in T_G^s(\overline{F}).$$

The "partial" inverse $T_G^s(F) \to N^1(F)$ therefore maps $diag(x, x', x^\sigma, (x')^\sigma)$ to x^σ/x'. The composite map $\pi : T_G \to T_G^s \to N^1$ is defined over F, where the first map is $Int(h^{-1})$, and it maps $\eta \in T_G(F)$ to $diag(x, x', x^\sigma, (x')^\sigma) \in T_G^s(\overline{F})$. Hence,

$$\pi : T_G(F) \to N^1(F), \qquad \pi(\eta) = x^\sigma/x'.$$

3. **Conclusion.** The cohomology class $[a] \in H^1(W_{L/F}, \hat{\mathbf{T}}_G)$ of Lemma 8.3 is the product of two classes $[a_1]$ and $[a_2]$, where a_1 is the cocycle

$$a_1(w) = diag(1, \chi_0(v_0(w))^{-2}, 1, \chi_0(v_0(w))^2) \in \hat{\mathbf{T}}_G, \qquad w \in W_{L/F},$$

which is the image of

$$b(w) = \chi_0(v_0(w))^2 \in Z^1(W_{L/F}, \hat{N}^1)$$

under the composition of the inclusion $\hat{N}^1$ into the second factor of $\hat{N} \times \hat{N}^1$ and the isomorphism with $\hat{\mathbf{T}}_G$, and where $a_2(w) = (-1)^{i(w)} \cdot id$. This is the image of the cocycle $((-1)^{i(w)}, (-1)^{i(w)})$ of $\hat{N}$ under the composite of the inclusion into the first factor of $\hat{N} \times \hat{N}^1$ and the isomorphism with $\hat{\mathbf{T}}_G$. It is a coboundary in $\hat{N}$ since $((-1)^{i(w)}, (-1)^{i(w)}) = (1, -1)^w (1, -1)^{-1}$. Hence,

$$[a] = [a_1].$$

4. For the map $\pi : T_G \to N \times N^1 \to N_L^1$ considered in step 2 we showed $a_T = \hat{\pi}_*([b])$ in step 3. Functoriality of Langlands reciprocity therefore implies $\langle a_T, \eta \rangle = \langle \hat{\pi}_*([b]), \eta \rangle = \langle b, \pi(\eta) \rangle$. Since $\langle b, . \rangle$ is the character χ_0 of $N^1(F)$ by Lemma 8.3,

$$\boxed{\Delta_{III_2}(\eta; (t_1, t_2)) = \chi_0(x^\sigma / x')}.$$

This proves Lemma 8.2.

For nonelliptic tori from M the Langlands–Shelstad factor becomes Δ_{IV} for our choice of the endoscopic datum $(M, {}^L M, s, \xi)$. This is left as an exercise for the reader. But one should keep in mind that such statements depend on the choice of the particular endoscopic datum $(M, {}^L M, s, \xi)$. See, e.g., Lemma 9.18, where this is illustrated for the split torus.

8.2.14 The Archimedean Case $F = \mathbb{R}$

$\Delta(\eta, t_1, t_2) = (\chi_{L/F} |.|)((x - x^\sigma)(x' - (x')^\sigma))|xx'|^{-1}$ for $x = t_1 t_2$ and $x' = t_1 t_2^\sigma$ by Corollary 8.1, where η is defined by

$$\eta = \begin{pmatrix} a & 0 & bD_0 & 0 \\ 0 & a' & 0 & b'D_0 \\ b & 0 & a & 0 \\ 0 & b' & 0 & a' \end{pmatrix},$$

and where $x = a + b\sqrt{D_0}$ and $x' = a' + b'\sqrt{D_0}$. The local transfer factor satisfies the global hypothesis: The product over all local places is 1 for global data D_0, t_1, t_2.

If we consider the problem of the stabilization of the trace formula as in Kottwitz [51], we assume that the adelic transfer factor satisfies the global hypothesis [51], p. 178. For this we normalize the Archimedean transfer factor

such that it is the one given by Langlands and Shelstad [60]. The only cases of
elliptic tori in the Archimedean case are the ones of case 1, studied above. The
nonelliptic case is rather easy, and will be skipped.

Remark 8.5. Using this normalization, we therefore do not a priori follow the conventions of Kottwitz [51], p. 184, when we define $\Delta_\infty = \Delta_\infty(\gamma_M, \gamma)$, nor do we
follow [51], p. 186, when we define $h_\infty = h_\infty^M$. Instead, having fixed our choice
$\Delta_\infty(\gamma_M, \gamma)$ of the Archimedean transfer factor, we are now forced to define h_∞^M in
such a way that it satisfies formula (7.4) in [51]:

$$SO_{(t_1, t_2)}(h_\infty^M) = \langle \beta(\eta), s_M \rangle \cdot \Delta_\infty(\eta, (t_1, t_2)) \cdot e(I) \cdot tr\xi_{\mathbb{C}}(\eta) \cdot vol^{-1}.$$

Our choice of the Archimedean transfer factor may therefore result in a change
of h_∞ defined in [51], p. 186, by some constant. Since $F = \mathbb{R}$, to determine this
constant of proportionality, it is enough to consider elliptic case I with $D_0 = -1$
for some suitable chosen regular semisimple elliptic $(t_1, t_2) \in \mathbb{C}^*$. In particular, we
may assume $|t_1| = |t_2| = 1$ in formula (7.4). Since $\chi_{L/\mathbb{R}}|x| = x$ we have

$$\Delta_\infty(\eta, t_1, t_2) = (x - \bar{x})(x' - \bar{x'})|xx'|^{-1}.$$

For $x = t_1 t_2$ and $x' = t_1 \bar{t_2}$ this becomes $(t_1 t_2 - (t_1 t_2)^{-1})((t_1/t_2) - (t_2/t_1))$ or
$(t_1^2 + 1 + t_1^{-2}) - (t_2^2 + 1 + t_2^{-2})$. Thus, for $|t_i| = 1$

$$\Delta_\infty(\eta, t_1, t_2) = \Big(tr(\tau_2 \otimes \tau_0) - tr(\tau_0 \otimes \tau_2)\Big)(t_1, t_2),$$

where τ_k denotes the representation of $SU(2, \mathbb{R})$ on $Symm^k(\mathbb{C}^2)$. Suppose the
representation $\xi_{\mathbb{C}}$ defining the coefficient system E_λ is fixed corresponding to holomorphic discrete series of weight $k_1 \geq k_2 \geq 3$. By the Archimedean endoscopic
character transfer this is related to elliptic holomorphic discrete series of weights
$r_1 = k_1 + k_2 - 2$ and $r_2 = k_1 - k_2 + 2$ for $Gl(2, \mathbb{R})^2$. This follows easily from [51],
p. 182ff. See, e.g., [58], p. 212f.

Lemma 8.4. *Consider the z-extension $M_1 = Gl(2, \mathbb{R})^2$ of M. Let $h(\pi_{r_1} \otimes \pi_{r_2})$ denote pseudocoefficients for (holomorphic) discrete series of weights r_1 and r_2 on M_1
in the sense of Clozel and Delorme [23]. Then with our choice of $\Delta_\infty(\eta, (t_1, t_2))$
under the assumptions made above the function*

$$h_\infty^{M_1} = -\Big(h(\pi_{r_1} \otimes \pi_{r_2}) - h(\pi_{r_2} \otimes \pi_{r_1})\Big), \qquad (r_1 > r_2)$$

satisfies formula (7.4) in [51].

Proof. From [51], p. 186, it follows that our $h_\infty^{M_1}$ is proportional to the one defined
in [51] up to some universal constant independent of the particular choice of $\xi_{\mathbb{C}}$. We
can therefore assume $\xi_{\mathbb{C}}$ to be the trivial representation; hence, $(k_1, k_2) = (3, 3)$ and
$(r_1, r_2) = (4, 2)$. By our assumptions $tr\xi_{\mathbb{C}}(\eta) = 1$ and $e(I) = 1$ holds in formula
(7.4). Because we replaced M by M_1, the notions of stable orbital integrals and
ordinary orbital integrals coincide. According to the formulas given in Sect. 3.4

$$SO^{M_1}_{(t_1,t_2)}\big(h(\pi_4 \otimes \pi_2) - h(\pi_2 \otimes \pi_4)\big)(t_1, t_2)$$

$$= vol^{-1} \cdot \Big[\big(-tr(\tau_2(t_1))\big) \cdot \big(-tr(\tau_0(t_2))\big) - \big(-tr(\tau_0(t_1))\big) \cdot \big(-tr(\tau_2(t_2))\big)\Big]$$

$$= vol^{-1} \cdot \Delta_\infty(\eta, (t_1, t_2)),$$

since $tr(\tau(t)) = tr(\tau^*(t))$ for $|t| = 1$. To show $SO_{(t_1,t_2)}(h^M_\infty) = \langle \beta(\eta), s_M \rangle \cdot \Delta_\infty(\eta, (t_1, t_2)) \cdot e(I) \cdot tr\xi_{\mathbb{C}}(\eta) \cdot vol^{-1}$, as required, it therefore suffices to show

$$\langle \beta(\eta), s_M \rangle = -1.$$

By its definition [51], pp. 167 and 182, $\beta(\eta) \in X^*(\hat{G}_\eta)^{\Gamma(\infty)}$ depends on η only via the centralizing elliptic torus $I = G_\eta \subseteq G$. For our choice of η the structure homomorphism $h \in X_\infty$ defining the Shimura variety has values in I. The structure homomorphism h defines an algebraic homomorphism $\mu_h : \mathbb{G}_m \to I$ as in [51], p. 167, which is defined over $\mathbb{C}$. Up to some conjugation in $G(\mathbb{C})$ we get our $\mu : \mathbb{G}_m \to T^s_G \hookrightarrow G$ defined over $\mathbb{C}$ such that

$$\mu(z) = diag(z, z, 1, 1).$$

This defines a cocharacter $\mu \in X_*(T^s_G)$ which, in the notation used for cocharacters of T^s_G in this chapter, corresponds to the element

$$(m_1, m_2, m'_1, m'_2) = (1, 0, 1, 0) \in X_*(T^s_G).$$

On the other hand,

$$s = s_M = diag(1, -1, 1, -1) \in \hat{\mathbf{T}}_G,$$

defined by the endoscopic datum (see page 216), corresponds under the isomorphism $\hat{\mathbf{T}}_G \cong \hat{T}^s_G$ specified earlier in this chapter to the element $(w_1, w_2, w'_1, w'_2) = (-1, 1, 1, -1) \in X^*(T^s_G) \otimes \mathbb{C}^* = \hat{T}^s_G$. Hence, the pairing $\langle \mu, (w_1, w_2, w'_1, w'_2) \rangle = w_1 w'_1$ gives the value $\langle \mu, s \rangle = -1$, which proves the lemma. $\square$

Chapter 9
Fundamental Lemma (Twisted Case)

In this chapter we show under conditions formulated in Sect. 9.2 that the fundamental lemma for standard endoscopy implies the fundamental lemma for twisted base change endoscopy. In fact, this is shown in this chapter for the unit elements of the Hecke algebras. Using global arguments involving the Selberg trace formula, it suffices to prove the fundamental lemma and twisted fundamental lemma in general assuming the fundamental lemma for unit elements and standard endoscopy for residue characteristic $p \geq p_0$. See Corollary 9.4 and Chap. 10. It also implies the Frobenius formula (see Lemma 9.7) used in the comparison of trace formulas in Chap. 3. This formula will be the clue to unravel the terms in the Kottwitz formula stated in Theorem 3.1 that appear in the form of the twisted orbital integrals $TO_\delta(\phi_n)$.

9.1 Restriction of Scalars (Notations)

Let E be an extension of the local field F of degree $[E : F] = n$. For a connected reductive group G' over E with L-group ${}^L G' = \hat{G}' \lhd W_E$, the L-group of the restriction of scalars $G = Res_{E/F}(G')$ is

$$ {}^L G = I_{W_E}^{W_F}(\hat{G}') \lhd W_F $$

from [12], p. 35 (Formula 2), where for a group X with W_E-action

$$ I_{W_E}^{W_F}(X) = \{f : W_F \to X \mid f(w'w) = w' \cdot f(w), \ w \in W_F, w' \in W_E\}. $$

W_F acts by $(r_w f)(x) = f(xw)$. X can be embedded W_E-equivariant into $I_{W_E}^{W_F}(X)$ by $x \mapsto f_x(w) = w \cdot x$ for $w \in W_E$, and $f_x(w) = 1$ otherwise.

Remark 9.1. $I_{W_E}^{W_F}(X) = X^*(Res_{E/F}(T'))$ if $X = X^*(T')$ for a torus T' [12].

R. Weissauer, *Endoscopy for GSp(4) and the Cohomology of Siegel Modular Threefolds*,   291
Lecture Notes in Mathematics 1968, DOI: 10.1007/978-3-540-89306-6_9,
© Springer-Verlag Berlin Heidelberg 2009

Suppose W_E is a normal subgroup of W_F of index n, with a cyclic factor group generated by $\sigma_F \in W_F$. Suppose X is a W_F module, whose W_E module structure is defined by restriction. Then $(\theta f)(x) = \sigma_F^{-1} f(\sigma_F x)$ satisfies $(\theta f)(w'w)w'(\theta f)(w)$ for $w' \in W_E$. Hence, θ acts on $I_{W_E}^{W_F}(X)$ and commutes with the action of W_F. Furthermore, $\theta^n = id$ and $I_{W_E}^{W_F}(X)^\theta = X$. In fact θ-invariant functions f in $I_{W_E}^{W_F}(X)$ satisfy $f(w'x) = w' \cdot f(x)$ for all $w' \in W_F$ since σ_F and W_E generate W_F. Hence, $f(w) = w \cdot f(1)$ for $w \in W_F$.

Abelian Case. If X is Abelian, the norm $f(x) \mapsto \sum_{i=1}^n (\theta^i f)(1) \in X$ is defined. This W_F-equivariant map annihilates $(1 - \theta)X$ and induces an isomorphism

$$I_{W_E}^{W_F}(X)_\theta \cong X$$

of W_E modules. In fact $x \mapsto f_x$ defines a W_E splitting; hence, the norm map is surjective. To show that it is an isomorphism, it is enough to show that the kernel is generated by the $f(x) - (\theta f)(x)$ – an easy exercise.

E/F Normal. If E/F is normal with cyclic Galois group $\Gamma_{E/F}$ generated by σ_F, the description of $^L G$ as an induced group given above defines $\theta : \, ^L G \to \, ^L G$ for $G = Res_{E/F}(G')$. For homomorphisms $G'_1 \to G'_2$ for reductive groups over F this θ is functorial, and defines

$$\Psi : \hat{G} \longrightarrow \hat{G}' \times \ldots \times \hat{G}',$$

$$\Psi(f) = \big(f(1), \theta f(1), \ldots, (\theta^{n-1} f)(1)\big).$$

Lemma 9.1. Ψ *is an isomorphism such that* $\Psi \circ \theta = \hat{\theta} \circ \Psi$*, where* $\hat{\theta}(g_1, \ldots, g_n) = (g_2, \ldots, g_n, g_1)$*, and* $\Psi(r_w(f)) = w \cdot \Psi(f)$ *for all* $w \in W_F$.

This gives a convenient description of L-groups for $\hat{G} \cong \prod_{i=1}^n \hat{G}'$.

Proof. Notice $\Psi(\theta f) = (\theta f(1), \theta^2 f(1), \ldots, f(1)) = \hat{\theta}(\Psi(f))$ and $\Psi(r_w(f)) = w \cdot \Psi(f) = (1 \triangleleft w)(\Psi(f) \triangleleft 1)(1 \triangleleft w^{-1})$. Since σ_F and W_E generate W_F, it is enough to check this for $w = \sigma_F$ and $w \in W_E$. In both cases this formula is obvious since $\Psi(r_w(f)) = (f(w), \theta f(w), \ldots, \theta^{n-1} f(w)) = (w(f(1)), w(\theta f(1)), \ldots, w(\theta^{n-1} f(1)))$ for $w \in W_E$ and $\Psi(r_{\sigma_F}(f)) = (f(\sigma_F), \theta f(\sigma_F), \ldots, \theta^{n-1} f(\sigma_F)) = (\sigma_F(\theta f(1)), \sigma_F(\theta^2(1)), \ldots, \sigma_F(f(1)))$. $\quad\square$

Tori. Suppose E/F is unramified. $X^*(Res_{E/F}(T')_\theta) = X^*(Res_{E/F}(T'))^\theta \cong X^*(T')$ holds for the character groups of the F-tori T' and $T = Res_{E/F}(T')$ by the remark above. There exist natural isomorphisms

$$Res_{E/F}(T')^\theta \cong T', \qquad Res_{E/F}(T')_\theta \cong T'.$$

The composite of the canonical quotient map $N_{T,\theta} : Res_{E/F}(T') \to Res_{E/F}(T')_\theta$ with the natural isomorphism $Res_{E/F}(T')_\theta \cong T'$ defines the map

$$Nm_{E/F} : Res_{E/F}(T') \to T'.$$

For $x \in Res_{E/F}(T')(\overline{F}) = T'(\bigoplus_1^n \overline{F})$ the map $Nm_{E/F}$ is given by the first projection $pr_1(\prod_0^{n-1} \theta^i(x)) \in T'(\overline{F})$. In particular, $x \in T(E)$, which corresponds to $x = (x_\sigma)_{\sigma \in \Gamma_{E/F}} \in T'(\bigoplus_1^n \overline{F})$, maps to $\prod_{\sigma \in \Gamma_{E/F}} x_\sigma \in T'(\overline{F})$. On the level of F-valued points the map $Nm_{E/F}$ induces a map

$$Norm_{E/F} : T'(E) \to T'(F)$$

such that $x \in T'(E)$ maps to $\prod_{\sigma \in \Gamma_{E/F}} \sigma(x)$. Hence, with the notations from below we have

Lemma 9.2. *As maps from $T(F) = T'(E)$ to $T'(F)$, the norms N_θ, N_{σ_F}, and $Norm_{E/F}$ coincide.*

9.1.1 Norm Maps

Besides the field norm $N_{E/F} : E^* \to F^*$, we distinguish the following norm maps:

- $N_{T,\theta} : Res_{E/F}(T') \to Res_{E/F}(T')_\theta$ (the abstract norm).
- $Nm_{E/F} : Res_{E/F}(T') \to Res_{E/F}(T')_\theta \cong T'$.
- $Norm_{E/F} : T'(E) \to T'(F)$.
- $norm : Res_{E/F}(T') \to T' \cong T_M$, the composite of the map $Nm_{E/F}$ with the admissible isomorphism $T' \cong T_M$.
- $N_\theta(\delta) = \delta\theta(\delta)\ldots\theta^{n-1}(\delta) = (\delta\theta)^n$.
- $N_{\sigma_F}(\delta) = \delta\sigma_T(\delta)\ldots\sigma_T^{n-1}(\delta) = (\delta\sigma_F)^n\sigma_F^{-n}$.

Obviously $N_{\sigma_F}(\delta) = N_\theta(\delta)$ for any $\delta \in G(F) = G'(E)$.

9.2 Endoscopy for Twisted Base Change

Let F be a local non-Archimedean field of characteristic zero and E/F an unramified field extension of degree n. Let G be the Weil restriction $G = Res_{E/F}(G')$ of a connected reductive unramified quasisplit group G' over F. Assume G_{der} to be simply connected. Let B' be a Borel group of G', defined over F, and let T' be a maximal F-torus in B'. Then G is an algebraic group over F, with Borel subgroup $B = Res_{E/F}(B')$ containing $T = Res_{E/F}(T')$. The action of the Frobenius automorphism σ_F generating $\Gamma_{E/F}$ on $G'(E) = G(F)$ is induced by an algebraic F-automorphism θ of G.

General Remarks. Then, following [54], (2.1), consider twisted base change endoscopy for $(G, \theta) = (G, \theta, 1)$. Several constructions of [54] are simplified in this case. In particular:

1. G is quasisplit and the automorphism θ respects the pair (B, T); hence, $G = G^*, \theta = \theta^*$ and $\psi = id$ in the sense of [54], (1.2). The fixed group $T^\theta = T_d$ is a maximal F-split torus in G. The derived group of G is simply connected.
2. The restricted root system $R_{res}(G, T)$ can be identified with the root system $R(G^\theta, T^\theta) = R(G', T')$. In particular, it is reduced, and all its roots are of type R_1 in the sense of [54], (1.3).
3. The dual group $\hat{G}$ is $\prod_{i=1}^{n} \hat{G}'$. The Galois group acts as specified via Ψ. The action factorizes over the maximal unramified quotient of the Galois group. The Frobenius σ_F acts like the dual automorphism

$$\sigma_F(x_1, \ldots, x_n) = (\sigma_F(x_2), \ldots, \sigma_F(x_n), \sigma_F(x_1)).$$

Furthermore, $\hat{\theta}(x_1, \ldots, x_n) = (x_2, \ldots, x_n, x_1)$. We write

$$\iota : \hat{G}' \hookrightarrow \prod_{\nu=1}^{n} \hat{G}' = \hat{G}$$

for the diagonal map.

Standard Endoscopy. Fix a standard endoscopic datum $(M, {}^L M, s, \xi)$ for the group G'. Similarly to [51], p. 179, we make the following.

Assumptions.

- $s \in Z(\hat{M})^\Gamma$ and $Int(s) \circ \xi = \xi$.
- ξ is unramified. We view $\hat{\xi} : \hat{M} \hookrightarrow \hat{G}$ as an inclusion map, and assume $\xi(1 \triangleleft w) = c(w) \triangleleft w$ for $c(w) \in \hat{M}$.

In addition, we assume the existence of a z-pair (M_1, ξ_{M_1}) ([54], pp. 20–24) such that:

- $pr : M_1 \twoheadrightarrow M$ is an unramified z-extension. In particular, $M_{1,der}$ is simply connected, and M_1 is unramified.
- $\xi_{M_1} : {}^L M \to {}^L M_1$ is an L-homomorphism, of the form $\xi_{M_1} = {}^L pr$.

Finally we choose:

- An unramified 1-cocycle $d(w) \in Z^1(W_F, Z(\hat{M}_1))$ such that $d(\sigma_F) = \lambda^{-1} \in Z(\hat{M}_1)^\Gamma$.

The parameter λ may be chosen to be 1. However, in Chap. 10 it will turn out to be convenient to choose $\lambda \neq 1$ such that $\lambda^n = s^{-1}$. This is the reason for introducing the parameter λ.

Although the endoscopic datum is defined by $\mathcal{H} = {}^L M$, the approach of Kottwitz and Shelstad dictates choosing a z-pair if M_{der} is not simply connected ([54], (2.2) and (5.5)). The choice of a z-pair allows us to replace M by the group M_1, whose derived group is simply connected. For simplicity of exposition, therefore, we usually assume $M = M_1$ and $\xi_{M_1} = {}^L id$ in the following, although the results formulated can be easily extended to hold under the assumptions above. In fact, for the crucial computation of the transfer factor and for the matching conditions, the influence of the choice of the z-pair is explained on page 310 and in Sect. 9.8.

The Twisted Datum $({}^L M, M, s_E, \xi_E)$. Under the assumptions above we associate with the standard datum $(M, {}^L M, s, \xi)$ for G' a twisted endoscopic datum $({}^L M, M, s_E, \xi_E)$ for $(G, \theta, 1)$ in the sense of [54], (2.1). For $E = F$ the new datum is isomorphic, but is not the same as the one we started with. Put

$$s_E = (1, \ldots, 1, s) \in \prod_{\nu=1}^{n} \hat{G}'$$

and define $\xi_E : {}^L M \to {}^L G$ by

$$\xi_E(1 \times \sigma_F) = s_E \lhd \xi(\sigma_F),$$

$$\xi_E(h \times 1) = \big(\xi(h), \xi(h), \ldots, \xi(h)\big) \lhd 1.$$

Using λ, we attach a new z-pair $({}^L M_1, \xi_{M_1}^E)$ for the datum $({}^L M, M, s_E, \xi_E)$, where the z-extension is the same but the L-homomorphism $\xi_{M_1}^E : {}^L M \to {}^L M_1$ is twisted by the cocycle d:

$$\xi_{M_1}^E(\sigma_F) = \lambda^{-1} \cdot \xi_{M_1}(\sigma_F).$$

Notice that for our ξ_E the parameter s coincides with s^{-1} in [51] owing to the inverse normalization of the transfer factors in [51], p. 178, and that in the situation considered in [51] $\lambda = 1$.

If $M_1 = M_{der}$ is simply connected, we could ignore z-pairs to simplify the notation, and equivalently define

$$s_E = (1, \ldots, 1, s) \in \prod_{\nu=1}^{n} \hat{G}'$$

and $\xi_E : {}^L M \to {}^L G$ by

$$\xi_E(1 \times \sigma_F) = \iota(\lambda) s_E \lhd \xi(\sigma_F),$$

$$\xi_E(h \times 1) = \big(\xi(h), \xi(h), \ldots, \xi(h)\big) \lhd 1.$$

Obviously the above definitions determine an endoscopic datum, i.e., the properties [54], (2.1.1)–(2.1.4b), are satisfied: It is clear that s_E is $\hat{\theta}$-semisimple. Furthermore, the fixed group $Cent(\hat{G}, Int(s_E \circ \theta))$ is the s-centralizer $Cent(\hat{G}', Int(s)) \subseteq \hat{G}'$ diagonally embedded into $\hat{G} = \prod_{i=1}^{n} \hat{G}'$. Therefore, $\xi_E(\hat{M}) = Cent(\hat{G}, Int(s_E \circ \theta))^0$. Finally $Int(s_E) \circ \theta \circ \xi_E = a'_E \cdot \xi_E$ since $s \in Z(\hat{M})^\Gamma$. Here $a'_E = (1, \ldots, 1, 1) = 1$, where $Int(s) \circ \xi = a' \cdot \xi$.

a-data and χ-data. For maximal F-tori T' of G' the tori $T = Res_{E/F}(T')$ are θ-invariant maximal F-tori. The abstract norms $Nm_{E/F} : T \to T_\theta \cong T' (\cong T_M)$ were defined already. They define admissible isomorphisms. We can choose a-data and χ-data according to our identifications $R_{res}(G, T) = R(G^\theta, T^\theta) = R(G', T')$ to be the same for T_M, T' and T_M, T.

9.3 Matching

Suppose first $M = M_1$ is simply connected. For sufficiently regular semisimple elements $\gamma = t_M \in M(F)$ consider the orbital integral

$$O_\gamma^M(f^M) = \int_{M_\gamma(F)\backslash M(F)} f^M(h^{-1}\gamma h)dh.$$

The centralizer M_γ of γ is a maximal torus. The orbital integrals coincide with the orbital integrals considered in [54], (5.5), in the situation $M = M_1$ ([54], (2.1) and (2.2)), since $z_\sigma = 1$ ([54], (3.2)) and $\theta_M = Int(\gamma_0) = id$ ([54], (5.4)) and $\gamma_0 = 1$ with the notation in [54] By the regularity assumption

$$SO_{t_M}(g) = \sum_{g'} O_{g'}^M(g).$$

The summation extends over representatives of the conjugacy classes in the stable conjugacy class of $t_M \in M(F)$.

Matching Conditions. We say that functions f on $G(F)$ and f^M on $M(F)$ are matching if for the Kottwitz–Shelstad transfer factor $\Delta(t_M, \delta)$

$$SO_{t_M}(f^M) = \sum_\delta \Delta(t_M, \delta)O_{\delta\theta}(f)$$

holds for all strongly G-regular $t_M \in M(F)$ where

$$O_{\delta\theta} = \int_{Cent_\theta(\delta, G(F))} f(g^{-1}\delta\theta(g))dg.$$

The sum over δ (which might be empty) extends over representatives for the θ-conjugacy classes in $G(F)$ of elements $\delta \in G(F)$, whose "norm" is t_M. See [54], (5.5.1). This means that δ is contained in the $\overline{F}$-conjugacy class of $\mathcal{A}_{M/G}(\delta_M)$ ([54], (3.3.3)), where $\mathcal{A}_{M/G}$ is the composite of $Cl_{ss}(M) \to T_M/\Omega_M \to (T_G)_\theta/\Omega^\theta$ and the inverse of the isomorphism $Cl_{ss}(G, \theta) \to (T_G)_\theta/\Omega^\theta$ induced by the abstract norm map $N_{T,\theta} : T \to T_\theta$. See [54], p. 26ff.

M_{der} not Simply Connected. To drop the assumption $M_1 = M$ we need to introduce the z-pair (M_1, ξ_{M_1}) to formulate now the matching condition between functions f on $G(F)$ and f^{M_1} on $M_1(F)$, where f_{M_1} is supposed to be a function with compact support on $M_1(F)$ modulo the center $Z_{M_1}(F)$, such that

$$f^{M_1}(zm_1) = \lambda_{M_1}(z)^{-1} f^{M_1}(m_1)$$

holds for all $z \in Z_1(F) \subseteq Z_{M_1}(F)$, and Z_1 is the kernel of the z-extension $pr : M_1 \to M$. For details on this and for the precise formulation of the matching conditions in this generality we refer to [54], pp. 70–72. For our purposes, however, it is important to keep in mind that the character λ_{M_1} depends on the chosen z-pair. In our situation, even in the simply connected case $M_1 = M$, if we replace $\xi_{M_1} = {}^L id : {}^L M \to {}^L M_1$ by a twist $\tilde{\xi}_{M_1} = d \cdot \xi_{M_1}$ with a 1-cocycle d, then the (unramified) character λ_{M_1} of $M_1(F)$ is determined by d via Langlands reciprocity: If (f, f^{M_1}) match for ξ_{M_1}, then $(f, \tilde{f}^{M_1})$ match for $\tilde{\xi}_{M_1}$, where $\tilde{f}^{M_1}(m_1) = \lambda_{M_1}^{-1}(m_1) f^{M_1}(m_1)$ and λ_{M_1} is determined by $\lambda = \lambda_{M_1}(\sigma_F) \in Z_{M_1}^\Gamma$.

Notation. Put $O_t^{G',\kappa}(f') = \sum_{t''} \kappa(inv(t,t'')) \cdot O_{t''}^{G'}(f')$, where the sum extends over representatives of the conjugacy classes in the stable conjugacy class of $t \in G'(F)$, and write

$$\Delta(t_M, t'') = \Delta(M, G', t_M, t) \cdot \kappa(inv(t, t'')),$$

$$\Delta(t_M, \delta'') = \Delta(M, G, t_M, \delta) \cdot \kappa(inv(\delta, \delta''))$$

using [54], Theorem 5.1.D (see also pp. 54–55, 77, and 89), and similarly put $O_{\delta\theta}^{G,\kappa}(f) = \sum_{\delta''} \kappa(inv(\delta, \delta'')) \cdot O_{\delta''\theta}^{G}(f)$ for the invariants $inv(\delta, \delta'')$ and $inv(t, t'')$ defined as in Kottwitz [46]. The sum runs over representatives of the θ-conjugacy classes in the stable θ-conjugacy class of δ. Here we assume $\delta \in G(F) = G'(E)$ such that $N_\theta(\delta)$ is conjugate to $t \in G'(F)$ in $G'(\overline{F})$. If such δ does not exist in $G(F)$, then t is not a norm and we put $O_{\delta\theta}^{G,\kappa}(f) = 0$.

9.4 The Twisted Fundamental Lemma (Unit Element)

By assumption there exist hyperspecial maximal compact subgroups $K \subseteq G(F)$, $K' \subseteq G'(F)$, and $K_M \subseteq M(F)$. We choose K by "extension of scalars" from K', granting that $K = K_G, K' = K'_{G'}$ satisfy conditions (a)–(c) in [46], p. 240. We write $1_K, 1_{K'}$, and 1_{K_M} for the corresponding characteristic functions.

Assumption. We assume now that the fundamental lemma for the standard endoscopic datum $(M, {}^L M, s, \xi)$ holds for unit elements. In other words, we assume that $1_{K'}$ and 1_{K_M} are matching functions

$$\Delta(M, G', t_M, t) O_t^{G',\kappa}(1_{K'}) = SO_t^M(1_{K_M}).$$

Under this assumption, we use [46] to compare with the twisted situation. In fact Kottwitz proved in [46], p. 241, the formula $O_t^{G'}(1_{K'}) = O_{\delta\theta}^G$, provided that $t \leftrightarrow \delta$

holds in the notation in [46] or that $t \leftrightarrow \delta$ holds in the notations used later. This is equivalent to equations (A) and (B) below. In fact Kottwitz proved more.

Theorem 9.1. *The correspondence* $t \overset{M}{\leftrightarrow} \delta$ *(for any fixed M, see the notation below) induces a bijection from the set of conjugacy classes of* $t \in G'(F)$ *such that* $O_t^{G'}(1_{K'}) \neq 0$ *to the set of θ-conjugacy classes of* $\delta \in G(F) = G'(E)$ *such that* $O_{\delta\theta}^G(1_K) \neq 0$. *Moreover, it implies* $t = N_\theta(\delta)$ *and equality of orbital integrals* $O_t^{G'}(1_{K'}) = O_{\delta\theta}^G(1_K)$.

The assumption regarding the standard fundamental lemma and Theorem 9.1 imply for $t \leftrightarrow \delta$

$$\Delta(M, G', t_M, t)\Delta^\kappa(G', G, t, \delta)O_{\delta\theta}^{G,\kappa}(1_K) = \Delta(M, G', t_M, t)O_t^{G',\kappa}(1_{K'}) = SO_t^M(1_{K_M}).$$

Here $t \in T_{G'}(F)$ is the image of $t_M \in T_M(F)$ under the admissible embedding, and $t = norm(\delta^*), \delta^* \in T_G(\overline{F})$ for regular t. The factor

$$\Delta^\kappa(G', G, t, \delta)$$

is not a transfer factor in the usual sense. It will be defined only when $O_{\delta\theta}^{G,\kappa}(1_K) \neq 0$ holds. In this case it is implicitly determined by the normalizations made in Theorem 9.1 or equivalently Proposition 1 in [46]. This will be explained in detail below. We first observe that the formula for the orbital integrals, stated above, immediately implies

Lemma 9.3. *Suppose the standard fundamental lemma holds for the unit elements* $1_{K_M}, 1_{K'}$. *Then the twisted base change fundamental lemma for the unit elements* $1_{K_M}, 1_K$, *i.e., the matching of* 1_{K_M} *and* 1_K *for sufficiently regular elements t, is equivalent to the following statement: The twisted endoscopic transfer factor is given by the formula*

$$\Delta(M, G, t_M, \delta) = \Delta(M, G', t_M, t)\Delta^\kappa(G', G, t, \delta)$$

whenever $O_{\delta\theta}^{G,\kappa}(1_K) \neq 0$ *holds.*

In the following we compute the twisted transfer factor $\Delta(M, G, t_M, \delta)$ following [54]. Lemma 9.17 and Theorems 9.2 and 9.3 imply that $\Delta(M, G, t_M, \delta)/\Delta$ (M, G', t_M, t) is equal to $\Delta^\kappa(G', G, t, \delta)$. Hence, we have

Lemma 9.4. *Suppose the standard fundamental lemma holds for the unit elements* $1_{K_M}, 1_{K'}$. *Then the twisted base change fundamental lemma holds for the unit elements* $1_{K_M}, 1_K$.

9.4.1 The Condition $t \leftrightarrow \delta$

We fix a unimodular matrix

$$M = \begin{pmatrix} b & a \\ 1 & n \end{pmatrix} \in Sl(2, \mathbb{Z}), \qquad n = [E : F].$$

There is little chance of confusing this with the endoscopic group M. Let $\mathbf{L}$ be the completion of F^{un}, the maximal unramified extension of F. Consider $t \in G'(F)$ and $\delta \in G'(E)$. The elements t and σ_F generate a subgroup in $G'(\mathbf{L}) \times \langle \sigma_F \rangle$ isomorphic to $\mathbb{Z}^2$. The same applies for $\delta\sigma_F$ and σ_F^n. We may ask whether these two subgroups $\mathbb{Z}^2$ are conjugate under some $c \in G'(\mathbf{L})$; more precisely, such that $Int(c) \circ M$ also maps generators to the generators:

$$(A) \quad Int(c)(t^a \sigma_F^n) = \sigma_F^n, \qquad (B) \quad Int(c)(t^b \sigma_F) = \delta\sigma_F.$$

If (A) and (B) hold, then we write $t \overset{M}{\leftrightarrow} \delta$. For a fixed matrix M we then write $t \leftrightarrow \delta$ as in [46]. However, we consider different choices of the matrix M. For every choice of M, there exists an identity between κ-orbital integrals on $G'(F)$ and twisted κ orbital integrals on $G'(E) = G(F)$, as stated in Theorem 9.1.

The Preferred Choice. Assume $b = 1$ and $a = n - 1$ (then the other choices are of type $a' \in a + n\mathbb{Z}$). From now on we reserve the notation $t \leftrightarrow \delta$ for this particular choice.

Lemma 9.5. *With this choice, for elements δ in the support of $O_{\delta\theta}^{G,\kappa}(1_K)$ the previously defined factor $\Delta^\kappa(G', G, t, \delta)$ is uniquely characterized by the following two properties:*

(i) $\Delta^\kappa(G', G, t, \delta) = 1$ *if* $t \leftrightarrow \delta$.
(ii) $\Delta^\kappa(G', G, t', \delta') = \kappa(inv(t', t))\Delta^\kappa(G', G, t, \delta)\kappa(inv(\delta, \delta'))$.

Proof. The first statement follows from Theorem 9.1. For the second property notice the following. For a fixed maximal torus T in G, define $\mathcal{D}(T) = Ker(H^1(F, T) \to H^1(F, G))$ as a subset of $H^1(F, T)$. For $[\xi] \in \mathcal{D}(T)$ we have $\xi(\sigma) = A^{-1}A^\sigma$, $A \in G(\overline{F})$. Then $T_\xi = ATA^{-1}$ is a torus in G defined over F. The map $\phi = \phi_{T_\xi T} : T \to T_\xi$ defined by $t \mapsto AtA^{-1}$ is F-rational, i.e., $\phi(\sigma(t)) = \sigma(\phi(t))$. It defines an F-isomorphism between the tori T and T_ξ. Furthermore, (ξ, A) defines a hypercohomology class in $H^1(F, T \to G)$. Let $[\eta] \in \mathcal{D}(T_\xi)$ be a class with torus $(T_\xi)_\eta$ and map $\phi_{(T_\xi)_\eta T_\xi} : T_\xi \to (T_\xi)_\eta$ defined by $\eta(\sigma) = B^{-1}B^\sigma$, $B \in G(\overline{F})$. Then the composite map $\phi_{(T_\xi)_\eta T_\xi}\phi_{T_\xi T} : T \to (T_\xi)_\eta$ maps t to $BAtA^{-1}B^{-1}$. This composite map is the map $\phi_{T_\zeta T}$ attached to $\zeta(\sigma) = A^{-1}B^{-1}B^\sigma A^\sigma = A^{-1}\eta(\sigma)A\xi(\sigma)$. Therefore, $(\phi_{T_\xi T})_*^{-1}(\eta)(\sigma) = A^{-1}\eta(\sigma)A$ implies

$$[\zeta] = (\phi_{T_\xi T})_*^{-1}([\eta]) + [\xi].$$

In other words,

$$inv(T, T_\zeta) = inv(T, T_\xi) + (\phi_{T_\xi T})_*^{-1}(inv(T_\xi, T_\zeta)).$$

Similarly, $\phi_{T_\xi T}(\mathcal{D}(T)) = \phi_{T_\xi T}(\xi) + \mathcal{D}(T_\xi)$. However, in the p-adic case the sets $\mathcal{D}(T)$ are subgroups of $H^1(F, T)$. This implies $\phi_{T_\xi T}(\mathcal{D}(T)) = \mathcal{D}(T_\xi)$. Hence, in the p-adic case we can tacitly identify the groups $\mathcal{D}(T_\xi)$ with a particular one, say, $\mathcal{D}(T)$, via the isomorphisms $\phi_{T_\xi T}$. This allows us to transfer characters κ on $\mathcal{D}(T)$ to characters κ_ξ on the different groups $\mathcal{D}(T_\xi)$. For $\kappa = \langle ., s \rangle$ this implies

$$\kappa(inv(T, T_\zeta)) = \kappa(inv(T, T_\xi))\kappa_\xi(inv(T_\xi, T_\zeta)).$$

For regular stably conjugate elements one therefore gets the analogous formula

$$\kappa(inv(t, t'')) = \kappa(inv(t, t'))\kappa_\xi(inv(t', t'')).$$

Hence, $O_t^\kappa(f) = \kappa(inv(t, t'))O_{t'}^\kappa(f)$. Since the transfer factor $\Delta^\kappa(G', G, t, \delta)$ has to be consistent with the transformation property of the orbital integral, this proves (ii). In fact the corresponding property also holds for the change from δ to δ' by a similar argument. This proves the lemma. $\square$

9.4.2 The Class $\alpha(t; \delta)$

In a different context, Kottwitz introduced a transfer factor which is easier to define than the transfer factors of Langlands and Shelstad. For this we need the elements $\alpha(t; \delta)$. To define

$$\alpha(t; \delta) \in X^*(\hat{T}_{G'}^\Gamma), \qquad t \in T_{G'}(F)$$

for $T_{G'} = T'$ suppose $\delta \in G(F) = G'(E)$ – then $\mathbf{N}_\theta(\delta) = \mathbf{N}_{\sigma_T}(\delta)$ – and suppose

$$\mathbf{N}_\theta(\delta) = \mathbf{N}_{\sigma_F}(\delta) = \delta \cdot \sigma_F(\delta) \ldots \cdot \sigma_F^{n-1}(\delta) = dtd^{-1}, \qquad d \in G'(\overline{F}).$$

Then by a theorem of Steinberg there exists an element $c \in G'(\mathbf{L})$, where $\mathbf{L}$ is the completion of the maximal unramified extension F^{un} of F, such that

$$\mathbf{N}_{\sigma_F}(\delta) = \delta \cdot \sigma_F(\delta) \ldots \cdot \sigma_F^{n-1}(\delta) = ctc^{-1}, \qquad c \in G'(\mathbf{L}).$$

This implies

$$b = c^{-1}\delta\sigma_F(c) \in T_{G'}(\mathbf{L}), \qquad T_{G'} = G'_t.$$

In fact, b is uniquely defined by the given data up to σ_F-conjugacy in $T_{G'}(\mathbf{L})$. Hence, there is a uniquely determined class $[b]$ in $B(T_{G'}) = H^1(\langle \sigma_L \rangle, T_{G'}(\mathbf{L}))$.

For an arbitrary F-torus T one has a canonical isomorphism [47] $\nu : X_*(T)_\Gamma \to B(T)$ for $B(T) = B_F(T) = T(\mathbf{L})/(1 - \sigma_F)$. The inverse image of b under this isomorphism defines

$$\alpha(t; \delta) = \nu^{-1}([b]) \in X_*(T_{G'})_\Gamma.$$

Elementary Characters. For an arbitrary F-torus T the composition

$$T(F) \longrightarrow B(T) \xrightarrow{\nu^{-1}} X_*(T)_\Gamma$$

is a homomorphism. It is not difficult to see that its kernel $K(T)$ is a subgroup of finite index in the maximal compact subgroup ${}^0T$ of $T(F)$. This is clear for anisotropic tori S because then $B(S)$ is finite. In general, consider a maximal split torus T_d in T. Then $S = T/T_d$ is anisotropic. The claim is easily reduced to the split case by the exact sequence $B(T_d) \to B(T) \to B(S) \to 0$.

A character $\chi : T(F) \to \mathbb{C}^*$ is called *elementary* if it is trivial on $K(T)$. Every elementary character is continuous. We get a functorial group homomorphism

$$\hat{T}^\Gamma \to Hom_{elm}(T(F), \mathbb{C}^*) \subseteq Hom_{cont}(T(F), \mathbb{C}^*)$$

from the above map $T(F) \to X_*(T)_\Gamma$, and the now to be defined pairing between $\hat{T}^\Gamma$ and $X_*(T)_\Gamma$.

Pairings. Since $X^*(\hat{T}^\Gamma) = X_*(T)_\Gamma$ there is a canonical pairing

$$\langle ., . \rangle : \quad X_*(T)_\Gamma \times \hat{T}^\Gamma = X_*(T)_\Gamma \times (\mathbb{C}^* \otimes X^*(T))^\Gamma \to \mathbb{C}^*,$$

or

$$\langle ., . \rangle : \quad B(T) \times \hat{T}^\Gamma \to \mathbb{C}^*.$$

This being said, it is clear how $\langle \alpha(t; \delta), s \rangle$ is defined: $\alpha(t; \delta) \in X_*(T_{G'})_\Gamma$ can be evaluated against $s \in \hat{T}^\Gamma_{G'}$.

In fact, the pairing $\langle ., . \rangle$ induces an isomorphism

$$\hat{T}^\Gamma \cong Hom(X_*(T)_\Gamma, \mathbb{C}^*),$$

since for a Γ-module A with trivial Γ-action, $Hom(Y, A)^\Gamma$ is the set of homomorphisms $f : Y \to A$ satisfying $f(\gamma^{-1}y) = f(y), \gamma \in \Gamma$. This is equivalent to $f(I) = 0$ for $I = Ker(Y \to Y_\Gamma)$. But the homomorphisms $f : Y_\Gamma \to A$ are the homomorphisms $f : Y \to A$ such that $f(I) = 0$.

Explicit Description of ν. Choose a lift $\alpha \in X_*(T)$ of $\overline{\alpha} \in X_*(T)_{\Gamma_F}$. Consider a splitting field K/L of the extension of T to L. There is a natural map $Norm_{K/L} : T(K) \to T(L)$ induced from $Nm_{K/L} : Res_{K/L}(T) \to T$. Then $\nu(\overline{\alpha}) = [Norm_{K/L}(\alpha(\pi_K))]$ in $B(T) = T(L)/(1 - \sigma_F)$. This is the description of $\alpha(t; \delta)$, given in [47], (2.5).

Hence, for an unramified finite extension E/F the diagram

$$
\begin{array}{ccc}
X_*(T)_{\Gamma_F} & \xrightarrow{\ \nu\ } & T(\mathbf{L})/(1 - \sigma_F) \\
\uparrow & & \uparrow \\
X_*(T)_{\Gamma_E} & \xrightarrow{\ \nu\ } & T(\mathbf{L})/(1 - \sigma_E)
\end{array}
$$

commutes. Therefore, we have

Lemma 9.6. *Let E/F be an unramified extension, and $s \in (\hat{T}')^{\Gamma_F} \subseteq (\hat{T})^{\Gamma_E}$. For $\delta \in T'(E) \subseteq T'(\mathbf{L})$ let $[\delta]_F \in B_F(T')$ and $[\delta]_E \in B_E(T')$ denote the corresponding classes for T' considered as a torus over F and E, respectively . Then $\langle [\delta]_F, s \rangle_F = \langle [\delta]_E, s \rangle_E$.*

9.4.3 The Frobenius Fundamental Lemma

From the study of the trace formula in the context of Shimura varieties, Kottwitz was led to introducing a slightly different kind of twisted fundamental lemma. In this form it appears naturally during the comparison of trace of the Frobenius endomorphism on the cohomology ([51], (7.2) and (7.3)) with the trace of certain Hecke correspondences which generalize the Eichler–Shimura correspondence.

This new formula is different from the twisted fundamental lemma in two respects:

1. It is required only for a certain specific spherical function $f = \phi_n$ on $G(F) = G'(E)$, which is the characteristic function of the double coset $K_G \mu_h^{-1}(\pi_E) K_G$, where μ_h is defined from the datum h of the Shimura variety [51], p. 173.
2. It involves a different transfer factor defined by a factor $\alpha(t; \delta)$.

Frobenius Formula. This formula was postulated by Kottwitz [51], p. 180, to obtain the crucial formula (7.2) in [51]. It is a variant of the twisted fundamental lemma for a special spherical function denoted ϕ_j in [49], p. 173 (in fact owing to a slightly different normalization we actually have to deal with the function ϕ_n defined in [65], p. 202). It is the characteristic function of the double coset $GSp(4, E)diag(p, p, 1, 1)GSp(4, E)$. In our notation $[E : F] = n$ denotes the index $j = [F : E]$ in [51], and the notation for E and F is reversed. In fact, the following lemma follows from loc. cit.:

Lemma 9.7 (Frobenius Formula). *The twisted fundamental lemma for $(M, {}^L M, s_E, \xi_E)$ for the Kottwitz spherical function ϕ_n, $n = [E : F]$ and the*

- **Assertion $A(E, f)$.** $\Delta(M, G, t_M, \delta) = \Delta(M, G', t_M, t) \langle \alpha(t; \delta), \lambda^n s \rangle$ *holds*[1] *whenever $O_{\delta\theta}^{G,\kappa}(f) \neq 0$*

for the function $f = \phi_n$ both together imply formula (7.2) in [51].

In fact it is enough to have this formula in the setting of [51], (7.2), i.e., under our assumptions $s \in Z(\hat{M})^{\Gamma}$ and the assumption that the embedding $\xi : {}^L M \to {}^L G'$ is unramified, in particular, G' and G are then unramified over F.

Concerning the Proof of the Frobenius Formula (7.2). Assertion $A(E, f)$, formulated in Lemma 9.7, is not unrelated to the twisted fundamental lemma for f. To the contrary, assertion $A(E, f)$ for the unit element $f = 1_K$ implies the twisted

---

[1] The notation for and the conditions imposed on t, δ as in Lemma 9.3 for all $\delta \in G(F)$ that satisfy $O_{\delta\theta}^{G,\kappa}(f) \neq 0$.

fundamental lemma for the unit element $f = 1_K$, provided the fundamental lemma for standard endoscopy holds for the unit element $1_{K'}$ (see Lemmas 9.3 and 9.8).

Corollary 9.1. *Assertion $A(E, 1_K)$ and the standard fundamental lemma for the unit element imply the twisted fundamental lemma for the unit element.*

Remark 9.2. The standard fundamental lemma for the unit element (for residue characteristic different from 2) implies by global methods the standard fundamental lemma [35]. This should also be true for the twisted fundamental lemma. For the case $G' = GSp(4), G = Res_{E/F}(G')$, see Chap. 10. More generally, see [113]. The twisted fundamental lemma and Lemma 9.17 and Theorem 9.3 imply $A(E, \phi_n)$. Then the twisted fundamental lemma and $A(E, \phi_n)$ imply the Frobenius formula. See Corollary 9.4. In Sect. 9.6 assertion $A(E, f)$ turns out to be a kind of explicit Langlands reciprocity law for elementary characters of tori.

Lemma 9.8. $\langle \alpha(t; \delta), s \rangle = \Delta^\kappa(G', G, t, \delta)$ *for all δ with* $O_{\delta\theta}^{G,\kappa}(1_K) \neq 0$.

Proof. By Lemma 9.5 it is enough to show:

(i) $\kappa(\alpha(t; \delta)) = 1$ for $t \leftrightarrow \delta$ and $O_{\delta\theta}^{G,\kappa}(1_K) \neq 0$.
(ii) $\langle \alpha(t'; \delta'), s \rangle = \langle inv(t', t), s \rangle \langle \alpha(t; \delta), s \rangle \langle inv(\delta, \delta'), s \rangle$. $\square$

Proof of (ii). See [51], p. 168, for the dependence on (t, t') and [51], p. 170. For example, for $t' \in T_\xi, t \in T$ with $T_\xi = ATA^{-1}, t' = ATA^{-1}$, then $b' = (cA^{-1})^{-1}$ $\delta\sigma_F(cA^{-1}) \in T_\xi(\mathbf{L})$, where A can be chosen in $G'(\mathbf{L})$. Thus, $b' = \phi_{T_\xi T}(b)$ $A\sigma_F(A^{-1}) = \phi_{T_\xi T}(b) inv(T_\xi, T)$. Hence, $\alpha(t'; \delta) = (\phi_{T_\xi T})_*(\alpha(t; \delta)) inv(T_\xi, T)$ because $inv(T_\xi, T) = -\phi_{T_\xi T}(inv(T, T_\xi)) = [A\sigma_F(A^{-1})]$. Similarly for a change of δ. The proof of statement (i) is given in Sect. 9.4.4. $\square$

9.4.4 The Support of $O_{\delta\theta}^{G,\kappa}$

Recall that the condition $t \leftrightarrow \delta$ for our preferred choice ($b = 1$, $a = n - 1$) is equivalent to equations (A) and (B) stated on page 299. Kottwitz [46], p. 241, showed that formulas (A) and (B) are equivalent to formulas (A') and (B')

$$(A') \quad t^{n-1} = c^{-1} \sigma_F^n(c), \qquad (B') \qquad t = c^{-1} \delta\sigma_F(c)$$

or (see [46], p. 242) are equivalent to formulas (C') and (D')

$$(C') \quad \mathbf{N}_{\sigma_F}(\delta) = \delta \cdot \sigma_F(\delta) \dots \cdot \sigma_F^{n-1}(\delta) = ctc^{-1}, \quad (D') \quad (\delta\sigma_F)^{-a} \sigma_F^a = c\sigma_F(c)^{-1}.$$

Lemma 9.9. *For $t \in G'(F)$ and $\delta \in G'(E)$ the following statements are equivalent:*

1. *$t \leftrightarrow \delta$ for our preferred choice $a = n - 1, b = 1$.*
2. *Equations (B') and (C') hold for some $c \in G'(\mathbf{L})$.*

Proof. Condition (1) is equivalent to (A) and (B) and these imply (A'), (B'), (C'), (D'). Conversely, since (A), (B) and (A'), (B') are equivalent, it is enough to show that (B') and (C') imply (A'). $N = \mathbf{N}_{\sigma_F}(\delta) \in G'(E)$ since $\delta \in G'(E)$. $t^n = \mathbf{N}_{\sigma_F}(t)$ since $t \in G'(F)$. Hence, (B') implies $t^n = \mathbf{N}_{\sigma_F}(t) = c^{-1}\mathbf{N}_{\sigma_F}(\delta)\sigma_F^n(c) = c^{-1}\sigma_F^n(c)\sigma_F^n(c) = c^{-1}\sigma_F^n(c)\sigma_F^n(c^{-1}Nc)$. Since $c^{-1}Nc = t \in G'(F)$ by (C'), therefore, $t^n = c^{-1}\sigma_F^n(c)t$. Dividing by t proves (A'). $\square$

Corollary 9.2. *For $t \in T'(F), \delta \in G'(E)$ satisfying (C') the following assertions are equivalent:*

1. $\alpha(t; \delta) = [t] \in B(T')$ *via the identification* $\nu : B(T') \cong X_*(T')_\Gamma$.
2. $t \leftrightarrow \delta$ *for our preferred choice.*

Proof. If $t \leftrightarrow \delta$ then condition (B') implies $b = t$ for $b = c^{-1}\delta\sigma_F(c)$; hence, $\alpha(t; \delta) = [t]$. Suppose conversely $\alpha(t; \delta) = [t]$ holds. Since we are free to make a change $c \mapsto cc_0, c_0 \in T'(\mathbf{L})$, condition (B') is $[t] = \alpha(t; \delta)$ in the quotient $B(T')$. By Lemma 9.9 this implies $t \leftrightarrow \delta$. $\square$

Proof of Lemma 9.8, statement (i). $t \leftrightarrow \delta$ is equivalent to (B') and (C') (Lemma 9.9). These imply $\alpha(t; \delta) = [t] \in B(T_{G'})$ (Corollary 9.2). But $O_t^{G',\kappa}(1_{K'}) \neq 0 \Leftrightarrow O_{\delta\theta}^{G,\kappa}(1_K) \neq 0$ (Theorem 9.1). Therefore, statement (i) of Lemma 9.8 follows from $\square$

Lemma 9.10 (support). $O_t^{G',\kappa}(1_{K'}) \neq 0$ *implies* $\kappa([t]) = 1$.

Proof. Suppose $t \leftrightarrow \delta$, say, for the preferred choice. According to Theorem 9.1 one can find t' such that $\delta \leftrightarrow t'$ for a modified choice of the parameter $a' = a + kn$. Then by Proposition 2 in [46], this implies $inv(t, t') = [t^{-k}] \in B(T')_{tor} = H^1(F, T')$. For $k = -1$, $[t] = inv(t, t')$. Now by Theorem 9.1 $O_t^{G',\kappa}(1_{K'}) = O_{\delta\theta}^{G,\kappa}(1_K) = O_{t'}^{G',\kappa'}(1_{K'}) \neq 0$. Since $\kappa(inv(t, t'')) = \kappa(inv(t, t'))\kappa'(inv(t', t''))$, where $\kappa' \circ \phi = \kappa$ for $\phi : G'_t \cong G'_{t'}$, this implies $O_t^{G',\kappa}(1_{K'}) = \kappa(inv(t, t'))O_{t'}^{G',\kappa}(1_{K'}) \neq 0$. Hence, $\kappa(inv(t, t')) = \kappa([t]) = 1$, which proves Lemma 9.10. $\square$

Questions. Suppose the fundamental lemma holds for the unit elements $1_{K_M}, 1_{K'}$. Then $O_t^{G',\kappa}(1_{K'}) \neq 0$ for all regular $t \in \psi(T_M(F) \cap K_M)$ for an admissible embedding $\psi : T_M \cong T' \subseteq G'$. Then, e.g., by the last lemma, $\psi(T_M \cap K_M)$ is contained in the kernel of the character $\kappa \circ [.] : T'(F) \cap K' \to \mathcal{D}(T') \to \mathbb{C}^*$. Under what conditions is it equal to the kernel of this character? Does $\kappa([t]) = 1$ for $t \in T'(F) \cap K'$ imply $O_t^{G',\kappa}(1'_K) \neq 0$?

9.5 The Norm Map and Stable Conjugacy

Let the situation be as in Sect. 9.2. Let $\psi : T_M \hookrightarrow T' \subseteq G'$ be an admissible embedding attached to the given standard endoscopic datum such that $T = Res_{E/F}(T')$

is a maximal torus in G. There is a canonical isomorphism $T_\theta \cong T'$, which defines the abstract norm map $Nm_{E/F} : T \to T'$.

For $\gamma_M \in T_M(F)$ let t denote its image in $T'(F)$ and γ (or γ_0) its image in $T_\theta(F)$. Choose $\delta^* \in T(\overline{F})$

$$Nm_{E/F}(\delta^*) = t \in T'(F),$$

where t is supposed to be *regular*. Notice this is equivalent to condition (1) $N_{T,\theta}(\delta^*) = \gamma$ in [54], p. 63. Let $t_i \in G'(F)$ denote representatives of the finitely many F-conjugacy classes of elements in $G'(F)$ stably conjugate to t. Then $inv(t, t_i) \in \mathcal{D}(G, T')$. Furthermore, choose representatives $\delta_i \in G(F)$ of the finitely many θ-conjugacy classes in $G(F)$ of elements in $G(F)$ stably θ-conjugate to δ^*. This set may be empty. Then

$$\delta^* = g_i \delta_i \theta(g_i)^{-1}, \qquad g_i \in G_{sc}(\overline{F})$$

defines

$$v_i(\sigma) = g_i \sigma(g_i)^{-1} \in T_{sc}(\overline{F})$$

such that $\sigma(\delta^*) = v_i(\sigma)^{-1}\delta^*\theta(v_i(\sigma))$. Then $\sigma(\delta^*) = \delta^* \mod (1 - \theta)T$. Hence, we can find $\lambda \in T(\overline{F})$ depending on i, such that $\sigma(\delta^*) = \delta^*\lambda^{1-\theta}$. Then $(v_i(\sigma)\lambda)^{-1}\delta^*\theta(v_i(\sigma)\lambda) = \delta^*$, which implies $v_i(\sigma)\lambda \in T^\theta(\overline{F})$. Hence, $v_i(\sigma) \in T(\overline{F})$ satisfies

$$\delta^*\sigma(\delta^*)^{-1} = v_i(\sigma)^{1-\theta}.$$

So for $\pi : T_{sc} \to T$ the pair $(v_i(\sigma)^{-1}, \delta^*)$ defines a hypercocycle in $Z^1(\Gamma_F, T_{sc} \overset{(1-\theta)\pi}{\longrightarrow} T)$.

Definition 9.1. Its class (denoted $inv(t, \delta)$ in [54], p. 63, if $T_1 = T$ in the notation in loc. cit.)

$$\mathbf{V}_i = [(v_i(\sigma)^{-1}, \delta^*)] \in H^1(F, T_{sc} \overset{(1-\theta)\pi}{\longrightarrow} T)$$

is independent of the choice of $\delta^* \in T(\overline{F})$ with $Nm_{E/F}(\delta^*) = t$, since another choice $\delta^* w^{1-\theta}, w \in T(\overline{F})$ defines a change by a hypercoboundary. Suppose $T_1 \neq T$. Then $\mathbf{V} = [(V_i(\sigma)^{-1}, \delta_1^*)]$ for $\delta_1^* \in T_1(\overline{F})$ with $\delta_1^* \mapsto t_1 \in T_{M_1}(F)$ for a lift $t_1 \in T_{M_1}(F)$ of $t \in T_M(F)$ similarly defines a cohomology class $inv(t_1, \delta) \in H^1(F, T_{sc} \to T_1)$.

Definition 9.2. Let

$$\mathbf{inv}_\theta(t, \delta_i) = [(v_i(\sigma)^{-1}, \delta^*)] \in H^1(F, T \overset{(1-\theta)}{\longrightarrow} T)$$

be the image of $\mathbf{V}_i$ induced from the chain complex map (π, id)

$$
\begin{array}{ccc}
T_{sc} & \xrightarrow{(1-\theta)\pi} & T \\
\pi \downarrow & & \downarrow id \\
T & \xrightarrow{(1-\theta)} & T
\end{array}
$$

Some Exact Sequences. Since $T^\theta = T'$ and $T_\theta \cong T'$, one has exact sequences

$$
0 \longrightarrow V \xrightarrow{\ i\ } T \xrightarrow{Nm_{E/F}} T' \longrightarrow 0
$$

$$
0 \longrightarrow T' \longrightarrow T \xrightarrow{1-\theta} V \longrightarrow 0
$$

for $(1-\theta)T = V$ and the inclusion map $i : V \to T$.

The Rational Case. Suppose δ^* can be chosen in $T(F)$ such that $Nm_{E/F}(\delta^*) = t$. Then the invariants

$$
inv(\delta^*, \delta_i) \in Ker(H^1(F, T^\theta) \to H^1(F, G)) = \mathcal{D}(T, \theta)
$$

are defined, and parameterize the conjugacy classes within the stable conjugacy class. Then we write $\delta^* = \delta$.

Lemma 9.11. *In the rational case* $(v_i(\sigma)^{-1}, \delta^*) = (v_i(\sigma)^{-1}, 1) \cdot (1, \delta^*)$ *is a product of two hypercocycles. The cohomology class of the first hypercocycle is the image of* $inv(\delta^*, \delta_i)$ *under the chain complex map* $\varphi : (T^\theta \to 0) \to (T \xrightarrow{1-\theta} T)$ *defined as the composite of the following vertical chain complex maps:*

$$
\begin{array}{ccc}
T & \xrightarrow{1-\theta} & T \\
id \uparrow & & \uparrow i \\
T & \xrightarrow{1-\theta} & V \\
\uparrow & & \uparrow \\
T^\theta & \longrightarrow & 0
\end{array}
$$

In the derived category the first map is a quasi-isomorphism. Since $T_\theta = T/V$, one has a distinguished triangle

$$
(T^\theta \to 0) \to (T \xrightarrow{1-\theta} T) \to (0 \to T_\theta) \to .
$$

Together with the second distinguished triangle, defined by the maps $(0, id)$ and $(id, 0)$,

$$
(0 \to T) \to (T \xrightarrow{1-\theta} T) \to (T \to 0) \to
$$

this defines the exact sequences

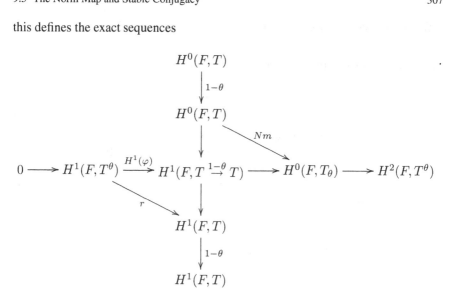

The maps r and Nm making the triangles in the diagram commutative are induced from the inclusion $T^\theta \hookrightarrow T$ and from the norm $Nm_{T,\theta} : T \to T_\theta$, respectively.

In the twisted base change situation we may identify T^θ and T_θ with T'.

Lemma 9.12. *Then the maps Nm and r become, respectively, the norm*

$$Norm_{E/F} : T(F) = T'(E) \to T'(F),$$

and the restriction map of Galois cohomology

$$res : H^1(F, T') \to H^1(E, T').$$

Proof. The first statement is obvious. For the second notice $H^1(F, T) \cong H^1(E, T')$ by Shapiro's lemma. This isomorphism is the composition of two maps, the restriction map $H^1(F, T) \to H^1(E, T/E)$ and the projection on one component of $T/E = \prod_1^n T'$. Since the restriction map is functorial, the map induced from the inclusion $T' = T^\theta \hookrightarrow T$ factors over the restriction map $H^1(F, T') \to H^1(E, T')$. The inclusion $T'(E) \to T(E) = \prod_{i=1}^n T'(E)$ composed with the projection on the one component is the identity map. This determines r as claimed, and completes the proof. $\square$

Corollary 9.3. *Suppose $res : H^1(F, T') \to H^1(F, T)^\theta$ is surjective. Then we are in the rational case.*

Proof. Obvious. $\square$

9.6 Comparison of the Transfer Factors Δ and Δ^E

In this section we compute the quotient of transfer factors

$$\frac{\Delta(M, G, t_M, \delta)}{\Delta(M, G', t_M, t)}$$

since it is relevant for the twisted fundamental lemma for the unit element (Lemma 9.3). For this let the situation be as in Sect. 9.2. In particular, G is quasisplit over F and θ fixes a F-splitting $(B, T_0, \{x_\alpha\})$ of G. Therefore, $\psi = id, u(\sigma) = 1, m = id, z = 1$ in the notation in [54]. Furthermore, $G^\theta = G'$ and $T^\theta = T'$ and $T_\theta \cong T'$ (not induced by the natural map $T^\theta \to T_\theta$). Then

Lemma 9.13.
$$\frac{\Delta(M, G, t_M, \delta)}{\Delta(M, G', t_M, t)} = \frac{\Delta_{III}(M, G, t_M, \delta)}{\Delta_{III}(M, G', t_M, t)}.$$

Proof. This is an easy consequence of the definition. Since the definitions are cumbersome, we give details of the argument for the convenience of the reader. In our quasisplit situation Δ is a product of four factors:

1. The first factor $\Delta_I(\gamma_M, \delta)$ is of cohomological nature. It is attached to a class $\lambda_{\{a_\alpha\}}(T^x)$ in the Galois cohomology group $H^1(F, T^x)$, where T^x is the intersection $T_{sc} \cap G^x$. G^x is the group of fixed points of the lift θ_{sc} to G_{sc}, where G_{sc} is the simply connected covering of G_{der}. It is defined by the induced a-data via $R(G^x, T^x) \subseteq R_{res}(G, T)$ (see [54], (4.2)). But $G^x = G^{\theta_{sc}}_{sc} = G'_{sc}$ and $T^x = T^{\theta_{sc}}_{sc} = T'_{sc}$. So the group T^x and the class $\lambda_{\{a_\alpha\}}(T^x)$ do not depend on E, and for $E = F$ it is the same class as for the original endoscopic setting we started from. The factor $\Delta_I(\gamma_M, \delta)$ is the value of the Tate–Nakayama pairing of this class against a class $s_{T,\theta}$ in $\pi_0((\hat{T}^x)^\Gamma)$ constructed from s_E or s_F, respectively, or s. It is defined to be the image of $s_E \in \hat{\mathbf{T}}_M \subseteq \hat{\mathbf{T}}_G$ under the maps $\hat{\mathbf{T}}_G \to \hat{\mathbf{T}}_{G_{ad}} \to (\hat{\mathbf{T}}_{G_{ad}})_{\theta_{ad}} \to \pi_0((\hat{T}^x)^\Gamma)$. Note $\hat{T}^x = (T^{\hat{\theta}_{sc}}_{sc}) \cong ((\hat{T})_{ad})_{\theta_{ad}}$. In our case the map $\hat{\mathbf{T}}_{G_{ad}} \to (\hat{\mathbf{T}}_{G_{ad}})_{\theta_{ad}}$ maps the class of $s_E = (1, \ldots, s)$ to the image of $s = s_F$ in $\hat{\mathbf{T}}_{G'_{ad}}$. Therefore, again the class in $\pi_0((\hat{T}^x)^\Gamma)$ does not depend on E. For $E = F$ it is the same as for the original endoscopic datum we started from. Therefore,

$$\Delta^E_I(\gamma_M, \delta) = \Delta_I(\gamma_M, t).$$

2. For Δ_{II} remember that all roots are of type R_1. Therefore,

$$\Delta^E_{II}(\gamma_M, \delta) = \prod_{\alpha_{res}} \chi_{\alpha_{res}} \left(\frac{N\alpha(\delta^*) - 1}{a_{\alpha_{res}}} \right).$$

The sum is over the Galois orbits of restricted roots which do not come from M. See [54], Lemma 4.3.A. Furthermore,

$$N\alpha = \sum_{i=0}^{l_\alpha - 1} \theta^i \alpha.$$

Here $l_\alpha = n$ is the length of the Galois orbit of α. In our case $N\alpha(\delta^*) = \alpha_{res}(N_\theta(\delta^*)) = \alpha_{res}(t)$, where $t \in T'(F) \cong T_\theta(F)$ is the image of $\gamma_M \in T_M(F)$ under the admissible embedding $T_M \to T_\theta$. Observe $\delta^* \in T(\overline{F})$ has norm $Nm_{E/F}(\delta^*) = t \in T'(F)$. This is equivalent to (3.3.4) in [54]. In our case m, defined in [54], (3.1), is the identity. We get

$$\Delta_{II}(t, \gamma_M) = \Delta_{II}^E(\delta, \gamma_M) = \prod_\alpha \chi_\alpha \left(\frac{\alpha(t) - 1}{a_\alpha} \right).$$

The product is over the Galois orbits of roots in $R(G', T')$ which do not come from M. This is obviously independent of E.

3. The fourth factor ([54], (4.5.A)) is treated similarly:

$$\Delta_{IV}^E(\gamma_M, \delta) = \prod_{\alpha_{res}} |N\alpha(\delta) - 1|_F^{1/2}.$$

The product is over all restricted roots not from M. They are of type R_1. The same argument as above gives

$$\Delta_{IV}(\gamma_M, t) = \Delta_{IV}^E(\gamma_M, \delta) = \prod_\alpha |\alpha(t) - 1|_F^{1/2}.$$

The product is over the Galois orbits of roots in $R(G', T')$ which do not come from M. It is obviously independent of the choice of E. This completes the proof of the lemma. $\square$

9.6.1 The Factor $\Delta_{III}(t, \delta)$

According to the definition of Kottwitz and Shelstad, this factor is defined by the value $\langle \mathbf{V}, \mathbf{A} \rangle$ of a hypercohomology pairing of the classes of the following two hypercohomology cocycles

$$\mathbf{V} = [(v(\sigma), \delta^*)] \in H^1(F, T_{sc} \overset{(1-\theta)\pi}{\longrightarrow} T),$$

$$\mathbf{A} = [(a_T^{-1}, s)] \in H^1(W_F, \hat{T} \overset{\hat{\pi}(1-\hat{\theta})}{\longrightarrow} (\hat{T})_{ad}).$$

See [54], p. 63. In fact, in loc. cit. it is enough to consider the easier case where U is T_{sc} and where S is $T_1 = T$ as explained in [54], p. 63 (quasisplit case for $\theta = \theta^*$) or [54], p. 38 (general case). The class $\mathbf{V}$ was already defined on page 305. First assume that the z-pair is $(M_1, \xi_{M_1}) = (M, {}^L id)$. The class $\mathbf{A}$ is defined by an endoscopic datum $({}^L M, M, s, \xi)$ of (G, θ) as follows. By transport of structure (twisted action) it is represented by a hypercocycle (a_T^{-1}, s) with values in the complex $\hat{\mathbf{T}}_G \to (\hat{\mathbf{T}}_G)_{ad}$, where:

1. $s \in (\hat{\mathbf{T}}_G)_{ad}$ is the image of the element $s \in \hat{\mathbf{T}}_G$, defined by the underlying endoscopic datum, under the map $\hat{\mathbf{T}}_G \to (\hat{\mathbf{T}}_G)_{ad}$.
2. a_T is the 1-cocycle in $Z^1(W_F, \hat{\mathbf{T}}_G)$ such that $\xi_{T_M} = a_T \cdot \xi_1$, where $\xi_{T_M} = \xi_\chi^M \circ \xi$ and $\xi_1 = \iota \circ \check{\xi}_1$ are defined by the first and the second row of the following diagram:

$$
\begin{array}{ccccc}
{}^L T_M & \xrightarrow{\;\xi_\chi^M\;} & {}^L M = \mathcal{H} & \xrightarrow{\;\xi\;} & {}^L G \\
{\cong}\big\downarrow & & & & \big\| \\
{}^L(T_\theta) & \xrightarrow{\;\check{\xi}_1\;} & {}^L((\hat{G}^{\hat\theta})^0) = {}^L G' & \xrightarrow{\;\iota\;} & {}^L G
\end{array}
$$

The left vertical map is induced from the admissible isomorphism $T_\theta \cong T_M$. The map ξ_χ^M is defined by the chosen χ-data induced for M, the second map ξ is part of the underlying endoscopic data. The map $\check{\xi}_1$ is defined by the induced χ-data. The precise definition only involves constructions in the group of fixed points $(\hat{G}^{\hat\theta})^0 \subseteq \hat{G}^{\hat\theta}$. See [54], p. 40. In our case of the twisted base change endoscopy $\hat{G}^{\hat\theta}$ is connected and ${}^L(\hat{G}^{\hat\theta}) = {}^L G'$. For twisted base change endoscopy the last map ι is defined by the "diagonal" embedding $\hat{\iota} : \hat{G}' \to \hat{G} = \prod_{i=1}^n \hat{G}'$. In the case of standard endoscopy the map $\xi_1 = \xi_\chi^G$, as well as ξ_χ^M, is defined as in Langlands and Shelstad [60], (2.6).

The description above applies both for the standard endoscopic datum $(M, {}^L M, s, \xi)$ attached to $(G, 1)$, and for the twisted base change datum $(M, {}^L M, s_E, \xi_E)$ attached to (G, θ). We denote the corresponding classes $\mathbf{A}$ and $\mathbf{A}^E$, and similarly $\mathbf{V}$ and $\mathbf{V}^E$.

Now let us drop the assumption that $(M_1, \xi_{M_1}) = (M, {}^L id)$. Then the cocycle a_T is determined by the following more complicated commutative diagram (see [54], pp. 43–44), which collapses to the diagram above in the special case $(M_1, \xi_{M_1}) = (M, {}^L id)$:

$$
\begin{array}{ccccccccc}
{}^L T_M & \xrightarrow{\;{}^L norm\;} & {}^L T_G & \xleftarrow{\;\beta\;} & \mathcal{U} & \xrightarrow{\;\alpha_0\;} & {}^L T_{M_1} & \xleftarrow{\;{}^L pr\;} & {}^L T_M \\
{\scriptstyle\hat\xi_1}\big\downarrow & {\scriptstyle\xi_1}\searrow & {\scriptstyle\xi_1'}\big\downarrow & & \big\downarrow & & {\scriptstyle\xi_{T_M}'}\big\downarrow & & {\scriptstyle\xi_\chi^M}\big\downarrow \\
{}^L((\hat{G}^{\hat\theta})^0) & \xrightarrow{\;\iota\;} & {}^L G & \xleftarrow{\;\xi\;} & \mathcal{H} & \xrightarrow{\;\xi_{M_1}\;} & {}^L M_1 & \xleftarrow{\;{}^L pr\;} & {}^L M
\end{array}
$$

Here $norm : T_G \twoheadrightarrow T_M$ and $pr : M_1 \twoheadrightarrow M$ and $pr : T_{M_1} \twoheadrightarrow T_M$, respectively. The dotted vertical arrows are certain canonical extensions defined in loc. cit. The group $\mathcal{U} \subseteq \mathcal{H}$ defined in loc. cit. contains $\hat{T}_M$ such that $\mathcal{U}/\hat{T}_M = W_F$. The L-homomorphism $\mathcal{U} \to {}^L(T_G \times T_{M_1})$, defined by $(\beta, \frac{1}{\alpha_0})$, is trivial on $\hat{T}_M$ and induces the desired cocycle $a_T : W_F = \mathcal{U}/\hat{T}_M \to {}^L(T_G \times T_{M_1})/\hat{T}_M = {}^L T_{M_1}$. The hypercocycle (a_T^{-1}, s) defines a cohomology class (loc. cit., p. 63):

$$\mathbf{A}_0 \in H^1(W_F, \hat{T}_1 \to (\hat{T})_{ad}).$$

Similarly, $(v(\sigma)^{-1}, \delta_1^*)$, now for $\delta_1^* \in T_1(F)$ with image $t_1 \in T_{M_1}(F)$, defines a class in $H^1(F, T_{sc} \to T_1)$. Here $T_1 = T \times_{T_M} T_{M_1}$ is the Cartesian product induced by $(norm, pr)$.

9.6.2 The Quotient $\Delta^E_{III}/\Delta_{III}$

Since all constructions involving the χ-data do not change in the passage from F to E, the only sensitive dependence on E in the definition of the factor Δ^E_{III} comes from the map ξ^E. For simplicity we concentrate on the case $M_1 = M$. To extend the formulations and proofs to the case $M_1 \neq M$ is an easy exercise, which is left to the reader. We only indicate at the end of this chapter (Sect. 9.8) how this is done in the "rational" case, since this is the only case that is relevant for our applications.

Standard Endoscopy. We start with the case $(M, {}^L M, s, \xi)$.

For $\xi^M_\chi(w) = b(w)w$ with $b(w) \in \hat{M}$ we get $\xi \circ \xi^M_\chi(w) = \xi(b(w))\xi(w)$. Since $\xi|\hat{M}$ is the inclusion map $\hat{M} \subseteq \hat{G}$, we can write this in the form $\xi \circ \xi^M_\chi(w) = b(w)\xi(w)$. The defining equation $\xi \circ \xi^M_\chi(w) = a_{T'}(w)\xi^{G'}_\chi(w)$ therefore gives

$$b(w)\xi(w) = a_{T'}(w)\xi^{G'}_\chi(w).$$

Suppose $\xi(w) = c(w)w$. Our assumptions $Int(s) \circ \xi = \xi$ and $s \in Z(\hat{M})^\Gamma$ imply $sc(w) = c(w)s$. Hence, $c(w) \in (\hat{G}')_s = \hat{M}$ from [54], (2.1.4b).

Twisted Case. Now the case $(M, {}^L M, s_E, \xi_E)$ for the z-pair $(M, {}^L id)$.

Now $\xi_E \circ \xi^M_\chi(w) = \xi_E(b(w)w) = \iota(b(w))\xi_E(w)$ since $\xi_E|\hat{M}$ is the diagonal embedding $\iota : \hat{M} \hookrightarrow \hat{G} = \prod_{i=1}^n \hat{G}'$. Since $\check{\xi}_1 = \xi^{G'}_\chi$ in our case, we obtain the defining equation

$$\iota(b(w))\xi_E(w) = a^E_T(w)\iota(\xi^{G'}_\chi)(w)$$

for $a^E_T(w)$. Hence, by comparison with $\iota(b(w))\iota\xi(w) = \iota(a_{T'}(w))\iota(\xi^{G'}_\chi(w))$

$$\iota(b(w))^{-1} \cdot \iota\xi(w)\xi_E(w)^{-1} \cdot \iota(b(w))^{-1} = \iota(a_{T'}(w)) \cdot a^E_T(w)^{-1} \in \hat{\mathbf{T}}_G.$$

ξ and ξ_E are unramified; hence $\iota\xi(w)\xi_E(w)^{-1}$ is unramified and is uniquely determined by $\iota\xi(\sigma_F)\xi_E(\sigma_F)^{-1} = (\lambda, \ldots, \lambda, s\lambda)^{-1} \in \iota(Z(\hat{M})^\Gamma)$. Since $b(w) \in \hat{M}$ commutes with this cocycle, we obtain

Lemma 9.14. *The 1-cocycle* $a^E_T(w) = \iota(a_{T'}(w)) \cdot \tau^E(w)$ *with values in* $\hat{\mathbf{T}}_G$ *is a product of 1-cocycles, where* $\tau^E(w)$ *is the unramified 1-cocycle determined by*

$$\tau^E(\sigma_F) = (\lambda, \ldots, \lambda, s\lambda) \in \hat{\mathbf{T}}_G.$$

Remark 9.3. For $M_1 \neq M$ then similarly $\tau^E(\sigma_F) = (s_E, \lambda^{-1})$ mod $\hat{T}_M$ in $(\hat{\mathbf{T}} \times \hat{\mathbf{T}}_{M_1})/\hat{\mathbf{T}}_M = \hat{\mathbf{T}}_1$.

The corresponding hypercocycle is also a product of hypercocycles in $Z^1(W_F, \hat{T}_G \to (\hat{T}_G)_{ad})$

$$\mathbf{A}^E = (\iota(a_{T'}(w))^{-1}\tau^E(w)^{-1}, s_E) = (\iota(a_{T'}(w))^{-1}, 1) \cdot (\tau^E(w)^{-1}, s_E)$$

and $\mathbf{A} = (a_{T'}(w)^{-1}, s) = (a_{T'}(w)^{-1}, 1) \cdot (1, s)$, respectively, for standard endoscopy.

The Hypercocycle $(\tau^E(w)^{-1}, s_E)$. Concerning the twisted Galois actions, transported from $\hat{T}_G, (\hat{T}_G)_{ad} \cong \widehat{(T_{sc})}$, the pair $(\tau^E(w)^{-1}, s_E)$

$$\tau^E(w) = (\lambda, \dots, \lambda, s\lambda) \in \hat{\mathbf{T}}_G,$$

$$s_E = (1, \dots, 1, s) \in (\hat{\mathbf{T}}_G)_{ad} = \hat{\mathbf{T}}_G/Z(\hat{G})$$

not only defines a hypercohomology cocycle in $Z^1(W_F, \hat{\mathbf{T}}_G \xrightarrow{1-\theta} \hat{\mathbf{T}}_{G_{ad}})$, but also in

$$Z^1(W_F, \hat{\mathbf{T}}_G \xrightarrow{1-\theta} \hat{\mathbf{T}}_G).$$

Notice $(\hat{T}_G)_{ad} = \hat{T}_G/Z(\hat{G}) \cong (\widehat{(T_G)_{sc}})$ (see [12], p. 44). Our assumption $s \in Z(\hat{M})^\Gamma$ implies that $s \in \hat{\mathbf{T}}_M = \hat{\mathbf{T}}_{G'}$ is invariant under conjugation by elements in $\hat{M}$ and is invariant under W_F (with the L-action on $\hat{M}$). Hence, $s \in \mathbf{T}_M$ is also invariant under the twisted action, transported from $\hat{T}_M$, since this only involves conjugations in $\hat{M}$. Similarly for the twisted action on $\hat{\mathbf{T}}_{G'}$ (transported from T'). Therefore, $s \in \hat{T}'$ is invariant. Hence,

$$(\tau^E)^{-1}(w) = (\lambda^{-1}, \dots, \lambda^{-1}, \lambda^{-1}s^{-1}) \in \hat{\mathbf{T}}_G,$$

$$s_E = (1, \dots, 1, s) \in \hat{\mathbf{T}}_G$$

defines a hypercocycle in $Z^1(W_F, \hat{\mathbf{T}}_G \xrightarrow{1-\theta} \hat{\mathbf{T}}_G)$ since $(1 - \theta)\tau^E(\sigma_F)^{-1} = (1, \dots, 1, s, s^{-1}) = \sigma_F(s_E)s_E^{-1}$ in $\hat{\mathbf{T}}_G$. Again we used $\sigma_F(s) = s$ for the twisted or untwisted action on $\hat{\mathbf{T}}_{G'}$.

Compatibility with Pullbacks ([54], p. 138, Lemma A.3.B). For $K^\bullet = (A \to B)$ and $\hat{K}^\bullet = \hat{B} \to \hat{A}$ one has an exact sequence

$$0 \to ((\hat{B})^\Gamma)^0 \to H^1(W_F, \hat{K}^\bullet) \to Hom_{cont}(H^1(F, K^\bullet), \mathbb{C}^*) \to 0$$

which is functorial with respect to chain complex homomorphisms $\phi : K_1^\bullet \to K_2^\bullet$. Furthermore, for $\mathbf{A} \in H^1(W_F, \hat{K}_1^\bullet)$ of the form $\mathbf{A} = \hat{\phi}_*(\mathbf{X})$ the following holds:

$$\langle \mathbf{V}, \mathbf{A} \rangle = \langle \mathbf{V}, \hat{\phi}_*(\mathbf{X}) \rangle = \langle \phi_*(\mathbf{V}), \mathbf{A} \rangle.$$

An Application. For the chain complex homomorphism $\phi = (id, Nm_{E/F})$

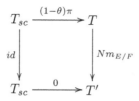

we have

$$\iota(a_{T'}(w)^{-1}, 1) = \hat{\phi}_*(a_{T'}(w)^{-1}, 1), \tag{9.1}$$

$$\phi_*(\mathbf{V}) = [(v_i(\sigma)^{-1}, t)]. \tag{9.2}$$

Hence,

$$\langle \mathbf{V}, \iota[(a_{T'}(w)^{-1}, 1)] \rangle = \langle [(v_i(\sigma), t)], [(a_{T'}(w)^{-1}, 1)] \rangle.$$

Note $H^1(F, A \xrightarrow{0} B) = H^0(F, B) \oplus H^1(F, A)$ and $H^1(W_F, \hat{B} \xrightarrow{0} \hat{A}) = H^0(W_F, \hat{A}) \oplus H^1(W_F, \hat{B})$. From [54], (A.3.13) and (A.3.14), it follows that the hypercohomology pairing degenerates

$$\langle [(a, b)], [(\hat{b}, \hat{a})] \rangle = \langle a, \hat{a} \rangle \langle b, \hat{b} \rangle_L^{-1}$$

into a pairing between $H^0(F, B)$ and $H^1(W_F, \hat{B})$ (inverse of the Langlands reciprocity pairing) and a pairing between $H^1(F, A)$ and $H^0(W_F, \hat{A})$ (Tate–Nakayama pairing). Hence,

$$\langle [(v_i(\sigma)^{-1}, t)], [(a_{T'}(w)^{-1}, 1)] \rangle = \langle a_{T'}(w), t \rangle_L \langle 1, v_i(\sigma)^{-1} \rangle = \langle [(a_{T'}(w)], t \rangle_L$$

$$= \Delta_{III_2}(\gamma_M, t) = \Delta_{III_2}(\gamma_M, t) \Delta_{III_1}(\gamma_M, t) = \Delta_{III}(\gamma_M, t)$$

in accordance with Langlands and Shelstad [60], (3.5). The factor $\Delta_{III_1}(\gamma_M, t)$ is trivial by our assumptions. t is the image of γ_M under an admissible embedding. By Lemma 9.14, therefore, we have

Lemma 9.15. $\frac{\Delta_{III}(M, G, t_M, \delta)}{\Delta_{III}(M, G', t_M, t)} = \Delta_{III}^E / \Delta_{III} = \langle [(v_i(\sigma)^{-1}, \delta^*)], [(\tau^E)^{-1}, s_E)] \rangle.$

Since $(t^E)^{-1}, s_E)$ also defines a hypercocycle in $Z^1(W_F, \hat{\mathbf{T}}_G \xrightarrow{1-\theta} \hat{\mathbf{T}}_{G_{ad}})$, we can consider the pullback with respect to the chain complex homomorphism $\phi = (\pi, id)$

$$\begin{array}{ccc}
T_{sc} & \xrightarrow{(1-\theta) \circ \pi} & T \\
{\scriptstyle \pi} \downarrow & & \downarrow {\scriptstyle id} \\
T & \xrightarrow{1-\theta} & T
\end{array}$$

By the compatibility of the pairing with pullback, we therefore obtain from Lemmas 9.13 and 9.15

Theorem 9.2. *For the pairing between* $H^1(W_F, \hat{T} \xrightarrow{1-\hat{\theta}} \hat{T})$ *and* $H^1(F, T \xrightarrow{1-\theta} T)$

$$\frac{\Delta(M, G, t_M, \delta)}{\Delta(M, G', t_M, t)} = \langle \mathbf{inv}_\theta(t, \delta_i), [(\tau^E)^{-1}, s_E)] \rangle.$$

Theorem 9.2 implies that the assertion $A(E, f)$ – relevant for the Frobenius formula – is equivalent to the assertion that

$$\langle \mathbf{inv}_\theta(t, \delta_i), [(\tau^E)^{-1}, s_E)] \rangle = \langle \alpha(t; \delta_i), s \rangle$$

holds whenever $O^G_{\delta\theta}(f) \neq 0$.

Remark 9.4. Both $\langle \alpha(t; \delta_i), s \rangle$ and $\langle [(\tau^E)^{-1}, s_E)], \mathbf{inv}(t, \delta_i) \rangle$ have the same transformation property if $\delta_i \in G(F)$ is replaced by $\delta \in G(F)$ within the same stable θ-conjugacy class. Kottwitz and Shelstad [54], Theorem 5.1.D, gave $\Delta(M, G, t, \delta) = \Delta(M, G, t, \delta_i) \langle inv(\delta_i, \delta), s \rangle$. On the other hand, in the proof of Lemma 9.8 we have already seen $\alpha(t; \delta) = \alpha(t; \delta_i) \langle inv(\delta_i, \delta), s \rangle$. Therefore, if Theorem 9.2 holds for one δ_i, it holds for all δ_i in the stable θ-conjugacy class.

The Rational Case. Suppose there exists $\delta^* \in T(F)$ such that $Nm_{E/F}(\delta^*) = t \in T'(F)$. Then it is enough to prove Theorem 9.2 for this $\delta_i = \delta^*$, denoted now simply as δ. By definition of c (on page 303) we can choose $c = 1$, and obtain

$$\alpha(t; \delta) = [c^{-1} \cdot \delta \cdot \sigma_F(c)] = [\delta] \in B(T') = T'(\mathbf{L})/(1 - \sigma_F)T'(\mathbf{L}).$$

Furthermore, $\mathbf{inv}(t; \delta) = [(1, \delta)] \in H^1(W_F, T \xrightarrow{1-\theta} T)$ according to Sect. 9.5 since $g_i = 1$. The desired assertion $\langle \mathbf{inv}_\theta(t, \delta_i), [(t^E)^{-1}, s_E)] \rangle = \langle \alpha(t; \delta_i), s \rangle$ therefore boils down to proving the formula

$$\langle t^E, \delta \rangle_L = \langle [\delta], s \rangle$$

since

$$\langle \mathbf{inv}_\theta(t, \delta), [(\tau^E)^{-1}, s_E)] \rangle = \langle \tau^E, \delta \rangle_L$$

by (A.3.13) in [54]. On the right we have the Langlands reciprocity pairing for the F-torus T, and on the left the pairing between $B(T')$ and $\hat{T}'^\Gamma$ for the F-torus T'.

The pairing $\langle [\delta], s \rangle$ on the left depends on E only from the way that $\delta \in T'(E) \subseteq T'(\mathbf{L})$. By Lemma 9.6 we can replace F by E and T' by its base field extension T'/E since $s \in \hat{T}'^{\Gamma_F} \subseteq \hat{T}'^{\Gamma_E}$. To compare both sides we also compute the right side by the Langlands pairing of the torus T'/E using Shapiro's lemma.

Lemma 9.16. *For* $T = Res_{E/F}(T')$ *the Langlands reciprocity pairing between* $H^1(W_F, \hat{T})$ *and* $T(F)$ *is compatible with Shapiro's lemma, i.e., base extension from* F *to* E.

In Other Words: Under the canonical isomorphism $H^1(W_F, \hat{T}) \cong H^1(W_E, \hat{T}')$ defined by Shapiro's lemma, and the canonical isomorphism $T(F) = T'(E)$, the Langlands reciprocity pairing for the F-torus $T = Res_{E/F}$ pairing between $H^1(W_F, T)$ and $T(F)$ coincides with the Langlands reciprocity pairing between $H^1(W_E, T'/E)$ and $T'(E)$ for the E-torus T'.

Proof. Shapiro's isomorphism $H^q(V, I) \cong H^q(U, A)$ for induced modules I, where U is a subgroup of V of finite index, is defined as the composite $\pi \circ res$ of two maps

$$H^q(V, I) \xrightarrow{res} H^q(U, I) \xrightarrow{H^q(\pi)} H^q(U, A) ,$$

where $\pi : I = Ind(U, V, A) = \bigoplus_{g \in V/U} A \to A$ is the U-equivariant projection on the component of the $g = 1$ coset and res is the cohomological restriction map [66], p. 68. We may replace W_F by $W_{K/F}$ for a normal splitting field extension K/F of T containing E. Let $\Gamma = \Gamma_{K/F}$ denote its Galois group. The Langlands reciprocity law is the statement that the corestriction map cor in the diagram below is an isomorphism. This follows from the following diagram by the 5-lemma.

$$0 \to H_T^{-1}(\Gamma, Hom(C, \hat{T})) \to Hom(C, \hat{T})_\Gamma \xrightarrow{N_\Gamma} Hom(C, \hat{T})^\Gamma \to H_T^0(\Gamma, Hom(C, X^*(\hat{T})), \mathbb{C}^*)$$

$$\downarrow{\cup u} \qquad\qquad \downarrow{cor} \qquad\qquad \downarrow{id} \qquad\qquad \downarrow{\cup u}$$

$$0 \longrightarrow H^1(\Gamma, \hat{T}) \xrightarrow{Inf} H^1(W_{K/F}, \hat{T}) \xrightarrow{res} H^1(\overline{K}^*, \hat{T})^\Gamma \xrightarrow{trans} H^2(\Gamma, \hat{T}),$$

where $C = \overline{K}^*$ [63], p. 130 and formula (8.6.1). Since this diagram is functorial in T, for the compatibility of this isomorphism with Shapiro's lemma it is therefore enough to check that all the other maps in this diagram except cor are compatible with the cohomological restriction maps from W_F to W_E. For this observe that the first and the fourth vertical map is an isomorphism ([63], Lemma 8.3) induced by the cup product with the fundamental class. Since cup products are compatible with restrictions [63], p. 4 and (0.1.5), and since the fundamental class u is compatible with restrictions by the Tate–Nakayama theorem [63], Theorem 0.2, the lemma follows. $\square$

Shapiro's Lemma. We apply Lemma 9.16 to compute $\langle \tau^E, \delta \rangle_L$ for $\delta \in H^0(F, T) = T(E) = H^0(E, T'/E)$. The image $res(\tau^E) \in H^1(W_E, \hat{T})$ of $\tau^E \in H^1(W_F, \hat{T})$ is the unramified cocycle determined by $\sigma_E = (\sigma_F)^n \mapsto N_{\sigma_F}((\lambda, \ldots, \lambda, s\lambda)) = (\lambda^n s, \lambda^n s, \ldots, \lambda^n s)$. Hence, the projection $H^1(\pi)$ gives the class of the unramified cocycle $\tau^{\lambda^n s} \in H^1(W_E, \hat{T}')$ defined $\sigma_E \mapsto \lambda^n s$. Thus, the value $\langle \tau^E, \delta \rangle_L$ of the Langlands pairing for the F-torus T is the same as the value $\langle \tau^{\lambda^n s}, \delta \rangle_L$ of the Langlands pairing of the E-torus T'/E obtained by base change from T' over F.

Lemma 9.17. *Suppose we are in the rational case $t = Nm_{E/F}(\delta)$ for $\delta \in T(F) = T'(E)$. Then*

$$\boxed{\langle \mathbf{inv}_\theta(t, \delta), [(t^E)^{-1}, s_E)] \rangle = \langle \tau^{\lambda^n s}, \delta \rangle_L}$$

for the Langlands pairing $\langle .,.\rangle_L$ of the torus T' over E between δ and the class of the unramified cocycle $\tau^{\lambda^n s}(\sigma_F) = \lambda^n s \in \mathbf{T}_{G'}$ in $H^1(W_E, \hat{T}'_G)$.

$$\boxed{\langle \alpha(t; \delta), \lambda^n s \rangle = \langle [\delta], \lambda^n s \rangle}$$

for $[\delta] \in B_E(T') = T'(\mathbf{L})/(1 - \sigma_E)T'(\mathbf{L})$ and $\lambda^n s \in (\hat{T}')^{\Gamma_E}$.

The Main Conclusion. The desired assertion $A(E, f)$ of the Frobenius formula (Lemma 9.7) follows from the following

Assertion $A(T', E)$. *The elementary character in $Hom_{cont}(T'(E), \mathbb{C}^*)$ defined by $T'(E) \ni \delta \mapsto \langle [\delta], \lambda^n s \rangle \in \mathbb{C}^*$ coincides with the class of the unramified cocycle $\tau^{\lambda^n s}$ in $H^1(W_F, \hat{T}')$ under the Langlands reciprocity map $Hom_{cont}(T'(E), \mathbb{C}^*) \cong H^1(W_E, \hat{T}')$ for the torus T' over the local field E.*

In fact, the support condition $O^{G,\kappa}_{\delta\theta}(f) \neq 0$ implies $Nm_{E/F}(\delta) = t \in T'(F)$, provided the twisted fundamental lemma holds for f; hence, we are in the rational case without restriction of generality, so we can apply Lemma 9.17. Notice the support condition of the twisted fundamental lemma holds true for $f = 1_K$ by Theorem 9.1; hence, in particular, Lemma 9.17 implies that the twisted base change fundamental lemma holds for the unit elements. But this implies that the twisted base change fundamental lemma holds for all spherical functions (see Chap. 10); hence, the support condition holds for spherical f. Thus, for the proof of $A(E, f)$ we can assume that we are in the rational case so that assertion $A(E, f)$ follows from $A(T', E)$ by Lemma 9.17.

Before we proceed to prove assertion $A(T', E)$ in Theorem 9.3, let us state the consequence.

Corollary 9.4. *The fundamental lemma for standard endoscopy implies the fundamental lemma for the twisted base change, and the Frobenius formula.*

9.6.3 The Explicit Reciprocity Law

Consider the category of tori over a p-adic field F. There are natural transformations

$$\psi_i : \hat{T}^\Gamma \to H^1(W_F, \hat{T})$$

between the functors $T \mapsto \hat{T}^\Gamma$ and $T \mapsto H^1(W_F, \hat{T})$ defined, respectively, by

$$\psi_1 : \hat{T}^\Gamma \to Hom_{elm}(T(F), \mathbb{C}^*) \subseteq Hom_{cont}(T(F), \mathbb{C}^*) \cong H^1(W_F, \hat{T}),$$

where the last isomorphism is the Langlands reciprocity map (for the definition of the first map and the notion of elementary characters, see page 300), and

$$\psi_2 : \hat{T}^\Gamma = H^1(\mathbb{Z}, \hat{T}^\Gamma) \overset{infl}{\to} H^1(W_F, \hat{T}^\Gamma) \to H^1(W_F, \hat{T}).$$

Here inflation is with respect to the unramified quotient map $ord : W_F \to \mathbb{Z}$, and the second map is induced from the inclusion map $\hat{T}^\Gamma \subseteq \hat{T}$.

Theorem 9.3. $\psi_1 = \psi_2$.

Proof. See Theorem 11.1. $\square$

The statement $\psi_1 = \psi_2$ for F replaced by E and T replaced by T' is just assertion $A(T', E)$ stated above. Hence, Theorem 9.3 implies Corollary 9.4.

9.7 Twisted Transfer Factors for Split Tori

For split tori choose trivial a-data and trivial χ-data: $\chi_\lambda = 1$ for all restricted roots and $a_\lambda = 1$ for the positive restricted roots and $a_\lambda = -1$ for the negative restricted roots. Using trivial a-data and χ-data, we get

Lemma 9.18. *Suppose G' and M are F-split groups. Then for the maximal split torus $T = Res_{E/F}(T')$ in the Borel group B the transfer factor is*

$$\Delta(\gamma, \delta) = \tau(\delta) \cdot \prod_\alpha |\alpha(t) - 1|_F^{1/2}.$$

The product is over all restricted roots which do not come from M for $\delta \in T(F), t \in T_\theta(F) \cong T'(F)$ and $\gamma \in T_M(F)$ related by the norm and the underlying isomorphism $T_\theta \cong T_M$, respectively. τ is an unramified character of $T(F)$ determined by $\lambda^n s \in (\hat{T})^\Gamma$ via Langlands reciprocity.

Proof. The restricted root system $R_{res}(G, T_G)$ can be identified with the root system $R(G^\theta, T_G^\theta)$. In other words, the root system of the torus T restricted to $T^\theta = T'$ can be identified with the root system of the group G'. $\square$

The First Factor $\Delta_I(\gamma, \delta)$. This factor is of cohomological nature [54], (4.2), i.e., attached to the Galois cohomology group $H^1(F, T^x)$, where T^x is the intersection $T_{sc} \cap G^x$. G^x is the group of fixed points of the lift θ_{sc} to G_{sc}, where G_{sc} is the simply connected covering of G_{der}. In our case $G_{der} = G$ is simply connected; hence, $G^x = G'_{der}$. Thus, T^x has vanishing cohomology by Shapiro's lemma. Hence, $\Delta_I(\gamma, \delta) = 1$.

The Second Factor $\Delta_{II}(\gamma, \delta)$. It is also trivial for split tori. It is attached to the restricted roots which do not come from M. It is 1 since trivial χ-data were chosen [54], Lemma 4.3.A.

The Fourth Factor ([54], Lemma 4.5.A). It is

$$\Delta_{IV}(\gamma,\delta) = \prod_{\alpha_{res}} |N\alpha(\delta^*) - 1|_{\overline{F}}^{1/2}.$$

The product is over all restricted roots not from M, which are all of type R_1, and

$$N\alpha = \sum_{i=0}^{l_\alpha-1} \theta^i \alpha,$$

where $l_\alpha = n$ is the length of the Galois orbit of α. In our case $N\alpha(\delta^*) = \alpha_{res}(N_\theta(\delta^*)) = \alpha_{res}(t)$, where $t \in T_\theta(F)$ is the image of $\gamma \in T_M(F)$ under the admissible embedding $T_M \to T_\theta$. Notice that $\delta^* \in T(\overline{F})$ satisfies $N_\theta(\delta^*) = t \in T_\theta(F)$, which corresponds to $\gamma \in T_M$ under the fixed isomorphism $T_\theta(F) \cong T_M(F)$. Furthermore, observe that $\delta^* = g\delta\theta(g)^{-1}$ for some $g \in G_{sc}(\overline{F})$ using the notation in [54], (3.3.4) since in our case m, defined as in [54], (3.1), is the identity.

Claim 9.1. $\mathbf{V} \in H^1(F, T_{sc} \overset{(1-\theta)\pi}{\to} T)$ is the image of the class $[D] = \delta^* \in H^0(F,T)$ under the natural map $j : H^0(F,T) \to H^1(F, T_{sc} \to T)$ (defined in [54], (A.3.13)).

Proof. This follows from the long exact sequence of hypercohomology together with $H^1(F, T_{sc}) = H^1(F, T_{der}) = 1$ (Hilbert 90 and Shapiro's lemma). In other words, we can assume $v(\sigma) = 1$ by modifying $g \in G_{sc}(\overline{F})$ from the left by an appropriate element in $T_{sc}(\overline{F})$. After this modification $[(v(\sigma), \delta^*)] = [(1, \delta^*)]$. The hypercocycle condition implies $\sigma(\delta^*) = \delta^*$, $\delta^* \in T(F)$. Therefore, $\mathbf{V} = j([D])$ is the image of $D = \delta^* \in H^0(F,T)$. Furthermore, $\delta^* \in G(F)$ is θ-conjugate over F to $\delta \in G(F)$ since $v(\sigma) = 1$ implies $g \in G_{sc}(F)$. $\square$

The Element A ([54], p. 39). It is represented by a pair (a_T^{-1}, s_E).

Recall that the endoscopic embedding ${}^L M \to {}^L G$ is unramified, and $\lambda, s \in Z(\hat{M})^\Gamma$. For the Frobenius element σ_F we get $\xi_{T_M}(1 \times \sigma_F) = \xi(1 \times \sigma_F) = \iota(\lambda) \cdot s_E \lhd \sigma_F$. Also $\xi_1(\sigma) = 1 \cdot n(\sigma) \times \sigma = 1 \lhd \sigma$ holds, since the χ-data are trivial and the torus is split such that $n(\sigma_T) = 1$ (in the notation in [54, 60]). This implies

$$a_T(\sigma_F) = \iota(\lambda) \cdot s_E = (\lambda, \dots, \lambda, \lambda \cdot s) \in \hat{\mathbf{T}}_G.$$

a_T satisfies the compatibility equation $(1 - \hat{\theta})a_T(\sigma_F) = s_E^{-1}\sigma_F(s_E) = (1, \dots, 1, s, s^{-1})$, and hence defines a hypercocycle.

The factor $\Delta_{III}(\gamma, \delta)$ is obtained by a hypercohomology pairing evaluated at these two classes $j([D])$ and $\mathbf{A}$. Using (A.3.13) in [54], we get

$$\langle j([D]), \mathbf{A} \rangle = \langle [D], \hat{i}(\mathbf{A}) \rangle_L^{-1},$$

where $\langle .,. \rangle_L : H^0(F,T) \times H^1(W_F, \hat{T}) \to \mathbb{C}^*$ is the Langlands reciprocity pairing, and where $\hat{i} : H^1(W_F, \hat{T} \to \hat{T}_{sc}) \to H^1(W_F, \hat{T})$ is the natural map. In our case

$\hat{i}(\mathbf{A}) = \hat{i}(a_T^{-1}, s_E) = [a_T^{-1}]$. Therefore,

$$\Delta_{III}(\gamma, \delta) = \langle \delta, a_T \rangle_L.$$

To Compute this Number. E/F is unramified; hence, so is the class of a_T. So we may proceed following Borel [12], Sects. 9.5 and 6.3. Let T_d denote the maximal F-split torus in T. The Langlands pairing defines an isomorphism $H^1(W_F, \hat{T}) \cong Hom_{cont}(T(F), \mathbb{C}^*)$ functorial in T. It is shown in [12] that unramified classes can be evaluated by passage to $T_d \hookrightarrow T$. In other words, let $\chi_0 : T(F) \to \mathbb{C}^*$ denote the unramified character corresponding to the cohomology class $[a_T] \in H^1(W_F, \hat{T}) \cong Hom_{cont}(T(F), \mathbb{C}^*)$ via Langlands reciprocity. Then the restriction $\chi_0 : T(F) \to \mathbb{C}^*$ to $T_d(F)$ corresponds to the cocycle which maps σ to the image of $a_T(\sigma)$ under the map $\hat{T} = \mathbb{C}^* \otimes X^*(T) \to \hat{T}_d = \mathbb{C}^* \otimes X^*(T_d)$. Since $X^*(T) \to X^*(T_d)$ is $(x_1, \ldots, x_n) \mapsto \sum x_i$, the canonical map $\hat{T} \to \hat{T}_d$ dual to the restriction map is the product over the components $\hat{T} = \prod_{i=1}^n \hat{T}_d \to \hat{T}_d$. Thus, the image of $a_T(\sigma_F) = \lambda s_E$ is the product of the components of $a_T(\sigma_F)$ for $s \in \hat{T}_d$. In other words, the image is represented by the unramified cocycle whose value at the Frobenius σ_F is $\lambda^n \cdot s$. This proves Lemma 9.18.

9.8 The Case $M_1 \neq M$

The main results of the last sections Lemmas 9.17 and 9.18 and Corollary 9.4 extend to the case $M_1 \neq M$. For this it is enough to consider the rational case where $\delta^* = \delta \in T_G(F)$ exists with $t = Nm_{E/F}(\delta)$. For $M_1 \neq M$ consider the Cartesian diagrams defining $\hat{T}_1 = (\hat{T}_G \times \hat{T}_{M_1})/\hat{T}_M$ with (s_E, λ^{-1}) mod $\hat{T}_M$ in $\hat{T}_1$

$$
\begin{array}{ccc}
T_1(F) & \longrightarrow & T_G(F) \ni \delta^* \\
\downarrow & & \downarrow {\scriptstyle norm} \\
t_1 \in T_{M_1}(F) & \xrightarrow{\ pr\ } & T_M(F) \ni t
\end{array}
\qquad
\begin{array}{ccc}
\hat{T}_1 & \longleftarrow & \hat{T}_G = \prod_{\nu=1}^n \hat{T}' \\
\uparrow & & \uparrow {\scriptstyle \iota} \\
\hat{T}_{M_1} & \xleftarrow{\ \hat{pr}\ } & \hat{T}_M = \hat{T}'
\end{array}
$$

The z-extension $pr : M_1 \to M$ defines a corresponding z-extension $Res_{E/F}(pr) :$ $\tilde{M} := Res_{E/F}(M_1) \to Res_{E/F}(M)$, and hence a surjection $Res_{E/F}(pr) : \tilde{T} := Res_{E/F}(T_{M_1}) \to T_G = Res_{E/F}(T') \cong Res_{E/F}(T_M)$ factorizing over T_1

$$
\begin{array}{ccccc}
\tilde{\delta} \in \tilde{T}(F) & \xrightarrow{\ \mu\ } & T_1(F) & \longrightarrow & T_G(F) \ni \delta^* \\
& {\scriptstyle norm} \searrow & \downarrow & & \downarrow {\scriptstyle norm} \\
& & t_1 \in T_{M_1}(F) & \xrightarrow{\ pr\ } & T_M(F) \ni t
\end{array}
$$

Notice $\tilde{T}(F) \rightarrow T(F)$ is also a surjection. Therefore, we find $\tilde{\delta} \in \tilde{T}(F)$ with image $\delta \in T(F)$. It defines $t_1 \in T_{M_1}(F)$ over $t \in T_M(F)$. Notice t_1 is uniquely determined by t up to some element in $Z_1(F)$. Hence, the invariant $inv(t_1, \delta)$ is defined as the class of the cocycle $(1, \delta_1)$ in $image(H^0(F, T_1) \rightarrow H^1(F, T_{sc} \rightarrow T_1))$. In fact this class is already in the image of the composed map

$$H^0(F, \tilde{T}) \rightarrow H^0(F, T_1) \rightarrow H^1(F, T_{sc} \rightarrow T_1).$$

By functoriality the hypercohomology pairing $< inv(t_1, \delta), \mathbf{A}_0 >$ between $inv(t_1, \delta)$ and $\mathbf{A}_0$ can therefore be evaluated as in the proof of Lemma 9.17,

$$< inv(t_1, \delta), \mathbf{A}_0 > = \langle \tilde{\delta}, \tilde{a}_T \rangle_L,$$

where now $\tilde{a}_T$ is the image class of a_T under $H^1(\hat{\mu}) : H^1(W_F, \hat{T}_1) \rightarrow H^1(W_F, \hat{\tilde{T}})$. Notice

$$\hat{\mu} : \hat{T}_1 \rightarrow \hat{\tilde{T}} = \prod_{\nu=1}^{n} \hat{T}_{M_1}$$

maps the unramified cocycle whose value at σ_F is (s_E, λ^{-1}) mod $\hat{T}_M$ (see Remark 9.3, where we now identify $\mathbf{T}$ with T, etc. by transport of structure) to the unramified cocycle whose value at σ_F is $(\lambda, \ldots, \lambda, s\lambda) \in \prod_1^n \hat{T}_{M_1}$. The remaining computations for the proof of Lemmas 9.17 and 9.18 and Corollary 9.4 are now completely analogous to the case $M_1 = M$, except that $\hat{T}_G$ has to replaced by $\hat{\tilde{T}}$ and (t, δ) has to be replaced by $(t_1, \tilde{\delta})$.

Chapter 10
Reduction to Unit Elements

The reduction of the fundamental lemma for arbitrary spherical Hecke operators to the case of the unit elements of the spherical Hecke algebra for the case of twisted endoscopy, although formulated only for $G' = GSp(4)$ and base change in this chapter, holds for twisted endoscopy in greater generality. However, to avoid technical considerations, we restrict ourselves here to the special case. The general case will be considered elsewhere [113].

10.1 The Unramified Endoscopic Lift

Concerning endoscopic data, in particular the choice of an L-embedding $\xi : {}^L M \to {}^L G$, we use the notation and conventions in Sects. 6.3 and 8.1. For this choice the corresponding unramified endoscopic lift, which depends on the chosen endoscopic datum, is described in Lemma 4.25.

As in Chap. 6 and Sect. 8.1 fix the embedding $\xi : {}^L M \hookrightarrow {}^L G$, and tori and Borel subgroups $B_M \subseteq M$ and $B = B_G \subseteq G$. In particular, B_M is the Borel group defined by the matrices

$$\gamma = \begin{pmatrix} t_1 & * \\ 0 & t_1' \end{pmatrix} \times \begin{pmatrix} t_2 & 0 \\ * & t_2' \end{pmatrix}$$

modulo the elements $(t \cdot id, t^{-1} \cdot id)$ for $t \in F^*$. T_M^s is the split torus of diagonal matrices in B_M of the form $t = (diag(t_1, t_1'), diag(t_2, t_2'))$, modulo matrices of the form $(diag(t, t), diag(t^{-1}, t^{-1}))$ for $t \in F^*$. The positive roots R_M^+ are the characters $\chi_1(t) = t_1/t_1'$ and $\chi_2(t) = t_2'/t_2$. $T_G^s \subseteq B_G \subseteq GSp(4)$ denotes the split torus of diagonal matrices, and elements of B_G have the form

$$\delta = \begin{pmatrix} x_1 & 0 & * & * \\ * & x_2 & * & * \\ 0 & 0 & x_1' & * \\ 0 & 0 & 0 & x_2' \end{pmatrix}.$$

R. Weissauer, *Endoscopy for GSp(4) and the Cohomology of Siegel Modular Threefolds*, Lecture Notes in Mathematics 1968, DOI: 10.1007/978-3-540-89306-6_10, © Springer-Verlag Berlin Heidelberg 2009

An admissible isomorphism ϕ between the two split tori T_M^s and T_G^s was defined in Lemma 8.1:

$$\phi : \left(diag(t_1, t_1'), diag(t_2, t_2')\right) \bmod F^* \mapsto diag(t_2' t_1', t_! t_2', t_1 t_2, t_2 t_1'),$$

with inverse $\phi^{-1} : diag(x_1, x_2, x_1', x_2') \mapsto diag(x_2, x_1) \times diag(x_1'/x_2, 1) \bmod F^*$. Under ϕ the unramified characters of the two tori correspond to each other. The character

$$(\xi, \eta) \times (\xi', \eta')\left(diag(t_1, t_1') \times diag(t_2, t_2')\right) = \xi(t_1)\eta(t_1')\xi'(t_2)\eta'(t_2'),$$

where $\xi\eta = \xi'\eta'$, corresponds to $diag(x_1, x_2, x_1', x_2') \mapsto \xi(x_2)\eta(x_1)\xi'(x_1'/x_2)\eta'(1)$ or, using the notation in Sect. 4.7, to

$$(\chi_1 \boxtimes \chi_2 \boxtimes \chi)\left(diag(x_1, x_2, x_1', x_2')\right) = \chi_2(x_1)\chi_1(x_2)\chi(x_1 x_1').$$

Hence, $\eta = \chi\chi_2$, $\xi/\xi' = \chi_1$, $\xi' = \chi$ and $\chi = \xi'$, $\chi\chi_1 = \xi$, $\chi\chi_2 = \eta$, $\chi\chi_1\chi_2 = \eta'$ by comparison. The corresponding map from the set of isomorphism classes of unramified principal series representations of M

$$Ind_{\left(\begin{smallmatrix} * & * \\ 0 & * \end{smallmatrix}\right)}^{Gl(2,F)}(\xi, \eta) \boxtimes Ind_{\left(\begin{smallmatrix} * & 0 \\ * & * \end{smallmatrix}\right)}^{Gl(2,F)}(\xi', \eta')$$

to the set of isomorphism classes of unramified principal series representations

$$\chi_1 \times \chi_2 \triangleleft \chi = Ind_{B(F)}^{G(F)}(\chi_1 \boxtimes \chi_2 \boxtimes \chi),$$

maps ξ, η, ξ', η' to $\chi_1 = \xi/\xi', \chi_2 = \eta/\xi', \chi = \xi'$. This describes the unramified endoscopic lift (Lemma 4.25).

Satake Parameters. To reformulate this in terms of the Satake parameters set,

$$\alpha = \xi(\pi_F), \ \beta = \eta(\pi_F), \quad \alpha' = \xi'(\pi_F), \ \beta' = \eta'(\pi_F), \ \text{with } \alpha\beta = \alpha'\beta'$$

and $\alpha_0 = \chi(\pi_F), \alpha_1 = \chi_1(\pi_F), \alpha_2 = \chi_2(\pi_F)$.

Then $\alpha_0 = \alpha'$, $\alpha_0\alpha_1 = \alpha$, $\alpha_0\alpha_2 = \beta$, $\alpha_0\alpha_1\alpha_2 = \beta'$, and the above correspondence between isomorphism classes of principal series representations comes from the L-embedding, which under the isomorphism ψ defined in Sect. 8.1 maps

$$\begin{pmatrix} \alpha & 0 \\ 0 & \beta \end{pmatrix} \times \begin{pmatrix} \alpha' & 0 \\ 0 & \beta' \end{pmatrix} \in \hat{T}_M^s$$

to

$$\begin{pmatrix} \alpha & 0 & 0 & 0 \\ 0 & \beta' & 0 & 0 \\ 0 & 0 & \beta & 0 \\ 0 & 0 & 0 & \alpha' \end{pmatrix} = \begin{pmatrix} \alpha_0\alpha_1 & 0 & 0 & 0 \\ 0 & \alpha_0\alpha_1\alpha_2 & 0 & 0 \\ 0 & 0 & \alpha_0\alpha_2 & 0 \\ 0 & 0 & 0 & \alpha_0 \end{pmatrix} \in \hat{\mathbf{T}}_G \subseteq {}^L G$$

in $\hat{\mathbf{T}}_G = \hat{\mathbf{T}}_M \subseteq {}^L M \subseteq {}^L G$. This image is uniquely defined in $\hat{\mathbf{T}}_G$ up to the action of the Weyl group. We may interchange α_1 and α_2 and leave α_0 unchanged. Or we may replace α_0 by $\alpha_0\alpha_2$ and α_2 by α_2^{-1} and leave α_1 unchanged.

These two substitutions generate the Weyl group $\mathbf{W}_G$ of $\hat{\mathbf{T}}_G$. The matrix in $\hat{\mathbf{T}}_G$ above corresponds to $(\alpha_0\alpha_2, \alpha_0, \alpha_1, 1) \in \hat{T}_G^s$ under the canonical isomorphism $\hat{\mathbf{T}}_G \cong \hat{T}_G^s$ (see page 274), in accordance with $\chi_2\chi(x_1)\chi_1(x_2)\chi(x_1') = (\alpha_0\alpha_2)^{v(x_1)}\alpha_1^{v(x_2)}\alpha_0^{v(x_1')}1^{v(x_2')}$.

10.2 Twisted Transfer Factors for the Split Torus in M_1

For the proof of the fundamental lemma for the twisted endoscopic datum $(^LM, M, s_E, \xi_E)$ and the z-pair (M_1, ξ_{M_1}) we follow the method of Labesse [57] and consider elementary functions. For the matching of elementary functions on M_1 and G it is enough, owing to support properties of their orbital integrals, to know the transfer factors for the split torus $T_{M_1}^s$ of M_1.

Suppose $(^LM, M, s_E, \xi_E)$ and the z-pair (M_1, ξ_{M_1}) are defined as in Sect. 9.2, where $M_1 = Gl(2)^2$ and $pr : M_1 \to M$ is the map already considered in Chap. 6.4, so that $\xi_{M_1} = {}^Lpr$ is the canonical inclusion, i.e., we make the

Assumption. $\lambda = 1$ (in the sense of Sect. 9.2).

Fix maximal split tori in M and $\hat{M}$ and in M_1 and $\hat{M}_1$ together with a Borel subgroup as in Chap. 6. Consider the θ-stable torus $T_G = Res_{E/F}(D)$, where D is the split E-torus of diagonal matrices in $GSp(4, E)$. Consider its dual $\hat{T}_G$ and choose an embedding into $\hat{G}$ with Borel group $\hat{B}$ as in Chap. 6. The norm map induces an isomorphism $norm : (T_G^s)_\theta \cong T_M^s$. In fact $(\hat{T}_G)^{\hat{\theta}} \cong \hat{T}_M^s$ as in [54], Chap. 3, where the extended $norm : T_G(F) \to T_M^s(F)$ is obtained by the following $norm : T_G(F) \to T_{M_1}^s(F)$ by composition with the projection $pr : T_{M_1}^s \to T_M^s$

$$norm : \quad \delta = \begin{pmatrix} x_1 & 0 & * & * \\ * & x_2 & * & * \\ 0 & 0 & x_1' & * \\ 0 & 0 & 0 & x_2' \end{pmatrix} \mapsto \gamma = \begin{pmatrix} N_{E/F}(x_2) & * \\ 0 & N_{E/F}(x_1) \end{pmatrix} \times \begin{pmatrix} N_{E/F}(x_1'/x_2) & 0 \\ * & 1 \end{pmatrix},$$

for x_1, x_1', x_2, x_2' in E satisfying the relation $x_1 x_1' = x_2 x_2'$. Stars indicate zeros, but as before the choice of the Borel subgroup. For $\delta = diag(x_1, x_2, x_1', x_2')$ in $G(F) = GSp(4, E)$ put $t = diag(N_{E/F}(x_1), N_{E/F}(x_2), N_{E/F}(x_1'), N_{E/F}(x_2'))$ in $T'(F) \subseteq G'(F) = GSp(4, F)$.

Since $M_1 \neq M$, the matching conditions ([54], (5.5.1)) involve smooth functions f on $G(F)$ with compact support, respectively smooth functions f^{M_1} on $M_1(F)$ with compact support modulo $Z_1(F)$ such that $f^{M_1}(zm) = \lambda_{M_1}(z)^{-1}f^{M_1}(m)$ holds for all $z \in Z_1(F)$, where Z_1 is the kernel of the projection $pr : M_1 \to M$. The matching condition is

$$\sum_{\delta'} \Delta(\delta_H, \delta') \int_{G^{\delta'\theta}(F)\backslash G(F)} f(g^{-1}\delta'\theta(g))dg/dt'$$

$$= \sum_{\delta'_M} \int_{M^{\delta'_M}(F)\backslash M(F)} f^{M_1}(m^{-1}\delta'_M m)dm/dt$$

for suitably normalized measures (see [54], (5.5.1)). The second sum is over all representatives δ'_M of $M(F)$-conjugacy classes in the $M(\overline{F})$-conjugacy class of $\delta_M \in M_1(F)$ for some fixed strongly G-regular semisimple elliptic element δ_M. The first sum extends over all representatives δ' for the θ-conjugacy classes under $G(F)$ of elements δ' in $G(F)$, whose norm is δ_M (if the first sum is over the empty set, this means that the second sum has to vanish). The norm condition means that δ is contained in the $\overline{F}$-conjugacy class of $\mathcal{A}_{M/G}(\delta_M)$ ([54], (3.3.3)), where $\mathcal{A}_{M/G}$ is the composite of $Cl_{ss}(M) \rightarrow T_M/\Omega_M \rightarrow (T_G)_\theta/\Omega^\theta$ and the inverse of the isomorphism $Cl_{ss}(G, \theta) \rightarrow (T_G)_\theta/\Omega^\theta$ induced by the abstract norm map $N_{T,\theta} : T \rightarrow T_\theta$. See [54], p. 25ff.

The transfer factor $\Delta(\delta_M, \delta')$ for elements $\delta_M \in T^s_{M_1}(F)$ in the split torus is given by

Lemma 10.1. *For* $\delta \in T_G(F)$ *in the maximal split diagonal torus* $T_G \cong Res_{E/F}(\mathbb{G}_m^3)$ *and corresponding* $\gamma \in T_{M_1}(F)$ *the transfer factor is*

$$\Delta(\gamma, \delta) = \tau(\delta) \cdot \prod_\alpha |\alpha(t) - 1|_F^{1/2}.$$

The product is over all restricted roots which do not come from M *and*

$$\tau(\delta) = (\chi_0 \boxtimes \chi_0 \boxtimes \chi_0)(\delta),$$

where χ_0 *is the nontrivial unramified quadratic character of* F^*.

Proof. Apply Lemma 9.18. The character τ corresponds via unramified Langlands reciprocity to $s \in \hat{T}_G$, which is given by the Satake parameters $\alpha_0 = \alpha_1 = \alpha_2 = -1$ since $\lambda = 1$. Therefore, $\tau = \chi_1 \boxtimes \chi_2 \boxtimes \chi$ for $\chi = \chi_1 = \chi_2$ equal to χ_0. $\square$

Remark 10.1. For some unramified character χ_{2n} of F^* with $\chi_0(x) = \chi_{2n}(N_{E/F}(x))$ the unramified character $(g_1, g_2) \mapsto \chi_{2n}(det(g_2))$ of $M_1(F)$ coincides with the character $\lambda_{M_1}^{-1}$ on the subgroup of $Z_1(F) \subseteq M_1(F)$ of elements which are in the image of the norm $Z_1(E) \rightarrow Z_1(F)$. In fact $\lambda_{M_1} = 1$ since $\xi_{M_1} = {}^L id$ and $\mathcal{H} = {}^L M$ [54], p. 23.

For other conjugacy classes of tori we need not compute these transfer factors explicitly since for them elementary function have vanishing orbital integrals.

10.3 The Twisted Spherical Lift

Let us express Lemma 10.1 in terms of an unramified endoscopic lift

$$\Pi : \mathcal{E}_{un}\big(M_1(F), Z_1(F)\big) \rightarrow \mathcal{E}_{un}\big(GSp(4, E)\big)$$

for $M_1 = GL(2, F)^2$ and $M = M_1/Z_1$. For

$$\begin{pmatrix} \alpha & 0 \\ 0 & \beta \end{pmatrix} \times \begin{pmatrix} \alpha' & 0 \\ 0 & \beta' \end{pmatrix} \times \sigma_F \quad \in {}^L M_1$$

the twisted base change for the datum $(M, {}^L M, s_E, \xi_E)$ and the z-pair $(M_1, {}^L id)$ defines a corresponding element in ${}^L Res_{E/F}(GSp(4))$

$$s_E \cdot (g, g, \ldots, g) \lhd \sigma_F, \qquad g = \begin{pmatrix} \alpha & 0 & 0 & 0 \\ 0 & \beta' & 0 & 0 \\ 0 & 0 & \beta & 0 \\ 0 & 0 & 0 & \alpha' \end{pmatrix} \in {}^L \mathbf{T_G}.$$

By Langlands duality this corresponds to an unramified character of the maximal split subtorus $T_d(F)$ of $T_G(F)$. Its Langlands dual in ${}^L T_d \cong {}^L T'$ is given by (see the argument on page 319)

$$\begin{pmatrix} \alpha^n & 0 & 0 & 0 \\ 0 & -(\beta')^n & 0 & 0 \\ 0 & 0 & \beta^n & 0 \\ 0 & 0 & 0 & -(\alpha')^n \end{pmatrix} \times \sigma_F.$$

In other words, the corresponding unramified character lift maps the isomorphism class of the spherical constituent $\sigma = (\sigma_1, \sigma_2)$ of the unramified principal series

$$Ind_{\begin{pmatrix} * & * \\ 0 & * \end{pmatrix}}^{Gl(2,F)}(\xi, \eta) \otimes Ind_{\begin{pmatrix} * & 0 \\ * & * \end{pmatrix}}^{Gl(2,F)}(\xi', \eta')$$

of $M_1(F)$ to the isomorphism class of the spherical constituent of the unramified principal series representations

$$\chi_1 \times \chi_2 \lhd \chi,$$

where

$$\chi_1 = (\frac{\xi}{\xi'} \circ N_{E/F}) \cdot \chi_0, \ \chi_2 = (\frac{\eta}{\xi'} \circ N_{E/F}) \cdot \chi_0, \ \chi = (\xi' \circ N_{E/F}) \cdot \chi_0.$$

Here χ_0, as above, is the unique nontrivial unramified quadratic character of F^*. The characters are considered on the diagonal torus T_d with "entries" in F^*. In the case $E = F, n = 1$ this differs from the "untwisted" lift by the multiplication with the character $\chi_0 \boxtimes \chi_0 \boxtimes \chi_0$ or the change $\alpha_0 \mapsto -\alpha_0, \alpha_1 \mapsto -\alpha_1$ and $\alpha_2 \mapsto -\alpha_2$.

Hecke Algebra. The preceding lift Π from the isomorphism classes of irreducible spherical representations of $\mathcal{E}_{un}(M_1(F), Z_1)$, whose central character is trivial on Z_1, to the isomorphism classes of irreducible spherical representations of $\mathcal{E}_{un}(Res_{E/F}(GSp(4)))$

$$\Pi : \mathcal{E}_{un}(M_1(F), Z_1(F)) \to \mathcal{E}_{un}(GSp(4, E))$$

defines a homomorphism between Hecke algebras $\phi \mapsto f$

$$\mathcal{H}(GSp(4, E), GSp(4, o_E)) \to \mathcal{H}(M_1(F), K_{M_1}; Z_1(F)),$$

where f for $\phi \in \mathcal{H}(GSp(4,E), GSp(4, o_E))$ is defined by the identity

$$tr\ (\sigma_1 \times \sigma_2)(f)\ =\ tr\ \Pi(\sigma_1 \times \sigma_2)(\phi)$$

for all spherical representations $\sigma_1 \times \sigma_2$.

10.3.1 The Frobenius Formula

Let $\phi = \phi_n$ be the characteristic function of the double coset considered in the Frobenius formula (Lemma 9.7). The spherical function ϕ_n on $GSp(4,E)$ is characterized by the property $tr\ \pi(\phi_n)\ =\ q_E^{3/2}(\alpha_0 + \alpha_0\alpha_1 + \alpha_0\alpha_2 + \alpha_0\alpha_1\alpha_2)$ for all irreducible spherical representations π of $GSp(4,E)$ with Satake parameters $\alpha_0, \alpha_1, \alpha_2$ [52], Theorem 2.1.3. Notice $q_E = p^n$.

Lemma 10.2. *Let $f^{(n)}$ be the spherical function on $Gl(2,F)$ characterized by the property $tr\ \sigma(f^{(n)}) = q_E^{3/2}(\alpha^n + \beta^n)$ for all irreducible spherical representations σ of $Gl(2,F)$ with Satake parameters α, β. Then $\phi_n^{M_1} = f^{(n)} \otimes 1_{Gl(2,o_F)} - 1_{Gl(2,o_F)} \otimes f^{(n)}$ defines a spherical function on $M_1(F) = Gl(2,F)^2$ which satisfies*

$$tr\ (\sigma_1 \times \sigma_2)(\phi_n^{M_1}) = tr\ \Pi(\sigma_1 \times \sigma_2)(\phi_n).$$

Proof. For irreducible spherical representations σ_1, σ_2 of $Gl(2,F)$ with common central character, given by the Satake parameters α, β and α', β', respectively, we have

$$tr\ (\sigma_1 \times \sigma_2)(\phi_n^{M_1}) = q_E^{3/2}(\alpha^n + \beta^n - (\alpha')^n - (\beta')^n).$$

The lift $\Pi(\sigma_1 \times \sigma_2)$ has Satake parameters determined by $\alpha_0\alpha_1 = \alpha^n$, $\alpha_0\alpha_1\alpha_2 = -(\beta')^n$, $\alpha_0\alpha_2 = \beta^n$, and $\alpha_0 = -(\alpha')^n$ as shown above. Therefore, again $tr\ \Pi(\sigma_1 \times \sigma_2) = q_E^{3/2}(\alpha^n + \beta^n - (\alpha')^n - (\beta')^n)$, which proves the lemma. $\square$

$\tilde{\phi}_n^{M_1}(g) = \int_{Z_1(F)} \phi_n^{M_1}(gz)dz$, as a spherical function in $\mathcal{H}(M_1(F), K_{M_1}; Z_1)$, has similar properties for spherical representations $\sigma = (\sigma_1, \sigma_2)$ with $\omega_{\sigma_1} = \omega_{\sigma_2}$. Once the twisted fundamental lemma is proven, it is the matching function for ϕ_n in the Frobenius formula. For an explicit formula for $f^{(n)}$ see Casselman [18], Lemma 5.3.1.

10.3.2 The Modified Twisted Lift Π_{mod}

For the proof of the twisted fundamental lemma it is convenient to modify the z-pair by choosing now $\lambda \neq 1$ such that $\lambda^n s = 1$ for the rest of this chapter, since this simplifies formulas for the proof of the twisted fundamental lemma. For this modified z-pair (see page 295) we have

Lemma 10.3. *For* $\delta \in T_G(F)$ *in the maximal split diagonal torus* $T_G \cong Res_{E/F}(\mathbb{G}_m^3)$ *and corresponding* $\gamma \in T_{M_1}(F)$ *the transfer factor is*

$$\Delta(\gamma, \delta) = \prod_\alpha |\alpha(t) - 1|_F^{1/2}.$$

In practice this yields the following.

Modified Endoscopic Lift Π. Let τ_E denote the square of unramified character $\tau_{2n}(g_1, g_2) = \chi_{2n}(det(g_2))$ of $M_1(F)$. Let $\mathcal{E}(M_1(F), \tau_E)$ denote the set of isomorphism classes of irreducible admissible representations $\sigma = (\sigma_1, \sigma_2)$ of $M_1(F)$ such that $\omega_{\sigma_2} = \omega_{\sigma_1} \tau_E$. Let $\mathcal{E}_{un}(M_1(F), \tau_E)$ denote the subset defined by the spherical representations. Then we have bijections

$$\mathcal{E}(M(F)) = \mathcal{E}(M_1(F), Z_1(F)) \cong \mathcal{E}(M_1(F), \tau_E),$$

where the second map is given by $\sigma = (\sigma_1, \sigma_2) \mapsto (\sigma_1, \sigma_2 \otimes \tau_{2n})$. Similarly, one has isomorphisms of Hecke algebras

$$\mathcal{H}(M) = \mathcal{H}(M_1; Z_1) \cong \mathcal{H}(M_1; \tau_E),$$

where the second isomorphism is defined by $f(g) \mapsto \chi_{2n}(g) \cdot f(g)$. Orbital integrals thereby change by a "relative" transfer factor $\Delta_{MM_1}(g) = \chi_{2n}(g)$.

   Composed with the twisted endoscopic lift $\Pi : \mathcal{E}_{un}(M_1(F), \tau_E) \to \mathcal{E}_{un}(GSp(4, E))$ described in the last section this defines a surjective lift Π_{mod} from the isomorphism classes of irreducible spherical representations of $\mathcal{E}_{un}(M_1(F), \tau_E)$ to the isomorphism classes of irreducible spherical representations of $\mathcal{E}_{un}(Res_{E/F}(GSp(4)))$

$$\Pi_{mod} : \mathcal{E}_{un}(M_1(F), \tau_E) \to \mathcal{E}_{un}(GSp(4, E)).$$

On the level of Satake parameters Π_{mod} is given by

$$\begin{pmatrix} \alpha & 0 \\ 0 & \beta \end{pmatrix} \times \begin{pmatrix} \alpha' & 0 \\ 0 & \beta' \end{pmatrix} \times \sigma_F \mapsto \begin{pmatrix} \alpha^n & 0 & 0 & 0 \\ 0 & \beta'^n & 0 & 0 \\ 0 & 0 & \beta^n & 0 \\ 0 & 0 & 0 & \alpha'^n \end{pmatrix} \times \sigma_F$$

The unramified lift Π has finite cardinality of $2 \cdot n^3$ or less. There exists a homomorphism between Hecke algebras

$$b : \mathcal{H}(GSp(4, E), GSp(4, o_E)) \to \mathcal{H}(M_1(F), K_{M_1}, \tau_E),$$

where $b(f)$ for $f \in \mathcal{H}(GSp(4, E), GSp(4, o_E))$ is characterized by the identities

$$tr\,(\sigma_1 \times \sigma_2)(b(f)) = tr\,\Pi_{mod}(\sigma_1 \times \sigma_2)(f)$$

for spherical representations $\sigma_1 \times \sigma_2$.

10.3.3 Elementary Functions

For the split torus $T_{M_1} = T^s_{M_1} \subseteq M_1$ and the maximal split θ-stable torus $T_G \subseteq G$ of diagonal matrices let $T^-_{M_1} \subseteq T^s_{M_1}(F)$ and $T^-_G \subseteq T_G(F)$, respectively, denote the subsets of strictly contractive elements with respect to the fixed Borel subgroups. That is, for $\delta = diag(x_1, x_2, x'_1, x'_2) \in T_G(F)$ and $x_i, x'_i \in E^*$ we have

$$\delta \in T^-_G \text{ iff } \left|\frac{x_2}{x_1}\right| < 1 \text{ and } \left|\frac{x_1}{x'_1}\right| < 1.$$

Similarly, $\gamma = diag(t_1, t'_1) \times diag(t_2, t'_2) \in T^-_{M_1}$ iff $|t_1/t'_1| < 1$ and $|t'_2/t_2| < 1$. The abstract norm is $N_{T,\theta} : T_G \to (T_G)_\theta$. The latter is identified with T_M through the choice of an admissible isomorphism $(T_G)_\theta \cong T_M$. The composite $norm$: $T_G \to T_M$ maps $\delta \in T_G(F)$ to the element represented by the diagonal matrix

$$\gamma = diag(N_{E/F}(x_2), N_{E/F}(x_1)) \times diag(N_{E/F}(x'_1/x_2), 1).$$

This map extends to a map $norm : T_G(F) \to T^s_{M_1}(F)$. Notice $T^s_{M_1} = norm(T_G) \cdot Z_1$ and $norm(T_G) \cap Z_1 = \{1\}$. Obviously, since $\frac{x_2}{x'_1} = \frac{x_2}{x_1}\frac{x_1}{x'_1}$, we have

Lemma 10.4. $norm$: $T^-_G \longrightarrow T^-_{M_1}$ and $norm$: $(T^-_G)^\# \longrightarrow T^-_{M_1}$, where $\#$ denotes the reflection in the Weyl group W_G of the torus T_G, which maps $diag(x_1, x_2, x'_1, x'_2)$ to $diag(x'_1, x_2, x_1, x'_2)$. Notice $W^\theta_G = W_{G'} = \langle W_M, \# \rangle = \langle W_{M_1}, \# \rangle$ and t is a norm if and only if $t^\#$ is a norm.

Definition 10.1. For $\delta \in T^-_G$ the elementary function $g_{\delta\theta}(x)$ is defined as the characteristic function of the set of elements $k^{-1}\delta a\theta k$ for $k \in K = GSp(4, o_E)$ and a in the maximal compact subgroup ^0T_G of $T_G(F)$. Similarly, for $t \in T^-_{M_1}$, the elementary function $g_t(x)$ is defined to be $\tau_E(z)$ if $x = zk^{-1}tak$ for $k \in K = Gl(2, o_F)^2$, a in the maximal compact subgroup $^0T^s_{M_1}$ of $T^s_M(F)$, and $z \in Z_1(F)$, and is defined to be zero otherwise.

The twisted orbital integrals of the elementary function $g_{\delta\theta}$ and the stable orbital integral of the elementary function g_t vanish for elements not in the (twisted) conjugacy classes of elements in $T_G(F)$ $T^s_M(F)$, respectively. Suppose $norm(\delta) = t$. Then, as in [57], Lemmas 1–3, for a suitable choice of measures one obtains from Lemma 10.3

Lemma 10.5. For $\delta \in T^-_G$ and $t = norm(\delta) \in T^-_{M_1}$, the normalized elementary functions $f_\delta(x) = \frac{g_{\delta\theta}(x)}{D_{G,\theta}(\delta)}$ and $f_t(x) = \frac{g_t(x)}{D_{M_1}(t)}$ are matching functions for the endoscopic datum $(M, {}^LM, s_E, \xi_E)$ and the modified z-pair (M_1, ξ_{M_1}) given by $\lambda^n s = 1$ (see page 295).

Furthermore, for an admissible representation π of $M_1(F)$ of finite length such that $\pi(z) = \chi_E(z)^{-1} \cdot id$ the

Trace Identity.

$$Tr\big(f_t(x), \pi\big) \;=\; \int_{^0T_{M_1}} D_{M_1}(ta)\chi_\pi(ta)da$$

$$= \delta_{B_{M_1}}^{-1/2}(t) \int_{^0T_{M_1}} \chi_{\pi(B_{M_1})}(ta)da \;=\; \big(r_{B_{M_1}M_1}(\pi)\big)^{^0T_{M_1}}(t)$$

holds, again for measures chosen suitably. This follows from the Weyl integration formula, the Deligne–Casselman formula for characters, and the formula $D_{M_1}(x) = \delta_{B_{M_1}}^{-1/2}(x)$, which holds for elements $x \in T_{M_1}^-$.

Twisted Version. For admissible θ-stable representations Π of $G(F)$ of finite length the following holds

$$Tr\big(f_{\delta\theta}(x) \lhd \theta, \Pi\big) \;=\; \int_{^0T_G} D_{G,\theta}(\delta a)\chi_\Pi(a\delta \lhd \theta)da$$

$$= \delta_{B_G}^{-1/2}(N_{E/F}(\delta)) \int_{^0T_G} \chi_{\Pi(B)}(a\delta \lhd \theta)da \;=\; \big(r_{BG}(\Pi)\big)^{^0T_G}(\delta \lhd \theta).$$

This follows from the twisted Weyl integration formula, the twisted character formula proved by Clozel, and the formula $D_{G,\theta}(\delta) = \delta_B(Norm_{E/F}(\delta))^{-1/2}$, which holds for $\delta \in T_G^-$. For the last formula use [54], (4.5), and the fact that the restricted roots are of type R_1.

   Both traces vanish unless the representations π and Π, respectively, have a nontrivial fixed vector under a Iwahori subgroup. Such representations, if irreducible, are constituents of unramified induced representations $Ind(\lambda)$ and $Ind(\tilde{\lambda})$, respectively, for some $\tilde{\lambda} = \lambda \circ N_{E/F}$, where $Ind(.)$ means the unitary normalized induced representation of the principal series. See [57], pp. 526–527.

Shintani Identity. Finally notice the following well-known identity:

$$r_{BG}\Big(Ind(\tilde{\lambda})\Big)(\delta \lhd \theta) = r_{B'G'}\Big(Ind(\lambda)\Big)(N_{E/F}(\delta)),$$

where the left side is the character of the Jacquet module of the principal series representation $Ind(\tilde{\lambda})$ of $G(F) = GSp(4, E)$ extended to a representation of the semidirect product $GSp(4, E) \times \langle \theta \rangle$ so that θ fixes the spherical vector. The right-hand side is the corresponding character of the Jacquet module of the principal series representation $Ind(\lambda)$ of $G'(F) = GSp(4, F)$.

   These facts together imply

Lemma 10.6. *For unramified characters* χ *of* $T_M^s(F)$ *and* $t = norm(\delta)$

$$Tr\big(f_t, Ind_{B_{M_1}(F)}^{M_1(F)}(\chi)\big) + Tr\big(f_{t\#}, Ind_{B_{M_1}(F)}^{M_1(F)}(\chi)\big) = Tr\big(f_{\delta\theta}\lhd\theta, Ind_{B_G(F)}^{G(F)}(\chi \circ norm_{E/F})\big).$$

Proof. By the trace identity and its twisted version stated above, the assertion of the lemma is equivalent to the corresponding identity $\sum_{w \in W_{M_1}} \chi(w(t)) + \sum_{w \in W_{M_1}} \chi(w(t^{\#})) = \sum_{w \in W_G^\theta}(\chi \circ norm_{E/F})(w(\delta))$ on the level of Jacquet modules, which follows from $W_G^\theta = \langle W_{M_1}, \theta \rangle$. $\square$

10.3.4 Spherical Projectors

Both groups M_1 and G are of adjoint type. This means, following [57], that for every finite set $\mathcal{E} \subseteq \mathcal{E}(G(F))$ of inequivalent irreducible admissible representations Π of $G(F)$ containing a spherical irreducible representation Π_0, one can find a finite $\mathbb{C}$-linear combination $f(x) = \sum_i c_i \cdot f_{\delta_i \theta}(x)$ of elementary functions on $G(F)$ such that $tr\, \Pi(f \vartriangleleft \theta) = 0$ holds for all $\Pi \in \mathcal{E}$ unless $\Pi \cong \Pi_0$, and such that $tr\, \Pi(f \vartriangleleft \theta) = 1$ holds for $\Pi \cong \Pi_0$. This is shown in [57] via the construction of projectors onto a "maximal exponent" of the Jacquet module.

Then by the way this is constructed, it can also be achieved for the modified endoscopic lift $\Pi_{mod} : \mathcal{E}(M_1(F), \tau_E) \to \mathcal{E}(G(F))$ that the corresponding matching function

$$f^{M_1}(x) = \sum_i c_i \cdot f_{norm(\delta_i)}(x)$$

on $M_1(F)$ has the property that

- $tr\, \pi(f^{M_1}) = 0$ holds for all π in a fixed subset $\mathcal{E}_1 \subseteq \mathcal{E}(M_1(F), \chi_E)$ unless π is spherical and $\pi \in \Pi_{mod}^{-1}(\Pi_0)$
- and such that $tr\, \pi(f^{M_1}) = 1$ holds for $\Pi_{mod}(\pi) \cong \Pi_0$

However, using Lemma 10.5 this implies

Lemma 10.7. *Given finite sets $\mathcal{E}$ and $\mathcal{E}_1$ and a spherical $\Pi_0 \in \mathcal{E}$, there exist matching functions f and f^{M_1}, which are linear combinations of elementary functions, such that for all for $\pi \in \mathcal{E}_1$ and all for $\Pi \in \mathcal{E}$:*

1. $tr\, \pi(f^{M_1}) = 1$ or 0 depending on whether π is spherical and lifts to Π_0 or not.
2. $tr\, \Pi(f) = 1$ or 0 depending on whether Π is spherical and lifts to Π_0 or not.

10.4 The (Twisted) Fundamental Lemma

The fundamental lemma (for θ-semisimple and strongly θ-regular elements) in the twisted endoscopic setting for $(M, {}^L M, s_E, \xi_E)$ and all relevant tori is now a consequence of the twisted trace formula, its stabilization in [54], and the fundamental lemma for the unit element of the spherical Hecke algebras at almost all places. See [20, 57]. We only sketch the argument. We assume that the untwisted fundamental lemma holds for the unit elements at almost all places, and for all relevant tori.

In fact the arguments in [57] carry over almost verbatim:

Step 1. Suppose we have a fixed non-Archimedean field F_v with place v. The local twisted base change datum $(M, {}^L M, s_E, \xi_E)$ for $G(F_v)$ attached to the endoscopic datum $(M, {}^L M, s, \xi)$ for $G'(F_v)$ together with the z-pair (M_1, ξ_{M_1}) (as in Sect. 10.3) defines the modified twisted unramified character lift Π_{mod} from isomorphism classes of unramified irreducible representations in $\mathcal{E}(M_1(F_v), \tau_E)$ to isomorphism classes of unramified irreducible representations of $G(F_v)$

$$\pi_v \mapsto \Pi_{mod}(\pi_v).$$

Let b be the corresponding homomorphism of spherical Hecke algebras. Suppose that the fundamental lemma holds for unit elements for almost all non-Archimedean local places for the standard endoscopic datum $(M, {}^L M, s, \xi)$ of G'. Then the twisted fundamental for the datum $(M, {}^L M, \sigma_E, \xi_E)$ also holds for unit elements (Corollary 9.4). Fix now a spherical function f_v^0 of $G(F_v)$ and some sufficiently regular $t_{10} \in M_1(F)$ with corresponding $t_0 \in M(F_v)$ and $\delta_0 \in G(F_v)$ (if the latter exists). Then by Kazhdan's argument [44], using the global trace formula and the fundamental lemma for unit elements at almost all places, there exist local character sums involving a finite number of irreducible admissible representations π_v of $M_1(F_v)$ and Π_v of $G(F_v)$ at the fixed place v and complex numbers $b(\pi_v), b(\Pi_v)$

$$\sum_{\Pi_v} b(\Pi_v) \cdot tr \; \Pi_v(f_v \times \theta),$$

$$\sum_{\pi_v} b(\pi_v) \cdot tr \; \pi_v(f_v^{M_1})$$

with the following two properties:

1. Both sums are equal for matching elementary functions $(f_v, f_v^{M_1})$ with respect to the twisted local endoscopic datum $(M, {}^L M, \sigma_E, \xi_E)$.
2. The identities

$$\sum_{\Pi_{mod}(\pi_v) \cong \Pi_v} b(\pi_v) \; = \; b(\Pi_v)$$

for all unramified irreducible Π_v imply the fundamental lemma for $(M, {}^L M, \sigma_E, \xi_E)$ at v, i.e., the matching of $(f_v, b(f_v))$ for the fixed spherical functions $f_v = f_v^0$ at the fixed $\delta_0 \in G(F_v)$ and $t_{10} \in M_1(F_v)$.

The sums above involving $b(\Pi_v), b(\pi_v)$ are constructed as in [57] by choosing a number field F which realizes F_v as a completion. One constructs a cyclic extension E of F of degree n, which is unramified at v and totally real at the Archimedean place. Put $G = Res_{E/F}(G')$ globally. Without restriction of generality δ can be assumed to be a global element in $G(F)$. τ_E can be extended to a idele class character. One constructs adelic matching functions (f, f^{M_1}), which are products of local matching functions. At two suitably chosen places $w \neq v$ the functions $(f_w, f_w^{M_1})$

are chosen to be matching cuspidal functions, which is done by an easy adaption of the argument of Hales [35], Sect. 5, to the twisted case. At a finite number of places matching functions $(f_v, f_v^{M_1})$ are constructed with regular support by the implicit function theorem in such a way that all global κ-orbital integrals vanish except a fixed one belonging to the given endoscopic datum at v. Furthermore, at some places w the support of f_w is chosen to be elliptic. At the fixed place v the functions f_v and $f_v^{M_1}$ are allowed to be arbitrary elementary or spherical functions. However, f^v is constructed such that all global κ-orbital integrals $O_{\delta\theta}^\kappa(f)$ for $f = f_v f^v$ of the global trace formula vanish except $O_{\delta\theta}^\kappa(f)$ if f_v is chosen to be the fixed spherical function f_v^0. Similar properties can be achieved for fixed $t_1 \in M_1(F)$ lying above $t = Nm_{E/F}(\delta)$ and its stable orbital integral. (If for $t_1 \in M_1(F)$ there exists no $\delta \in G(F)$, one has to modify the arguments accordingly.) Then the simple regular elliptic global trace formula gives a formula $O_{\delta\theta}(f_v f^v) = \sum_{\Pi_v} b(\Pi_v) tr \, \Pi_v(f_v \times \theta)$ for $f_v = f_v^0$, and similarly for the simple regular elliptic global trace formula for M_1. This gives property (2) except that the sums may be over infinitely many representations, but are absolutely convergent. In fact, if f^w is now fixed at the non-Archimedean places $w \neq v$ and the support of f_w is fixed at the Archimedean place, the trace formula involves only irreducible cuspidal global representations Π with $\Pi_w^{K_w} \neq 0$ for all non-Archimedean places $w \neq v$, where K_w is a suitably chosen compact open group depending only on f^w. If f_v is assumed to be the spherical function f_v^0 or elementary at v, then the level of the automorphic forms Π at v can also be fixed by the Iwahori group K_v. Concerning the Archimedean places, first choose a fixed matching pair $(f_\infty, f_\infty^{M_1})$. Notice that for $f_\infty = \prod_{\nu=1}^n g_\infty$ on $\prod_1^n G'(\mathbb{R})$ this amounts to the matching of $(g_\infty^{*n}, f_\infty^{M_1})$ of the n-fold convolution product g_∞^{*n} of g_∞ on $G'(\mathbb{R})$ with $f_\infty^{M_1}$ on $M_1(\mathbb{R})$ with respect to standard endoscopy. Property (2) is then achieved from the trace formula. However, we can assume that we are still allowed to modify f_∞ as long as we do not change the support of f_∞. Similarly for the matching function $f_\infty^{M_1}$. For instance, we can modify f_∞ by applying multipliers of the convolution Hecke algebra as in Sect. 5.5. This allows us to separate the global trace identities into identities involving sums over packets with finitely many irreducible Archimedean representations. By fixing the level groups the finiteness of the number of irreducible automorphic cuspidal representations with fixed Archimedean components and fixed level and central character ensures, that we can assume without restriction of generality that the sums extend over finitely many irreducible cuspidal automorphic representations. Hence, only finitely many local representations intervene in (1). The equality (1) for matching elementary functions $(f_v, f_v^{M_1})$, on the other hand, follows from the stabilization of the simple regular elliptic trace achieved by Kottwitz and Shelstad [54], 7.4. *Step 2.* This being said, we now argue as in [57]. By property (1) the identity

$$\sum_{\pi_v} b(\pi_v) \cdot tr \, \pi_v(f_v^M) = \sum_{\Pi_v} b(\Pi_v) \cdot tr \, \Pi_v(f_v \times \theta)$$

holds for matching elementary functions $(f_v, f_v^{M_1})$. Recall these sums are finite and involve only representations of Iwahori level at v. Hence, they involve only a finite number of such representations. Therefore, by Lemma 10.7 there exists a pair $(f_v, f_v^{M_1})$ of matching linear combination of elementary functions, for which the last trace identity becomes

$$\sum_{\Pi_{mod}(\pi_v) \cong \Pi_v} b(\pi_v) = b(\Pi_v).$$

From property (2) the fundamental lemma for the fixed spherical function f_v^0 and the fixed δ_0 follows.

Step 3. Now vary f_v^0 and the fixed δ_0, which proves the fundamental lemma.

Chapter 11
Appendix on Galois Cohomology

Let F be a local non-Archimedean field of characteristic zero. Let π_F denote a prime element. Let F^{un} be the maximal unramified extension field of F in the algebraic closure $\overline{F}$ of F. Let $I = Gal(\overline{F}/F^{un})$ be the inertia group. Let $\sigma = \sigma_F$ denote the arithmetic Frobenius in $\Gamma_F = Gal(F^{un}/F)$. Let $W_F \subseteq \Gamma_F$ denote the Weil group of F. Notice $I \subseteq W_F$, and W_F/I is generated by σ_F.

For a connected reductive group G over F, let $\hat{G}$ denote the algebraic group over $\mathbb{C}$, which is the connected component of the Langlands L-group of G. Consider the category of reductive groups over F, whose morphisms are the group homomorphism $G \to H$, which are defined over F. The Borovoi fundamental group $\pi(G)$ is a functor from the category of reductive groups over F to the category of finitely generated Abelian groups with Γ_F-action. The Borovoi fundamental group is defined as follows. For tori T over F we have a canonical isomorphism $\pi(T) = Hom_{\overline{F}-alg}(\mathbb{G}_m, T)$, and $\hat{T} = \pi(T) \otimes_{\mathbb{Z}} \mathbb{C}^*$ with induced Γ_F-action. For semisimple G over F we have a canonical isomorphism $\pi(G) = \pi_1(G)$ (algebraic fundamental group). In general, let G_{sc} be the simply connected covering of the derived group G_{der} of G. Choose maximal F-tori $T_{sc} \to T_{der} \to T$. Then $\pi(G) = \pi(T)/image(\pi(T_{sc}))$.

The Langlands reciprocity law is a functorial isomorphism, denoted LR,

$$\Psi(T) \cong \Phi(T)$$

between the functors $\Psi(T) = Hom_{cont}(T(F), \mathbb{C}^*)$ and $\Phi(T) = H^1_{cont}(W_F, \hat{T})$ on the category of tori over F [12], Sect. 9.

Let $\Psi_{un}(T)$ denote the subgroup of $\Psi(T)$ defined by all unramified characters $\psi : T(F) \to \mathbb{C}^*$, where ψ is called unramified if ψ is trivial on the maximal compact subgroup of $T(F)$. If T is an unramified torus over F, i.e., splits over F^{un}, then $T(F^{un}) \cong K \times \pi(T)$ as a Γ_T-module, where K is a maximal compact subgroup of $T(F^{un})$. Furthermore, for unramified T over F we have a commutative diagram

R. Weissauer, *Endoscopy for GSp(4) and the Cohomology of Siegel Modular Threefolds*, 335
Lecture Notes in Mathematics 1968, DOI: 10.1007/978-3-540-89306-6_11,
© Springer-Verlag Berlin Heidelberg 2009

$$\Psi_{un} \cong Hom(\pi(T), \mathbb{C}^*) \xrightarrow{LR} infl(H^1_{cont}(\sigma_F^{\mathbb{Z}}, \hat{T}))$$

$$\Psi(T) \xrightarrow{LR} H^1_{cont}(W_F, \hat{T})$$

obtained by the inflation map $infl : H^1(W_F/I, \hat{T}^I) = H^1(W_F/I, \hat{T})$. Notice $\hat{T}^I = \hat{T}$ for unramified F-tori.

Let $\Psi_{elm}(T)$ denote the subgroup of $\Psi(T)$, which consists of all elementary characters $\psi : T(F) \to \mathbb{C}^*$. A character ψ is called elementary if $\psi \in \Psi(T)$ corresponds under the Langlands reciprocity map to a cocycle $a_\psi \in \Phi(T) = H^1_{cont}(W_F, \hat{T})$, which is the inflation of a cocycle of W_F/I with values in $\hat{T}^I$.

Example 11.1. Suppose $T = Res_{L/F}(\mathbb{G}_m)$, where L/F is a tame quadratic extension field. Then $T(F) \cong \{x \in L^* \mid N_{L/F}(x) = 1\}$ is a subgroup of o_L^*. In this case $\Psi_{un}(T) = \{1\}$ and $\Psi_{elm}(T) = \{1, \psi\}$, where $\psi : T(F) \to \{\pm 1\}$ is the unique nontrivial character obtained by the composition of the inclusion $T(F) \hookrightarrow o_L^*$ and the projection $o_L^* \to o_L^*/(1 + \pi_L o_L)$.

We say a torus over F is tame over F if it splits over a tame field extension of F. Let T be a torus defined over F. Let T_{tame} denote its maximal tame F-subtorus. Let $T_{un} \subseteq T_{tame}$ denote its maximal unramified F-subtorus. Then the following statements can be obtained by direct calculation [82]:

- $\Psi_{elm}(T)/\Psi_{un}(T) \cong \Psi_{elm}(T/T_{un})$ is a finite Abelian group.
- If T is unramified over F, then $\Psi_{un}(T) = \Psi_{elm}(T)$.
- If T is tame over F, and if T_{un} is trivial, then a character $\psi : T(F) \to \mathbb{C}^*$ is elementary if and only if ψ is trivial on the subgroup of topologically unipotent elements in $T(F)$ (i.e., the maximal pro-p-subgroup of $T(F)$).
- If T is tame over F, let Q be the quotient torus T/T_{un}. Then

$$\Psi_{elm}(T) \cong \Psi_{un}(T_{un}) \oplus \Psi_{elm}(Q).$$

- In general, $\Psi_{elm}(T) \cong \Psi_{elm}(T)[p^\infty] \times \Psi_{elm}(T_{tame})$.

From these statements it follows that to determine the elementary characters of $T(F)$ it is enough to determine the p-primary component $\Psi_{elm}(T)[p^\infty]$ of $\Psi_{elm}(T)$. I do not know of any direct description which allows us to determine whether a character of $T(F)$ is in $\Psi_{elm}(T)[p^\infty]$. However, one can give the following indirect characterization.

11.1 Explicit Reciprocity Law for Elementary Characters

Let G be a connected reductive group over F, and let $\pi(G)$ be its Borovoi fundamental group. Let us define the *characteristic* maps

$$\lambda_G : G(F) \to (\pi(G)_I)^{\Gamma_F}.$$

Let L denote the completion of F^{un}. The set $B(G)$ of σ_F-conjugacy classes of $G(L)$ admits a map $B(G) \to \pi(G)_{\Gamma_F}$ introduced by Kottwitz [47], (2.4.1), for tori. It is easy to check that the composed map

$$G(L) \to G(L)/(\sigma_F - conjugacy) = B(G) \to \pi(G)_{\Gamma_F}$$

is a surjective map which is functorial in G (Kottwitz–Borovoi). Since we have this map for all unramified extension fields of F, we obtain from these maps a functorial map

$$G(F^{un}) \to \pi(G)_I,$$

which commutes with the natural action of Γ_F/I. Considering the fixed groups under Γ_F, we obtain a map

$$\lambda_G : G(F) \to (\pi(G)_I)^{\Gamma_F},$$

which we call the characteristic map. From [48, 65, 82], we obtain

Theorem 11.1. *Let G be a connected reductive group over F. Then the characteristic map*

$$\lambda_G : G(F) \to (\pi(G)_I)^{\Gamma_F}$$

is a group homomorphism with the following properties:

- λ_G *is surjective.*
- λ_G *vanishes on the image of $G_{sc}(F)$.*
- *Functoriality, i.e., for a group homomorphism $H \to G$ of connected reductive F-groups we get a commutative diagram of homomorphisms*

$$
\begin{array}{ccc}
H(F) & \longrightarrow & G(F) \\
\lambda_H \downarrow & & \downarrow \lambda_G \\
(\pi(H)_I)^{\Gamma_F} & \longrightarrow & (\pi(G)_I)^{\Gamma_F}.
\end{array}
$$

- *Extends to a surjective, continuous, Γ_F-equivariant homomorphism*

$$\lambda_G : G(F^{un}) \to \pi(G)_I.$$

- *(A version of Steinberg's theorem.) The isomorphism $H^1(F, G) \cong H^1_{un}(F, G(F^{un}))$ induces an isomorphism*

$$H^1(\lambda_G) : H^1(F, G) \cong H^1_{un}(F, G(F^{un})) \to H^1(F, \pi(G)_I) \cong \pi(G)_{\Gamma_F}.$$

- *Assume $G = G_{sc}/\kappa$ is connected and semisimple. Then λ_G is the composite of the following maps*

$$G(F) \xrightarrow{\delta} H^1(F, \kappa) \xrightarrow{\cong} H^1(F, \pi(G)')' \xrightarrow{infl'} (\pi(G)_I)^{\Gamma_F},$$

where the first map is the coboundary map induced by $G = G_{sc}/\kappa$, the second isomorphism is the duality isomorphism of local Galois cohomology, and the last map is the dual of the inflation morphism.

- *Let G be connected and semisimple, and quasisplit over F of adjoint type. Let T be a maximal split F-torus in G, and $T_c(F)$ a maximal compact subgroup of $T(F)$. Then λ_G induces isomorphisms*

$$G_{sc}(F) \backslash G(F) \cong image(T_{sc}(F)) \backslash T(F)/T_c(F) \cong (\pi(G)_I)^{\Gamma_F}.$$

- *Suppose G is connected and semisimple. Then λ_G is trivial on any hyperspecial maximal compact subgroup K of $G(F)$.*
- *Suppose T is a torus over F. Let $^0T(F^{un})$ be the kernel of the map λ_G : $T(F^{un}) \to \pi(T)_I$. Then this group is acyclic: $H^\nu_{un}(F, {}^0T(F^{un})) = 0$ for all $\nu \geq 1$. Hence, $H^\nu(\lambda_T)$ induces isomorphisms for all $\nu \geq 1$*

$$H^\nu(F, T) \cong H^\nu_{un}(F, T(F^{un})) \xrightarrow[\sim]{H^\nu(\lambda_T)} H^\nu_{un}(F, \pi(T)_I) .$$

- *For an F-torus T there is an exact sequence*

$$0 \longrightarrow {}^0T(F) \longrightarrow T(F) \xrightarrow{\lambda_T} (\pi(T)_I)^{\Gamma_F} \longrightarrow 0$$

such that (i) λ_T is the universal elementary character of $T(F)$ and (ii) $^0T(F)$ is the intersection $\bigcap_{F'} N_{F'/F}(T(F'))$ over all unramified finite extension fields F'/F ("the Galois connected component").

Concerning the Last Statement. That λ_T is the universal elementary character of $T(F)$ means that for any character $\phi : (\pi(T)_I)^{\Gamma_F} \to \mathbb{C}^*$ the composed map $\psi = \phi \circ \lambda_T$ is an elementary character $\psi \in \Psi_{elm}(T)$ of $T(F)$, and that conversely any elementary character of $T(F)$ is uniquely obtained from such a character ϕ of $(\pi(T)_I)^{\Gamma_F}$ in this way.

Chapter 12
Appendix on Double Cosets

We now discuss a double coset decomposition for the symplectic group GSp $(2n, F)$, which in the case $n = 2$ was found by Schröder [81]. Let F be a local non-Archimedean field of residue characteristic not equal to 2, let $\mathfrak{o}_F$ be its ring of integers, and let π_F denote a prime element. Let $G(F) = GSp(2n, F) \subseteq Gl(2n, F)$ be the group of symplectic similitudes. Hence, $g \in G(F)$ iff $g'Jg = \lambda(g) \cdot J$ for a scalar $\lambda(g) \in F^*$, where

$$J = \begin{pmatrix} 0 & E \\ -E & 0 \end{pmatrix}$$

and where E denotes the unit matrix. Then $g \in G(F) \iff (g')^{-1} \in G(F) \iff g' \in G(F)$ and $J' = J^{-1} = -J \in G(F)$. Let $G(\mathfrak{o}_F) = GSp(2n, \mathfrak{o}_F)$ denote the group of all unimodular symplectic similitudes.

Centralizers. For $n = i + j$ and $i \leq j$ put

$$s = diag(E^{(i,i)}, -E^{(j,j)}, E^{(i,i)}, -E^{(j,j)}) \in G(F).$$

The connected component of the centralizer $H = (G_s)^0$ of s is a maximal connected reductive subgroup of G. $H(F)$ is isomorphic to the subgroup of all matrices (g_1, g_2) in $GSp(2i, F) \times GSp(2j, F)$ with similitude factor $\lambda(g_1) = \lambda(g_2)$

$$1 \to H(F) \to GSp(2i, F) \times GSp(2j, F) \to F^* \to 1.$$

The Matrices $g(e_1, \ldots, e_i)$. Let denote $g(e_1, \ldots, e_i)$ the upper triangular matrix

$$g(D) = \begin{pmatrix} E & S \\ 0 & E \end{pmatrix}, \qquad S = \begin{pmatrix} 0^{(i,i)} & D \\ D' & 0^{(j,j)} \end{pmatrix},$$

defined by $D = \left(diag(\pi_F^{e_1}, \ldots, \pi_F^{e_i}) , 0^{(i,j-i)} \right)$, where we assume $e_\nu \in \mathbb{Z}$ and

$$e_1 \leq e_2 \ldots \leq e_i \leq \infty.$$

R. Weissauer, *Endoscopy for GSp(4) and the Cohomology of Siegel Modular Threefolds*, Lecture Notes in Mathematics 1968, DOI: 10.1007/978-3-540-89306-6_12, © Springer-Verlag Berlin Heidelberg 2009

Theorem 12.1. *The matrices $g(e_1, \ldots, e_i)$, for $e_1 \leq e_2 \cdots \leq e_i \leq \infty$ and $e_\nu < 0$ or $e_\nu = \infty$ for all $\nu = 1, \ldots, i$, define a system of representatives for the double cosets*

$$H(F) \setminus G(F) / G(\mathfrak{o}_F).$$

Remark 12.1. An alternative choice would have been $g(e_1, \ldots, e_i)$ with $e_1 \leq \ldots \leq e_i \leq 0$. Using this representatives one obtains the following corollary.

Corollary 12.1. *Let T be the diagonal torus in H or in G. Then there exists an element $r \in G(F)$ such that the set of conjugates $\{trt^{-1} \mid t \in T(F)\}$ of r contains a complete set of representatives of $H(F) \setminus G(F) / G(\mathfrak{o}_F)$.*

For instance, one can choose $r = g(0, \ldots, 0) \in G(\mathfrak{o}_F)$. For $\mathbf{D} = diag(D, E, D^{-1}, E) \in T(F)$ and $D = diag(\pi_F^{e_1}, \ldots, \pi^{e_i})$ then $\mathbf{D}r\mathbf{D}^{-1} = g(e_1, \ldots, e_i)$.

The proof of the theorem requires some preparation. In the following we always assume that D satisfies $e_1 \leq \cdots \leq e_\nu$ and $e_{\nu+1} = \cdots = e_i = \infty$ for some $\nu \leq i$:

1. **The parabolic subgroups P_s.** There is a parabolic subgroup $P = P_s$ of G with Levi component L in $H = H_s$. Let $P = L \cdot N$, where N is the unipotent radical. Then Iwasawa decomposition $G(F) = L(F) \cdot N(F) \cdot G(\mathfrak{o}_F)$ allows us to choose representatives $g(M, N, U, V) \in N(F)$ of the form

$$g(M, N, U, V) = \begin{pmatrix} E & M & U & N \\ 0 & E & * & V \\ 0 & 0 & E & 0 \\ 0 & 0 & -M' & E \end{pmatrix}.$$

   Notice $g(0, 0, 0, V)g(M, N, U, 0) = g(M, N, U, V)$ and $V = V' = V^{(j,j)}$ is symmetric. Since $g(0, 0, 0, V) \in H(F)$ we can assume $V = 0$ and therefore write $g(M, N, *) = g(M, N, *, 0)$ for the representative

$$g(M, N, *) = \begin{pmatrix} E & M & * & N \\ 0 & E & N' & 0 \\ 0 & 0 & E & 0 \\ 0 & 0 & -M' & E \end{pmatrix}.$$

   For the moment $M, N \in Hom_F(F^j, F^i)$ are still arbitrary.
2. Notice $g(M_1, N_1, U_1) \cdot g(M_2, N_2, U_2) = g(M_1 + M_2, N_1 + N_2, U_1 + U_2 + M_1 \cdot N_2' - N_1 \cdot M_2')$; hence, $g(M, N, U) \cdot g(0, 0, \tilde{U}) = g(0, 0, \tilde{U}) \cdot g(M, N, U) = g(M, N, \tilde{U} + U)$ and $g(0, 0, U) \in H(F)$. $S = U - M \cdot N'$ is symmetric. For symmetric $S = S'$ now $g(0, 0, S) \in H(F)$. Hence, we can choose the representatives in the form

$$g(M, N) = g(M, N, M \cdot N').$$

Since $g(M_1, N_1)g(M_2, N_2) = g(M_1+M_2, N_1+N_2, M_1N_1'+M_2N_2'+M_1N_2'-N_1M_2') = g(M_1+M_2, N_1+N_2, (M_1+M_2)(N_1+N_2)' - M_2N_1' - N_1M_2') = g(0, 0, -M_2N_1' - N_1M_2') \cdot g(M_1+M_2, N_1+N_2)$

$$H(F) \cdot g(M_1, N_1)g(M_2, N_2) \cdot G(\mathfrak{o}_F) = H(F) \cdot g(M_1+M_2, N_1+N_2) \cdot G(\mathfrak{o}_F).$$

Hence, $(M, N) \in Hom_F(F^{2j}, F^i)$ can be modified within the double coset by adding an arbitrary element from $Hom_{\mathfrak{o}_F}(\mathfrak{o}_F^{2j}, \mathfrak{o}_F^i)$.

3. For $A = A^{(i,i)}$ and $B = B^{(i,i)}$ we later consider the special cases

$$M = (A, 0), \qquad N = (B, 0).$$

We then simply write $g(A, B)$ or $g(A, B, *)$ instead of $g(M, N)$ and $g(M, N, *)$, respectively. The formulas above are valid with A, B in place of M, N. For

$$A = \begin{pmatrix} A_0 & 0 \\ A_1 & A_2 \end{pmatrix}, \qquad B = \begin{pmatrix} B_0 & 0 \\ 0 & B_1 \end{pmatrix}$$

and $k \times k$-matrices A_0, B_0, and $k < i$ and integral matrices A_1, A_2, B_1 step (2) allows us to replace the matrices A_1, A_2 by zero and B_1 by the unit matrix, without changing the double coset.

4. Next, for $U_i \in Gl(i, \mathfrak{o}_F)$ we obtain equivalent representatives $g(M, N, *)$ and $g(U_i \cdot M, U_i \cdot N, *)$ by conjugation with $diag(U_i, E, (U_i')^{-1}, E)$.

5. On the $i \times 2j$-matrices in $Hom_F(F^{2j}, F^i)$ the elements $g \in Sp(2j, \mathfrak{o}_F)$ act by multiplication from the right

$$(\tilde{M}, \tilde{N}) = (M, N) \cdot g^{-1}.$$

$g(M, N, *)$ and $g(\tilde{M}, \tilde{N}, *)$ define the same double coset. It suffices to show this for generators g of $Sp(2j, \mathfrak{o}_F)$. For this notice $w_j \cdot g(M, N, 0) \cdot w_j^{-1} = g(N, -M, *)$ and $u_T \cdot g(M, N, 0) \cdot u_T^{-1} = g(M, N - MT, *)$ for the generators (see [28], Satz A.5.4)

$$w_j = \begin{pmatrix} E & 0 & 0 & 0 \\ 0 & 0 & 0 & E \\ 0 & 0 & E & 0 \\ 0 & -E & 0 & 0 \end{pmatrix}$$

and

$$u_V = \begin{pmatrix} E & 0 & 0 & 0 \\ 0 & E & 0 & V \\ 0 & 0 & E & 0 \\ 0 & 0 & 0 & E \end{pmatrix}.$$

For integral symmetric V these are contained in the intersection of $G(\mathfrak{o}_F)$ and $H(F)$. Hence, we may choose our representatives (A, B) in

$$Gl(i, \mathfrak{o}_F) \backslash Hom_F(F^{2j}, F^i) / Sp(2j, \mathfrak{o}_F),$$

where these, in addition, may be modified by elements from $Hom_{\mathfrak{o}_F}(\mathfrak{o}_F^{2j}, \mathfrak{o}_F^i)$.

12.1 Reduction to Standard Type

We say (M, N) are of *standard* type if

$$(M, N) = \Big((A, 0), (B, 0)\Big)$$

for an $i \times i$-diagonal matrix $B = diag(\pi_F^{e_1}, \ldots, \pi_F^{e_i})$ and a nilpotent $i \times i$-lower triangular matrix A such that:

(a) $e_1 \le e_2 \le \ldots \le e_i \le 0$.
(b) $B^{-1} \cdot A$ is an integral matrix.

The Reduction. We now construct elements in $Sp(2j, \mathfrak{o}_F) \times Gl(i, \mathfrak{o}_F)$ which transform a given $(M, N) \in Hom_F(F^{2j}, F^i)$ into standard type. For this temporarily replace (M, N) by $(N, -M)$ (using conjugation by w_j as in step (5)), and then replace the resulting matrix by its transpose in

$$Hom_F(F^i, F^{2j}).$$

By this $Sp(2j, \mathfrak{o}_F)$ now acts from the left and $Gl(i, \mathfrak{o}_F)$ acts from the right. Our argument now proceeds using induction. Start with an arbitrary matrix in $Hom_F(F^i, F^{2j})$. We say it is of *weak r-standard type* if it is of the form

$$\begin{pmatrix} B_r & * \\ 0 & * \\ -A_r' & * \\ 0 & * \end{pmatrix},$$

where $r \le i \le j$ and $B_r = diag(\pi_F^{e_1}, \ldots, \pi_F^{e_r})$ and $e_1 \le \ldots \le e_r \le \infty$, such that A_r' is a strict upper triangular $r \times r$-matrix such that:

(a) π^{e_r} divides the greatest common divisor (gcd) π^e of all entries of the matrix denoted by a star.
(b) π^{e_ν} divides all entries of the νth column for $1 \le \nu \le r$.

If, in addition, the shape is

$$\begin{pmatrix} B_r & 0 \\ 0 & * \\ -A_r' & * \\ 0 & * \end{pmatrix}$$

we say the matrix is *partially* of r-standard type.

By elimination of the right upper block a representative of weak partial r-standard type can be transformed to become partially of r-standard type. Use right multiplication with some element in $Gl(i, \mathfrak{o}_F)$ to clear the first r rows of the dotted area by adding columns. This does not change condition (b), since $e_1 \le \ldots e_r$

and e_r is less than or equal to the gcd of the remaining columns (beginning from $r + 1$). Since we add multiples of $\pi_F^{e-e_\nu} \lambda, \lambda \in \mathfrak{o}_F$ times the νth column ($\nu \leq r$), we add terms in $\pi_F^e \mathfrak{o}_F$ as follows from condition (b). Therefore, the gcd of the back columns will not be changed by this procedure.

The Induction Step. For a matrix partially of $(r - 1)$-standard type consider the columns beginning from the rth column. By right multiplication with a permutation matrix in $Gl(i, \mathfrak{o}_F)$ one can achieve the gcd π_F^e of all these columns already being the gcd of the entries of the rth column vector $v \in F^{2j}$. The first $(r-1)$-entries of v are zero since the matrix we started with was partially of $(r - 1)$-standard type, and since the permutations of columns beginning from the rth column do not change the property such that the upper entries of these columns vanish.

Now our modifications will only involve multiplications with elements in $G(\mathfrak{o}_F)$ from the left. This changes the columns beginning from the $(r + 1)$th. In particular, this may destroy the property that the first $(r - 1)$-coordinates of these columns vanish. The given matrix is of the form

$$\begin{pmatrix} B_{r-1} & 0 & * \\ 0 & * & * \\ -A'_{r-1} & * & * \\ 0 & * & * \end{pmatrix}$$

such that the gcd π_F^e of the "middle" rth column divides the gcd of all columns beginning from the $(r + 1)$th. This property is preserved under multiplication with substitutions from $G(\mathfrak{o}_F)$. Hence, in principle, we can concentrate on the first r columns since it is enough to bring our representative into a form of weak partial r-standard type. We therefore temporarily ignore all columns beginning from the $(r + 1)$th column.

A suitable symplectic transformation of an embedded $Sp(2(j - r + 1), \mathfrak{o}_F)$ by multiplication from the left allows us to make all coordinates of v be zero, except the rth and the $(j + 1), \ldots, (j + r - 1)$th coordinate entries. By this the first $(r - 1)$-columns of our representative will not be changed. In addition we can achieve the rth coordinate entry of v being a power π_F^f of the prime element. For this notice that the unimodular symplectic matrices act transitively on primitive vectors ([28], Hilfssatz A.5.2).

After this the matrix is almost of weak partial r-standard type, being of the form

$$\begin{pmatrix} B_r & * \\ 0 & * \\ -A'_r & * \\ 0 & * \end{pmatrix}$$

such that (a) is satisfied. We are done if the rth coordinate entry π_F^f of the rth column is equal to π_F^e. If it is not, then $e < f$. Then there exists ν with $1 \leq \nu < r$ such that the gcd of the rth column is realized at the $(j + \nu)$th coordinate entry. It then remains to bring the gcd of column v to the "top." Left multiplication by a symplectic unimodular substitution – on the standard basis w_i of F^{2j} given by $w_\mu \mapsto w_\mu$ for $\mu \neq j+\nu, j+r$ and by $w_{j+\nu} \mapsto w_{j+\nu}+w_r$ and $w_{j+r} \mapsto w_{j+r}+w_\nu$ – has no effect on the lower half, i.e., A'_r will not be changed. Also the zero blocks on the left side will not be changed. The matrix B_r, on the other hand, will be modified. Since the rth line of $-A'_r$ is zero, only the last line of B_r will be changed – in fact by addition of the νth line of $-A'_r$. Let $x_1, \ldots, x_r$ denote the new entries. For example, $x_r = \pi_F^{e_r} + \pi_F^e = \varepsilon \cdot \pi^e$ $(\varepsilon \in \mathfrak{o}_F^*)$.

Next the modified B_r will again be diagonalized by left multiplication by a unimodular symplectic matrix of the form $diag(U, E, (U')^{-1}, E)$, where

$$
U = \begin{pmatrix}
1 & 0 & \cdots & 0 & 0 \\
0 & 1 & \cdots & 0 & 0 \\
\cdot & & & & \cdot \\
\cdot & & & & \cdot \\
\cdot & & & & \cdot \\
0 & 0 & \cdots & 1 & 0 \\
y_1 & y_2 & & y_{r-1} & \varepsilon^{-1}
\end{pmatrix}.
$$

This transforms $-A'_r$ into $-(U')^{-1} \cdot A'_r = -A'_r$ (the rth column of A'_r is zero) and transforms B_r into $U \cdot B_r$. For suitable y_ν the matrix $U \cdot B_r$ will become a diagonal matrix with the entries $diag(\pi_F^{e_1}, \ldots, \pi_F^{e_{r-1}}, \pi_F^e)$, provided $y_\nu \cdot \pi_F^{\varepsilon_\nu} = -\varepsilon^{-1}x_\nu$ holds. By condition (b) of the matrix of partial $(r-1)$-standard type we started with, such y_ν can be chosen in $\mathfrak{o}_F$. This implies $U \in Gl(r, \mathfrak{o}_F)$ and $diag(U, E, (U')^{-1}, E) \in G(\mathfrak{o}_F)$. This shows that our new matrix is now of weak partial r-standard type such that $e_r = e$, and it is a representative in the double coset of the matrix we started from. This completes the proof of the induction step.

Iterating this i times, we can get a matrix of partial i-standard type. Reverse transposition and reverse conjugation by w_j therefore gives an equivalent matrix replacing (M, N), which now is almost of standard type. It is of the form $(M, N) = ((A, 0), (B, 0))$ for $B = diag(\pi_F^{e_1}, \ldots, \pi_F^{e_i})$ and a lower triangular matrix A, whose diagonal is zero, and such that $e_1 \leq e_2 \leq \ldots \leq e_i \leq \infty$. Choose k to be maximal such that $e_k < 0$. By step (3) we can assume without restriction of generality $e_\nu = 0$ for $\nu > k$. Then B^{-1} is defined, and $B^{-1}A$ is an integral matrix. So we have a matrix of standard type.

Summary. *There exist representatives of the double cosets $H(F) \setminus G(F)/G(\mathfrak{o}_F)$ of the form $g = g((A, 0), (B, 0))$, such that:*

- $B = diag(\pi_F^{e_1}, \ldots, \pi_F^{e_i})$ *is a diagonal invertible $i \times i$-matrix with $e_1 \leq \ldots e_i \leq 0$.*
- *A is a lower triangular matrix.*
- *B^{-1} has integral entries.*
- *The lower $i \times i$-triangular matrix $B^{-1} \cdot A$ has integral entries.*
- *$B^{-1} \cdot A'$ is an $i \times i$-matrix with integral entries.*

12.2 The Quadratic Embedding

The matrix $\Lambda_s = s \cdot J = J \cdot s$ is skew-symmetric

$$\Lambda_s = \begin{pmatrix} 0 & 0 & E & 0 \\ 0 & 0 & 0 & -E \\ -E & 0 & 0 & 0 \\ 0 & E & 0 & 0 \end{pmatrix}.$$

For $g \in G(F)$ the conditions $g \in H(F)$ and $\lambda(g)^{-1} g' \cdot \Lambda_s \cdot g = \Lambda_s$ are equivalent; since $\lambda(g)^{-1} \cdot (g')^{-1} = JgJ^{-1}$ and $J^{-1}\Lambda_s = s$, the first equation is equivalent to $s \cdot g = g \cdot s$.

Consequence. $Elm(H(F) \cdot g) = \lambda(g)^{-1} g' \cdot \Lambda_s \cdot g$ *defines an injection*

$$Elm : H(F) \setminus G(F) \hookrightarrow \Lambda^2(F^{2n})$$

of the cosets $H(F) \setminus G(F)$ into the vector space $\Lambda^2(F^{2n})$ of skew-symmetric $2n$-matrices.

Remark 12.2. The quadratic form $q(\Lambda) = Trace(\Lambda \cdot J \cdot \Lambda \cdot J)$ defines a nondegenerate symmetric bilinear form on $\Lambda^2(F^{2n})$ such that $q(\lambda(g)^{-1} g' \cdot \Lambda \cdot g) = q(\Lambda)$ holds for all $g \in G(F)$.

Notation. We write $Elm(A, B)$ for the matrix $Elm(g(A, B, AB'))$. Then $Elm(A, B)$ is a skew-symmetric matrix contained in the symplectic group $Sp(2n, F)$.

By definition Λ_s and g are both contained in $G(F) = GSp(2n, F)$. In all that follows, we may therefore restrict ourselves to the case $i = j$ since $Elm(A, B)$ is in $Sp(2i, F) \times Sp(2(j - i), F)$, and its "component" is in $Sp(2(j - i), F)$ is $J = J^{(j-i,j-i)}$.

Assumption. For simplicity of notation we therefore assume from now on $j = i$, without restriction of generality.

Then

$$Elm(A, B) = \begin{pmatrix} 0 & X \\ -X' & \mathcal{A} \end{pmatrix}$$

defined by $n \times n$-block matrices $X = \begin{pmatrix} E & 0 \\ 2A' & -E \end{pmatrix}$ and $\mathcal{A} = \begin{pmatrix} 2(B \cdot A' - A \cdot B') & -2 \cdot B \\ 2 \cdot B' & 0 \end{pmatrix}$.

Remark 12.3. The skew-symmetric matrix $\mathcal{A}$ is invertible since B is invertible.

So there are matrices Z and $\tilde{\mathcal{A}}$ such that

$$\begin{pmatrix} 0 & X \\ -X' & \mathcal{A} \end{pmatrix} = \begin{pmatrix} E & Z \\ 0 & E \end{pmatrix} \begin{pmatrix} \tilde{\mathcal{A}} & 0 \\ 0 & \mathcal{A} \end{pmatrix} \begin{pmatrix} E & 0 \\ Z' & E \end{pmatrix}.$$

Notice $X = Z \cdot \mathcal{A}$, $\tilde{\mathcal{A}} = -Z \cdot \mathcal{A} \cdot Z'$, and $Z = X \cdot \mathcal{A}^{-1}$, $X' = -\mathcal{A} \cdot Z'$, $\tilde{\mathcal{A}} = Z \cdot X' = X \cdot \mathcal{A}^{-1} \cdot X'$.

Corollary 12.2. *The $(n \times n)$-matrix Z is symmetric. Hence,*

$$g(Z) = \begin{pmatrix} E & 0 \\ Z & E \end{pmatrix} \in G(F).$$

Proof. $Z = X \cdot \mathcal{A}^{-1}$ satisfies $Z = Z'$ if $-\mathcal{A}' \cdot Z \cdot \mathcal{A} = \mathcal{A} \cdot X$ is symmetric. Since

$$\mathcal{A} \cdot X = \begin{pmatrix} 2 \cdot (B \cdot A' - A \cdot B') & -2 \cdot B \\ 2 \cdot B' & 0 \end{pmatrix} \begin{pmatrix} E & 0 \\ 2 \cdot A' & -E \end{pmatrix}$$

$$= \begin{pmatrix} 2B \cdot A' - 2A \cdot B' - 4B \cdot A' & 2B \\ 2B' & 0 \end{pmatrix}$$

$$= \begin{pmatrix} -2 \cdot (B \cdot A' + A \cdot B') & 2 \cdot B \\ 2 \cdot B' & 0 \end{pmatrix}$$

is symmetric, Z is also symmetric. $\square$

It follows that

Fact. $g(Z)' \cdot Elm(A, B) \cdot g(Z) = \begin{pmatrix} \tilde{\mathcal{A}} & 0 \\ 0 & \mathcal{A} \end{pmatrix}$, *where* $\tilde{\mathcal{A}} = (\mathcal{A}')^{-1} = -\mathcal{A}^{-1}$ *and*

$$\mathcal{A} = \begin{pmatrix} 2(BA' - AB') & -2 \cdot B \\ 2 \cdot B' & 0 \end{pmatrix} = -\mathcal{A}'.$$

Formula for Z. $\mathcal{A}$ is invertible by assumption. Since $Elm(A, B)$ and $g(Z)$, and hence also $g(Z)'$, are symplectic matrices, we have $\tilde{\mathcal{A}} = (\mathcal{A}')^{-1}$. Notice that

$$\begin{pmatrix} E & -A \\ 0 & E \end{pmatrix} \begin{pmatrix} 0 & -2 \cdot B \\ 2 \cdot B' & 0 \end{pmatrix} \begin{pmatrix} E & 0 \\ -A' & E \end{pmatrix}$$

$$= \begin{pmatrix} -2A \cdot B' & -2 \cdot B \\ 2 \cdot B' & 0 \end{pmatrix} \begin{pmatrix} E & 0 \\ -A' & E \end{pmatrix}$$

$$= \begin{pmatrix} -2(A \cdot B' - B \cdot A') & -2 \cdot B \\ 2 \cdot B' & 0 \end{pmatrix} = \mathcal{A}.$$

Hence,

$$-2 \cdot Z = -2 \cdot X \cdot \mathcal{A}^{-1} = \begin{pmatrix} E & 0 \\ 2 \cdot A' & -E \end{pmatrix} \begin{pmatrix} E & 0 \\ A' & E \end{pmatrix} \begin{pmatrix} 0 & B \\ -B' & 0 \end{pmatrix}^{-1} \begin{pmatrix} E & A \\ 0 & E \end{pmatrix}$$

$$= \begin{pmatrix} E & 0 \\ A' & -E \end{pmatrix} \begin{pmatrix} 0 & -(B')^{-1} \\ B^{-1} & 0 \end{pmatrix} \begin{pmatrix} E & A \\ 0 & E \end{pmatrix}$$

$$= \begin{pmatrix} 0 & -(B')^{-1} \\ -B^{-1} & -A' \cdot (B')^{-1} \end{pmatrix} \begin{pmatrix} E & A \\ 0 & E \end{pmatrix}$$

$$= \begin{pmatrix} 0 & -(B')^{-1} \\ -B^{-1} & -B^{-1} \cdot A - A' \cdot (B')^{-1} \end{pmatrix} .$$

Since we have shown that we can assume the representative to be of standard type, the matrices B^{-1} and $B^{-1}A$ are integral; hence, Z is also integral. Therefore, we have

Fact. *The symplectic matrix $g(Z)$ is contained in $G(\mathfrak{o}_F)$.*
The injection Elm already defined induces an injection elm

$$\boxed{elm: \ H(F) \backslash G(F)/G(\mathfrak{o}_F) \ \hookrightarrow \ \Lambda^2(F^{2n})/G(\mathfrak{o}_F)} .$$

A Consequence. Suppose (M, N) is of standard type. Consider the double coset of $g(M, N) = g(A, B)$. Its image $elm(A, B)$ in $\Lambda^2(F^{2n})/G(\mathfrak{o}_F)$ is represented by the symplectic block matrix

$$diag(\tilde{\mathcal{A}}, \mathcal{A}) = diag(-\mathcal{A}^{-1}, \mathcal{A}).$$

12.3 Elementary Divisors

We dispose over another obvious map

$$\Lambda^2(F^{2n})/G(\mathfrak{o}_F) \ \to \ \Lambda^2(F^{2n})/\big(Gl(2n, \mathfrak{o}_F) \times \mathfrak{o}_F^*\big).$$

Here $(h, \varepsilon) \in Gl(2n, \mathfrak{o}_F) \times \mathfrak{o}_F^*$ acts on $\Lambda^2(F^{2n})$ by $\Lambda \mapsto \varepsilon \cdot h' \cdot \Lambda \cdot h$. For this we may consider the general case $i \leq j$, and we then claim

Lemma 12.1. *The composed map*

$$\mathcal{L}: \ H(F) \backslash G(F) / G(\mathfrak{o}_F) \ \longrightarrow \ \Lambda^2(F^{2n})/\big(Gl(2n, \mathfrak{o}_F) \times \mathfrak{o}_F^*\big),$$

which maps $H(F)gG(\mathfrak{o}_F)$ to the orbit of $\lambda(g)^{-1}g'\Lambda_s g$, is an injection.

We say two skew-symmetric invertible matrices in $\Lambda^2(F^m)$ are equivalent if there exists a unimodular matrix h in $Gl(m, \mathfrak{o}_F)$ such that $\Lambda_1 = h' \cdot \Lambda_2 \cdot h$. Concerning the orbits (right side of the map in the last lemma) recall the result of Frobenius:

(A) Λ_1 and Λ_2 are equivalent if and only if they have the same elementary divisors (understood in the usual sense).

(B) The product of the first k-elementary divisors (in the usual sense) is the gcd of all $k \times k$-minors.

(C) $\varepsilon \cdot \Lambda$ and Λ are equivalent for any $\varepsilon \in \mathfrak{o}_F^*$.

Hence, the orbits $\Lambda^2(F^m)/(Gl(m, \mathfrak{o}_F) \times \mathfrak{o}_F^*)$ are described by the elementary divisors.

Proof of Lemma 12.1. Without restriction of generality we again assume $i = j$. Then the skew-symmetric $(n \times n)$-matrix $\mathcal{A}$ can be brought into the following Frobenius standard form by a suitable unimodular transformation $U \in Gl(n, \mathfrak{o}_F)$:

$$U' \cdot \mathcal{A} \cdot U = diag\left(\begin{pmatrix} 0 & \pi_F^{a_1} \\ -\pi_F^{a_1} & 0 \end{pmatrix}, \ldots, \begin{pmatrix} 0 & \pi_F^{a_i} \\ -\pi_F^{a_i} & 0 \end{pmatrix}\right),$$

where $a_1 \leq \ldots \leq a_i$. These symplectic elementary divisors are determined by the elementary divisors of the matrix $U'\mathcal{A}U$ (in the usual sense), which are $\pi_F^{a_1}, \pi_F^{a_1}, \pi_F^{a_2}, \pi_F^{a_2}, \cdots$.

The diagonalizing matrix U defines

$$g = diag((U')^{-1}, U) \in Sp(2n, \mathfrak{o}_F) \subseteq Gl(2n, \mathfrak{o}_F).$$

The symplectic $2n \times 2n$-matrix $diag(\tilde{\mathcal{A}}, \mathcal{A}) = diag((\mathcal{A}')^{-1}, \mathcal{A})$ will be transformed by $g \in G(\mathfrak{o}_F)$ into the "symplectic normal form"

$$diag\left(\begin{pmatrix} 0 & \pi_F^{-a_1} \\ -\pi_F^{-a_1} & 0 \end{pmatrix}, \ldots, \begin{pmatrix} 0 & \pi_F^{-a_i} \\ -\pi_F^{-a_i} & 0 \end{pmatrix}, \begin{pmatrix} 0 & \pi_F^{a_1} \\ -\pi_F^{a_1} & 0 \end{pmatrix}, \ldots, \begin{pmatrix} 0 & \pi_F^{a_i} \\ -\pi_F^{a_i} & 0 \end{pmatrix}\right).$$

This symplectic normal form defines the same coset in $\Lambda^2(F^{2n})/G(\mathfrak{o}_F)$ as the matrices $diag(\tilde{\mathcal{A}}, \mathcal{A})$ and $elm(M, N)$.

Claim 12.1. $a_i \leq -a_i$. In other words, the exponents of the elementary divisors of $diag(\tilde{\mathcal{A}}, \mathcal{A})$, in increasing order, are the numbers

$$a_1, a_1, a_2, a_2, \ldots, a_i, a_i, -a_i, -a_i, \ldots, -a_1, -a_1.$$

(In the general case $j > i$ there are $n - 2i$ additional zeros in the middle.) Hence, the elementary divisors of $diag(\tilde{\mathcal{A}}, \mathcal{A})$ uniquely determine the exponents $a_1 \leq a_2 \leq \cdots \leq a_i$ of the symplectic Frobenius normal form of $\mathcal{A}$, as defined above. This immediately implies the lemma, provided the claim $a_i \leq 0$ holds. To show this claim, notice $\pi_F^{-a_i} = det(\mathcal{A})^{-1} \cdot gcd(\Lambda^{n-1}(\mathcal{A}))$ and $det(\mathcal{A})^{-1} \cdot \Lambda^{2i-1}(\mathcal{A}) = \mathcal{A}^{-1}$. Hence, $\pi_F^{-a_i} = gcd(\mathcal{A}^{-1})$ is the first elementary divisor of $\mathcal{A}^{-1}$. Thus, to prove the claim, it suffices to show that $\mathcal{A}^{-1}$ is an integral matrix. Since

$$\mathcal{A} = -2 \begin{pmatrix} B & 0 \\ 0 & E \end{pmatrix} \begin{pmatrix} G & E \\ -E & 0 \end{pmatrix} \begin{pmatrix} B' & 0 \\ 0 & E \end{pmatrix}$$

for the matrix $G = B^{-1}A - (B^{-1}A)'$, we get

$$\mathcal{A}^{-1} = -\frac{1}{2} \begin{pmatrix} (B')^{-1} & 0 \\ 0 & E \end{pmatrix} \begin{pmatrix} 0 & -E \\ E & G \end{pmatrix} \begin{pmatrix} B^{-1} & 0 \\ 0 & E \end{pmatrix}.$$

Since B^{-1} and G are integral, $\mathcal{A}^{-1}$ is integral, which proves the lemma. $\square$

Proof of Theorem 12.1. By Lemma 12.1 it suffices to show that for $(e_1, \ldots, e_i)$, subject to the conditions stated in Theorem 12.1, the elementary divisors of the matrices

$$Elm\Big(g(e_1, \ldots, e_i)\Big) = \begin{pmatrix} 0 & 0 & E & 0 \\ 0 & 0 & 0 & -E \\ -E & 0 & 0 & -2 \cdot D \\ 0 & E & 2 \cdot D' & 0 \end{pmatrix}$$

determine $(e_1, \ldots, e_i)$ uniquely such that every possible constellation of elementary divisors – as determined above – is realized by some $Elm(g(e_1, \ldots, e_i))$. This, however, is rather obvious. The elementary divisors of $Elm(g(e_1, \ldots, e_i))$ are $\pi_F^{e_1}, \pi_F^{e_1}, \ldots, \pi_F^{e_r}, \pi_F^{e_r}, \ldots$, where $r \leq i$ is chosen to be maximal such that $e_r < 0$. The following elementary divisors are pairs of 1 and then followed by the inverse numbers $\pi_F^{-e_r}, \ldots, \pi_F^{-e_1}$ (in fact notice it is enough to consider minors in the right lower $n \times n$-block). This implies that the representatives $g(e_1, \ldots, e_i)$ uniquely represent the double cosets $H(F) \backslash G(F) / G(\mathfrak{o}_F)$, which proves the theorem. $\square$

Remark 12.4. In fact we have now also determined the image of the map $\mathcal{L}$. It consists of all orbits which contain a matrix in Frobenius normal form with exponents which satisfy

$$a_1 \leq \cdots \leq a_i \leq -a_i \leq \cdots - a_1.$$

12.4 The Compact Open Groups

Now fix some representative $g(D)$ as in Theorem 12.1. For simplicity assume $i = j$. Recall $D = D'$. Then

$$H_D = H(F) \cap g(D)G(\mathfrak{o}_F)g(D)^{-1}$$

is a compact open subgroup of $H(F)$. For

$$h = \begin{pmatrix} \alpha_1 & 0 & \beta_1 & 0 \\ 0 & \alpha_2 & 0 & \beta_2 \\ \gamma_1 & 0 & \delta_1 & 0 \\ 0 & \gamma_2 & 0 & \delta_2 \end{pmatrix}$$

in $H(F)$ we have the symplectic conditions $\alpha_i' \delta_i - \gamma_i' \beta_i = \lambda \cdot E, \lambda \in \mathfrak{o}_F^*$, $\alpha_i' \gamma_i = \gamma_i' \alpha_i$, $\gamma_i \delta_i' = \delta_i' \beta_i$. Furthermore, h is contained in H_D if and only if

$$g(D)^{-1} \cdot h \cdot g(D) = \begin{pmatrix} \alpha_1 & -D\gamma_2 & -D\gamma_2 D' + \beta_1 & -D\delta_2 + \alpha_1 D \\ -D'\gamma_1 & \alpha_2 & -D'\delta_1 + \alpha_2 D' & -D'\gamma_1 D + \beta_2 \\ \gamma_1 & 0 & \delta_1 & \gamma_1 D \\ 0 & \gamma_2 & \gamma_2 D' & \delta_2 \end{pmatrix}$$

is contained in $G(\mathfrak{o}_F)$. Then $\alpha_i, \gamma_i, \delta_i$ and $D'\gamma_1, D\gamma_2$ and $\gamma_2 D'$ and $\gamma_1 D$ are integral, and $det(h) \in \mathfrak{o}_F^*$ (first integrality conditions). Furthermore, we have the four congruence conditions (*) modulo integral matrices:

$$\beta_1 \equiv D\gamma_2 D', \qquad \beta_2 \equiv D'\gamma_1 D,$$

$$D\delta_2 \equiv \alpha_1 D, \qquad D'\delta_1 \equiv \alpha_2 D'.$$

Since D^{-1} is integral, and hence $D^{-1}\beta_1, \beta_1(D')^{-1}, (D')^{-1}\beta_2, \beta_2 D^{-1}$ and $D^{-1}\alpha_1 D, D'\delta_1(D')^{-1}, D\delta_2 D^{-1}, (D')^{-1}\alpha_2 D'$ are necessarily integral (second integrality conditions). We reformulate the integrality conditions by introducing the integral skew-symmetric matrix

$$\Lambda_D = \begin{pmatrix} 0 & D^{-1} \\ -D^{-1} & 0 \end{pmatrix}.$$

Define

$$GSp(\Lambda_D) = \left\{ h \in Gl(F^{2i}) \mid h'\Lambda_D h = \lambda \cdot \Lambda_D, \ \lambda \in F^* \right\}.$$

Notice $diag(E, -E) \in GSp(\Lambda_D)$ and $J \in GSp(\Lambda_D)$; hence,

$$I = \begin{pmatrix} 0 & E \\ E & 0 \end{pmatrix} \in GSp(\Lambda_D),$$

and, therefore, $g = \left(\begin{smallmatrix} a & b \\ c & d \end{smallmatrix}\right) \in GSp(\Lambda_D) \iff g^I = IgI = \left(\begin{smallmatrix} d & c \\ b & a \end{smallmatrix}\right) \in GSp(\Lambda_D)$.

Also notice that $g_k \in GSp(\Lambda_D)$ holds for the two matrices ($k = 1, 2$)

$$g_k := \begin{pmatrix} a_k & b_k \\ c_k & d_k \end{pmatrix} = \begin{pmatrix} 0 & D \\ E & 0 \end{pmatrix} \begin{pmatrix} \alpha_k & \beta_k \\ \gamma_k & \delta_k \end{pmatrix} \begin{pmatrix} 0 & D \\ E & 0 \end{pmatrix}^{-1} = \begin{pmatrix} D\delta_k D^{-1} & D\gamma_k \\ \beta_k D^{-1} & \alpha_k \end{pmatrix}.$$

All the integrality conditions stated above when put together express the fact that both matrices g_k and $diag(D, D)^{-1} g_k diag(D, D)$ are integral matrices (for $k = 1, 2$) with equal similitude factor in $\mathfrak{o}_F^*$. If $\Gamma = (\mathfrak{o}_F)^{2i}$ denotes the standard lattice in F^{2i}, then Λ_D defines a skew-symmetric pairing $\langle ., . \rangle_D$ on Γ by $\langle v, w \rangle_D = v'\Lambda_D w$. The dual lattice $\Gamma^* = \{w \in F^{2i} \mid \langle w, \Gamma \rangle_D \in \mathfrak{o}_F\}$ is $\Gamma^* = diag(D, D)(\Gamma)$. The matrices in $GSp(\Lambda_D)$, which preserve the lattice Γ, define a compact open subgroup $Aut(\Gamma, \Lambda_D) \subseteq GSp(\Lambda_D)$. Each $g \in Aut(\Gamma, \Lambda_D)$ preserves the dual lattice Γ^*. Hence, $g(\Gamma^*) = \Gamma^*$ or $diag(D, D)^{-1} \cdot g \cdot diag(D, D) \in Aut(\Gamma, \Lambda_{D^{-1}})$. This shows that the first and second integrality conditions are equivalent to

$$g_1, g_2 \in Aut(\Gamma, \Lambda_D).$$

This defines an injective homomorphism

$$H_D \hookrightarrow Aut(\Gamma, \Lambda_D) \times Aut(\Gamma, \Lambda_D),$$

induced from the injection of $H(F)$ into $GSp(\Lambda_D) \times GSp(\Lambda_D)$, which is defined by

$$h \mapsto (k_1, k_2) = (g_1, g_2^I).$$

Now $\Gamma \subseteq \Gamma^*$ since Λ_D is integral. Hence, we get a homomorphism

$$1 \to K_D \to Aut(\Gamma, \Lambda_D) \to Aut(\Gamma^*/\Gamma)$$

with kernel, say, K_D. Obviously $g \in K_D \Longleftrightarrow (g - id)diag(D, D)$ is integral. For $\mathcal{G} = M_{i,i}(\mathfrak{o}_F) \cdot D^{-1} \subseteq M_{i,i}(\mathfrak{o}_F)$ the above four congruences (*) are equivalent to $b_1 \equiv c_2$, $c_1 \equiv b_2$, $d_1 \equiv a_2$, and $a_1 \equiv d_2$ modulo $\mathcal{G}$. In other words, the conditions (*) mean $(g_1 - g_2^I)diag(D, D)$ is integral, or $(id - g_1^{-1}g_2^I)diag(D, D)$ is integral. In other words, we get the condition

$$g_1 \equiv g_2^I \bmod K_D,$$

or $k_1 \equiv k_2 \bmod K_D$. Hence, H_D is isomorphic to the group of all $(k_1, k_2) \in Aut(\Gamma, \Lambda_D)^2$ such that $k_1 \equiv k_2 \bmod K_D$. Since K_D is a normal subgroup of $Aut(\Gamma, \Lambda_D)$, this proves that H_D is isomorphic to the semidirect product $K_D \triangleleft Aut(\Gamma, \Lambda_D)$

$$1 \to K_D \to H_D \to Aut(\Gamma, \Lambda_D) \to 1.$$

Example 12.1. For $D = d \cdot E$ we have $GSp(\Lambda_D) = GSp(2i, F)$, and $Aut(\Gamma, \Lambda_D) = G(\mathfrak{o}_F)$ such that K_D is the principal congruence subgroup of level d.

12.5 The Twisted Group $\tilde{H}$

Whereas in the last section we considered $i \leq j$, we now have to restrict ourselves to the special case $n = i + j$, where $i = j$. In this special case the normalizer N_s of the subgroup $H = G_s$ of $G = GSp(2n)$ is not connected. This now allows us to define Galois twists $\tilde{H}$ of the group H considered in the last section. For $i = j$ the centralizer in the adjoint group of the element s (defined at the beginning of this chapter) is nontrivial. The element

$$w = \begin{pmatrix} 0 & E & 0 & 0 \\ E & 0 & 0 & 0 \\ 0 & 0 & 0 & E \\ 0 & 0 & E & 0 \end{pmatrix} \in G(\mathfrak{o}_F)$$

in G_s generates $N_s/G_s \cong \mathbb{Z}/2\mathbb{Z}$. We have $ws = -sw$, $w^2 = id$, and $wJ = Jw$.

Suppose K/F is a quadratic field extension, and σ is the generator of the Galois group of this extension. Since $H^1(F, GSp(2n))$ is trivial, there exists an $g_0 \in GSp(2n, K)$ such that

$$g_0^{-1}\sigma(g_0) = w, \qquad \sigma(g_0) = g_0 \cdot w.$$

This condition determines the coset $G(F) \cdot g_0$ uniquely. Then the group $\tilde{H} = g_0 H g_0^{-1}$ is invariant under σ, and defines a form $\tilde{H}$ of H over F together with an embedding $\tilde{H} \hookrightarrow G$ defined over F. Notice for the norm-1-subgroup N^1 in $Res_{K/F}(\mathbb{G}_m)$

$$1 \to \tilde{H} \to Res_{K/F}(GSp(2i)) \to N^1 \to 1,$$

where the morphism on the right is $g \mapsto (\sigma(\lambda(g))/\lambda(g)$. Let us make some choice, i.e., $g_0 = diag(1, \frac{\alpha^2}{2}, \alpha^2, 2) \cdot diag(U, (U')^{-1})$ for $U = \left(\begin{smallmatrix} 1/2 & 1/2 \\ -\alpha & \alpha \end{smallmatrix}\right)$, where $K = F(\alpha)$ and $A^{-1} = \alpha^2 \in F^*$. We can assume that the valuation is $v_K(\alpha^2) = 0$ or -1 depending on whether K/F is unramified or not since the residue characteristic is different from 2. Then $\tilde{H}$ becomes the subgroup of all the matrices η defined on page 239.

The Map $\tilde{\mathcal{L}}$. Now consider the commutative diagram

$$
\begin{array}{ccc}
\tilde{H}(F) \backslash G(F)/G(\mathfrak{o}_F) & \longrightarrow & g_0 H(K) g_0^{-1} \backslash G(K)/G(\mathfrak{o}_K) \\
\tilde{\mathcal{L}} \downarrow & & \downarrow L(g_0^{-1}) \\
\Lambda^2(K^{2n})/(Gl(2n, \mathfrak{o}_K) \times \mathfrak{o}_K^*) & \xleftarrow{\;\;\mathcal{L}\;\;} & H(K) \backslash G(K)/G(\mathfrak{o}_K)
\end{array}
$$

where the upper map is induced by the scalar extension maps, the right vertical bijection is $\tilde{H}(K)gG(\mathfrak{o}_K) \mapsto H(K)g_0^{-1}gG(\mathfrak{o}_K)$, and the lower horizontal injective map is defined as in Lemma 12.1, but now for the field K instead of F. The left vertical map is the composition of the other maps $\tilde{\mathcal{L}}(\tilde{H}(F)gG(\mathfrak{o}_F)) = \mathcal{L}(H(K)g_0^{-1}gG(\mathfrak{o}_K)) = $ orbit of $\lambda(g_0^{-1}g)^{-1} \cdot (g_0^{-1}g)'\Lambda_s(g_0^{-1}g) = $ orbit of $\lambda(g)^{-1} \cdot g'\Lambda_s g$ for

$$
\tilde{\Lambda}_s = \begin{pmatrix} 0 & 0 & 0 & -\alpha^{-1}E \\ 0 & 0 & -\alpha E & 0 \\ 0 & \alpha E & 0 & 0 \\ \alpha^{-1}E & 0 & 0 & 0 \end{pmatrix}.
$$

The Image of $\tilde{\mathcal{L}}$. The F-rational element $diag(E, A \cdot E, E, E) \times 1$ transforms $\tilde{\Lambda}_s$ to $-\alpha^{-1} \cdot J$ within its orbits. Hence, the image of $\tilde{\mathcal{L}}$ is contained in the image of the $Gl(2n, F) \times F^*$ orbit of the matrix $-\alpha^{-1} \cdot J$. Therefore,

$$image(\tilde{\mathcal{L}}) \subseteq \big(v_K(\alpha), \ldots, v_K(\alpha)\big) + v_K(F^*)^n \subseteq v_K(K^*)^n,$$

considered as a subset of the n exponents, which define the Frobenius normal form of the skew-symmetric $2n \times 2n$-matrix. On the other hand, this image is contained in the image of the lower horizontal map $\mathcal{L}$, which is the set $\{(a_1, \ldots, a_i, -a_i, \ldots, -a_1) \in v_K(K^*)^n \mid a_1 \leq \cdots a_\nu \leq -a_\nu \leq \cdots - a_1\}$.

We claim that $image(\tilde{\mathcal{L}})$ is the intersection of these two sets. We show that every element of this intersection is some $\tilde{\mathcal{L}}(\tilde{H}(F)\tilde{g}(D)G(\mathfrak{o}_F))$. Notice

$$\tilde{\mathcal{L}}(\tilde{H}(F)\tilde{g}(D)G(\mathfrak{o}_F)) = \text{orbit of } \begin{pmatrix} E & T \\ 0 & E \end{pmatrix}' \tilde{\Lambda}_s \begin{pmatrix} E & T \\ 0 & E \end{pmatrix} = \text{orbit of } \begin{pmatrix} 0 & M \\ -M' & TM - M'T \end{pmatrix}$$

for $M = \begin{pmatrix} 0 & M_0^{-1} \\ M_0 & 0 \end{pmatrix}$, $M_0 = -\alpha E$. For $T = \begin{pmatrix} A^{-1} \cdot D & 0 \\ 0 & 0 \end{pmatrix}$ this gives the orbit of

$$\begin{pmatrix} 0 & 0 & 0 & -\alpha^{-1}E \\ 0 & 0 & -\alpha E & 0 \\ 0 & \alpha E & 0 & -\alpha D \\ \alpha^{-1}E & 0 & \alpha D & 0 \end{pmatrix}.$$

Obviously for $e = v_K(\alpha) \in \{0, -\frac{1}{2}\}$ and $D = A^{-1} \cdot diag(\pi_F^{e_1}, \cdots, \pi_F^{e_i})$ as in Theorem 12.1 the first $2i$-elementary divisors of this matrix are $(e + e_1, e + e_1, e + e_2, e + e_2, \ldots, e + e_i, e + e_i, *, \ldots, *)$ (arises from the lower right block, since αD gives rise to the minors with the highest order denominator). This suffices to prove the claim with the representatives $\tilde{g}(D) = \begin{pmatrix} E & T \\ 0 & E \end{pmatrix}$, where $T = \begin{pmatrix} A^{-1} \cdot D & 0 \\ 0 & 0 \end{pmatrix}$ for D as above.

Galois Descent. From the last argument it follows that every $g \in G(F)$ can be written in the form $g = h\tilde{g}(D)k_0^{-1}$ for some $h \in \tilde{H}(K)$, some $k_0 \in G(\mathfrak{o}_K)$, and some D as in Theorem 12.1. Then $\sigma(h)\tilde{g}(D)\sigma(k^{-1}) = h\tilde{g}(D)k^{-1}$ implies $\tilde{g}(D)^{-1}h^{-1}\sigma(h)\tilde{g}(D) = k^{-1}\sigma(k)$ or

$$b(\sigma) = k^{-1}\sigma(k) \in H_D(K) := \left(\tilde{g}(D)^{-1}\tilde{H}(K)\tilde{g}(D) \right) \cap G(\mathfrak{o}_K).$$

Suppose the 1-cocycle $b(\sigma) = k^{-1}\sigma(k) \in H_D(K)$ is a 1-coboundary $b(\sigma) = y^{-1}\sigma(y)$ for some $y \in H_D(K)$. Then $\tilde{k} = yk^{-1} \in G(\mathfrak{o}_F)$ and $g = (h\tilde{g}(D)y^{-1}\tilde{g}(D)^{-1}) \cdot \tilde{g}(D) \cdot (yk^{-1}) = \tilde{h} \cdot \tilde{g}(D) \cdot \tilde{k}$. Since $g, \tilde{k} \in G(F)$ and $\tilde{g}(D) \in G(F)$, $\tilde{h} \in \tilde{H}(F)$ and

$$g \in \tilde{H}(F) \cdot \tilde{g}(D) \cdot G(\mathfrak{o}_F).$$

The Obstruction. For $n = 2$ the group $H_D(K)$ is isomorphic to the group $Gl(2, R)^0$, where R is the ring $\mathfrak{o}_F \otimes_{\mathfrak{o}_F} \mathfrak{o}_K(i)$ (see page 248). Here σ acts by its natural action on the first factor, and is trivial on the second factor. By Shapiro's lemma the class $[b(\sigma)]$ in the cohomology $H^1(\langle\sigma\rangle, H_D(K))$ is trivial if this holds for its image in the quotient $H^1(\langle\sigma\rangle, Gl(2, \mathfrak{o}_K/(\pi_F^i))^0$. Now it is easy to show that the fiber over the trivial element under the reduction map

$$H^1\left(\langle\sigma\rangle, Gl(2, \mathfrak{o}_K/(\pi_F^i))^0\right) \to H^1\left(\langle\sigma\rangle, Gl(2, \mathfrak{o}_K/(\pi_K))^0\right)$$

is trivial. This is easily shown by induction on i. If K/F is unramified, the cohomology set $H^1\left(\langle\sigma\rangle, Gl(2, \mathfrak{o}_K/(\pi_K))^0\right)$ is trivial. If K/F is ramified, this is not the case. But in the ramified case, for the vanishing of the obstruction classes $[b(\sigma)]$

it is therefore sufficient that the image of the 1-cocycle $b(\sigma)$ is trivial in the quotient group $Gl(2, \mathfrak{o}_K/(\pi_K))^0$ of $H_D(K)$. However, this follows from Lemma 7.5 since the reduction of $b(\sigma) = k^{-1}\sigma(k) \in G(\mathfrak{o}_K)$ in $G(\mathfrak{o}_K/(\pi_K))$ is trivial. Notice $k^{-1}\sigma(k) = k^{-1}k = 1$ in $G(\mathfrak{o}_K/(\pi_K))$.

Injectivity of $\tilde{\mathcal{L}}$. In general, if all the above-defined cohomology obstructions $[b(\sigma)]$ in $H^1(\langle\sigma\rangle, H_D(K))$ are trivial, we obtain

$$G(F) = \bigcup_D \tilde{H}(F) \cdot \tilde{g}(D) \cdot G(\mathfrak{o}_F).$$

Furthermore, since $\tilde{\mathcal{L}}(H(F) \cdot \tilde{g}(D_1) \cdot G(\mathfrak{o}_F)) = \tilde{\mathcal{L}}(H(F) \cdot \tilde{g}(D_2) \cdot G(\mathfrak{o}_F))$ implies $D_1 = D_2$ as shown above, we even conclude

Theorem 12.2. *The matrices $\left(\begin{smallmatrix} E & T \\ 0 & E \end{smallmatrix}\right)$ for $T = \left(\begin{smallmatrix} D & 0 \\ 0 & 0 \end{smallmatrix}\right)$ and $D = A^{-1} \cdot diag(\pi_F^{e_1},$ $\cdots, \pi_F^{e_i})$, where $e_1 \leq \cdots \leq e_i$ and $e_\nu < 0$ or $e_\nu = \infty$ for $\nu = 1, \ldots, i$, define inequivalent representatives and for $n = 2$ a full system of representatives for the double cosets*

$$\tilde{H}(F) \setminus G(F)/G(\mathfrak{o}_F).$$

In general, consider the matrix group $H' = Res_{K/F}(GSp(\Lambda_D))^0$. For an integral extension ring $\mathfrak{O}$ of $\mathfrak{o}_F$ with fraction field L consider the subset of $H'(L) = GSp(\Lambda_D)^0(L \otimes_F K)$ defined by all block matrices $g = \left(\begin{smallmatrix} A & B \\ C & D \end{smallmatrix}\right)$ for which A, B, C, and D are matrices of the form

$$X \otimes 1 + Y \cdot D^{-1} \otimes \sqrt{A}^{-1},$$

such that X and Y are $i \times i$-matrices with coefficients in $\mathfrak{O}$. In fact, this subset defines a subgroup. For $\mathfrak{O} = \mathfrak{o}_K$ this group is isomorphic to $H_D(K)$, and σ acts on these matrices by $\sigma(X \otimes 1 + YD^{-1} \otimes \sqrt{A}^{-1}) = \sigma(X) \otimes 1 + \sigma(Y)D^{-1} \otimes \sqrt{A}^{-1}$, via its natural action on $\mathfrak{O} = \mathfrak{o}_K$. Notice that the coefficients of the matrices g are in $\mathfrak{O} \otimes_{\mathfrak{o}_F} \mathfrak{o}_K \subseteq L \otimes_F K$.

References

1. Arthur J. A trace formula for reductive groups I: Terms associated to classes in $G(\mathbb{Q})$. *Duke Math. J.*, 45:911–952, 1978.
2. Arthur J. On elliptic tempered representations. *Acta Math.*, 171, 1993.
3. Arthur J. On some problems suggested by the trace formula, in Lie Group Representations II. *Springer Lecture Notes in Math.*, 1041, 1984.
4. Arthur J. The L^2-Lefschetz numbers of Hecke operators. *Invent. Math.*, 97:257–290, 1989.
5. Arthur J. The invariant trace formula I, local theory. *J. AMS 1*, 323–383, 1988.
6. Arthur J. The invariant trace formula II, global theory. *J. AMS 1*, 501–554, 1988.
7. Arthur J. Trace formula in invariant form. *Ann. Math.*, 114:1–74, 1981.
8. Arthur J., Clozel L. Simple Algebras, Base Change, and the Advanced Theory of the Trace Formula. *Annals of Mathematics Studies*, 120, Princeton University Press, Princeton, 1989.
9. Arthur J., Gelbart S. Lectures on automorphic L-functions, in L-functions and Arithmetic, in: Durham Symposium, edited by Coates J., Taylor M.J., July 1989 *Lond. Math. Soc. Lect. Notes*, 153:1–61, 1991.
10. Bernstein J.N. Le "centre" de Bernstein, in travaux en cours, Representations des groupes reductifs sur un corps local. *Hermann*, 1984.
11. Bernstein J.N., Deligne P., Kazhdan D. Trace Paley–Wiener theorem for reductive p-adic groups. *J. Anal. Math.*, 47:180–192, 1986.
12. Borel A. Automorphic L-functions. *Proc. Symp. Pure Math.*, XXXIII, 1977, part 2.
13. Borel A. Stable real cohomology of arithmetic groups II. *Collected Papers*, III:650–684, Springer, Berlin, 1983.
14. Borel A., Harish-Chandra. Arithmetic subgroups of algebraic groups. *Ann. Math.*, 75: 485–535, 1962.
15. Borel A., Wallach N. Continuous cohomology, discrete subgroups, and representations of reductive groups. Princeton University Press, Princeton, 1980.
16. Cartier P. Representations of p-adic groups. *Proc. Symp. Pure Math.*, XXXIII:111–155, 1979, part 1.
17. Casselman W. Characters and Jacquet modules. *Math. Ann.*, 230:101–105, 1977.
18. Casselman W. The Hasse–Weil ζ-function of some moduli varieties of dimension greater than one. *Proc. Symp. Pure Math.*, XXXIII, 1977, part 2.
19. Chai C.L., Faltings G. Degenerations of Abelian Varieties, Ergebnisse d.Math. und ihrer Grenzgebiete. *3.Folge Band 22*, 1990.
20. Clozel L. The fundamental lemma for stable base change. *Duke Math. J.*, 61:255–302, 1990.
21. Clozel L. The Schwartz space of a reductive p-adic group, in Harmonic analysis on reductive groups, edited by Barker, Sally. *Progr. Math.*, 101:101–121, 1991.
22. Clozel L., Delorme P. Le théorème de Paley–Wiener invariant pour les groupes de Lie réductifs. *Invent. Math.*, 77:427–453, 1984.

23. Clozel L., Delorme P. Pseudo-coefficients et cohomologie des groupes de Lie réductifs reels. *C.R. Acad. Sci. Paris*, 300:385–387, serie I, 1985.

24. Deligne P. Varietes de Shimura: Interpretation modulaire, et techniques de construction de modeles canoniques. *Proc. Sym. Pure Math.* 33:247–290, 1979, part II.

25. Deligne P., Kazhdan D., Vigneras M.-F. Representations des algebres centrales simples p-adiques, Representations des groupes réductifs sur un corps local. *Hermann*, 1984.

26. Flicker Y.F. Matching of orbital integrals on $Gl(4)$ and $GSp(2)$ Mem. Am. Math. Soc., 137, 1999.

27. Franke J. Harmonic analysis in weighted L_2-spaces. Ann. Sci. Ecole Norm. Sup., 31: 181–279, 1998.

28. Freitag E. Siegelsche Modulformen *Grundlehren der mathematischen Wissenschaften*, Springer, Berlin, Band 254, 1983.

29. Gelbart S., Jacquet H. Forms of Gl(2) from the analytic point of view, in: Automorphic Forms, Representations, and L-functions. *Proc. Symp. Pure Math., AMS*, 23:213–251, 1979, part 1.

30. Goresky M., Harder G., MacPherson R. Weighted cohomology. *Invent. Math.*, 116:139–213, 1994.

31. Goresky M., Kottwitz R., MacPherson R. Discrete series characters and the Lefschetz formula for Hecke operators. *Duke Math. J.*, 89:477–554, 1997.

32. Goresky M., MacPherson R. The topological trace formula. *J. Reine Angew. Math.*, 560: 77–150, 2003.

33. Goresky M., MacPherson R. Lefschetz numbers of Hecke correspondences. *The zeta functions of Picard modular surfaces*, 1992.

34. Goresky M., MacPherson R. Local contribution to the Lefschetz fixed point formula. *Invent. Math.*, 111:1–33, 1993.

35. Hales T.C. On the fundamental lemma for standard endoscopy: Reduction to unit elements. *Can. J. Math.*, 47:974–994, 1995.

36. Hales T.C. Shalika germs on $GSp(4)$. *Asterisque*, 171–172:195–256, 1989.

37. Harder G. A Gauss–Bonnet formula for discrete arithmetically defined groups. *Ann. Sci. Ecole Norm. Sup.*, 4:409–455, 1971.

38. Harder G. General purpose of the (topological) trace formula. Preprint.

39. Harder G. Letter to Goresky and MacPherson on the topological trace formula. *Springer Lecture Notes in Math.*, 1562.

40. Harder G. Eisensteinkohomologie und die Konstruktion gemischter Motive. *Springer Lecture Notes in Math.*, 1562, Springer, Berlin, 1993.

41. Howe R., Piatetski-Shapiro I.I. Some examples of automorphic forms on Sp_4. *Duke Math. J.*, 50(1):55–106, 1983.

42. Jacquet H., Langlands R.P. Automorphic Forms on GL(2). *Springer Lecture Notes in Math., 114*, 1970.

43. Kaiser C. Ein getwistetes fundamentales Lemma. Dissertation Bonn, 1997.

44. Kazhdan D. Cuspidal geometry of p-adic groups. *J. Anal. Math.*, 47:1–36, 1986.

45. Kostant B. Lie algebra cohomology and the generalized Borel–Weil theorem. *Ann. Math.* 74(2):329–387, 1961.

46. Kottwitz R.E. Base change for the unit elements of Hecke algebras. *Compos. Math.*, 60: 237–250, 1986.

47. Kottwitz R.E. Isocrystals with additional structure. *Compos. Math.*, 56:201–220, 1985.

48. Kottwitz R.E. Isocrystals with additional structure. II. *Compos. Math.*, 109:255–339, 1997.

49. Kottwitz R.E. Rational conjugacy classes in reductive groups. *Duke Math. J.*, 49:785–806, 1982.

50. Kottwitz R.E. Tamagawa numbers. *Ann. Math.*, 127:629–646, 1988.

51. Kottwitz R.E. Shimura varieties and λ-adic representations, in: Ann Arbor Proceedings, edited by Clozel L., Milne J.S., July 1988. *Perspectives of Mathematics*, Academic, London, pp. 161–209, 1990.

52. Kottwitz R.E. Shimura varieties and twisted orbital integrals. *Math. Ann.*, 269:287–300, 1984.

53. Kottwitz R.E. Stable trace formula: Elliptic singular terms. *Math. Ann.*, 275:629–646, 1986.
54. Kottwitz R.E., Shelstad D. Foundations of twisted endoscopy. *Asterisque*, 255, 1999.
55. Kudla S., Rallis S. Ramified degenerate principal series representations for $Sp(n)$. *Isr. J. Math.*, 78:209–256, 1992.
56. Kudla S.S., Rallis S., Soudry D. On the degree 5 L-function for $Sp(2)$. *Invent. Math.*, 107, 1992.
57. Labesse J.P. Fonctions elementaires et lemma fondamental pour le changement de base stable. *Duke Math. J.*, 61:519–530, 1990.
58. Langlands R.P. Automorphic representations, Shimura varieties, and motives. Ein Märchen. *Proc. Symp. Pure Math., AMS*, 23:205–246, 1979, part 2.
59. Langlands R.P. Les débuts d'une formule des traces stables. *Pub. Math. de l'Univ. Paris VII*, 13, 1982.
60. Langlands R.P., Shelstad D. On the definition of transfer factors. *Math. Ann.*, 278:219–271, 1987.
61. Langlands R.P., Shelstad D. Descent for transfer factors. Grothendieck Festschrift, vol II. *Progr. Math.*, 87:485–563, 1990.
62. Laumon G. Sur la cohomologie à supports compacts des variétés de Shimura pour $GSp(4)_{\mathbb{Q}}$. Preprint, 7390:77, Paris 1994, and *Compos. Math.*, 105:267–359, 1996.
63. Milne J.S. Arithmetic duality theorems. *Perspectives of Mathematics*, Academic, London, p. 1, 1986.
64. Milne J.S. The conjecture of Langlands and Rapoport for Siegel modular varieties. *Bull. Am. Math. Soc.*, 24:335–341, 1991.
65. Milne J.S. The points on a Shimura variety modulo a prime of good reduction, in: The zeta functions of Picard modular surfaces, edited by Langlands R.P., Ramakrishnan D.. *Les Publicationes CRM*, pp. 151–253, 1992.
66. Neukirch J. Klassenkörpertheorie. *Bibliographisches Institut Mannheim*, BI-Hochschultaschenbuch, 1969.
67. Oesterle J. Nombres de Tamagawa et groupes unipotentes en characteristique p. *Invent. Math.*, 78:13–88, 1984.
68. Ono T. On Tamagawa numbers, in: Algebraic groups and discontinuous subgroups *Proc. Symp. Pure Math.*, 9:122–132, 1966.
69. Piatetski-Shapiro I.I. On the Saito–Kurokawa Lifting. *Invent. Math.*, 71:309–338, 1983.
70. Piatetski-Shapiro I.I., Soudry D. L and ϵ factors for $GSp(4)$. *J. Fac. Sci. Univ. Tokyo, Sec. 1a*, 28, 1981.
71. Piatetski-Shapiro I.I., Soudry D. On a correspondence of automorphic forms on orthogonal groups of order five. *J. Math. Pures Appl.*, 66:407–436, 1987.
72. Deligne P. Cohomologie étale, Seminaire de Geometrie Algebrique du Bois-Marie. SGA $4\frac{1}{2}$, *Springer Lecture Notes in Math., 569*, Springer, Berlin, 1977.
73. Pink R. On the calculation of local terms in the Lefschetz–Verdier trace formula and its application to a conjecture of Deligne. Preprint, 1990.
74. Ribet K.A. Galois representations attached to eigenforms with nebentypus, in: Modular functions of one variable V. *Springer Lecture Notes in Math.*, 601:17–52, 1977.
75. Rodier F. Les representations des groupes p-adiques. *Bull. Soc. Math. France*, 116:15–42, 1988.
76. Rogawski J. Automorphic representations of unitary groups in three variables. *Annals of Mathematics Studies*, Princeton University Press, Princeton, p. 123, 1990.
77. Rogawski J. Representations of $Gl(n)$ and division algebras over a p-adic field. *Duke Math. J.*, 50:161–196, 1983.
78. Saito M. Representations unitaires des groupes symplectiques. *J. Math. Soc. Japan*, 24:232–251, 1972.
79. Sally P.J., Tadic M. Induced representations and classifications for GSp(2,F) and Sp(2,F). *Suppl. Bull. S.M.F., Tome*, 121:75–133, 1993, fasc. 1.
80. Schneider P., Stuhler U., Representation theory and sheaves on the Bruhat–Tits building. *Publications Mathématiques de l'IHÉS*, 85:97–191, 1997.

81. Schröder M. Zählen der Punkte mod p einer Shimuravarietät zu $GSp(4)$. Thesis, University of Mannheim, 1993.

82. Schröder M., Weissauer R. An explicit Langlands reciprocity law for unramified L-group homomorphisms. *Mannheimer Manuskripte*, 1997.

83. Schwermer J. On arithmetic quotients of the Siegel upper half space of degree two. *Compos. Math.*, 58:233–258, 1986.

84. Serre J.P. A course in arithmetic, Graduate Texts in Math. 7, Springer Verlag, Berlin.

85. Serre J.P. Arithmetic groups. *Collected papers*, III(1972–1984):503–534, 1986.

86. Serre J.P. Quelques applications du theoreme de densite de Chebotarev. *Collected papers*, Springer, Berlin, III.

87. Shahidi F. A proof of Langlands conjecture on Plancherel measures; complementary series for p-adic groups. *Ann. Math.*, 132:273–330, 1990.

88. Shahidi F. Langland's conjecture on Plancherel measures for p-adic groups. *Harmonic Analysis on Reductive Groups (Brunswick, Me, 1989), Progr. Math.*, vol 101, Birkhäuser, Boston, pp. 227–295, 1991.

89. Shahidi F. On nonvanishing of L-functions. *Bull. AMS (New Series)*, 2:442–464, 1980.

90. Shelstad D. Characters and inner forms of a quasi-split group over $\mathbb{R}$. *Compos. Math.*, 39: 11–45, 1979.

91. Shelstad D. L-indistinguishability for real groups. *Math. Ann.*, 259:385–430, 1982.

92. Shimura G. Introduction to the arithmetic theory of automorphic functions, Princeton University Press, Princeton, 1971.

93. Silberger A.J. Discrete Series and classification for p-adic groups I. *Am. J. Math.*, 103: 1241–1321, 1981.

94. Soudry D. A uniqueness theorem for representations of $GSO(6)$ and the strong multiplicity one theorem for generic representations of $GSp(4)$. *Isr. J. Math.*, 58(3), 1987.

95. Soudry D. Automorphic forms on GSp(4), in: Papers on Analysis, Number Theory and Automorphic L-functions part II, edited by Gelbart S., Howe R., Sarnak P. Weizmann Science Press, Israel, pp. 291–303, 1990.

96. Soudry D. The L and γ factors for generic representations of $GSp(4, k) \times Gl(2, k)$ over a local nonarchimedian field k. *Duke Math. J.*, 51(2):335–394, 1984.

97. Soudry D. The CAP representations of $GSp(4, \mathbb{A})$. *J. Reine Angew. Math.*, 383:87–108, 1988.

98. Springer T.A. Reductive groups. *Proc. Symp. Pure Math. AMS, Automorphic forms, Representations, and L-functions*, 23:3–29, 1979, part 1.

99. Springer T.A., Steinberg R. Conjugacy classes, in: Seminar on algebraic groups and related finite groups. *Springer Lecture Notes in Math.*, 131:167–266, 1970.

100. Tadic M. Representations of p-adic symplectic groups. Compos. Math., 90:123–181, 1994.

101. Tate J. Number theoretic background, *in Proc. Symp. in Pure Math.*, 33:3–26, 1979, part 2.

102. Tate J.T. Fourier analysis in number fields and Hecke's zeta-functions, in: Algebraic Number Theory, edited by Cassels-Fröhlich, Academic, London, 1967.

103. Tits J. Reductive groups over local fields. *Proc. AMS Symp. Pure Math.*, 33:29–69, 1979, part 1.

104. Vigneras M.F. Correspondances entre representations automorphes de $Gl(2)$ sur une extension quadratique de $GSp(4)$ sur $\mathbb{Q}$, conjecture locale de Langlands pour $GSp(4)$. *Contemp. Math.*, 53, 1986.

105. Waldspurger J.-L. Correspondences de Shimura et quaternions. *Forum Math.*, 3:219–307, 1991.

106. Waldspurger J.-L. Un exercice sur GSp(4,F) et les representations de Weil. *Bull. Soc. Math. France*, 115:35–69, 1987.

107. Waldspurger J.-L. Une formule des traces locale pour les algebres de Lie p-adiques. *J. Reine Angew. Math.*, 465:41–99, 1995.

108. Waldspurger J.-L. Le lemme fondamental implique le transfert. *Compos. Math.*, 105: 153–236, 1997.

109. Warner G. Harmonic Analysis on Semi-simple Liegroups *Grundlehren der Math. Wissenschaften*, Springer, London, pp. 188–189, 1972.

110. Weissauer R. Differentialformen zu Untergruppen der Siegelschen Modulgruppe zweiten Grades. *J. Reine Angew. Math.*, 391:100–156, 1988.
111. Weissauer R. Modular forms of genus 2 and weight 1. *Math. Z.*, 210:91–96, 1992.
112. Weissauer R. On certain degenerate eigenvalues of Hecke operators. *Compos. Math.*, 83: 127–145, 1992.
113. Weissauer R. Spectral approximation of twisted κ-orbital integrals. Preprint.
114. Weissauer R. The Picard group of Siegel modular threefolds. *J. Reine Angew. Math.*, 430:179–211, 1992.
115. Weissauer R. Four dimensional Galois representations, in: Formes Automorphes (II), Le cas du groupe GSp(4), edited by Tilouine J., Carayol H., Harris M., Vignéras M.F., *Astérisque 302*, pp. 67–150, 2005.
116. Weissauer R. Double cosets for classical groups in the unramified case. Preprint. www.mathi.uni-heidelberg.de/~weissaue/papers.html
117. Weselmann U. On double coset decompositions for the algebraic group G_2. Preprint. www.mathi.uni-heidelberg.de/~weselman

Index

Nomenclature

Lecture Notes in Mathematics

For information about earlier volumes
please contact your bookseller or Springer
LNM Online archive: springerlink.com

Vol. 1881: S. Attal, A. Joye, C.-A. Pillet, Open Quantum Systems II, The Markovian Approach (2006)

Vol. 1882: S. Attal, A. Joye, C.-A. Pillet, Open Quantum Systems III, Recent Developments (2006)

Vol. 1883: W. Van Assche, F. Marcellàn (Eds.), Orthogonal Polynomials and Special Functions, Computation and Application (2006)

Vol. 1884: N. Hayashi, E.I. Kaikina, P.I. Naumkin, I.A. Shishmarev, Asymptotics for Dissipative Nonlinear Equations (2006)

Vol. 1885: A. Telcs, The Art of Random Walks (2006)

Vol. 1886: S. Takamura, Splitting Deformations of Degenerations of Complex Curves (2006)

Vol. 1887: K. Habermann, L. Habermann, Introduction to Symplectic Dirac Operators (2006)

Vol. 1888: J. van der Hoeven, Transseries and Real Differential Algebra (2006)

Vol. 1889: G. Osipenko, Dynamical Systems, Graphs, and Algorithms (2006)

Vol. 1890: M. Bunge, J. Funk, Singular Coverings of Toposes (2006)

Vol. 1891: J.B. Friedlander, D.R. Heath-Brown, H. Iwaniec, J. Kaczorowski, Analytic Number Theory, Cetraro, Italy, 2002. Editors: A. Perelli, C. Viola (2006)

Vol. 1892: A. Baddeley, I. Bárány, R. Schneider, W. Weil, Stochastic Geometry, Martina Franca, Italy, 2004. Editor: W. Weil (2007)

Vol. 1893: H. Hanßmann, Local and Semi-Local Bifurcations in Hamiltonian Dynamical Systems, Results and Examples (2007)

Vol. 1894: C.W. Groetsch, Stable Approximate Evaluation of Unbounded Operators (2007)

Vol. 1895: L. Molnár, Selected Preserver Problems on Algebraic Structures of Linear Operators and on Function Spaces (2007)

Vol. 1896: P. Massart, Concentration Inequalities and Model Selection, Ecole d'Été de Probabilités de Saint-Flour XXXIII-2003. Editor: J. Picard (2007)

Vol. 1897: R. Doney, Fluctuation Theory for Lévy Processes, Ecole d'Été de Probabilités de Saint-Flour XXXV-2005. Editor: J. Picard (2007)

Vol. 1898: H.R. Beyer, Beyond Partial Differential Equations, On linear and Quasi-Linear Abstract Hyperbolic Evolution Equations (2007)

Vol. 1899: Séminaire de Probabilités XL. Editors: C. Donati-Martin, M. Émery, A. Rouault, C. Stricker (2007)

Vol. 1900: E. Bolthausen, A. Bovier (Eds.), Spin Glasses (2007)

Vol. 1901: O. Wittenberg, Intersections de deux quadriques et pinceaux de courbes de genre 1, Intersections of Two Quadrics and Pencils of Curves of Genus 1 (2007)

Vol. 1902: A. Isaev, Lectures on the Automorphism Groups of Kobayashi-Hyperbolic Manifolds (2007)

Vol. 1903: G. Kresin, V. Maz'ya, Sharp Real-Part Theorems (2007)

Vol. 1904: P. Giesl, Construction of Global Lyapunov Functions Using Radial Basis Functions (2007)

Vol. 1905: C. Prévôt, M. Röckner, A Concise Course on Stochastic Partial Differential Equations (2007)

Vol. 1906: T. Schuster, The Method of Approximate Inverse: Theory and Applications (2007)

Vol. 1907: M. Rasmussen, Attractivity and Bifurcation for Nonautonomous Dynamical Systems (2007)

Vol. 1908: T.J. Lyons, M. Caruana, T. Lévy, Differential Equations Driven by Rough Paths, Ecole d'Été de Probabilités de Saint-Flour XXXIV-2004 (2007)

Vol. 1909: H. Akiyoshi, M. Sakuma, M. Wada, Y. Yamashita, Punctured Torus Groups and 2-Bridge Knot Groups (I) (2007)

Vol. 1910: V.D. Milman, G. Schechtman (Eds.), Geometric Aspects of Functional Analysis. Israel Seminar 2004-2005 (2007)

Vol. 1911: A. Bressan, D. Serre, M. Williams, K. Zumbrun, Hyperbolic Systems of Balance Laws. Cetraro, Italy 2003. Editor: P. Marcati (2007)

Vol. 1912: V. Berinde, Iterative Approximation of Fixed Points (2007)

Vol. 1913: J.E. Marsden, G. Misiołek, J.-P. Ortega, M. Perlmutter, T.S. Ratiu, Hamiltonian Reduction by Stages (2007)

Vol. 1914: G. Kutyniok, Affine Density in Wavelet Analysis (2007)

Vol. 1915: T. Bıyıkoğlu, J. Leydold, P.F. Stadler, Laplacian Eigenvectors of Graphs. Perron-Frobenius and Faber-Krahn Type Theorems (2007)

Vol. 1916: C. Villani, F. Rezakhanlou, Entropy Methods for the Boltzmann Equation. Editors: F. Golse, S. Olla (2008)

Vol. 1917: I. Veselić, Existence and Regularity Properties of the Integrated Density of States of Random Schrödinger (2008)

Vol. 1918: B. Roberts, R. Schmidt, Local Newforms for GSp(4) (2007)

Vol. 1919: R.A. Carmona, I. Ekeland, A. Kohatsu-Higa, J.-M. Lasry, P.-L. Lions, H. Pham, E. Taflin, Paris-Princeton Lectures on Mathematical Finance 2004. Editors: R.A. Carmona, E. Çinlar, I. Ekeland, E. Jouini, J.A. Scheinkman, N. Touzi (2007)

Vol. 1920: S.N. Evans, Probability and Real Trees. Ecole d'Été de Probabilités de Saint-Flour XXXV-2005 (2008)

Vol. 1921: J.P. Tian, Evolution Algebras and their Applications (2008)

Vol. 1922: A. Friedman (Ed.), Tutorials in Mathematical BioSciences IV. Evolution and Ecology (2008)

Vol. 1923: J.P.N. Bishwal, Parameter Estimation in Stochastic Differential Equations (2008)

Vol. 1924: M. Wilson, Littlewood-Paley Theory and Exponential-Square Integrability (2008)

Vol. 1925: M. du Sautoy, L. Woodward, Zeta Functions of Groups and Rings (2008)

Vol. 1926: L. Barreira, V. Claudia, Stability of Nonautonomous Differential Equations (2008)

Vol. 1927: L. Ambrosio, L. Caffarelli, M.G. Crandall, L.C. Evans, N. Fusco, Calculus of Variations and Non-Linear Partial Differential Equations. Cetraro, Italy 2005. Editors: B. Dacorogna, P. Marcellini (2008)

Vol. 1928: J. Jonsson, Simplicial Complexes of Graphs (2008)

Vol. 1929: Y. Mishura, Stochastic Calculus for Fractional Brownian Motion and Related Processes (2008)

Vol. 1930: J.M. Urbano, The Method of Intrinsic Scaling. A Systematic Approach to Regularity for Degenerate and Singular PDEs (2008)

Vol. 1931: M. Cowling, E. Frenkel, M. Kashiwara, A. Valette, D.A. Vogan, Jr., N.R. Wallach, Representation Theory and Complex Analysis. Venice, Italy 2004. Editors: E.C. Tarabusi, A. D'Agnolo, M. Picardello (2008)

Vol. 1932: A.A. Agrachev, A.S. Morse, E.D. Sontag, H.J. Sussmann, V.I. Utkin, Nonlinear and Optimal

Control Theory. Cetraro, Italy 2004. Editors: P. Nistri, G. Stefani (2008)

Vol. 1933: M. Petkovic, Point Estimation of Root Finding Methods (2008)

Vol. 1934: C. Donati-Martin, M. Émery, A. Rouault, C. Stricker (Eds.), Séminaire de Probabilités XLI (2008)

Vol. 1935: A. Unterberger, Alternative Pseudodifferential Analysis (2008)

Vol. 1936: P. Magal, S. Ruan (Eds.), Structured Population Models in Biology and Epidemiology (2008)

Vol. 1937: G. Capriz, P. Giovine, P.M. Mariano (Eds.), Mathematical Models of Granular Matter (2008)

Vol. 1938: D. Auroux, F. Catanese, M. Manetti, P. Seidel, B. Siebert, I. Smith, G. Tian, Symplectic 4-Manifolds and Algebraic Surfaces. Cetraro, Italy 2003. Editors: F. Catanese, G. Tian (2008)

Vol. 1939: D. Boffi, F. Brezzi, L. Demkowicz, R.G. Durán, R.S. Falk, M. Fortin, Mixed Finite Elements, Compatibility Conditions, and Applications. Cetraro, Italy 2006. Editors: D. Boffi, L. Gastaldi (2008)

Vol. 1940: J. Banasiak, V. Capasso, M.A.J. Chaplain, M. Lachowicz, J. Miękisz, Multiscale Problems in the Life Sciences. From Microscopic to Macroscopic. Będlewo, Poland 2006. Editors: V. Capasso, M. Lachowicz (2008)

Vol. 1941: S.M.J. Haran, Arithmetical Investigations. Representation Theory, Orthogonal Polynomials, and Quantum Interpolations (2008)

Vol. 1942: S. Albeverio, F. Flandoli, Y.G. Sinai, SPDE in Hydrodynamic. Recent Progress and Prospects. Cetraro, Italy 2005. Editors: G. Da Prato, M. Röckner (2008)

Vol. 1943: L.L. Bonilla (Ed.), Inverse Problems and Imaging. Martina Franca, Italy 2002 (2008)

Vol. 1944: A. Di Bartolo, G. Falcone, P. Plaumann, K. Strambach, Algebraic Groups and Lie Groups with Few Factors (2008)

Vol. 1945: F. Brauer, P. van den Driessche, J. Wu (Eds.), Mathematical Epidemiology (2008)

Vol. 1946: G. Allaire, A. Arnold, P. Degond, T.Y. Hou, Quantum Transport. Modelling, Analysis and Asymptotics. Cetraro, Italy 2006. Editors: N.B. Abdallah, G. Frosali (2008)

Vol. 1947: D. Abramovich, M. Mariño, M. Thaddeus, R. Vakil, Enumerative Invariants in Algebraic Geometry and String Theory. Cetraro, Italy 2005. Editors: K. Behrend, M. Manetti (2008)

Vol. 1948: F. Cao, J-L. Lisani, J-M. Morel, P. Musé, F. Sur, A Theory of Shape Identification (2008)

Vol. 1949: H.G. Feichtinger, B. Helffer, M.P. Lamoureux, N. Lerner, J. Toft, Pseudo-Differential Operators. Quantization and Signals. Cetraro, Italy 2006. Editors: L. Rodino, M.W. Wong (2008)

Vol. 1950: M. Bramson, Stability of Queueing Networks, Ecole d'Eté de Probabilités de Saint-Flour XXXVI-2006 (2008)

Vol. 1951: A. Moltó, J. Orihuela, S. Troyanski, M. Valdivia, A Non Linear Transfer Technique for Renorming (2009)

Vol. 1952: R. Mikhailov, I.B.S. Passi, Lower Central and Dimension Series of Groups (2009)

Vol. 1953: K. Arwini, C.T.J. Dodson, Information Geometry (2008)

Vol. 1954: P. Biane, L. Bouten, F. Cipriani, N. Konno, N. Privault, Q. Xu, Quantum Potential Theory. Editors: U. Franz, M. Schuermann (2008)

Vol. 1955: M. Bernot, V. Caselles, J.-M. Morel, Optimal Transportation Networks (2008)

Vol. 1956: C.H. Chu, Matrix Convolution Operators on Groups (2008)

Vol. 1957: A. Guionnet, On Random Matrices: Macroscopic Asymptotics, Ecole d'Eté de Probabilités de Saint-Flour XXXVI-2006 (2009)

Vol. 1958: M.C. Olsson, Compactifying Moduli Spaces for Abelian Varieties (2008)

Vol. 1959: Y. Nakkajima, A. Shiho, Weight Filtrations on Log Crystalline Cohomologies of Families of Open Smooth Varieties (2008)

Vol. 1960: J. Lipman, M. Hashimoto, Foundations of Grothendieck Duality for Diagrams of Schemes (2009)

Vol. 1961: G. Buttazzo, A. Pratelli, S. Solimini, E. Stepanov, Optimal Urban Networks via Mass Transportation (2009)

Vol. 1962: R. Dalang, D. Khoshnevisan, C. Mueller, D. Nualart, Y. Xiao, A Minicourse on Stochastic Partial Differential Equations (2009)

Vol. 1963: W. Siegert, Local Lyapunov Exponents (2009)

Vol. 1964: W. Roth, Operator-valued Measures and Integrals for Cone-valued Functions and Integrals for Cone-valued Functions (2009)

Vol. 1965: C. Chidume, Geometric Properties of Banach Spaces and Nonlinear Iterations (2009)

Vol. 1966: D. Deng, Y. Han, Harmonic Analysis on Spaces of Homogeneous Type (2009)

Vol. 1967: B. Fresse, Modules over Operads and Functors (2009)

Vol. 1968: R. Weissauer, Endoscopy for GSp(4) and the Cohomology of Siegel Modular Threefolds (2009)

Vol. 1969: B. Roynette, M. Yor, Penalising Brownian Paths (2009)

Vol. 1970: R. Kotecký, Methods of Contemporary Mathematical Statistical Physics (2009)

Vol. 1971: L. Saint-Raymond, Hydrodynamic Limits of the Boltzmann Equation (2009)

Vol. 1972: T. Mochizuki, Donaldson Type Invariants for Algebraic Surfaces (2009)

Recent Reprints and New Editions

Vol. 1702: J. Ma, J. Yong, Forward-Backward Stochastic Differential Equations and their Applications. 1999 – Corr. 3rd printing (2007)

Vol. 830: J.A. Green, Polynomial Representations of GL_n, with an Appendix on Schensted Correspondence and Littelmann Paths by K. Erdmann, J.A. Green and M. Schoker 1980 – 2nd corr. and augmented edition (2007)

Vol. 1693: S. Simons, From Hahn-Banach to Monotonicity (Minimax and Monotonicity 1998) – 2nd exp. edition (2008)

Vol. 470: R.E. Bowen, Equilibrium States and the Ergodic Theory of Anosov Diffeomorphisms. With a preface by D. Ruelle. Edited by J.-R. Chazottes. 1975 – 2nd rev. edition (2008)

Vol. 523: S.A. Albeverio, R.J. Høegh-Krohn, S. Mazzucchi, Mathematical Theory of Feynman Path Integral. 1976 – 2nd corr. and enlarged edition (2008)

Vol. 1764: A. Cannas da Silva, Lectures on Symplectic Geometry 2001 – Corr. 2nd printing (2008)

LECTURE NOTES IN MATHEMATICS　　　　🦄 Springer

Edited by J.-M. Morel, F. Takens, B. Teissier, P.K. Maini

Editorial Policy (for the publication of monographs)

1. Lecture Notes aim to report new developments in all areas of mathematics and their applications - quickly, informally and at a high level. Mathematical texts analysing new developments in modelling and numerical simulation are welcome.

   Monograph manuscripts should be reasonably self-contained and rounded off. Thus they may, and often will, present not only results of the author but also related work by other people. They may be based on specialised lecture courses. Furthermore, the manuscripts should provide sufficient motivation, examples and applications. This clearly distinguishes Lecture Notes from journal articles or technical reports which normally are very concise. Articles intended for a journal but too long to be accepted by most journals, usually do not have this "lecture notes" character. For similar reasons it is unusual for doctoral theses to be accepted for the Lecture Notes series, though habilitation theses may be appropriate.

2. Manuscripts should be submitted either to Springer's mathematics editorial in Heidelberg, or to one of the series editors. In general, manuscripts will be sent out to 2 external referees for evaluation. If a decision cannot yet be reached on the basis of the first 2 reports, further referees may be contacted: The author will be informed of this. A final decision to publish can be made only on the basis of the complete manuscript, however a refereeing process leading to a preliminary decision can be based on a pre-final or incomplete manuscript. The strict minimum amount of material that will be considered should include a detailed outline describing the planned contents of each chapter, a bibliography and several sample chapters.

   Authors should be aware that incomplete or insufficiently close to final manuscripts almost always result in longer refereeing times and nevertheless unclear referees' recommendations, making further refereeing of a final draft necessary.

   Authors should also be aware that parallel submission of their manuscript to another publisher while under consideration for LNM will in general lead to immediate rejection.

3. Manuscripts should in general be submitted in English. Final manuscripts should contain at least 100 pages of mathematical text and should always include

   – a table of contents;
   – an informative introduction, with adequate motivation and perhaps some historical remarks: it should be accessible to a reader not intimately familiar with the topic treated;
   – a subject index: as a rule this is genuinely helpful for the reader.

   For evaluation purposes, manuscripts may be submitted in print or electronic form, in the latter case preferably as pdf- or zipped ps-files. Lecture Notes volumes are, as a rule, printed digitally from the authors' files. To ensure best results, authors are asked to use the LaTeX2e style files available from Springer's web-server at:

   ftp://ftp.springer.de/pub/tex/latex/svmonot1/ (for monographs).

Additional technical instructions, if necessary, are available on request from: lnm@springer.com.

4. Careful preparation of the manuscripts will help keep production time short besides ensuring satisfactory appearance of the finished book in print and online. After acceptance of the manuscript authors will be asked to prepare the final LaTeX source files (and also the corresponding dvi-, pdf- or zipped ps-file) together with the final printout made from these files. The LaTeX source files are essential for producing the full-text online version of the book (see www.springerlink.com/content/110312 for the existing online volumes of LNM).
   The actual production of a Lecture Notes volume takes approximately 12 weeks.

5. Authors receive a total of 50 free copies of their volume, but no royalties. They are entitled to a discount of 33.3% on the price of Springer books purchased for their personal use, if ordering directly from Springer.

6. Commitment to publish is made by letter of intent rather than by signing a formal contract. Springer-Verlag secures the copyright for each volume. Authors are free to reuse material contained in their LNM volumes in later publications: a brief written (or e-mail) request for formal permission is sufficient.

Addresses:

Professor J.-M. Morel, CMLA,
École Normale Supérieure de Cachan,
61 Avenue du Président Wilson, 94235 Cachan Cedex, France
E-mail: Jean-Michel.Morel@cmla.ens-cachan.fr

Professor F. Takens, Mathematisch Instituut,
Rijksuniversiteit Groningen, Postbus 800,
9700 AV Groningen, The Netherlands
E-mail: F.Takens@math.rug.nl

Professor B. Teissier, Institut Mathématique de Jussieu,
UMR 7586 du CNRS, Équipe "Géométrie et Dynamique",
175 rue du Chevaleret
75013 Paris, France
E-mail: teissier@math.jussieu.fr

For the "Mathematical Biosciences Subseries" of LNM:

Professor P.K. Maini, Center for Mathematical Biology,
Mathematical Institute, 24-29 St Giles,
Oxford OX1 3LP, UK
E-mail: maini@maths.ox.ac.uk

Springer, Mathematics Editorial I, Tiergartenstr. 17
69121 Heidelberg, Germany,
Tel.: +49 (6221) 487-8259
Fax: +49 (6221) 4876-8259
E-mail: lnm@springer.com